AF396006

Quantum Computing and Quantum Machine
Learning for Engineers and Developers

Jesse Van Griensven Thé • Roydon Andrew Fraser • Jose Rosas-Bustos

Quantum Computing and Quantum Machine Learning for Engineers and Developers

 Springer

Jesse Van Griensven Thé
Mechanical Engineering
University of Waterloo
Waterloo, ON, Canada

Roydon Andrew Fraser
Mechanical Engineering
University of Waterloo
Waterloo, ON, Canada

Jose Rosas-Bustos
Mechanical Engineering
University of Waterloo
Waterloo, ON, Canada

ISBN 978-3-031-98244-6   ISBN 978-3-031-98245-3   (eBook)
https://doi.org/10.1007/978-3-031-98245-3

This Springer imprint is published by the registered company Springer Nature Switzerland AG
The registered company address is: Gewerbestrasse 11, 6330 Cham, Switzerland

If disposing of this product, please recycle the paper.

Foreword

The quantum journey that began in university laboratories a generation ago is now racing toward commercial reality. At the University of Waterloo's Institute for Quantum Computing, we have watched, and often helped, this elegant theory into devices, algorithms, and companies. Yet one barrier stubbornly remains: working engineers still struggle to connect the machinery of qubits, gates, and error-prone hardware with the day-to-day problems they already know how to solve in classical code.

That is precisely the gap Professor Jesse Van Griensven, Professor Roydon Fraser, and José R. Rosas-Bustos close in *Quantum Computing and Quantum Machine Learning for Engineers*. They speak the language of engineering practicality while never compromising on scientific depth. From the first pages, readers encounter lucid explanations of superposition, entanglement, and the Schrödinger equation, but always coupled to tasks an engineer care about, such as signal processing, optimization, control, secure communication, and quantum machine learning.

The book's nine chapters march in a logical progression. Foundations come first, presenting quantum mechanics, core linear algebra tools, and the gate model. Then proceeding to the engineer's workhorses: data structures and encoding, circuits, optimization, and finally, the quantum AI frontier, elaborating on quantum support vector machine, quantum neural networks, and hybrid classical-quantum algorithms. Each chapter ends exactly where the next must begin, and every concept appears again in code you can run directly from the authors' public repository.

Along the way you will meet topics such as amplitude and angle encoding schemes that pack classical datasets into qubits with mathematical elegance (Chap. 4), the quantum approximate optimization algorithm that already outperforms heuristic solvers on selected combinatorial problems (Chap. 6), and hybrid quantum-classical classifiers that hint at real-world quantum advantage in pattern recognition (Chap. 8). These are not toy demonstrations, they are the building blocks for tomorrow's secure networks, drug-discovery pipelines, and autonomous systems.

We are especially pleased to see the authors emphasize error-aware programming and hardware diversity. Quantum processors based on superconducting circuits, trapped ions, photonics, and spin qubits each receive balanced treatment, together with the noise models and mitigation strategies practitioners must master to move from PowerPoint to prototype. No matter which platform finally scales first, readers of this book will be ready.

Quantum engineering is becoming a multi-disciplinary team sport. Success now requires physicists, computer scientists, materials experts and, increasingly, software and systems engineers who can translate a customer's specification into a quantum workflow. With this book volume in hand, those engineers can join the field without detouring through a decade of theoretical physics.

Jesse, Roydon, and José have given our community a reference that is rigorous enough for graduate study yet practical enough for the design office. We expect to see it open on lab benches, in start-up boardrooms, and in lecture halls around the world. Most of all, we expect to meet the next generation of quantum innovators who will tell us this was the book that showed them where to begin.

Raymond Laflamme
Mike & Ophelia Lazaridis Professor of Physics, Founding Director, and Executive Director of the Institute for Quantum Computing, University of Waterloo from 2002 to 2017. He performed the world's first experimental demonstration of quantum error correction, a task necessary to make quantum computers resilient to noise. He also developed a blueprint for a quantum information processor using linear optics that is being implemented by many groups around the world. Prof. Laflamme received the Order of Canada for his significant scientific and leadership contributions to the country.

Mark Pecen
Mark Pecen serves as the CEO of Approach Infinity, Inc., a specialized advisory firm in Waterloo, Canada. As a recognized thought leader with over 35 years of experience in the ICT industry, Pecen has been quoted in the Wall Street Journal, Forbes, Bloomberg, Daily Telegraph, Cybersecurity Magazine, and others and is the author of numerous articles, papers, and book chapters. Pecen is a former board member of the Institute for Quantum Computing in Canada and is a co-founder of the European Telecommunication Standards Institute's Working Group for Quantum Safe Cryptography in France. He is a pioneer in wireless technology and the inventor of more than 100 fundamental patents in wireless communication, networking, and computing and is a graduate from the University of Pennsylvania, Wharton School of Business and the School of Engineering and Applied Sciences.

University of Waterloo, Waterloo, ON,   Raymond Laflamme
Canada
Approach Infinity, Waterloo, ON,   Mark Pecen
Canada

Preface

Quantum computing and quantum machine learning represent a revolutionary leap forward, promising immense computational power far beyond the capabilities of classical systems. The motivation behind this book is to provide professional engineers and software developers with practical insights, essential theory, and hands-on guidance necessary to harness quantum technologies for real-world applications effectively. This book addresses the growing demand among industry professionals for clear, applicable knowledge, bridging theoretical foundations with practical implementation.

The primary audience for this book includes professional engineers and software developers already familiar with classical computing, algorithmic design, and basic machine learning techniques. The content is tailored specifically for practitioners who seek to integrate quantum algorithms and quantum machine learning methods into their existing technology stack, enhancing computational efficiency and enabling innovative solutions to complex computational problems.

This book is structured into clear, concise chapters, each building logically upon the preceding ones:

Chapter 1: "Introduction to Quantum Computing" provides foundational knowledge, covering the evolution and core principles of quantum computing, essential linear algebra concepts, and the significance of quantum mechanics for engineers.

Chapter 2: "Quantum Computing Fundamentals" introduces wave-particle duality, uncertainty principles, Schrödinger equation, qubits, quantum gates, and quantum measurement concepts necessary for understanding quantum computations.

Chapter 3: "Quantum Circuits" discusses quantum programming languages and platforms, such as Qiskit, Google's Cirq, Microsoft's Q#, and quantum simulators, emphasizing quantum hardware architectures, error correction, and fault tolerance.

Chapter 4: "Quantum Algorithms" details critical quantum algorithms including Grover's search, quantum Fourier transform, Shor's factoring algorithm, and the variational quantum eigensolver (VQE), outlining their practical engineering applications.

Chapter 5: "Classical Artificial Intelligence" outlines core concepts in classical AI, including supervised and unsupervised learning, neural networks, optimization techniques, and their significance for engineers transitioning to quantum AI.

Chapter 6: "Quantum Data Structures" examines efficient data encoding methods crucial for quantum computing, highlighting quantum encoding techniques, noise mitigation, and challenges related to quantum data representation.

Chapter 7: "Quantum Optimization" explores quantum optimization methods like quantum approximate optimization algorithm (QAOA), quantum kernels, and quantum classifiers and compares quantum optimization to classical methods, detailing relevant engineering applications.

Chapter 8: "Introduction to Quantum AI" introduces quantum-enhanced AI methods, emphasizing quantum-enhanced algorithms, hybrid quantum-classical approaches, quantum data processing, and specific applications in finance, optimization, and generative modeling.

Chapter 9: "Quantum Neural Networks (QNNs)" discusses the motivation, evolution, and practical implementation of quantum neural networks, covering quantum perceptrons, quantum neurons, activation functions, training methods, and hybrid quantum-classical architectures.

To support practical engagement and continuous learning, this book includes robust online resources:

Codes: Comprehensive, practical code examples, simulations, and programming exercises to facilitate hands-on experimentation with quantum algorithms and machine learning models.

Errata: Regular updates and corrections to ensure content accuracy and reliability post-publication.

These supplementary materials can be accessed at our dedicated online resource page: https://aqtinstitute.org/quantum-computing-for-engineers.

We hope this book becomes a trusted guide for professional engineers and software developers, enabling them to effectively implement and innovate with quantum computing and quantum machine learning.

Waterloo, ON, Canada   Jesse Van Griensven Thé
Waterloo, ON, Canada   Roydon Andrew Fraser
Waterloo, ON, Canada   Jose Rosas-Bustos

Acknowledgments

This book required many hours away from our family and friends. This way, the authors would like to thank their wives, Cristiane Thé, Frances Fraser, and Aki Ikegami.

The content of this course was created for a doctoral program course in the Department of Mechanical Engineering and Mechatronics (robotics) at the University of Waterloo in Canada. We thank the course participants for the feedback on the content.

Contents

About the Authors

Jesse Van Griensven Thé obtained his Ph.D. in Computational Fluid Dynamics from the prestigious University of Waterloo. He is currently the CEO of EigenQ, a company that develops and commercializes quantum technologies, including post-quantum cryptography software and hardware. He teaches artificial intelligence and quantum computing at the doctoral level in the Department of Mechanical and Mechatronics Engineering (robotics) at the University of Waterloo. Prof. Jesse has received many international awards for his research in various fields and is the author and co-author of various patents, including two that cover new developments in the quantum Internet.

Roydon Andrew Fraser Graduated from Queen's University's, Canada, Engineering Physics, and Princeton University's, U.S.A., Mechanical and Aerospace Engineering programs, and is currently a professor of Mechanical and Mechatronics Engineering (MME) at the University of Waterloo, Canada. Fraser has over 150 journal publications and in 2024 was recognized by Clarivate as being among the top 0.1% of highly cited researchers. Fraser's vehicle research students have been recognized with over 30 external awards, while Fraser's teaching has been recognized with Teaching Excellence awards, NSF Long-Term Supervisor awards, and his past installment as the MME Teaching Chair.

Jose Rosas-Bustos is an accomplished inventor with multiple granted patents, trade secrets, and proprietary technologies spanning cybersecurity, infrastructure, and quantum technologies. He holds an MBA and serves as the CTO of EigenQ, where he leads the development of advanced quantum systems. José has delivered executive training, graduate-level instruction, and technical workshops on quantum technologies and cybersecurity across academic, corporate, and government institutions. He is currently pursuing a Ph.D. in Mechanical and Mechatronics Engineering at the University of Waterloo, focusing his research on quantum information processing and the quantum internet.

Chapter 1
Introduction to Quantum Computing

The fast development of quantum computing is changing the way we can compute and help us analyze problems. Quantum computing differs from classical computing in that it utilizes the principles of superposition, entanglement, and interference to simultaneously explore a large space of solutions. Through these principles, quantum computers can solve problems that classical systems cannot, with applications in optimization, simulations of quantum systems, and cryptography, among others. In order to innovate in the future, industries must understand and utilize quantum technology to remain competitive in tackling complicated problems.

This book helps engineers and researchers find their way in the world of quantum computing and quantum machine learning. This book combines theory, applications, and hands-on examples to give you a toolkit for designing and implementing quantum algorithms. This book discusses various transformative applications of quantum in the fields of engineering, machine learning, and science, from Shor's algorithm of integer factorization to quantum neural networks. If you're working on optimization problems, building fault-tolerant quantum systems, or changing the face of machine learning models, this book is perfect for the quantum revolution.

1.1 The Evolution of Computing: Classical Vs. Quantum

As the twentieth century progressed, computers followed Moore's law, which states that electronic processing power doubles every year. This allowed digital computers to be progressively faster, more efficient, and more capable. Computers that run on Boolean logic and binary operations have achieved an extraordinary sophistication level, solving problems in virtually every industry and discipline. Even with those improvements, some problems remain unsolvable by classical systems because they are too hard. Quantum computing is a fundamentally different approach. It employs quantum mechanics to solve problems that classical systems cannot solve efficiently. Classical computing and quantum computing are fundamentally different with

J. Van Griensven Thé et al., *Quantum Computing and Quantum Machine Learning for Engineers and Developers*, https://doi.org/10.1007/978-3-031-98245-3_1

respect to their concepts and computational capabilities. It is important to know when quantum computing is best for a particular problem or task or when classical computing is needed.

1.2   A Brief History of Quantum Computing

Quantum computing is an innovative discipline that employs the principles of quantum mechanics to perform computational tasks in ways that classical computers cannot. Its history spans theoretical explorations, mathematical formalizations, and advancements in experimental physics and engineering. This narrative traces the journey of quantum computing from its foundational principles to the milestones that shaped it as a scientific field.

The interaction between theoretical innovation and experimental ingenuity is demonstrated by the history of quantum computing. Quantum computing science has advanced remarkably quickly, from the fundamental ideas of quantum mechanics to the creation of hardware and the realization of quantum algorithms. This evolution represents a wider change in how we comprehend and use the basic laws of nature to solve challenging issues, in addition to reflecting developments in physics and computing.

1.2.1   The Foundations of Quantum Mechanics

The story of quantum computing begins with the development of quantum mechanics in the early twentieth century. Key discoveries during this period laid the groundwork for understanding the unique behaviors of particles at microscopic scales.

1. Max Planck introduced in 1900 the idea that energy is quantized, presenting the concept of energy packets or "quanta." This marked a departure from classical physics and set the stage for the quantum revolution. Note that Planck believed that this quantized energy was only a mathematical trick.
2. Albert Einstein's explanation of the photoelectric effect, in 1905, was the breakthrough that provided evidence of quantized energy levels and demonstrated the particle-like nature of light. His work inspired the development of the quantum framework.
3. Louis de Broglie, in 1924, advanced the understanding of wave-particle duality, showing that particles, such as electrons, exhibit both particle-like and wave-like behavior.
4. Erwin Schrödinger introduced his wave equation model in 1926. This equation describes how quantum states evolve over time, providing a mathematical framework for quantum systems. Note that Max Born provided the correct

understanding that Schrödinger's wave equation describes the probability of a particle's position.

5. Werner Heisenberg formulated the uncertainty principle in 1927, describing the inherent limits in measuring a particle's position and momentum simultaneously. This introduced probabilistic thinking into physics.

These discoveries, as summarized in Fig. 1.1, laid the theoretical foundation that quantum computing would later build upon.

1.2.2   Early Theoretical Connections to Computing

In the mid-twentieth century, theoretical physicists and computer scientists began exploring the implications of quantum mechanics for computation. This evolution is described below and represented in Fig. 1.2.

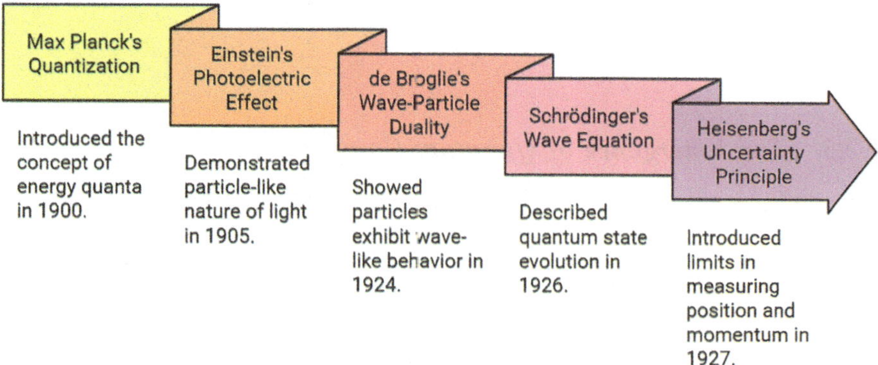

Fig. 1.1   Key developments in early quantum mechanics

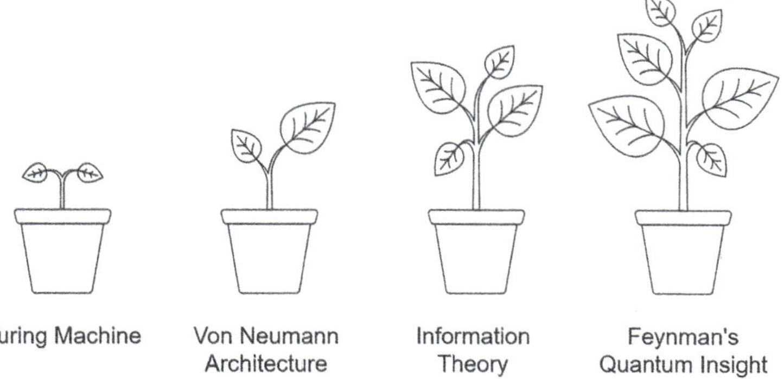

Fig. 1.2   Evolution from theory of computation to classical and quantum computers

1. Although not directly related to quantum mechanics, Alan Turing's development of the Turing machine, in 1936, established the theoretical basis of classical computation. It provided a benchmark against which future computational paradigms, including quantum computing, would be compared.
2. John Von Neumann's foundational work on computer architecture, known as Von Neumann Architecture, was first outlined in the "First Draft of a Report on the EDVAC," completed in June 1945. This was a revolutionary idea at the time, as it proposed that a computer's instructions (program) could be stored in the same memory as its data, enabling greater flexibility and efficiency.
3. In the 1950s and 1960s, Claude Shannon and others laid the foundation for information theory, including the concept of the bit, data transmission, encoding, and noise reduction. This influenced physicists to think about information processing in quantum systems.
4. In 1981, Richard Feynman argued in a lecture titled "Simulating Physics with Computers" that classical computers are inefficient at simulating quantum phenomena. Feynman's insights led to the realization that quantum computing could solve problems beyond the reach of classical computers, particularly in quantum simulations, cryptography, and optimization problems.

1.2.3 The Emergence of Quantum Computation

The late twentieth century saw the formalization of quantum computation, with significant contributions from pioneers who envisioned how quantum mechanics could be harnessed for computational tasks.

1. Paul Benioff, in 1980, was the first to formally describe a quantum Turing machine, providing a theoretical foundation for quantum computation. His work demonstrated that quantum systems could, in principle, perform universal computation.
2. Yuri Manin, in his 1980 book *Computable and Uncomputable*, independently proposed using quantum mechanics for computation, emphasizing its potential to solve problems beyond classical means.
3. In 1981, at a physics and computation conference, Richard Feynman proposed that quantum systems could be used to simulate other quantum systems more efficiently than classical computers. He famously remarked that "nature isn't classical" and suggested building computers that follow quantum laws.
4. In 1985, David Deutsch introduced the concept of quantum gates, analogous to classical logic gates, which operate on quantum bits (qubits). He also proposed a universal quantum computer that could perform any computation (Fig. 1.3).

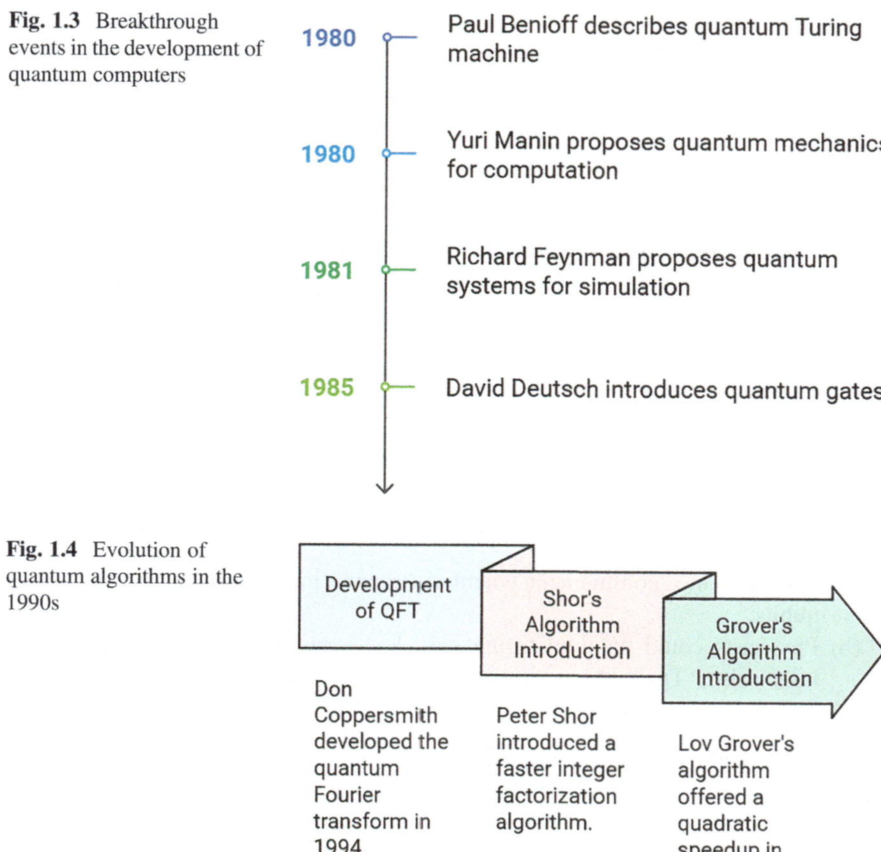

Fig. 1.3 Breakthrough events in the development of quantum computers

Fig. 1.4 Evolution of quantum algorithms in the 1990s

1.2.4   Quantum Algorithms: The Breakthroughs of the 1990s

The 1990s marked a turning point with the discovery of algorithms that showcased quantum computing's superiority for certain tasks. These algorithms highlighted the practical implications of quantum computing and underscored the need to develop functional quantum hardware. This is further explained in the text below and in Fig. 1.4.

1. Don Coppersmith developed the quantum Fourier transform (QFT) in 1994. The QFT is a quantum implementation of the discrete Fourier transform (DFT), which is a mathematical method that extracts frequency components from signals.
2. Peter Shor developed a quantum algorithm for integer factorization, which runs exponentially faster than the best-known classical algorithms. This posed a direct challenge to classical cryptographic systems like RSA, sparking significant interest in quantum computing. Note that this algorithm would not be possible without the preceding development of the quantum Fourier transform.

3. Lov Grover introduced an algorithm for searching unsorted datasets with a quadratic speedup over classical approaches in 1996. Though not as transformative as Shor's algorithm, it demonstrated the broader potential of quantum algorithms.

1.2.5   Development of Quantum Hardware

Building a quantum computer required the theoretical blueprint to be implemented into the physical world, and many interdisciplinary discoveries were made in the process. Here is a quick summary:

1. In 1995, Christopher Monroe and David Wineland achieved a milestone in quantum computing at the National Institute of Standards and Technology (NIST) with trapped ions by constructing one of the first quantum logic gates. The achievement was a testament to two findings:

   (a) Ions held in a confinement potential (trapped ion) could function as stable qubits.
   (b) Physicists could manipulate the particles' entangled quantum states using laser pulses. Trapped ions are an attractive candidate for scalability since they possess long coherence times and high fidelity.

2. In 1995, Andrew Steane and Peter Shor developed quantum error correction methods compensating for the fragile nature of quanta, which allowed for the construction of reliably functioning quantum machines.
3. 1999, John Clarke and his team began exploring superconducting circuits as qubits, which operated a Josephson junction to generate quantized energy levels. While the first qubits made from superconducting circuits were lossy and not manipulable, they paved the way for reasonably practical, scalable quantum processors in the following decades.
4. By 2000, experimentalists were starting to create basic quantum gates and even small-scale quantum circuits. The possibility had been proven, and excitement and anticipated advancements were on the rise.

In 2024, the five leading types of quantum computers are (i) superconducting, (ii) photonic, (iii) neutral atoms, (iv) trapped ions, and (v) quantum dot. Each technology hardware type has its own specific technological and engineering promises and challenges, but common to all are the challenges of scalability and noise limiting the size of problems that can be solved today on a quantum computer.

1.2.6 Academic and Industrial Momentum

The turn of the twenty-first century saw increased investment in quantum computing from academia, government, and private industry.

1. Government funding programs such as the U.S. National Quantum Initiative and similar initiatives in Europe and China underscored the strategic importance of quantum technology.
2. Industry involvement has become critical for the development of quantum computers. Companies like IBM, Google, Xanadu, and Rigetti began building quantum hardware, while others, like D-Wave, pursued quantum annealing as an alternative model.
3. In 2016, IBM launched the IBM Quantum Experience, allowing researchers worldwide to access quantum computers through the cloud. This democratized quantum computing and accelerated its adoption.
4. The release of open-source frameworks like Qiskit (IBM) and Cirq (Google) empowered developers to write quantum algorithms, fostering a global quantum development community.

1.3 Classical Computing Foundations

Classical computers use bits, which represent information in binary form as 0 or 1. Logical operations on these bits are carried out utilizing gates such as AND, OR, and NOT, which change their states to get the desired computing results. These operations are predictable and follow the Boolean algebra concepts.

A classical computer does computations in a sequential manner, following a set of well-defined processes. The Turing machine concept is based on this paradigm, which was codified by Alan Turing. A Turing machine manipulates symbols on tape in accordance with a set of rules, establishing the theoretical basis for classical computation.

Algorithmic efficiency is an important consideration in classical computing. Algorithms are evaluated based on their temporal complexity, which evaluates how calculation time varies with input size. Examples include:

1. **Sorting Algorithms**

    Quicksort, a widely used algorithm, operates with a best-case time complexity of $O(n \log n)$, where n is the number of elements to be sorted. Its divide-and-conquer approach makes it highly efficient for large datasets.
2. **Combinatorial Optimization**

    Problems like the traveling salesman problem (TSP) require finding the shortest path visiting a set of cities and returning to the starting point. A brute-force solution evaluates all $n!$ permutations of cities, resulting in a factorial time complexity $O(n!)$. Such scaling makes classical approaches infeasible for large n.

While classical computers excel at sequential, well-defined tasks like adding numbers, they struggle with problems requiring exponential exploration of possibilities. For example, factoring large integers, a critical task in cryptography, requires significant computational resources as the input size grows.

Classical computing is also constrained in its ability to perform parallel computations. Multi-core processors and distributed computing address this limitation to some extent, but they rely on duplicating classical hardware and dividing tasks rather than inherently altering the computational model.

1.4 Introduction to Linear Algebra for Quantum Computing

In both theoretical and practical applications, linear algebra knowledge is essential for solving linear equations, evaluating geometric transformations, and comprehending vector spaces. These ideas are essential to comprehending quantum computing. For example, the tensor product joins vector spaces, the vector or cross product computes orthogonal vectors, and the dot product assesses alignment. These ideas connect linear algebra to data science, quantum mechanics, and engineering applications.

1.4.1 Vector and Matrix Operations

Vectors and matrices are fundamental objects in linear algebra that allow us to represent and manipulate data efficiently. Their operations, such as addition, multiplication, and transformations, are the foundation for solving systems of equations, performing data analysis, and working in physics, engineering, and computer science.

This section will cover the most important operations on vectors and matrices with their formal definitions and properties. Table 1.1 summarizes these operations.

1.4.1.1 Vector Addition and Subtraction

Vectors of the same dimension can be added or subtracted by performing element-wise operations.

For two vectors $\mathbf{u} = [u_1, u_2, \ldots, u_n]$ and $\mathbf{v} = [v_1, v_2, \ldots, v_n]$, addition and subtraction are defined as:

$$\mathbf{u} + \mathbf{v} = [u_1 + v_1, u_2 + v_2, \ldots, u_n + v_n]$$

Table 1.1 Summary of vectors and matrices operations

Operation	Description	Output
Vector addition	Element-wise addition of vectors	Vector
Scalar multiplication	Scaling a vector/matrix by a scalar	Vector/ matrix
Matrix addition	Element-wise addition of matrices	Matrix
Matrix transpose	Flip rows into columns	Matrix
Matrix multiplication	Multiply compatible matrices	Matrix
Determinant	Scalar value indicating matrix properties	Scalar
Matrix inverse	Matrix that satisfies $AA^{-1} = I$	Matrix
Dot product	The sum of the products of corresponding entries of two vectors	Scalar
Cross product	A vector perpendicular to two given vectors in three-dimensional space	Vector
Tensor product	An operation on two vectors or matrices resulting in a higher-dimensional tensor	Tensor

$$\mathbf{u} - \mathbf{v} = \left[u_1 - v_1, u_2 - v_2, \ldots, u_n - v_n\right].$$

Properties

(a) Commutativity: $\mathbf{u} + \mathbf{v} = \mathbf{v} + \mathbf{u}$.
(b) Associativity: $(\mathbf{u} + \mathbf{v}) + \mathbf{w} = \mathbf{u} + (\mathbf{v} + \mathbf{w})$.

Python Code Example: Vector Addition and Subtraction

```python
#-------------------------------------------------------------
# Vector Addition and Subtraction
# Chapter 1 in the QUANTUM COMPUTING AND QUANTUM MACHINE LEARNING BOOK
#-------------------------------------------------------------
# Version 1.0
# Qiskit changes frequently.
# We recommend using the latest version from the book code repository at:
# https://aqtinitiative.org/quantum-computing-for-engineers
#
# (c) 2025 Jesse Van Griensven, Roydon Fraser, and Jose Rosas
# License: MIT - Citation required
#-------------------------------------------------------------
import numpy as np

# Define vectors
u = np.array([1, 2, 3])
v = np.array([4, 5, 6])

# Vector addition
add_result = u + v
```

```python
# Vector subtraction
sub_result = u - v

print("Vector Addition and Subtraction example")
print("Vector u:", u)
print("Vector v:", v)
print("Vector Addition:  ", add_result)
print("Vector Subtraction:", sub_result)

# Output:
# Vector Addition and Subtraction example
# Vector u: [1 2 3]
# Vector v: [4 5 6]
# Vector Addition:   [5 7 9]
# Vector Subtraction: [-3 -3 -3]
```

Figure 1.5 depicts the graphical addition of two vectors.

1.4.1.2 Scalar Multiplication

Multiplying a vector $\mathbf{u} = [u_1, u_2, ..., u_n]$ by a scalar $c \in \mathbb{R}$ scales each component of the vector:

$$c\mathbf{u} = [cu_1, cu_2, ..., cu_n]$$

Multiplications of a vector by a scalar are graphically represented in Fig. 1.6.

Python Code Example: Scalar Multiplication

```python
#---------------------------------------------------------------------
# Vector Scalar Multiplication
# Chapter 1 in the QUANTUM COMPUTING AND QUANTUM MACHINE LEARNING BOOK
#---------------------------------------------------------------------
# Version 1.0
# Qiskit changes frequently.
# We recommend using the latest version from the book code repository at:
# https://aqtinitiative.org/quantum-computing-for-engineers
#
# (c) 2025 Jesse Van Griensven, Roydon Fraser, and Jose Rosas
# License: MIT - Citation required
#---------------------------------------------------------------------
import numpy as np

# Define vectors
u = np.array([1, 2, 3])
v = np.array([4, 5, 6])
```

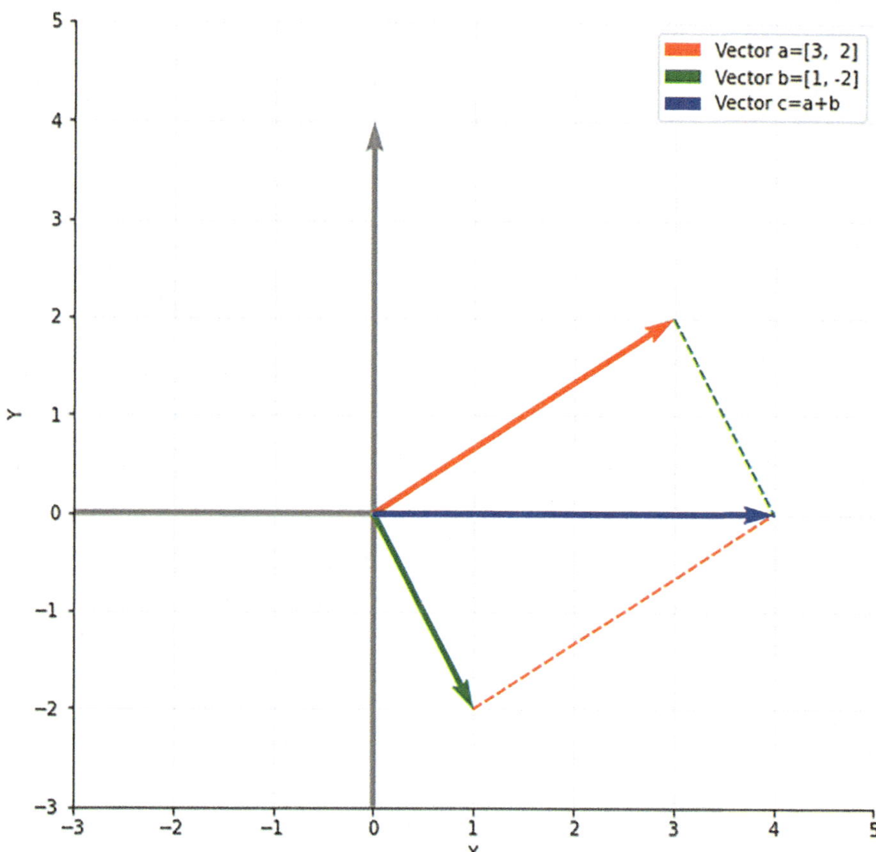

Fig. 1.5 Addition of vectors **a** and **b** (by the authors)

```
# Define Scalar
scalar = 3


# Execute multiplication operation
scaled_vector_u = scalar * u
scaled_vector_v = scalar * v

# Print Results
print("Vector u:", u)
print("Vector v:", v)
print("Scalar Multiplication u:", scaled_vector_u)
print("Scalar Multiplication v:", scaled_vector_v)

#Output:
# Vector u: [1 2 3]
```

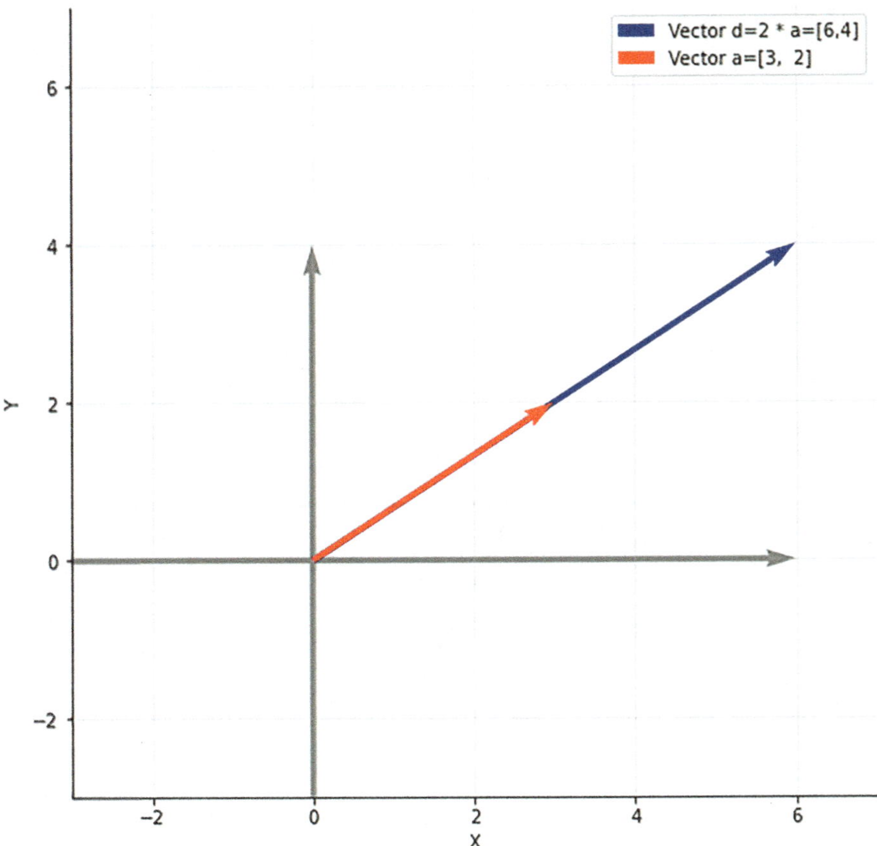

Fig. 1.6 Scalar multiplication of vector **a** by scalar 2 (by the authors)

```
# Vector v: [4 5 6]
# Scalar Multiplication u: [3 6 9]
# Scalar Multiplication v: [12 15 18]
```

1.4.1.3   Vector Norm (Magnitude)

The magnitude of a vector, also referred to as its length, corresponds to the Euclidean norm, which generalizes the Pythagorean theorem to n-dimensional space. This magnitude, called the "norm" of a vector $\mathbf{u} = [u_1, u_2, ..., u_n]$, is defined as:

$$\| \mathbf{u} \| = \sqrt{u_1^2 + u_2^2 + \cdots + u_n^2}$$

In Python, the norm can be calculated using `numpy.linalg.norm`.

Python Code Example: Vector Norm

```python
#-----------------------------------------------------------------
# Vector Norm
# Chapter 1 in the QUANTUM COMPUTING AND QUANTUM MACHINE LEARNING BOOK
#-----------------------------------------------------------------
# Version 1.0
# Qiskit changes frequently.
# We recommend using the latest version from the book code repository at:
# https://aqtinitiative.org/quantum-computing-for-engineers
#
# (c) 2025 Jesse Van Griensven, Rcydon Fraser, and Jose Rosas
# License: MIT - Citation required
#-----------------------------------------------------------------
import numpy as np

# Define vectors
u = np.array([1, 2, 3])
v = np.array([4, 5, 6])

# Execute NORM operation
norm_u = np.linalg.norm(u)
norm_v = np.linalg.norm(v)

# Print Results
print("Vector u:", u)
print("Vector v:", v)
print("Norm u:", norm_u)
print("Norm v:", norm_v)
# Output
# Vector u: [1 2 3]
# Vector v: [4 5 6]
# Norm u: 3.7416573
# Norm v: 8.7749643
```

1.4.1.4 Matrix Addition and Subtraction

Two matrices A and B of the same dimensions $m \times n$ can be added or subtracted element-wise:

$$A + B = \left[a_{ij} + b_{ij}\right], \quad A - B = \left[a_{ij} - b_{ij}\right]$$

1.4.1.5   Matrix Scalar Multiplication

Multiplying a matrix $A = [a_{ij}]$ by a scalar $c \in \mathbb{R}$ scales every element of the matrix:

$$cA = \left[ca_{ij} \right]$$

Python Code Example: Matrix Scalar Multiplication

```python
#-------------------------------------------------------------------
# Matrix Scalar Multiplication
# Chapter 1 in the QUANTUM COMPUTING AND QUANTUM MACHINE LEARNING BOOK
#-------------------------------------------------------------------
# Version 1.0
# Qiskit changes frequently.
# We recommend using the latest version from the book code repository at:
# https://aqtinitiative.org/quantum-computing-for-engineers
#
# (c) 2025 Jesse Van Griensven, Roydon Fraser, and Jose Rosas
# License: MIT - Citation required
#-------------------------------------------------------------------
import numpy as np

# Define the SYMPY routines we need
import sympy as sp
#-------------------------------------------------------------------
def sprint(Matrix):
    """ Prints a numpy Matrix in a nice format with sympy """
    # The input Matrix can be a Numpy or a Sympy Matrix
    SMatrix = sp.Matrix(Matrix)
    display( SMatrix )
    return
#-------------------------------------------------------------------

# Define Matrices A and B
A = np.array([[1, 2], [3, 4]])
B = np.array([[5, 6], [7, 8]])

scalar = 2

# Execute scalar multiplication
scaled_matrix_A = scalar * A
scaled_matrix_B = scalar * B

# Print Results
print("Matrix A:")
```

```
sprint(A)
print()

print("Matrix B:")
sprint(B)
print()

print("Scalar Multiplication A:")
sprint(scaled_matrix_A)
print()

print("Scalar Multiplication B:")
sprint(scaled_matrix_B)
print()
```

1.4.1.6   Matrix Transpose

The transpose of a matrix A flips its rows and columns:

$$A_{ij}^{T} = A_{ji}$$

For a matrix $A = \begin{bmatrix} 1 & 2 \\ 3 & 4 \end{bmatrix}$, its transpose is:

$$A^{T} = \begin{bmatrix} 1 & 3 \\ 2 & 4 \end{bmatrix}$$

Python Code Example: Matrix Transpose

```
#------------------------------------------------------------------
# Matrix Transpose
# Chapter 1 in the QUANTUM COMPUTING AND QUANTUM MACHINE LEARNING BOOK
#------------------------------------------------------------------
# Version 1.0
# Qiskit changes frequently.
# We recommend using the latest version from the book code repository at:
# https://aqtinitiative.org/quantum-computing-for-engineers
#
# (c) 2025 Jesse Van Griensven, Roydon Fraser, and Jose Rosas
# License: MIT - Citation required
#------------------------------------------------------------------
import numpy as np
```

```python
# Define the SYMPY routines we need
import sympy as sp
#-------------------------------------------------------------------
def sprint(Matrix):
    """ Prints a numpy Matrix in a nice format with sympy """
    # The input Matrix can be a Numpy or a Sympy Matrix
    SMatrix = sp.Matrix(Matrix)
    display( SMatrix )
    return
#-------------------------------------------------------------------

# Define Matrices A and B
A = np.array([[1, 2], [3, 4]])
B = np.array([[5, 6], [7, 8]])

# Execute the Transpose
transpose_A = A.T
transpose_B = B.T

# Print Results
print("Matrix A:")
sprint(A)
print()

print("Matrix B:")
sprint(B)
print()

print("Transpose of A:")
sprint(transpose_A)
print()

print("Transpose of B:")
sprint(transpose_B)
print()
```

1.4.1.7 Matrix Multiplication

The product of two matrices $A \in \mathbb{R}^{m \times n}$ and $B \in \mathbb{R}^{n \times p}$ is a matrix $C \in \mathbb{R}^{m \times p}$, where:

$$C_{ij} = \sum_{k=1}^{n} A_{ik}B_{kj}$$

Matrix multiplication is not commutative: $AB \neq BA$.

Python Code Example: Matrix Multiplication

```python
#-----------------------------------------------------------------
# Matrix Multiplication
# Chapter 1 in the QUANTUM COMPUTING AND QUANTUM MACHINE LEARNING BOOK
#-----------------------------------------------------------------
# Version 1.0
# Qiskit changes frequently.
# We recommend using the latest version from the book code repository at:
# https://aqtinitiative.org/quantum-computing-for-engineers
#
# (c) 2025 Jesse Van Griensven, Roydon Fraser, and Jose Rosas
# License: MIT - Citation required
#-----------------------------------------------------------------
import numpy as np

# Define the SYMPY routines we need
import sympy as sp
#-----------------------------------------------------------------
def sprint(Matrix):
    """ Prints a numpy Matrix in a nice format with sympy """
    # The input Matrix can be a Numpy or a Sympy Matrix
    SMatrix = sp.Matrix(Matrix)
    display( SMatrix )
    return
#-----------------------------------------------------------------

# Define Matrices A and B
A = np.array([[1, 2], [3, 4]])
B = np.array([[5, 6], [7, 8]])

Multi_A_B = np.dot(A, B)  # Matrix multiplication

# Print Results
print("Matrix A:")
sprint(A)
print()
```

```
print("Matrix B:")
sprint(B)
print()

print("Matrix Multiplication:")
sprint(Multi_A_B)
print()
```

1.4.1.8 Identity Matrix

The identity matrix I_n is a square $n \times n$ matrix with ones on the diagonal and zeros elsewhere:

$$
I_n = \begin{bmatrix} 1 & 0 & \cdots & 0 \\ 0 & 1 & \cdots & 0 \\ & & \ddots & \\ 0 & 0 & \cdots & 1 \end{bmatrix}
$$

For any matrix A, multiplying with I does not change A: $AI = IA = A$.

Python Code Example: Identity Matrix

```
#-------------------------------------------------------------------
# Matrix Multiplication
# Chapter 1 in the QUANTUM COMPUTING AND QUANTUM MACHINE LEARNING BOOK
#-------------------------------------------------------------------
# Version 1.0
# Qiskit changes frequently.
# We recommend using the latest version from the book code repository at:
# https://aqtinitiative.org/quantum-computing-for-engineers
#
# (c) 2025 Jesse Van Griensven, Roydon Fraser, and Jose Rosas
# License: MIT - Citation required
#-------------------------------------------------------------------
import numpy as np

I = np.identity(2)
print("Identity Matrix:\n", I)
```

1.4.1.9 Determinant of a Matrix

The determinant of a square matrix A is a scalar that provides information about the matrix, such as:

1. Whether the matrix is invertible ($\det(A) \neq 0$).
2. The scaling factor for a linear transformation.

For a 2×2 matrix A:

$$A = \begin{bmatrix} a & b \\ c & d \end{bmatrix}, \quad \det(A) = ad - bc$$

For a 3×3 matrix B:

$$B = \begin{bmatrix} a & b & c \\ d & e & f \\ g & h & i \end{bmatrix} = a\,det\begin{bmatrix} e & f \\ h & i \end{bmatrix} - b\,det\begin{bmatrix} d & f \\ g & i \end{bmatrix} + c\,det\begin{bmatrix} d & e \\ g & h \end{bmatrix}$$

The determinant of B, denoted as $\det(B)$, is calculated as:

$$\det(B) = a(ei - fh) - b(di - fg) + c(dh - eg)$$

Example Consider the matrix:

$$B = \begin{bmatrix} 1 & 2 & 3 \\ 0 & 4 & 5 \\ 1 & 0 & 6 \end{bmatrix}$$

Calculating its determinant:

$$\det(B) = 1(4 \cdot 6 - 5 \cdot 0) - 2(0 \cdot 6 - 5 \cdot 1) + 3(0 \cdot 0 - 4 \cdot 1)$$
$$\det(B) = 1(24 - 0) - 2(0 - 5) + 3(0 - 4)$$
$$\det(B) = 24 + 10 - 12 = 22$$

Since $\det(B) \neq 0$, matrix B is invertible.

Python Code Example: Determinant of A

```
#-------------------------------------------------------------------
# Matrix Multiplication
# Chapter 1 in the QUANTUM COMPUTING AND QUANTUM MACHINE LEARNING BOOK
#-------------------------------------------------------------------
# Version 1.0
# Qiskit changes frequently.
# We recommend using the latest version from the book code repository at:
# https://aqtinitiative.org/quantum-computing-for-engineers
#
# (c) 2025 Jesse Van Griensven, Roydon Fraser, and Jose Rosas
# License: MIT - Citation required
```

```
#--------------------------------------------------------------------
import numpy as np

# Define the SYMPY routines we need
import sympy as sp
from IPython.display import display

#----------------------------------- ----------------------------------
def sprint(Matrix, decimals=4):
  """ Prints a numpy Matrix in a nice format with sympy and specified
precision """
  # The input Matrix can be a Numpy or a Sympy Matrix
  SMatrix = sp.Matrix(Matrix)
  # Apply the rounding to each element
  SMatrix = SMatrix.applyfunc(lambda x: round(float(x), decimals) if
isinstance(x, (float, int, sp.Float)) else x)
  display(SMatrix)
  return
#----------------------------------- ----------------------------------

# Define Matrices A and B
A = np.array([[1, 2], [3, 4]])
B = np.array([[5, 6], [7, 8]])

det_A = np.linalg.det(A)
det_B = np.linalg.det(B)

# Print matrices and their determinants
print("Matrix A:")
sprint(A)
print()

print("Matrix B:")
sprint(B)
print()

print("Determinant of A: {:.4f}".format(det_A))
print("Determinant of B: {:.4f}".format(det_B))
```

1.4.1.10 Matrix Inverse

The inverse of a square matrix A is A^{-1}, such that:

$$AA^{-1} = A^{-1}A = I$$

A matrix has an inverse only if it is non-singular (i.e, $(\det(A) \neq 0)$).

Python Code Example: Matrix Inverse

```python
#--------------------------------------------------------------------
# Matrix Inverse
# Chapter 1 in the QUANTUM COMPUTING AND QUANTUM MACHINE LEARNING BOOK
#--------------------------------------------------------------------
# Version 1.0
# Qiskit changes frequently.
# We recommend using the latest version from the book code repository at:
# https://aqtinitiative.org/quantum-computing-for-engineers
#
# (c) 2025 Jesse Van Griensven, Roydon Fraser, and Jose Rosas
# License: MIT - Citation required
#--------------------------------------------------------------------
import numpy as np


#--------------------------------------------------------------------
def sprint(Matrix):
    """ Prints a numpy Matrix in a nice format with sympy """

# Define the SYMPY routines we need
import sympy as sp

    # The input Matrix can be a Numpy or a Sympy Matrix
    SMatrix = sp.Matrix(Matrix)
    display( SMatrix )
    return
#--------------------------------------------------------------------


# Define Matrices A and B
A = np.array([[1, 2], [3, 4]])
B = np.array([[5, 6], [7, 8]])

# Execute the Matrix Inversions
A_inv = np.linalg.inv(A)
B_inv = np.linalg.inv(B)

# Print Results
print("Matrix Inverse A:")
sprint(A_inv)
print()
```

```
print("Matrix Inverse B:")
sprint(B_inv)
print()

# Check if the inverse is correct. Multiplying a Matrix by its inverse
should yield the identity matrix
print("Checking A dot its inverse. We should get the identity matrix")
sprint(np.dot(A, A_inv))
```

1.4.2   Vector Spaces

A collection of objects known as vectors that meet specific algebraic requirements
makes up a vector space, a basic idea in linear algebra. Numerous applications in
mathematics, physics, and engineering require the knowledge of vector space
concepts.

1.4.2.1   Definition of Vector Space

A vector space V over a field F (e.g., the real numbers $\mathbb{R}$ or complex numbers $\mathbb{C}$) is a
set of elements (vectors) that satisfies the following properties:

1. **Closure Under Addition**
     For any two vectors $\mathbf{u}, \mathbf{v} \in V$, the sum $\mathbf{u} + \mathbf{v} \in V$.
2. **Closure Under Scalar Multiplication**
     For any $\mathbf{v} \in V$ and any scalar $c \in F$, the product $c\mathbf{v} \in V$.
3. **Properties of Vector Addition**

   (a) Commutativity: $\mathbf{u} + \mathbf{v} = \mathbf{v} + \mathbf{u}$.
   (b) Associativity: $(\mathbf{u} + \mathbf{v}) + \mathbf{w} = \mathbf{u} + (\mathbf{v} + \mathbf{w})$.
   (c) Existence of a zero vector 0: $\mathbf{v} + 0 = \mathbf{v}$.
   (d) Existence of additive inverses: For each $\mathbf{v}$, there exists $-\mathbf{v}$ such that $\mathbf{v} + (-\mathbf{v}) = 0$.

4. **Properties of Scalar Multiplication**

   (a) Distributivity: $c(\mathbf{u} + \mathbf{v}) = c\mathbf{u} + c\mathbf{v}$.
   (b) Compatibility: $(c_1 + c_2)\mathbf{v} = c_1\mathbf{v} + c_2\mathbf{v}$.
   (c) Associativity: $c_1(c_2\mathbf{v}) = (c_1 c_2)\mathbf{v}$.
   (d) Identity: $1\mathbf{v} = \mathbf{v}$, where 1 is the multiplicative identity.

Examples of Vector Spaces
1. Euclidean Space $\mathbb{R}^n$

The set of all n-dimensional real-valued vectors forms a vector space. For instance:

$$\mathbf{v} = \begin{bmatrix} v_1 \\ v_2 \\ \\ v_n \end{bmatrix} \quad \text{where } v_i \in \mathbb{R}$$

The corresponding **vector field** is the existence of a vector at every real value point in $\mathbb{R}^n$.

1. Polynomials

The set of all polynomials of degree less than or equal to n:

$$P_n = \{a_0 + a_1 x + \cdots + a_n x^n \mid a_i \in \mathbb{R}\}$$

2. Matrices

The set of all $m \times n$ matrices over $\mathbb{R}$ forms a vector space:

$$A = \begin{bmatrix} a_{11} & a_{12} & \cdots & a_{1n} \\ & & \ddots & \\ a_{m1} & a_{m2} & \cdots & a_{mn} \end{bmatrix}$$

3. Functions

The set of all continuous real-valued functions $f : \mathbb{R} \to \mathbb{R}$.

1.4.2.2   Subspaces

A subspace $W \subseteq V$ is a subset of a vector space V that is itself a vector space under the same addition and scalar multiplication. To verify that W is a subspace of V:

1. $0 \in W$ (contains the zero vector).
2. If $\mathbf{u}, \mathbf{v} \in W$, then $\mathbf{u} + \mathbf{v} \in W$ (closed under addition).
3. If $c \in F$ and $\mathbf{v} \in W$, then $c\mathbf{v} \in W$ (closed under scalar multiplication).

Example   The set of vectors in $\mathbb{R}^3$, where $x + y + z = 0$ forms a subspace of $\mathbb{R}^3$.

1.4.2.3   Linear Combinations and Span

A vector $\mathbf{v}$ is a linear combination of vectors $\mathbf{v}_1, \mathbf{v}_2, \ldots, \mathbf{v}_k$, if there exist scalars $c_1, c_2, \ldots, c_k$ such that:

$$\mathbf{v} = c_1\mathbf{v}_1 + c_2\mathbf{v}_2 + \cdots + c_k\mathbf{v}_k$$

The span of vectors $\mathbf{v}_1$, $\mathbf{v}_2$, ..., $\mathbf{v}_k$ is the set of all possible linear combinations:

$$\text{span}\{\mathbf{v}_1, \mathbf{v}_2, ..., \mathbf{v}_k\} = \left\{ \sum_{i=1}^{k} c_i\mathbf{v}_i \mid c_i \in \mathbb{R} \right\}$$

If the span of a set of vectors equals the entire vector space V, then those vectors **span** V.

Python Example Code: Vector Spaces

This is an example of verifying linear combinations and the span of vectors in Python using NumPy.

1. **Linear Combination Verification**

    The function checks if the target vector is a linear combination of the given basis vectors using `np.linalg.solve`.

2. **Span Demonstration**

    A linear combination of the basis vectors v_1 and v_2 is computed.

```python
#------------------------------------------------------------------
# Vector Spaces
# Chapter 1 in the QUANTUM COMPUTING AND QUANTUM MACHINE LEARNING BOOK
#------------------------------------------------------------------
# Version 1.0
# (c) 2025 Jesse Van Griensven, Roydon Fraser, and Jose Rosas
# License: MIT - Citation required
#------------------------------------------------------------------
import numpy as np
import sympy as sp
#------------------------------------ -----------------------------
def sprint(Matrix):
    """ Prints a numpy Matrix in a nice format with sympy """
    # The input Matrix can be a Numpy or a Sympy Matrix
    SMatrix = sp.Matrix(Matrix)
    display( SMatrix )
    return


#------------------------------------ -----------------------------
# Check if a vector is a linear combination of v1, v2, v3
def is_linear_combination(target, basis_vectors):
    # Solve the system Ax = b, where A = basis_vectors, b = target
    A = np.column_stack(basis_vectors)
    b = target
```

```python
    try:
        solution = np.linalg.solve(A, b)
        print("The target vector is a linear combination.")
        print("Solution (coefficients):", solution)
    except np.linalg.LinAlgError:
        print("The target vector is NOT a linear combination.")

    return
#------------------------------------ ----------------------------------

# Define vectors in R^3
v1 = np.array([1, 0, 0])
v2 = np.array([0, 1, 0])
v3 = np.array([0, 0, 1])

# Target vector
target_vector = np.array([1, 2, 3])
basis_vectors = [v1, v2, v3]

# Print Inputs
print("vectors in R^3")
print("v1: \n", v1)
print("v2: \n", v2)
print("v3: \n", v3)
print()
print("Target Vector: \n", target_vector)
print()
print("Basis Vector:")
sprint(basis_vectors)
print()

# Check if a vector is a linear combination of v1, v2, v3
is_linear_combination(target_vector, basis_vectors)

# Span demonstration: create a span of vectors
span_vectors = [v1, v2]
linear_combination = 2 * v1 + 3 * v2
print("Linear Combination of v1 and v2:", linear_combination)
```

1.4.3   Vector Basis and Coordinates

The concepts of vector basis and coordinates are fundamental in understanding the
structure of vector spaces. They provide a systematic way to represent vectors,

enabling operations such as linear transformations, projections, and changes of perspective within vector spaces.

These are the basis and coordinates key concepts for the vector spaces:

1. A basis is a set of linearly independent vectors that spans a vector space.
2. The coordinates of a vector are the unique scalars representing the vector as a linear combination of basis vectors.
3. A change of basis involves transforming vector coordinates from one basis to another using a change-of-basis matrix.

1.4.4 Basis of a Vector Space

A basis for a vector space V is a set of linearly independent vectors that span the entire space. Every vector in V can be written uniquely as a linear combination of the basis vectors.

A set of vectors $\{\mathbf{v}_1, \mathbf{v}_2, \ldots, \mathbf{v}_n\}$ is a basis for a vector space V if:

1. The vectors are linearly independent.
2. The vectors span the space V.

Formally, for any vector $\mathbf{v} \in V$, there exist unique scalars $c_1, c_2, \ldots, c_n$ such that:

$$\mathbf{v} = c_1 \mathbf{v}_1 + c_2 \mathbf{v}_2 + \cdots + c_n \mathbf{v}_n.$$

- The scalars $c_1, c_2, \ldots, c_n$ are called the coordinates of $\mathbf{v}$ with respect to the basis $\{\mathbf{v}_1, \mathbf{v}_2, \ldots, \mathbf{v}_n\}$.

Example of a Basis
1. Standard Basis for $\mathbb{R}^n$

   The standard basis for $\mathbb{R}^n$ consists of vectors with all coordinates equal 0 except one:

$$\mathbf{e}_1 = \begin{bmatrix} 1 \\ 0 \\ 0 \end{bmatrix}, \mathbf{e}_2 = \begin{bmatrix} 0 \\ 1 \\ 0 \end{bmatrix}, \ldots, \mathbf{e}_n = \begin{bmatrix} 0 \\ 0 \\ 1 \end{bmatrix}$$

   Every vector $\mathbf{v} \in \mathbb{R}^n$ can be expressed uniquely as:

$$\mathbf{v} = v_1 \mathbf{e}_1 + v_2 \mathbf{e}_2 + \cdots + v_n \mathbf{e}_n$$

   where $v_1, v_2, \ldots, v_n$ are the coordinates of $\mathbf{v}$.
2. Non-standard Basis

   Consider $\mathbb{R}^2$ with basis vectors $\mathbf{b}_1 = \begin{bmatrix} 1 \\ 1 \end{bmatrix}$ and $\mathbf{b}_2 = \begin{bmatrix} -1 \\ 1 \end{bmatrix}$. A vector $\mathbf{v} = \begin{bmatrix} 2 \\ 4 \end{bmatrix}$ can be expressed as:

$$\mathbf{v} = c_1 \mathbf{b}_1 + c_2 \mathbf{b}_2$$

To find the *two* scalars c_1 and c_2, we solve the *two* linearly independent coordinate equations given by:

$$c_1 \begin{bmatrix} 1 \\ 1 \end{bmatrix} + c_2 \begin{bmatrix} -1 \\ 1 \end{bmatrix} = \begin{bmatrix} 2 \\ 4 \end{bmatrix}$$

The equations corresponding to this multiplication and addition of basis vectors are linearly independent because the basis vectors are linearly combined and independent.

1.4.5 Coordinates of a Vector

The coordinates of a vector $\mathbf{v}$ with respect to a basis $B = \{\mathbf{b}_1, \mathbf{b}_2, ..., \mathbf{b}_n\}$ are the scalars $c_1, c_2, ..., c_n$ such that:

$$\mathbf{v} = c_1 \mathbf{b}_1 + c_2 \mathbf{b}_2 + \cdots + c_n \mathbf{b}_n$$

The vector $\mathbf{c} = \begin{bmatrix} c_1 \\ c_2 \\ \\ c_n \end{bmatrix}$ is called the coordinate vector of $\mathbf{v}$ with respect to the basis B. This allows any vector in a space to be represented uniquely as an ordered list or tuple of numbers.

Example

Let $\mathbf{b}_1 = \begin{bmatrix} 1 \\ 0 \end{bmatrix}$ and $\mathbf{b}_2 = \begin{bmatrix} 0 \\ 1 \end{bmatrix}$, which form the standard basis for $\mathbb{R}^2$. If $\mathbf{v} = \begin{bmatrix} 3 \\ 5 \end{bmatrix}$, then the coordinates of $\mathbf{v}$ with respect to $B = \{\mathbf{b}_1, \mathbf{b}_2\}$ are $[3, 5]$.

If the basis is changed to $\mathbf{b}_1' = \begin{bmatrix} 1 \\ 1 \end{bmatrix}$ and $\mathbf{b}_2' = \begin{bmatrix} 1 \\ -1 \end{bmatrix}$, the coordinates will differ. To find the new coordinates, we solve a system of linear equations.

1.4.6 Change of Basis

Changing the basis allows us to express a vector with respect to a different set of basis vectors. If a vector $\mathbf{v}$ has coordinates $\mathbf{c}$ in basis B and $\mathbf{d}$ in basis B', the relation between the two coordinate systems is given by a change-of-basis matrix P:

$$\mathbf{c} = \boldsymbol{P}\mathbf{d}$$

where $\boldsymbol{P}$ transforms coordinates from one basis to another.

To compute $\boldsymbol{P}$, express each new basis vector $\mathbf{b}'_j$ as a linear combination of the original basis $\mathbf{b}_i$.

Python Example Code: Finding Coordinates and Change of Basis
The following Python code uses NumPy to compute vector coordinates with respect to a non-standard basis and perform a basis change.

```python
#-----------------------------------------------------------------------
# Change of Basis
# Chapter 1 in the QUANTUM COMPUTING AND QUANTUM MACHINE LEARNING BOOK
#-----------------------------------------------------------------------
# Version 1.0
# (c) 2025 Jesse Van Griensven, Roydon Fraser, and Jose Rosas
# License: MIT - Citation required
#-----------------------------------------------------------------------
import numpy as np
import sympy as sp
#-------------------------------------- ----------------------------------
def sprint(Matrix):
    """ Prints a numpy Matrix in a nice format with sympy """
    # The input Matrix can be a Numpy or a Sympy Matrix
    SMatrix = sp.Matrix(Matrix)
    display( SMatrix )
    return
#-------------------------------------- ----------------------------------

# Define the original basis vectors
b1 = np.array([1, 1])
b2 = np.array([-1, 1])
B = np.column_stack((b1, b2))  # Basis matrix
Print("Basis Matrix")
sprint(B)
print()

# Define the vector v
v = np.array([2, 4])

# Solve for the coordinates of v in basis B
coordinates = np.linalg.solve(B, v)
print("Coordinates of v in the new basis B:", coordinates)
print()
```

```
# Define a new basis B'
b1_new = np.array([1, 0])
b2_new = np.array([0, 1])
B_new = np.column_stack((b1_new, b2_new))  # New basis matrix
print("New Basis Matrix")
sprint(B_new)
print()


# Change of basis matrix from B to B'
P = np.linalg.solve(B_new, B)
print("Change of basis matrix from B' to B:")
sprint(P)
print()


# Verify the transformation
v_new_basis = np.dot(P, coordinates)
print("Vector in new basis B':", v_new_basis)
```

1.4.7   "Products": Dot (Scalar), Vector (Cross), and Tensor

In linear algebra, different types of "products" are used to combine vectors, matrices, or tensors. These products help describe geometric, physical, and mathematical relationships between vectors and higher-dimensional objects. The most important products are:

1. Dot product (scalar product)
2. Cross product (vector product)
3. Tensor product

Each of these products has its unique properties, interpretations, and applications. Table 1.2 is a summary of these products.

Table 1.2 Summary of linear algebra products

Product	Definition	Result	Application
Dot product	$\mathbf{u} \cdot \mathbf{v}$	Scalar	Measure alignment, projections
Vector product	$\mathbf{u} \times \mathbf{v}$	Orthogonal vector to $(\mathbf{u},\mathbf{v})$ plane	Physics (torque, magnetic fields)
Tensor product	$\mathbf{u} \otimes \mathbf{v}$	Matrix/tensor	Quantum mechanics, ML weights

1.4.7.1   Dot Product (Scalar Product)

The dot product of two vectors $\mathbf{u}$ and $\mathbf{v}$ in $\mathbb{R}^n$ is defined as:

$$\mathbf{u} \cdot \mathbf{v} = \sum_{i=1}^{n} u_i v_i$$

The result of a dot product is a scalar (a single number), hence the name scalar product.

Geometric Interpretation

The dot product measures the alignment between two vectors. If θ is the angle between $\mathbf{u}$ and $\mathbf{v}$, then:

$$\mathbf{u} \cdot \mathbf{v} = \parallel \mathbf{u} \parallel \parallel \mathbf{v} \parallel \cos(\theta)$$

where $\|\mathbf{u}\|$ and $\|\mathbf{v}\|$ are the magnitudes (norms) of $\mathbf{u}$ and $\mathbf{v}$.

(a) If $\mathbf{u} \cdot \mathbf{v} > 0$: The vectors point in roughly the same direction.
(b) If $\mathbf{u} \cdot \mathbf{v} < 0$: The vectors point in opposite directions.
(c) If $\mathbf{u} \cdot \mathbf{v} = 0$: The vectors are orthogonal (perpendicular).

Example

For two vectors $\mathbf{u} = [1, 2, 3]$ and $\mathbf{v} = [4, -5, 6]$:

$$\mathbf{u} \cdot \mathbf{v} = (1)(4) + (2)(-5) + (3)(6) = 4 - 10 + 18 = 12$$

Figure 1.7 demonstrates the effect of a dot product as a projection.

1.4.8   Vector Product (Cross Product)

The vector product (cross product) of two vectors $\mathbf{u}$ and $\mathbf{v}$ in $\mathbb{R}^3$ results in a new vector $\mathbf{w}$, defined as:

$$\mathbf{w} = \mathbf{u} \times \mathbf{v}$$

where

$$\mathbf{w} = \begin{bmatrix} u_2 v_3 - u_3 v_2 \\ u_3 v_1 - u_1 v_3 \\ u_1 v_2 - u_2 v_1 \end{bmatrix}$$

An alternative way to understand the cross product is to expand vectors $\mathbf{u}$ and $\mathbf{v}$ in three-dimensional space:

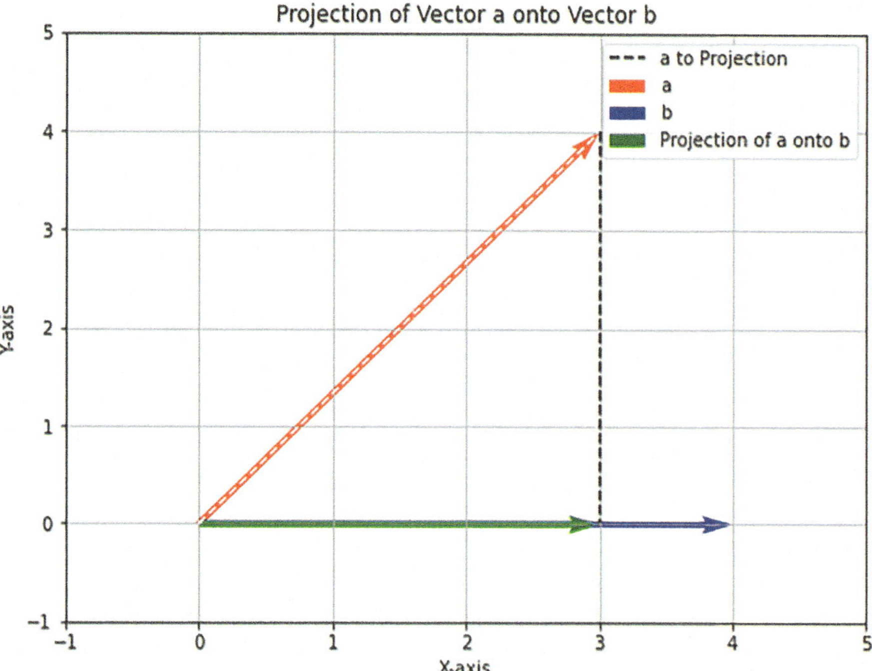

Fig. 1.7 Dot product as the projection of vector *a* onto vector *b*

$$\mathbf{u} = u_1\widehat{i} + u_2\widehat{j} + u_3\widehat{k}, \quad \mathbf{v} = v_1\widehat{i} + v_2\widehat{j} + v_3\widehat{k}$$

The cross-product $\mathbf{w} = \mathbf{u} \times \mathbf{v}$ results in a vector that is perpendicular to both $\mathbf{u}$ and $\mathbf{v}$. The components of $\mathbf{w}$ can be calculated using the following determinant:

$$w = u \times v = \begin{vmatrix} \widehat{i} & \widehat{j} & \widehat{k} \\ u_1 & u_2 & u_3 \\ v_1 & v_2 & v_3 \end{vmatrix}$$

Expand the determinant along the first row:

$$w = \widehat{i} \begin{vmatrix} u_2 & u_3 \\ v_2 & v_3 \end{vmatrix} - \widehat{j} \begin{vmatrix} u_1 & u_3 \\ v_1 & v_3 \end{vmatrix} + \widehat{k} \begin{vmatrix} u_1 & u_2 \\ v_1 & v_2 \end{vmatrix}$$

Notice the pattern similarity with the equation for calculating the determinant of a 3x3 matrix, as shown in Sect. 1.4.1.9. Calculating the minor determinants
For $\widehat{i}$:

$$\begin{vmatrix} u_2 & u_3 \\ v_2 & v_3 \end{vmatrix} = u_2 v_3 - u_3 v_2$$

For $\widehat{j}$:

$$\begin{vmatrix} u_1 & u_3 \\ v_1 & v_3 \end{vmatrix} = u_1 v_3 - u_3 v_1$$

For $\widehat{k}$:

$$\begin{vmatrix} u_1 & u_2 \\ v_1 & v_2 \end{vmatrix} = u_1 v_2 - u_2 v_1$$

Substitute the minor determinants into the expanded formula:

$$\mathbf{w} = (u_2 v_3 - u_3 v_2)\widehat{i} - (u_1 v_3 - u_3 v_1)\widehat{j} + (u_1 v_2 - u_2 v_1)\widehat{k}$$

Therefore, the cross-product vector $\mathbf{w}$ is:

$$\mathbf{w} = \begin{bmatrix} u_2 v_3 - u_3 v_2 \\ -(u_1 v_3 - u_3 v_1) \\ u_1 v_2 - u_2 v_1 \end{bmatrix}$$

The following is a list of cross product properties:

1. The result is orthogonal to both $\mathbf{u}$ and $\mathbf{v}$.
2. The magnitude of $\mathbf{u} \times \mathbf{v}$ is given by:

$$\| \mathbf{u} \times \mathbf{v} \| = \| \mathbf{u} \| \| \mathbf{v} \| \sin(\theta)$$

   where θ is the angle between $\mathbf{u}$ and $\mathbf{v}$.
3. The direction of $\mathbf{u} \times \mathbf{v}$ follows the right-hand rule.

Example

For $\mathbf{u} = [1, 0, 0]$ and $\mathbf{v} = [0, 1, 0]$:

$$\mathbf{u} \times \mathbf{v} = \begin{bmatrix} (0)(0) - (0)(1) \\ (0)(1) - (1)(0) \\ (1)(1) - (0)(0) \end{bmatrix} = \begin{bmatrix} 0 \\ 0 \\ 1 \end{bmatrix}$$

The cross-product magnitude of vector $\mathbf{u} = [3, 4]$ and vector $\mathbf{v} = [4, 0]$ is vector $\mathbf{w} = [0, 0, -16]$, which can be represented as an area, as shown in Fig. 1.8.

1.4.9 Tensor Product

The tensor product of two vectors $\mathbf{u} \in \mathbb{R}^m$ and $\mathbf{v} \in \mathbb{R}^n$ produces a matrix (or higher-dimensional tensor). It is denoted as $\mathbf{u} \otimes \mathbf{v}$ and is defined as:

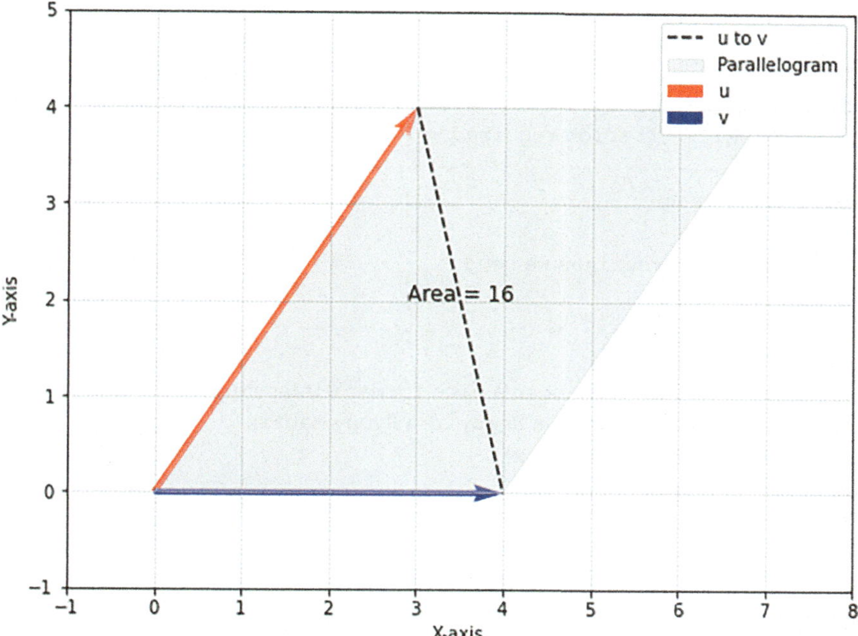

Fig. 1.8 Visualization of cross product magnitude (area of parallelogram)

$$\mathbf{u} \otimes \mathbf{v} = \begin{bmatrix} u_1 v_1 & u_1 v_2 & \cdots & u_1 v_n \\ u_2 v_1 & u_2 v_2 & \cdots & u_2 v_n \\ & & \ddots & \\ u_m v_1 & u_m v_2 & \cdots & u_m v_n \end{bmatrix}$$

The result of the tensor product is a **matrix** if the inputs are vectors, but it can be generalized to higher dimensions.

Tensor Product Properties
1. The tensor product is not commutative: $\mathbf{u} \otimes \mathbf{v} \neq \mathbf{v} \otimes \mathbf{u}$.
2. It is used to represent higher-order structures in physics and machine learning, such as quantum states and neural network weights.

Example
Let $\mathbf{u} = [1, 2]$ and $\mathbf{v} = [3, 4, 5]$. The tensor product $\mathbf{u} \otimes \mathbf{v}$ is:

$$\mathbf{u} \otimes \mathbf{v} = \begin{bmatrix} 1 \cdot 3 & 1 \cdot 4 & 1 \cdot 5 \\ 2 \cdot 3 & 2 \cdot 4 & 2 \cdot 5 \end{bmatrix} = \begin{bmatrix} 3 & 4 & 5 \\ 6 & 8 & 10 \end{bmatrix}$$

Python Example Code: Dot, Cross, and Tensor Products

```
#----------------------------------------------------------------
# Matrix Dot, Cross, and Tensor Products
```

```python
# Chapter 1 in the QUANTUM COMPUTING AND QUANTUM MACHINE LEARNING BOOK
#----------------------------------------------------------------------
# Version 1.0
# (c) 2025 Jesse Van Griensven, Roydon Fraser, and Jose Rosas
# License: MIT - Citation required
#----------------------------------------------------------------------
import numpy as np

# Define the SYMPY routines we need
import sympy as sp
#------------------------------------------- ---------------------------
def sprint(Matrix):
    """ Prints a numpy Matrix in a nice format with sympy """
    # The input Matrix can be a Numpy or a Sympy Matrix
    SMatrix = sp.Matrix(Matrix)
    display( SMatrix )
    return
#------------------------------------------- ---------------------------

u = np.array([1, 2, 3])
v = np.array([4, -5, 6])

# Compute dot product
dot_product = np.dot(u, v)
print("Dot Product:", dot_product)

# Compute cross product
cross_product = np.cross(u, v)
print("Cross Product:", cross_product)

# Compute tensor product
tensor_product = np.tensordot(u, v, axes=0)
print("Tensor Product:")
sprint(tensor_product)
```

1.4.10 Eigenvectors and Eigenvalues

Eigenvectors and eigenvalues are essential concepts in linear algebra that describe the behavior of linear transformations and help to simplify and analyze complex systems. They are fundamental in fields, such as engineering, computer science, and machine learning. Eigenvalues and eigenvectors have broad applications in science, physics, engineering, data analysis, machine learning, and quantum computing.

Table 1.3 summarizes the meaning and solution of eigenvectors and eigenvalues.

Table 1.3 Meaning and solution of eigenvectors and eigenvalues

Feature	Action	Solution
Eigenvalues	Scale the corresponding eigenvectors	Solving $\det(A - \lambda I) = 0$ yields the eigenvalue
Eigenvectors	Vectors that remain in the same direction under a linear transformation	Solving $(A - \lambda I)\mathbf{v} = 0$ gives the eigenvectors

1.4.10.1   Definition of Eigenvectors and Eigenvalues

Let A be an $n \times n$ square matrix. A nonzero vector $\mathbf{v}$ is called an eigenvector of A if there exists a scalar λ (the eigenvalue) such that:

$$A\mathbf{v} = \lambda \mathbf{v}$$

where

(a) $A\mathbf{v}$ represents the linear transformation of $\mathbf{v}$ by A.
(b) λ is the eigenvalue, a scalar that scales the vector $\mathbf{v}$.
(c) $\mathbf{v}$ is the eigenvector, which points in the same or opposite direction after transformation.

The special characteristic of an eigenvector $\mathbf{v}$ is that it is unchanged in direction during a transformation while its magnitude, the amount it stretches or shrinks, is scaled by λ.

1.4.10.2   Finding Eigenvalues and Eigenvectors

To find the eigenvalues and eigenvectors of a square matrix A, we solve using the following steps:

Step 1: Use Characteristic Equation to Solve for Eigenvalues
The eigenvalues are the solutions to the characteristic equation given by:

$$\det(A - \lambda I) = 0$$

where

(a) I is the identity matrix of the same size as A.
(b) λ is the eigenvalue.
(c) $A - \lambda I$ represents the matrix whose determinant must equal zero for non-trivial solutions.

Step 2: Use Eigenvalues to Solve for Eigenvectors
For each eigenvalue λ, we solve the equation

$$(A - \lambda I)\mathbf{v} = 0$$

This equation yields the eigenvector $\mathbf{v}$ corresponding to the eigenvalue λ. The solutions form a subspace of vectors that satisfy the condition.

Example

Consider the following matrix A:

$$A = \begin{bmatrix} 4 & -2 \\ 1 & 1 \end{bmatrix}$$

Step 1: Find the Eigenvalues

Find λ such that $\det(A - \lambda I) = 0$. Subtract λI from A:

$$A - \lambda I = \begin{bmatrix} 4 - \lambda & -2 \\ 1 & 1 - \lambda \end{bmatrix}$$

The determinant is:

$$\det(A - \lambda I) = (4 - \lambda)(1 - \lambda) - (-2)(1) = \lambda^2 - 5\lambda + 6 = 0$$

Solve the quadratic equation:

$$\lambda^2 - 5\lambda + 6 = 0$$

Thus, the eigenvalues are:

$$\lambda_1 = 3, \quad \lambda_2 = 2$$

Step 2: Find the Corresponding Eigenvectors

1. **For $\lambda = 3$**

    Substitute $\lambda = 3$ into $A - \lambda I$:

$$A - 3I = \begin{bmatrix} 1 & -2 \\ 1 & -2 \end{bmatrix}$$

    Solve $(A - 3I)\mathbf{v} = 0$. Let $\mathbf{v} = \begin{bmatrix} x \\ y \end{bmatrix}$. From the first row:

    $x - 2y = 0 => x = 2y$

    The eigenvector corresponding to $\lambda = 3$ is:

$$\mathbf{v}_1 = \begin{bmatrix} 2 \\ 1 \end{bmatrix}$$

2. **For $\lambda = 2$**

    Substitute $\lambda = 2$ into $A - \lambda I$:

$$A - 2I = \begin{bmatrix} 2 & -2 \\ 1 & -1 \end{bmatrix}$$

Solve $(A - 2I)\mathbf{v} = 0$. From the first row:

$2x - 2y = 0 => x = y$

The eigenvector corresponding to $\lambda = 2$ is:

$$\mathbf{v}_2 = \begin{bmatrix} 1 \\ 1 \end{bmatrix}$$

Python Code Example: Compute Eigenvalues and Eigenvectors

We can compute the eigenvalues and eigenvectors of a matrix.

```python
#----------------------------------------------------------------------
# Matrix Eigenvalues and Eigenvectors
# Chapter 1 in the QUANTUM COMPUTING AND QUANTUM MACHINE LEARNING BOOK
#----------------------------------------------------------------------
# Version 1.0
# Qiskit changes frequently.
# We recommend using the latest version from the book code repository at:
# https://aqtinitiative.org/quantum-computing-for-engineers
#
# (c) 2025 Jesse Van Griensven, Roydon Fraser, and Jose Rosas
# License: MIT - Citation required
#----------------------------------------------------------------------
import numpy as np

#----------------------------------------------------------------------
def sprint(Matrix, decimals=4):
    """ Prints a numpy Matrix in a nice format with sympy and specified
precision """
    # Define the SYMPY routines we need
    import sympy as sp
    # The input Matrix can be a Numpy or a Sympy Matrix
    SMatrix = sp.Matrix(Matrix)
    # Apply the rounding to each element
    SMatrix = SMatrix.applyfunc(lambda x: round(float(x), decimals) if
isinstance(x, (float, int, sp.Float)) else x)
    display(SMatrix)
    return
#----------------------------------------------------------------------
```

```
# Define the matrix
A = np.array([[4, -2],
       [1, 1]])

# Compute eigenvalues and eigenvectors
eigenvalues, eigenvectors = np.linalg.eig(A)

# Display Input
print("Matrix A:")
sprint(A)
print()

# Display results
print("Eigenvalues:\n", eigenvalues)
print()

print("Eigenvectors")
sprint(eigenvectors)
# Output:
# Matrix A:
# [[ 4 -2]
# [ 1 1]]
# Eigenvalues: [3. 2.]
# Eigenvectors:
# [[ 0.89442719 0.70710678]
# [ 0.4472136  0.70710678]]
```

1.4.11   Properties of Eigenvalues and Eigenvectors

The following list contains important properties of eigenvalues and eigenvectors.

1. **Trace and Eigenvalues**

    The sum of the eigenvalues of a matrix equals the trace of the matrix (sum of diagonal elements):

$$\text{trace}(A) = \sum_{i=1}^{n} \lambda_i$$

2. **Determinant and Eigenvalues**

    The product of the eigenvalues equals the determinant of the matrix:

$$\det(A) = \prod_{i=1}^{n} \lambda_i$$

3. Linearly Independent Eigenvectors

For distinct eigenvalues, the corresponding eigenvectors are linearly independent.

4. Diagonalizability

If a matrix A has n linearly independent eigenvectors, it can be diagonalized as follows:

$$D = P^{-1}AP$$

where D is a diagonal matrix formed by replacing the 1's in the identity matrix by the eigenvalues of A, and P is a matrix whose columns are the eigenvectors. Diagonalization is the process of finding D and P, a process that enables A to be represented by

$$A = PDP^{-1}$$

Which often makes subsequent calculations easier.

1.5   The Dirac Notation: Bra-Ket Formalism

The Dirac notation, also known as bra-ket notation, is a concise way to represent vectors in quantum mechanics. This subsection re-introduces some linear algebra, now employing the Dirac notation.

1.5.1   Bra ($\langle \phi |$)

A "bra" is the complex conjugate transpose (or Hermitian adjoint) of a ket:

$$\langle \phi | = \begin{bmatrix} \phi_1^* & \phi_2^* & \cdots & \phi_n^* \end{bmatrix}.$$

where the complex conjugate of a complex number $a + ib$ is $a - ib$, where a is the real part of the complex number, and b is the imaginary part.

1.5.2   Ket ($|\psi\rangle$)

A "ket" is a column vector representing a quantum state:

$$|\psi\rangle = \begin{bmatrix} \psi_1 \\ \psi_2 \\ \\ \psi_n \end{bmatrix}$$

1.5.3   Bra-Ket: Inner Product ($\langle\phi\,|\,\psi\rangle$)

The inner product between a bra and a ket produces a scalar:

$$\langle\phi\,|\,\psi\rangle = \phi_1^*\psi_1 + \phi_2^*\psi_2 + \cdots + \phi_n^*\psi_n$$

This is equivalent to the dot or scalar product of the vectors ϕ and ψ.

1.5.4   Bra-Ket: Outer Product ($|\psi\rangle\langle\phi|$)

The outer product produces a matrix:

$$|\psi\rangle\langle\phi| = \begin{bmatrix} \psi_1 \\ \psi_2 \\ \\ \psi_n \end{bmatrix} \begin{bmatrix} \phi_1^* & \phi_2^* & \cdots & \phi_n^* \end{bmatrix}$$

These constructs are critical for describing quantum states and quantum mechanics operations. This is equivalent to the tensor product of ψ and ϕ.

1.5.5   Bases and Basis States

In linear algebra, a basis is a set of linearly independent vectors that span a vector space. Every vector in the space can be uniquely expressed as a linear combination of these basis vectors.

1.5.5.1   Standard Basis

For $\mathbb{C}^n$, the standard basis consists of vectors:

$$| e_1 \rangle = \begin{bmatrix} 1 \\ 0 \\ 0 \end{bmatrix}, \quad | e_2 \rangle = \begin{bmatrix} 0 \\ 1 \\ 0 \end{bmatrix}, \quad \ldots, \quad | e_n \rangle = \begin{bmatrix} 0 \\ 0 \\ 1 \end{bmatrix}$$

1.5.5.2 Orthonormal Bases

In quantum computing, basis vectors are often orthonormal, satisfying:

(a) Orthogonality: $\langle e_i \mid e_j \rangle = 0$ if $i \neq j$.
(b) Normalization: $\langle e_i \mid e_i \rangle = 1$

1.5.5.3 Quantum Basis States

Quantum systems are described using a basis of states, often called computational basis states. For a single qubit, these are:

$$| 0 \rangle = \begin{bmatrix} 1 \\ 0 \end{bmatrix}, \quad | 1 \rangle = \begin{bmatrix} 0 \\ 1 \end{bmatrix}$$

These represent the two fundamental states available to a single qubit. By definition, a qubit can exist in only two orthonormal basis states, such as spin-up and spin-down, or 0 and 1.

A general qubit state is expressed as:

$$| \psi \rangle = \alpha | 0 \rangle + \beta | 1 \rangle, \quad \text{where } \alpha, \beta \in \mathbb{C} \text{ and } | \alpha |^2 + | \beta |^2 = 1$$

This expression represents a superposition of the basis states $|0\rangle$ and $|1\rangle$. When the qubit $|\psi\rangle$ is measured, the probability of finding it in state $|0\rangle$ is $|\alpha|^2$ and in state $|1\rangle$ is $|\beta|^2$.

1.5.6 Matrices and Operators

Operators in quantum mechanics are represented as matrices.

1.5.6.1 Matrix–Vector Multiplication

Given an operator A and a state vector $|\psi\rangle$, the result is:

$$A \,|\,\psi\rangle = \begin{bmatrix} a_{11} & a_{12} \\ a_{21} & a_{22} \end{bmatrix} \begin{bmatrix} \psi_1 \\ \psi_2 \end{bmatrix} = \begin{bmatrix} a_{11}\psi_1 + a_{12}\psi_2 \\ a_{21}\psi_1 + a_{22}\psi_2 \end{bmatrix}$$

where ψ_1 and ψ_2 are alternatives to α and β, respectively. While α and β are conventionally used to describe qubits, when larger quantum states need to be described, such as in a quantum registrar of multiple qubits, the ψ_1 and ψ_2 become advantageous.

1.5.6.2 Hermitian Operators

An operator A is Hermitian if $A^\dagger = A$. These are crucial operators because they correspond to observable quantities in quantum mechanics. A is a matrix, while the dagger superscript matrix $A^\dagger$ is the conjugate transpose of the operator A.

1.5.7 Change of Basis and Unitary Transformation

Changing the basis of a vector allows us to express it in a different reference frame. This is achieved using a unitary transformation represented by a matrix U:

$$|\psi'\rangle = U \,|\,\psi\rangle$$

The unitary operator preserves the inner product of the basis vector.

1.5.7.1 Example: Pauli Matrices

The Pauli matrices are fundamental in quantum mechanics and are often used to define basis transformations:

$$\sigma_x = \begin{bmatrix} 0 & 1 \\ 1 & 0 \end{bmatrix}, \quad \sigma_y = \begin{bmatrix} 0 & -i \\ i & 0 \end{bmatrix}, \quad \sigma_z = \begin{bmatrix} 1 & 0 \\ 0 & -1 \end{bmatrix}$$

Each Pauli matrix is a *unitary operator*. This property can be verified by multiplying the matrix by its conjugate transpose (Hermitian adjoint), which yields the identity matrix:

$$\sigma_k{}^\dagger \sigma_k = I \quad for\ k \in \{x, y, z\}$$

1.5.8 Quantum Measurements

Quantum measurements are described using projection operators. For example, measuring a qubit computationally involves the operators:

$$P_0 = |0\rangle\langle 0| = \begin{bmatrix} 1 & 0 \\ 0 & 0 \end{bmatrix}, \quad P_1 = |1\rangle\langle 1| = \begin{bmatrix} 0 & 0 \\ 0 & 1 \end{bmatrix}$$

If a state $|\psi\rangle = \alpha\,|0\rangle + \beta\,|1\rangle$ is measured via $P_0\,|\psi\rangle$, this yields $\alpha\,|0\rangle$ with a probability outcome for $|0\rangle$ of $|\alpha|^2$. Similarly, $P_1\,|\psi\rangle$ yields $\beta\,|1\rangle$ with a probability outcome for $|1\rangle$ of $|\beta|^2$.

1.6 Complex Numbers, Fourier Transform, and Hilbert Space

This introduction provides a foundation for understanding complex numbers, the Fourier transform, and Hilbert space. These are the three mathematical pillars essential for quantum computing. Complex numbers capture the amplitudes and phases of quantum states, the Fourier transform facilitates frequency-domain analysis and is central to quantum algorithms, and Hilbert space offers the structure to describe quantum systems. By mastering these concepts, engineers can build a strong foundation for exploring quantum computation and quantum mechanics.

1.6.1 Complex Numbers

Complex numbers extend real numbers by introducing the imaginary unit i, which is defined as $i^2 = -1$ or $i = \sqrt{-1}$. Complex numbers are fundamental in quantum mechanics because quantum states and operators often involve complex amplitudes.

A complex number z is expressed as:

$$z = a + bi, \quad \text{where } a, b \in \mathbb{R}$$

- a: The real part, denoted as $\mathrm{Re}(z)$.
- b: The imaginary part, denoted as $\mathrm{Im}(z)$.

Complex numbers are represented graphically in a unit circle, as shown in the following figure (developed by the authors) (Fig. 1.9).

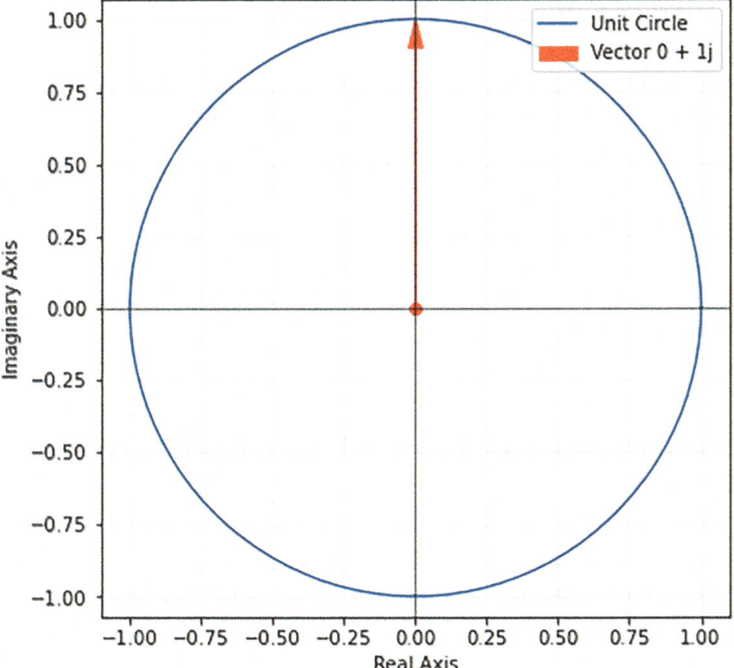

Fig. 1.9 Complex unit circle (developed by the authors)

Table 1.4 Complex number operations

Operation type	Execution
Addition	$z_1 + z_2 = (a_1 + a_2) + i(b_1 + b_2)$
Multiplication	$z_1 z_2 = (a_1 a_2 - b_1 b_2) + i(a_1 b_2 + a_2 b_1)$
Conjugate	The complex conjugate of $z = a + bi$ is $z^* = a - bi$
Modulus	$\lvert z \rvert = \sqrt{(a^2 + b^2)}$

1.6.1.1 Arithmetic Operations with Complex Numbers

Table 1.4 summarizes the main operations with complex numbers.

1.6.1.2 Magnitude and Phase

The magnitude (length or modulus) of z is:

$$\lvert z \rvert = \sqrt{a^2 + b^2}$$

The phase (or argument) θ is the angle z makes with the real axis:

$$\theta = \arg(z) = \tan^{-1}\left(\frac{b}{a}\right)$$

1.6.1.3 Polar Form

A complex number can also be represented in polar form using Euler's formula:

$$z = |z|\,e^{i\theta}, \quad \text{where } e^{i\theta} = \cos\theta + i\sin\theta$$

This representation is particularly useful in quantum mechanics for expressing quantum states and phase factors, as can be seen from Fig. 1.10 (obtained from the article in Wikipedia located at: https://en.wikipedia.org/wiki/Euler%27s_formula).

Python Code Example: Complex Numbers

```
#-------------------------------------------------------------------
# Complex Numbers
# Chapter 1 in the QUANTUM COMPUTING AND QUANTUM MACHINE LEARNING BOOK
#------------------- ------------------------------------------
--------------------------
# Version 1.0
# Qiskit changes frequently.
# We recommend using the latest version from the book code repository at:
# https://aqtinitiative.org/quantum-computing-for-engineers
#
# (c) 2025 Jesse Van Griensven, Roydon Fraser, and Jose Rosas
# License: MIT - Citation required
```

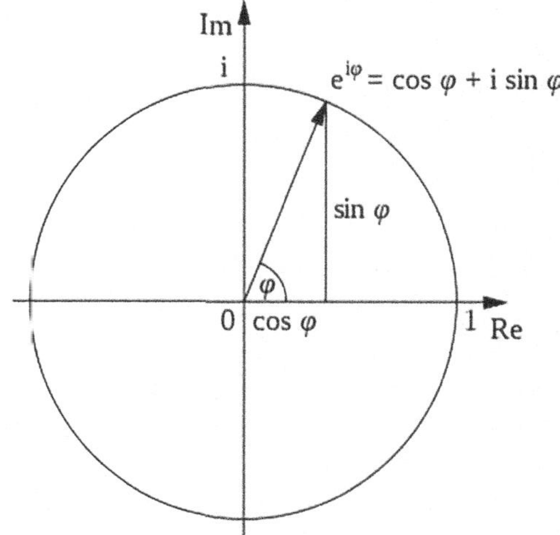

Fig. 1.10 Graphical representation of the Euler's formula (Wikipedia.com)

```
#-------------------------------------------------------------------
import numpy as np

# Define two complex numbers
complex_num1 = 3 + 4j  # 3 + 4i
complex_num2 = 1 - 2j  # 1 - 2i

# Addition of complex numbers
complex_sum = complex_num1 + complex_num2
# The real parts are added together, and the imaginary parts are added
together
# (3 + 1) + (4 - 2)i = 4 + 2i
print("Addition of complex numbers:")
print(f"{complex_num1} + {complex_num2} = {complex_sum}")
print()

# Multiplication of complex numbers
complex_product = complex_num1 * complex_num2
# Use the distributive property: (3 + 4i)(1 - 2i) = 3 - 6i + 4i - 8i^2
# Remember i^2 = -1, so it becomes: 3 - 2i + 8 = 11 - 2i

print("Multiplication of complex numbers:")
print(f"{complex_num1} * {complex_num2} = {complex_product}")
print()

# Conjugate of a complex number
# The conjugate of a complex number changes the sign of the imaginary part
# For example, the conjugate of 3 + 4i is 3 - 4i
complex_conjugate1 = complex_num1.conjugate()
complex_conjugate2 = complex_num2.conjugate()

print("Conjugate of complex numbers:")
print(f"Conjugate of {complex_num1} = {complex_conjugate1}")
print(f"Conjugate of {complex_num2} = {complex_conjugate2}")
```

1.6.2   Fourier Transform

The Fourier transform (FT) is a mathematical tool that decomposes a function into its frequency components. It plays a key role in quantum computing, particularly in the quantum Fourier transform (QFT), which underpins algorithms like Shor's.

1.6.2.1 Continuous Fourier Transform

The Fourier transform of a function $f(t)$ is:

$$F(\omega) = \int_{-\infty}^{\infty} f(t)e^{-i\omega t}\,dt$$

where

- $F(\omega)$ represents the frequency-domain function.
- ω is the angular frequency.

The inverse Fourier transform recovers $f(t)$ from $F(\omega)$:

$$f(t) = \frac{1}{2\pi} \int_{-\infty}^{\infty} F(\omega)e^{i\omega t}\,d\omega$$

1.6.2.2 Discrete Fourier Transform (DFT)

For discrete signals, the Fourier transform is represented as the DFT:

$$F(k) = \sum_{n=0}^{N-1} f(n)e^{-i2\pi kn/N}, \quad k = 0, 1, \dots, N-1$$

The quantum Fourier transform (QFT) is the quantum analog of the DFT, providing exponential speedup for many problems.

Python Code Example: Fourier Transform

```python
#------------------------------------------------------------------
# Chapter 1 Fourier Transform
# Chapter 1 in the QUANTUM COMPUTING AND QUANTUM MACHINE LEARNING BOOK
#------------------------------------------------------------------
# Version 1.0
# Qiskit changes frequently.
# We recommend using the latest version from the book code repository at:
# https://aqtinitiative.org/quantum-computing-for-engineers
#
# (c) 2025 Jesse Van Griensven, Roydon Fraser, and Jose Rosas
# License: MIT - Citation required
#------------------------------------------------------------------
import numpy as np
import matplotlib.pyplot as plt

# Define the time domain
time = np.linspace(0, 1, 100, endpoint=False)  # 1 second sampled at
100 Hz
```

```python
# Create a simple signal: a sine wave with a frequency of 5 Hz
frequency = 5 # Frequency in Hz, 5 cycles per second
signal = np.sin(2 * np.pi * frequency * time)

# Perform the Fourier Transform
frequency_domain = np.fft.fft(signal)  # Fast Fourier Transform
frequencies = np.fft.fftfreq(len(signal), d=time[1] - time[0])  #
Frequency bins

# Take the magnitude to represent the frequency domain
magnitude = np.abs(frequency_domain)

# Plot the original signal in the time domain
plt.figure(figsize=(12, 6))

plt.subplot(2, 1, 1)
plt.plot(time, signal)
plt.grid(alpha = 0.2)
plt.title("Time-Domain Signal (Sine Wave) - 5 Hertz - ")
plt.xlabel("Time (s)")
plt.ylabel("Amplitude")

# Plot the magnitude of the Fourier Transform in the frequency domain
plt.subplot(2, 1, 2)
plt.stem(frequencies[:len(frequencies)//2], magnitude[:len
(magnitude)//2], basefmt=" ", use_line_collection=True)
plt.title("Frequency-Domain Representation (Fourier Transform)")
plt.xlabel("Frequency (Hz)")
plt.ylabel("Magnitude")
plt.tight_layout()
plt.grid(alpha = 0.2)

plt.show()
```

The results from the code above are described in Fig. 1.11.

1.7 Hilbert Space

A Hilbert space is a mathematical framework that generalizes Euclidean space to potentially infinite dimensions. In quantum mechanics, it serves as the space of quantum states, where vectors represent states and operators act on these states. A Hilbert space $\mathcal{H}$ is a complete vector space equipped with an inner product, denoted $\langle \cdot, \cdot \rangle$, that satisfies the requirements expressed in Table 1.5.

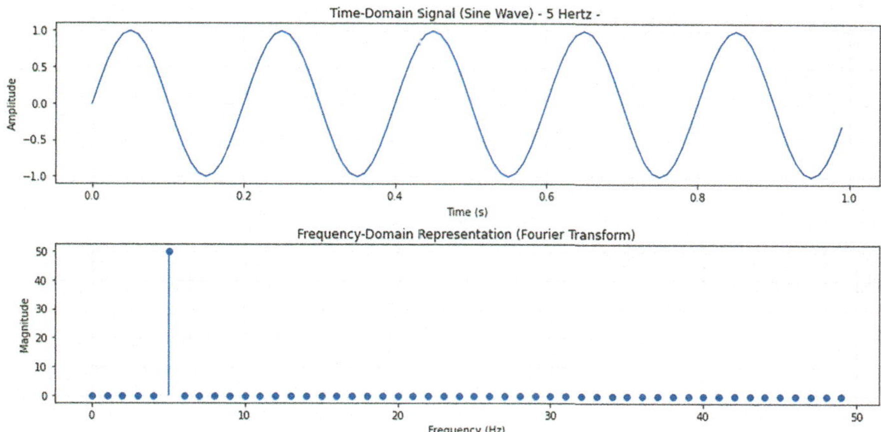

Fig. 1.11 Fourier analysis of a single mode at 5 Herz (by the authors)

Table 1.5 Hilbert space requirements

Hilbert space requirement	Definition
Conjugate symmetry	$\langle \phi, \psi \rangle = \langle \psi, \phi \rangle^*$
Linearity	$\langle a\phi + b\chi, \psi \rangle = a\langle \phi, \psi \rangle + b\langle \chi, \psi \rangle$
Positive definiteness	$\langle \phi, \phi \rangle \geq 0$, with equality only if $\phi = 0$

1.7.1   Quantum States in Hilbert Space

In quantum mechanics:

(a) Quantum states are vectors in a Hilbert space.
(b) Inner products represent overlaps between states:

$$\langle \phi \,|\, \psi \rangle = \sum_i \phi_i^* \psi_i$$

(c) Normalization: Quantum states satisfy $\langle \psi \,|\, \psi \rangle = 1$.

Example: Single-Qubit Hilbert Space
For a single qubit, the Hilbert space is $\mathbb{C}^2$. A general qubit state is:

$$|\psi\rangle = \alpha |0\rangle + \beta |1\rangle, \quad \text{with } |\alpha|^2 + |\beta|^2 = 1$$

1.7.2   Operators in Hilbert Space

Operators act on quantum states and are represented as matrices. Common types include those in Table 1.6.

1.7.2.1   Example: Pauli Operators

The Pauli matrices σ_x, σ_y, σ_z are examples of Hermitian and unitary operators:

$$\sigma_x = \begin{bmatrix} 0 & 1 \\ 1 & 0 \end{bmatrix}, \quad \sigma_y = \begin{bmatrix} 0 & -i \\ i & 0 \end{bmatrix}, \quad \sigma_z = \begin{bmatrix} 1 & 0 \\ 0 & -1 \end{bmatrix}$$

1.7.3   Basis in Hilbert Space

The choice of basis vectors in a Hilbert space determines how quantum states and operators are represented.

1.7.3.1   Orthonormal Basis

A set of vectors $\{|e_i\rangle\}$ is an orthonormal basis if:

$$\langle e_i \mid e_j \rangle = \delta_{ij}$$

where δ_{ij} is the Kronecker delta defined by $\delta_{ij} = \begin{cases} 1 & i=j \\ 0 & i \neq j \end{cases}$.

1.7.3.2   Change of Basis

Changing the basis in Hilbert space is achieved using unitary transformations:

$$|\psi'\rangle = U |\psi\rangle$$

Table 1.6 Operators in Hilbert space

Hilbert space operator	Description
Hermitian operators	$A = A^\dagger$, representing observables
Unitary operators	$U^\dagger U = I$, representing quantum gates
Projection operators	Represent measurements in quantum mechanics

1.7.4 The Quantum Paradigm

Quantum computing processes information using quantum bits, or qubits, which operate according to the principles of quantum mechanics. Unlike classical bits that can be in one of the two states (0 or 1), qubits can exist in a superposition of states:

$$|\psi\rangle = \alpha|0\rangle + \beta|1\rangle$$

where α and β are complex probability amplitudes that satisfy $|\alpha|^2 + |\beta|^2 = 1$.

Superposition enables a single qubit to represent multiple states simultaneously. For a system of n qubits, the total number of possible states is 2^n, compared to n states in a classical system. This exponential growth in representational capacity is the foundation of quantum parallelism, allowing quantum computers to evaluate multiple solutions simultaneously.

Entanglement is another cornerstone of quantum computing. When qubits become entangled, their states are correlated such that the measurement of one qubit instantaneously determines the state of the other, regardless of distance. This property is mathematically described using Bell states, such as:

$$|\Phi^+\rangle = \frac{1}{\sqrt{2}}(|00\rangle + |11\rangle)$$

Entanglement enables quantum computers to process information non-locally, which is essential for tasks like quantum teleportation and secure communication.

Quantum computing also leverages interference, the constructive or destructive combination of probability amplitudes. By carefully designing quantum algorithms, engineers can amplify the probability of correct solutions while suppressing incorrect ones. This principle is fundamental in algorithms like Grover's search algorithm, which provides a quadratic speedup for unstructured search problems.

The physical realization of qubits relies on diverse technologies, including superconducting circuits, trapped ions, and photonic systems. Each platform offers distinct advantages and faces specific challenges, such as coherence time limitations, gate fidelity, and scalability. Despite these challenges, advances in quantum hardware have enabled significant milestones—most notably, demonstrations of quantum supremacy, referring to tasks in which quantum systems outperform the most powerful classical supercomputers.

Quantum computers redefine computational complexity by enabling solutions to problems considered intractable for classical systems. For example:

1. **Integer Factorization**

   Shor's algorithm runs in polynomial time, $O((\log N)^3)$, for factoring integers compared to the exponential scaling of the best classical algorithms. This poses a direct threat to classical cryptographic schemes like RSA.

2. **Quantum Simulations**

   Quantum systems can efficiently simulate other quantum systems, solving Schrödinger's equation for large molecules. This capability has profound implications for material science and drug discovery.

3. **Optimization Problems**

   The quantum approximate optimization algorithm (QAOA) provides efficient solutions to combinatorial optimization problems by utilizing quantum superposition and interference:

$$|\psi\rangle = e^{-i\gamma H_c} e^{-i\beta H_m} |\psi_0\rangle$$

where H_c encodes the problem constraints and H_m facilitates exploration of the solution space.

Because of how quantum computers work, which includes superposition, parallelism, and entanglement, they can accomplish more than classical systems for many tasks. However, classical systems will still need to exist and operate in the real world. The advances that quantum systems will create will bring about a new dawn, especially in fields with vast computational needs.

Quantum computing is a different way of calculating solutions to problems. Superposition, entanglement, and interference get the potential solution universe going beyond the boundaries of previously known possibilities to answer some of the more complex inquiries. Yet until the rest of the engineering and scientific world can join the quantum realm, they need to know how to cross the divide from one side of the equation to the other. While quantum computing provides an alternative solution to calculating answers, it's the nuances that may take time to learn without instruction.

1.8   Core Principles: Superposition, Entanglement, and Interference

Quantum computing relies upon the principles of quantum mechanics, which govern the behavior of matter and energy at atomic and subatomic levels. Classical systems, for example, operate on a different set of governing standards. Therefore, investigating these three quantum phenomena, which are composed of superposition, entanglement, and interference existing within our universe, explains why it was historically developed and how it will revolutionize processing power.

1.8.1   Superposition

Superposition is a defining property of quantum systems, allowing particles to exist in multiple states simultaneously. In the context of quantum computing, a qubit, the basic unit of quantum information, can exist in a state that is a linear combination of $|0\rangle$ and $|1\rangle$:

$$|\psi\rangle = \alpha\,|0\rangle + \beta\,|1\rangle$$

where α and β are complex probability amplitudes satisfying $|\alpha|^2 + |\beta|^2 = 1$. These amplitudes represent the probabilities of measuring the qubit in the $|0\rangle$ or $|1\rangle$ state, respectively.

Superposition enables quantum computers to represent and process exponentially larger amounts of information compared to classical systems. For a classical bit, the state is either 0 or 1, while a system of n classical bits can represent one of $\mathbf{2^n}$ possible states at a given time. In contrast, an n-qubit quantum system can exist in a superposition of all $\mathbf{2^n}$ states simultaneously:

$$|\psi\rangle = \sum_{i=0}^{2^n - 1} c_i\,|i\rangle$$

where c_i are the complex amplitudes associated with each basis state $|i\rangle$.

This property underpins quantum parallelism, allowing quantum computers to evaluate multiple possibilities in parallel. For example, consider a database search problem, where N entries need to be checked to find a specific element. A classical algorithm would require $O(N)$ steps to search the database sequentially. In a quantum computer, superposition enables simultaneous evaluation of all N entries, dramatically reducing the computational effort.

Superposition also facilitates quantum algorithms like Shor's algorithm for integer factorization, which leverages the ability to perform parallel computations on superposed states to achieve exponential speedup. However, the power of superposition is only realized when combined with interference and measurement, which refine the results by amplifying correct solutions and suppressing incorrect ones.

Measurement in quantum mechanics collapses a superposed state into one of its basis states. For instance, measuring the state $|\psi\rangle = \alpha\,|0\rangle + \beta\,|1\rangle$ results in $|0\rangle$ with probability $|\alpha|^2$ and $|1\rangle$ with probability $|\beta|^2$. The act of measurement destroys the superposition, emphasizing the importance of designing quantum algorithms that minimize intermediate measurements.

1.8.2 Entanglement

Entanglement is a uniquely quantum phenomenon in which the states of two or more particles become correlated such that the state of one particle is dependent on the state of another, regardless of the physical distance between them. This property is critical for many quantum computing and communication protocols, enabling tasks that are impossible in classical systems.

Mathematically, an entangled state cannot be expressed as the product of individual states of the particles involved. For example, the Bell state:

$$|\Phi^-\rangle = \frac{1}{\sqrt{2}}(|00\rangle - |11\rangle)$$

is an entangled state of two qubits. Measuring the first qubit in the $|0\rangle$ or $|1\rangle$ state immediately determines the state of the second qubit, even if they are separated by large distances.

Entanglement has profound implications for quantum computing:

1. **Quantum Teleportation**

    Entanglement enables the transfer of quantum states from one location to another without physically transmitting the particle. This process involves three key steps: entanglement distribution, measurement, and classical communication. Quantum teleportation is represented mathematically by:

$$|\psi\rangle_A = \alpha|0\rangle + \beta|1\rangle, \quad |\Phi^+\rangle_{BC} = \frac{1}{\sqrt{2}}(|00\rangle + |11\rangle)$$

    Using these states, the quantum information in $|\psi\rangle_A$ can be transferred to qubit C.

2. **Quantum Key Distribution (QKD)**

    Protocols like BB84 and E91 rely on entanglement to establish secure cryptographic keys. In QKD, any attempt to intercept the entangled particles disturbs their states, alerting the parties involved to potential eavesdropping.

3. **Quantum Error Correction**

    Entanglement is used to encode logical qubits in multiple physical qubits, enabling the correction of errors without directly measuring the logical state. For example, the three-qubit code uses entanglement to protect against single-qubit errors:

$$|0_L\rangle = |000\rangle, \quad |1_L\rangle = |111\rangle$$

Entanglement is a key enabler of quantum supremacy, allowing quantum systems to solve problems like optimization and factorization with efficiencies unattainable by classical systems.

1.8.3 Interference

Interference is the phenomenon where quantum probability amplitudes combine, either constructively or destructively, to influence the likelihood of different outcomes. Constructive interference enhances the probability of desired solutions, while destructive interference suppresses incorrect ones. This principle is harnessed by quantum algorithms to achieve computational advantages.

The state of a quantum system can be expressed as:

$$|\psi\rangle = \sum_i c_i e^{i\theta_i} |i\rangle$$

where $e^{i\theta_i}$ represents the phase of each amplitude. The phase factors θ_i determine how amplitudes interfere with each other during quantum operations.

Interference is central to the operation of Grover's search algorithm, which provides a quadratic speedup for searching unsorted databases. The algorithm iteratively applies the Grover operator, which combines amplitude inversion and phase shifts to amplify the probability of the correct solution. After $O(\sqrt{N})$ iterations, the probability of measuring the desired state approaches unity.

Another example of interference is in the quantum Fourier transform (QFT), a key component of Shor's algorithm. The QFT maps a quantum state from the time domain to the frequency domain, enabling efficient computation of periodicity:

$$|\psi\rangle = \frac{1}{\sqrt{N}} \sum_{k=0}^{N-1} e^{2\pi i k x / N} |k\rangle$$

Interference between amplitudes ensures that only frequencies corresponding to the periodicity of the input function have significant probabilities.

Interference is also leveraged in quantum simulations, where the constructive combination of amplitudes represents physically meaningful solutions to quantum systems governed by Schrödinger's equation:

$$H |\psi\rangle = E |\psi\rangle$$

where the interference helps identify the ground-state energy E, crucial for applications in material science and chemistry and H is the Hamiltonian discussed more in Sect. 1.9.1 below, in Chap. 2, and later chapters. This is the time-independent form of Schrödinger's equation.

By carefully designing quantum circuits and manipulating phase factors, interference allows quantum computers to outperform classical systems in solving complex problems. This principle, combined with superposition and entanglement, forms the backbone of quantum computing, enabling transformative advances across disciplines.

1.9   Why Engineers Need to Understand Quantum Computing

Engineers and software developers will know where to apply quantum computing since it will solve problems that cannot be solved by classical systems. The way classical computing operates is bound to time and space, while quantum mechanical phenomena such as superposition, entanglement, and interference create an astronomical computing capacity. Experts who understand where quantum computing can be effectively applied will lead advancements in optimization, simulation, and machine learning. Therefore, engineers and developers must develop awareness of its application domains to remain at the forefront of their respective fields.

Quantum computing simplifies and speeds up everything from optimization and simulations to machine learning by solving many of the most complex problems facing today's specialists. The more professionals understand how quantum computing works, the more they'll be able to develop new applications in any engineering field, such as energy, healthcare, aerospace, and even manufacturing. For engineers and developers, quantum computing is a necessity to stay on top of contemporary, high-tech, high-value solutions to problems.

1.9.1   Advancements in Optimization

Optimization is a critical aspect of engineering, spanning fields, such as logistics, telecommunications, energy management, and structural design. Classical optimization methods struggle with problems that exhibit combinatorial complexity, where the number of possible solutions grows exponentially with the problem size.

Quantum computers provide a novel approach to optimization through algorithms, such as the quantum approximate optimization algorithm (QAOA). QAOA is designed to solve combinatorial optimization problems more efficiently than classical methods. The algorithm operates by encoding the problem into a cost Hamiltonian H_c and using a mixing Hamiltonian H_m to explore the solution space:

$$| \psi \rangle = e^{-i\beta H_m} e^{-i\gamma H_c} | \psi_0 \rangle$$

where $|\psi_0\rangle$ is the initial state, and β and γ are tunable parameters optimized to maximize the probability of finding the best solution.

The following is a sample of quantum optimization applications:

1. **Logistics and Supply Chain Management**

    Quantum algorithms optimize delivery routes, warehouse locations, and inventory management. For instance, solving the traveling salesman problem for large networks becomes feasible using QAOA, minimizing travel costs and time.

2. **Energy Optimization**

   Power grid management and renewable energy integration benefit from quantum optimization by balancing energy loads and minimizing losses. For example, optimizing the placement of solar panels and wind turbines increases efficiency.

3. **Aerospace and Automotive Design**

   Structural optimization in aerospace and automotive engineering, such as minimizing weight while maintaining strength, can leverage quantum techniques to evaluate a vast array of configurations rapidly.

By enhancing the ability to solve optimization problems, quantum computing enables engineers to design more efficient systems and processes.

1.9.2   Revolutionizing Simulations

Simulations are indispensable in engineering for predicting the behavior of physical systems under various conditions. Classical computers rely on numerical approximations to solve complex equations, such as partial differential equations, which can become computationally prohibitive for large systems. Quantum computers offer an exponential advantage in simulating quantum systems by directly solving equations that govern their behavior.

The core of quantum simulation lies in solving Schrödinger's equation, which describes the quantum state of a physical system:

$$H \, | \, \psi \rangle = E \, | \, \psi \rangle$$

where H is the Hamiltonian operator representing the system's energy, $| \psi \rangle$ is the wave function, and E is the energy eigenvalue. Quantum computers naturally simulate such systems by encoding the Hamiltonian into quantum circuits.

The following is a sample of quantum simulation applications:

1. **Material Science**

   Quantum simulations accelerate the discovery of new materials by modeling their quantum properties, such as conductivity, magnetism, and strength. For example, simulating high-temperature superconductors could lead to breakthroughs in energy transmission.

2. **Chemistry and Drug Discovery**

   Quantum computers model molecular interactions with unparalleled accuracy. This capability enables the design of drugs that target specific proteins or enzymes, reducing the time and cost of pharmaceutical development. For instance, simulating the binding affinity of a drug molecule to a target protein can be done efficiently in a quantum system.

3. **Environmental Engineering**

   Engineers use quantum simulations to study complex systems, such as atmospheric dynamics, ocean circulation, and pollutant behavior. These simulations help in designing strategies to mitigate climate change and manage natural resources effectively.

By revolutionizing simulations, quantum computing equips engineers with powerful tools to model, analyze, and optimize complex systems that were previously infeasible to study.

1.9.3 Enhancing Machine Learning

Machine learning (ML) is transforming engineering fields by enabling predictive modeling, pattern recognition, and decision-making in real time. Classical ML models require extensive computational resources for training, particularly when dealing with large datasets or high-dimensional feature spaces. Quantum computing introduces novel approaches to machine learning, offering faster and more accurate training of models.

Quantum machine learning (QML) utilizes quantum principles such as superposition and entanglement to process data more efficiently. For example, quantum-enhanced support vector machines (SVMs) use quantum kernels to map data into higher-dimensional spaces, improving classification accuracy.

1.9.4 Quantum Neural Networks (QNNs)

QNNs combine the architecture of classical neural networks with the computational advantages of quantum circuits. A QNN uses parameterized quantum circuits to optimize model weights efficiently. The gradient of the loss function L is computed using central differencing:

$$\frac{\partial L}{\partial \theta} = \frac{L\left(\theta + \frac{\pi}{2}\right) - L\left(\theta - \frac{\pi}{2}\right)}{2}$$

where θ represents the circuit parameters.

The following is a sample of quantum machine learning applications:

1. **Predictive Maintenance**

    In manufacturing and aerospace, QML models analyze sensor data to predict equipment failures, reducing downtime and maintenance costs.

2. **Autonomous Vehicles**

    Quantum-enhanced ML improves object detection, decision-making, and route optimization in self-driving cars. For instance, QNNs process high-dimensional sensor data faster than classical systems, enabling real-time responses.

3. **Medical Imaging**

   Quantum models enhance image recognition in radiology, allowing early detection of diseases such as cancer by analyzing complex patterns in medical images.

QML represents a significant leap in computational power, enabling engineers to develop smarter and more efficient systems.

1.10 Applications in Engineering Fields

Quantum computing offers transformative capabilities across a variety of engineering disciplines. Its ability to process and analyze data at unprecedented scales, optimize complex systems, and model physical phenomena positions it as a critical tool for advancing modern engineering. This section explores its applications in aerospace engineering, electrical engineering, civil engineering, artificial intelligence, machine learning, and energy systems.

Quantum computing's applications in engineering are vast, spanning fields from aerospace and electrical engineering to artificial intelligence and energy systems. Its ability to solve optimization problems, perform accurate simulations, and enhance machine learning positions quantum computing as a transformative technology that will redefine the engineering landscape.

1.10.1 Aerospace Engineering

The aerospace industry is characterized by its reliance on precise optimization, simulation, and design processes. Quantum computing introduces new possibilities for tackling the computational challenges in this field.

1.10.1.1 Fuel Efficiency Optimization

Fuel efficiency is one of the most critical considerations in aerospace engineering, directly impacting operational costs and environmental sustainability. Quantum optimization algorithms, such as the quantum approximate optimization algorithm (QAOA), optimize flight routes, engine designs, and fuel usage. These algorithms explore large solution spaces efficiently, reducing costs while maintaining safety standards.

1.10.1.2 Aerodynamic Simulations

Aerodynamic simulation involves solving the Navier–Stokes equations, which describe fluid flow. Classical methods, like computational fluid dynamics (CFD), are computationally intensive, especially for turbulent flows. Quantum computers, by leveraging quantum-enhanced solvers, can handle these simulations more efficiently:

$$\frac{\partial \rho}{\partial t} + \nabla \cdot (\rho \mathbf{u}) = 0$$

where ρ is the fluid density, and $\mathbf{u}$ is the velocity vector. These simulations enable engineers to design more efficient and stable aircraft and spacecraft.

1.10.1.3 Structural Optimization

Quantum algorithms support the optimization of structural components by balancing weight, strength, and material cost. For example, optimizing the layout of a spacecraft's internal structure to minimize material usage while maximizing robustness can benefit from quantum optimization techniques.

1.10.1.4 Case Study

A leading aerospace company utilized quantum algorithms to design fuel-efficient flight paths across multiple global hubs. The quantum solution reduced fuel consumption by 12% compared to classical methods, saving millions in operational costs.

1.10.2 Electrical Engineering

Electrical engineering is at the forefront of quantum technology development, particularly in designing quantum circuits and advancing fault-tolerant quantum devices.

The following is a sample of applications in electrical engineering:

1.10.2.1 Quantum Circuit Design

Quantum circuit design involves creating architectures that can execute quantum algorithms with minimal errors. Engineers use quantum computing frameworks, such as Qiskit and Cirq, to simulate and test circuit designs before implementing them on physical hardware. The accuracy and scalability of these designs are critical for ensuring reliable quantum computation.

1.10.2.2   Fault-Tolerant Devices

Fault tolerance is a major challenge in quantum computing, requiring additional qubits to correct errors. Electrical engineers develop error-correcting codes, such as surface codes, to encode logical qubits:

$$|0_L\rangle = |000\rangle, \quad |1_L\rangle = |111\rangle$$

These codes mitigate decoherence and gate errors, enhancing the stability of quantum systems.

1.10.2.3   Power Grid Optimization

Quantum algorithms optimize power grid management by minimizing energy losses and balancing supply with demand. Quantum systems model power flows with greater precision than classical methods, enabling efficient integration of renewable energy sources.

1.10.2.4   Case Study

A research group used quantum simulations to design an energy-efficient superconducting qubit system, reducing energy loss in quantum circuit operations by 20%.

1.10.3   Civil Engineering

Civil engineering involves large-scale resource allocation, project planning, and infrastructure design, all of which are computationally intensive tasks. Quantum computing provides innovative solutions to these challenges.

1.10.3.1   Resource Allocation

Quantum algorithms optimize the allocation of resources in large projects, such as urban construction and infrastructure development. QAOA finds efficient solutions for scheduling tasks, allocating equipment, managing labor costs, and reducing project delays and budget overruns.

1.10.3.2   Traffic Flow Optimization

Quantum computers analyze and optimize traffic flow in urban environments, reducing congestion and improving transportation efficiency. Quantum-enhanced simulations model complex traffic networks, accounting for variables such as vehicle density, signal timing, and route preferences.

1.10.3.3   Structural Design

Quantum algorithms assist in designing earthquake-resistant buildings by optimizing the placement of structural reinforcements and evaluating stress distribution under dynamic loads. These simulations rely on solving equations governing elasticity and material properties.

1.10.3.4   Case Study

A metropolitan transit authority used quantum algorithms to optimize the construction schedule of a new subway line. The approach reduced project time by 18% while staying within budget constraints.

1.10.4   Artificial Intelligence and Machine Learning

Quantum computing accelerates the training and accuracy of machine learning (ML) models, revolutionizing predictive analytics and decision-making in engineering applications.

1.10.4.1   Faster Model Training

Quantum-enhanced learning algorithms utilize quantum parallelism to accelerate the training of ML models. For instance, quantum support vector machines (QSVMs) use quantum kernels to map data into high-dimensional feature spaces, improving classification accuracy:

$$K(x, x') = |\langle \psi(x) | \psi(x') \rangle|^2$$

1.10.4.2 Improved Accuracy in Predictions

Quantum neural networks (QNNs) optimize model parameters with higher efficiency than classical methods, reducing training time while improving model performance. Engineers and developers employ QNNs to analyze sensor data, predict equipment failures, and optimize maintenance schedules.

1.10.4.3 Applications in Engineering

(a) **Robotics**
Quantum-enhanced ML algorithms improve object recognition and path planning for autonomous robots in manufacturing and construction.
(b) **Predictive Analytics**
Quantum algorithms enhance predictive analytics for risk assessment, enabling engineers to anticipate and mitigate potential failures in complex systems.

1.10.4.4 Case Study

A robotics company integrated QNNs into its autonomous systems, achieving a 30% improvement in obstacle detection accuracy and a 25% reduction in computation time.

1.10.5 Energy Systems

The energy sector benefits significantly from quantum computing through advancements in grid optimization, energy storage, and material discovery.

1.10.5.1 Electric Grid Optimization

Quantum computers model and optimize power grids by solving large-scale optimization problems involving load balancing, network stability, and renewable energy integration. Quantum simulations account for real-time fluctuations in energy supply and demand, minimizing operational inefficiencies.

1.10.5.2 Battery Materials Discovery

Quantum simulations enable the discovery of new battery materials with higher energy densities and longer lifespans. These simulations solve quantum mechanical equations to predict the behavior of materials at the atomic level, accelerating the development of next-generation batteries.

1.10.5.3   Renewable Energy Systems

Quantum computing improves the efficiency of solar and wind energy systems by optimizing the placement and configuration of energy harvesting devices. Engineers and developers can apply quantum algorithms to maximize energy output while minimizing costs.

1.10.5.4   Case Study

A renewable energy firm used quantum simulations to design a new lithium-air battery, achieving a 15% increase in energy density compared to conventional lithium-ion batteries.

1.11   Overview of the Book Structure

This book provides a structured exploration of quantum computing and quantum artificial intelligence tailored for engineers and software developers. The book contests are described below:

Chapter 1	Introduction to Quantum Computing
Chapter 2	Quantum Computing Fundamentals
Chapter 3	Quantum Circuits
Chapter 4	Quantum Algorithms
Chapter 5	Classical Artificial Intelligence
Chapter 6	Quantum Data Structures
Chapter 7	Quantum Optimization
Chapter 8	Introduction to Quantum AI
Chapter 9	Quantum Neural Networks

Exercise Questions: Introduction to Quantum Computing
  1. **Classical vs. Quantum Computing**

     (a) Explain how quantum computing differs from classical computing in terms of problem-solving capability.
     (b) Provide an example of a problem that is intractable for classical computers but manageable for quantum computers.

  2. **Foundations of Quantum Mechanics**

(a) Describe Max Planck's contribution to quantum mechanics and explain the significance of quantized energy levels.

(b) What is the Heisenberg uncertainty principle, and how does it introduce probabilistic thinking into physics?

3. **Early Theoretical Connections**

(a) Discuss the role of Alan Turing's Turing machine in the development of classical computation. How does it relate to quantum computation?

(b) Explain how Claude Shannon's information theory influenced the exploration of information processing in quantum systems.

4. **Emergence of Quantum Computation**

(a) Summarize Richard Feynman's proposal regarding quantum systems and their computational applications.

(b) What is a quantum gate, as introduced by David Deutsch? How does it compare to a classical logic gate?

5. **Quantum Algorithms**

(a) What is the quantum Fourier transform, and why is it important for quantum algorithms?

(b) Explain how Grover's algorithm demonstrates a quadratic speedup over classical search methods.

6. **Development of Quantum Hardware**

(a) Discuss the significance of Christopher Monroe and David Wineland's demonstration of quantum logic gates using trapped ions.

(b) What role does quantum error correction play in building reliable quantum systems?

7. **Linear Algebra Foundations**

(a) Define a vector space and explain how it is used to represent quantum states in quantum computing.

(b) Using Dirac notation, express the inner product of two quantum states and explain its significance.

8. **Superposition and Entanglement**

(a) Write the mathematical expression for a superposed quantum state and explain the concept of quantum parallelism.

(b) Provide an example of an entangled state and describe its importance in quantum teleportation.

9. **Quantum Computing Principles**

(a) What are the three core principles of quantum mechanics utilized in quantum computing? Briefly describe each.

(b) How does interference contribute to the success of quantum algorithms like Grover's search?

10. **Mathematical Tools for Quantum Computing**

    (a) Explain the polar representation of complex numbers and its relevance to quantum mechanics.
    (b) Describe the role of the Fourier transform in quantum algorithms like Shor's.

11. **Engineering Applications of Quantum Computing**

    (a) Identify two optimization problems that can benefit from quantum computing and describe how quantum methods improve them.
    (b) How do quantum simulations aid in drug discovery and material science?

12. **The Role of Engineers in Quantum Computing**

    (a) Why is it essential for engineers to understand quantum computing, and how does it impact their role in technological innovation?
    (b) Discuss one application of quantum machine learning in engineering fields, such as robotics or predictive maintenance.

Additional Bibliography

1. C.H. Bennett, G. Brassard, Quantum cryptography: Public key distribution and coin tossing, in *Proceedings of IEEE International Conference on Computers, Systems and Signal Processing*, (IEEE, 1984), pp. 175–179
2. D. Broglie, Louis., Recherches sur la théorie des quanta (PhD Thesis). Ann. Phys. **10**, 3 (1925)
3. D. Deutsch, Quantum theory, the church–Turing principle and the universal quantum computer. Proc. R. Soc. A **400**(1818), 97–117 (1985)
4. M.H. Devoret, R.J. Schoelkopf, Superconducting circuits for quantum information: An outlook. Science **339**(6124), 1169–1174 (2013)
5. A. Einstein, Über einen die Erzeugung und Verwandlung des Lichtes betreffenden heuristischen Gesichtspunkt. Ann. Phys. **17**, 132–148 (1905)
6. E. Farhi, J. Goldstone, S. Gutmann, A quantum approximate optimization algorithm. arXiv (2014) arXiv:1411.4028 [quant-ph]
7. R.P. Feynman, Simulating physics with computers. Int. J. Theor. Phys. **21**(6–7), 467–488 (1982)
8. L.K. Grover, A fast quantum mechanical algorithm for database search, in *Proceedings of the Twenty-Eighth Annual ACM Symposium on Theory of Computing*, (ACM, 1996), pp. 212–219
9. A.S. Holevo, Bounds for the quantity of information transmitted by a quantum communication channel. Probl. Inf. Transm. **9**(3), 177–183 (1973)
10. M.W. Johnson et al., Quantum annealing with manufactured spins. Nature **473**(7346), 194–198 (2011)
11. E. Knill, R. Laflamme, Theory of quantum error-correcting codes. Phys. Rev. A **55**(2), 900–911 (1997)
12. S. Majidy, C. Wilson, R. Laflamme, *Building Quantum Computers—A Practical Introduction* (Cambridge Press, 2025)

13. C. Monroe et al., Demonstration of a fundamental quantum logic gate. Phys. Rev. Lett. **75**(25), 4714–4717 (1995)
14. M.A. Nielsen, I.L. Chuang, *Quantum Computation and Quantum Information* (Cambridge University Press, 2000)
15. M. Planck, Über das Gesetz der Energieverteilung im Normalspektrum. Ann. Phys. **4**, 553–563 (1901)
16. P.W. Shor, Algorithms for quantum computation: Discrete logarithms and factoring, in *Proceedings of the 35th Annual Symposium on Foundations of Computer Science*, (IEEE, 1994), pp. 124–134
17. P.W. Shor, Scheme for reducing decoherence in quantum computer memory. Phys. Rev. A **52**(4), R2493–R2496 (1995)
18. A. Turing, On computable numbers, with an application to the Entscheidungsproblem. Proc. Lond. Math. Soc. **42**, 230–265 (1936)
19. V. Neumann, John., *First Draft of a Report on the EDVAC* (Moore School of Electrical Engineering, University of Pennsylvania, 1945)

Chapter 2
Intro to Quantum Mechanics and Computing

Quantum mechanics provides the theoretical foundation for understanding the behavior of particles at atomic and subatomic scales. Its principles underpin the operation of quantum computing and other quantum technologies. This section explores the fundamental concepts of wave-particle duality and the uncertainty principle, emphasizing their mathematical representations and practical implications.

The principles of wave-particle duality and the uncertainty principle form the bedrock of quantum mechanics, providing insights into the behavior of particles at the quantum level. These principles are not only foundational for theoretical physics but also have practical applications in engineering fields, including nanotechnology, cryptography, and photonic device design. Understanding these concepts is essential for leveraging quantum mechanics in the development of advanced technologies.

2.1 Wave-Particle Duality

Wave-particle duality is a foundational property of quantum mechanics, encapsulating the dual nature of matter and energy as both discrete particles and continuous waves. This principle is critical for interpreting quantum phenomena such as interference, diffraction, and quantization in quantum systems.

2.1.1 Historical Context

The concept of wave-particle duality emerged from experiments in the early twentieth century. The double-slit experiment is a classic demonstration of this principle. When light or electrons pass through two closely spaced slits, an interference pattern characteristic of waves appears on the detection screen. However, when particles are observed individually, they behave like discrete entities.

© The Author(s), under exclusive license to Springer Nature Switzerland AG 2025

J. Van Griensven Thé et al., *Quantum Computing and Quantum Machine Learning for Engineers and Developers*, https://doi.org/10.1007/978-3-031-98245-3_2

The experiment's results reveal that particles exhibit wave-like behavior when not measured but appear as discrete particles upon observation. This duality is mathematically described by the wave function $\Psi(x, t)$, which encapsulates all possible states of a quantum system.

2.1.2 The Equation for Wave Function

The wave function $\Psi(x, t)$ is given by:

$$\Psi(x, t) = Ae^{i(kx - \omega t)}$$

where

- A is the amplitude of the wave.
- k is the wave number.
- ω is the angular frequency.
- t and x are time and position, respectively.

The wave function is complex, and its physical interpretation requires the calculation of the probability density, which is the modulus squared of $\Psi(x, t)$:

$$|\Psi(x, t)|^2 = \Psi^*(x, t)\Psi(x, t).$$

This probability density determines the likelihood of finding a particle at position x and time t.

2.1.3 Probability Density and Interference Patterns

In the double-slit experiment, the interference pattern arises due to the superposition of wave functions from the two slits. The combined wave function at a point on the screen is:

$$\Psi_{\text{total}} = \Psi_1 + \Psi_2$$

The intensity is:

$$I = |\Psi_{\text{total}}|^2 = |\Psi_1|^2 + |\Psi_2|^2 + 2|\Psi_1||\Psi_2|\cos\phi$$

where ϕ is the phase difference between the waves from the two slits. The term $\mathbf{2|\Psi_1||\Psi_2|\cos\phi}$ represents the interference contribution, leading to bright and dark fringes on the screen.

2.1.4 Implications for Engineers

Understanding wave-particle duality is critical for designing and optimizing nano-scale devices and photonic circuits. The following are engineering examples:

(a) **Nanoscale Transistors**

Quantum effects like tunneling and wave interference affect the behavior of electrons in modern transistors, necessitating precise control over device dimensions.

(b) **Photonic Devices**

Devices such as lasers and photonic waveguides exploit the wave-like nature of light to guide and manipulate photons. Engineers use principles of wave interference to design these components for optimal performance.

2.2 The Uncertainty Principle

The uncertainty principle, introduced by Werner Heisenberg, asserts that certain pairs of non-commuting observables, a prime example being position and momentum, cannot be simultaneously measured with arbitrary precision. This principle reflects the inherent limitations of quantum systems and holds profound implications for both theoretical understanding and practical application.

2.2.1 Statement of Heisenberg's Uncertainty Principle

The mathematical expression of the uncertainty principle is:

$$\Delta x \Delta p \geq \frac{\hbar}{2}$$

where

Δx is the uncertainty in position.
Δp is the uncertainty in momentum.
$\hbar$ is the reduced Planck's constant ($\hbar = h/2\pi$).

This uncertainty principle inequality implies that the product of uncertainties in position and momentum is bounded by a fundamental limit, making it impossible to simultaneously know both properties with absolute precision.

2.2.2 Physical Interpretation

The uncertainty principle arises from the wave-like nature of particles. Position x and momentum p are conjugate variables, and their relationship is governed by Fourier transforms. A localized wave packet (representing precise position) requires a broad range of momentum components, while a narrow momentum distribution leads to a delocalized wave packet.

Mathematically, the uncertainty principle is a consequence of the non-commuting nature of the position ($\widehat{x}$) and momentum ($\widehat{p}$) operators:

$$[\widehat{x}, \widehat{p}] = \widehat{xp} - \widehat{px} = i\hbar$$

2.2.3 Implications for Measurement

In quantum mechanics, measurement collapses a particle's wave function to a specific state, introducing uncertainty in conjugate properties. For example:

(a) Measuring a particle's position precisely ($\Delta x \to 0$) increases the uncertainty in its momentum ($\Delta p \to \infty$).
(b) Conversely, precise momentum measurement leads to high uncertainty in position.

2.2.4 Applications of the Uncertainty Principle

2.2.4.1 Quantum Cryptography

The uncertainty principle underpins the security of quantum cryptographic protocols like quantum key distribution (QKD). In QKD, any attempt to eavesdrop on a quantum communication channel introduces measurable disturbances, alerting the communicating parties.

2.2.4.2 Nanotechnology

In nanoscale devices, quantum confinement effects arise due to the uncertainty principle. For instance:

- **Quantum Dots:** These nanostructures confine electrons in all three spatial dimensions, resulting in discrete energy levels.
- **Scanning Tunneling Microscopes (STMs):** The principle helps explain electron tunneling, which allows STMs to image surfaces at atomic resolution.

2.2.4.3   Atomic and Molecular Spectroscopy

The uncertainty principle governs the linewidth of spectral lines. A shorter-lived excited state corresponds to a broader spectral line due to increased energy uncertainty (ΔE).

2.2.4.4   Particle Accelerators

Engineers designing particle accelerators account for uncertainty in particle position and momentum to achieve precise beam focusing and control.

2.2.5   Mathematical Formulation

Consider a particle confined within a one-dimensional box of length L. The uncertainty in position is approximately $\Delta x = L$. The uncertainty in momentum is given by:

$$\Delta p \geq \frac{\hbar}{2\Delta x} = \frac{\hbar}{2L}$$

This relationship illustrates the trade-off between spatial confinement and momentum spread, which is critical in designing quantum devices.

2.3   The Schrödinger Equation

The Schrödinger equation is one of the foundational equations of quantum mechanics. Developed by Erwin Schrödinger in 1925, it describes how the quantum state of a physical system evolves over time. In non-relativistic quantum mechanics (i.e., energies and speeds much lower than those requiring relativistic corrections), it provides essential insights into the behavior of particles at the atomic and subatomic scales. Note that the Dirac equation is a more modern version of Schrödinger's since it includes relativistic effects.

In one spatial dimension, the time-dependent Schrödinger equation can be written as:

$$i\hbar \frac{\partial}{\partial t} \Psi(x, t) = \left[-\frac{\hbar^2}{2m} \frac{\partial^2}{\partial x^2} + V(x) \right] \Psi(x, t)$$

where

i is the imaginary unit ($\sqrt{-1}$).

$\hbar$ ("h-bar") is the reduced Planck's constant, defined by $\hbar = \frac{h}{2\pi}$.

$\Psi(x, t)$ is the wave function of the system, which encodes the system's quantum state.

m is the mass of the particle (for a single particle).

$V(x)$ is the potential energy as a function of position x.

H is the Hamiltonian operator represented by the terms in straight brackets on the right-hand side of the equation above $H = \left[-\frac{\hbar^2}{2m} \frac{\partial^2}{\partial x^2} + V(x) \right]$.

For higher dimensions (e.g., three-dimensional space), x becomes a vector $\mathbf{r} = (x, y, z)$, and the spatial derivative term becomes the Laplacian (i.e., ∇^2).

The Schrödinger equation gives us the probability amplitudes for finding particles in specific states. Solving it provides allowed energy levels (e.g., the discrete "quantum" states of an electron in a hydrogen atom). Almost every concept in non-relativistic quantum mechanics, such as electron orbitals, quantum tunneling, discrete energy levels, and the uncertainty principle, can be connected back to solutions of the Schrödinger equation (and its extensions or approximations). Its probabilistic interpretation was a major departure from classical determinism. Rather than specifying precise trajectories, quantum mechanics speaks in terms of probabilities, reflecting a deeper limit to how precisely one can "know" or "measure" certain physical quantities.

It is usually not feasible to solve the Schrödinger equation precisely for real-world scenarios. Consequently, approximate solutions are obtained using methods like the simple finite difference method. To simplify, this section presents and solves the 1D Schrödinger equation. The solution principle is the same for a 2D and 3D version of this equation. The 1D time-independent Schrödinger equation for quantum systems is:

$$-\frac{\hbar^2}{2m} \frac{d^2\psi}{dx^2} + V(x)\psi(x) = E\psi(x)$$

To further simplify the presentation, let's set $m = 1$ and $h = 1$. This reduces the above equation to:

$$-\frac{1}{2} \frac{d^2\psi(x)}{dx^2} + V(x)\psi(x) = E\,\psi(x)$$

One needs to approximate the second derivative of the Ψ in relation to x. Employing a central difference, which is second-order accurate, for the second derivative, $d^2\Psi/dx^2$ is approximated by:

$$\frac{\partial^2 \psi}{\partial x^2} = \frac{\psi_{i+1} - 2\psi_i + \psi_{i-1}}{\Delta x^2}$$

This will transform the above equation to:

$$-\frac{1}{2}\left(\frac{\psi_{j+1} - 2\psi_j + \psi_{j-1}}{h^2}\right) + V_j\psi_j = E\psi_j$$

One must discretize the spatial dimension, from zero to X_N, into small sections, which I will call Δx. The number of segments in the modeling domain is $N = X_N/\Delta x$. One must also define the conditions on the boundaries of the modeling domain. The boundary conditions will be set to zero at both ends. That is:

$$\Psi_0 = 0$$

$$\Psi_N = 0$$

Applying the following parameters on the 1D gridded modeling domain:

- $X_0 = 0.0$ (meter)
- $X_N = 0.1$ (meter)
- $N = 100$ (number of gridded segments on the modeling domain)
- By selecting only the first three states, one gets the following solution presented in the figure below.

Python Code Example: Solving the Schrodinger Equation

```python
#---------------------------------------------------------------------
# Code to Solve the Schrodinger Equation
# Chapter 2 in the QUANTUM COMPUTING AND QUANTUM MACHINE LEARNING BOOK
#---------------------------------------------------------------------
# Version 1.0
# Qiskit changes frequently.
# We recommend using the latest version from the book code repository at:
# https://aqtinitiative.org/quantum-computing-for-engineers

# (c) 2025 Jesse Van Griensven, Roydon Fraser, and Jose Rosas
# License: MIT - Citation required
#---------------------------------------------------------------------
import numpy as np
import matplotlib.pyplot as plt


#---------------------------------------------------------------------
def Vpot(x):
    """ Potential Energy Function """
    return x**2
#---------------------------------------------------------------------
```

```python
# Spatial grid parameters
x0 = 0.0 # [meter]
xN = 0.1 # [meter]
N = 100 # Number of segments - Used to define the delta x (dx)

# Create x grid
x = np.linspace(x0, xN, N)

# Define step size
dx = x[1] - x[0]

# Define the number of "waves" (States) we want to obtain in the solution
States = 4

# T: Kinetic Energy matrix

T = np.zeros((N-2)**2).reshape(N-2,N-2)

# Account for the boundary condition
for i in range(N-2):
  for j in range(N-2):
    if i==j:
      T[i,j]= -2
    elif np.abs(i-j)==1:
      T[i,j]=1
    else:
      T[i,j]=0

# V is the Potential Energy matrix
V = np.zeros((N-2)**2).reshape(N-2, N-2)

for i in range(N-2):
  for j in range(N-2):
    if i==j:
      V[i,j] = Vpot(x[i+1])
    else:
      V[i,j] = 0

# Hamiltonian Matrix
# $ \LARGE H = -\frac{T}{2 \cdot dx^2} + V$
#

# Hamiltonian matrix - Computing in vectorized form
H = -T / (2*dx**2) + V
```

```python
# Solve to obtain the eigenvalues and eigenvectors
Eigval, Eigvec = np.linalg.eig(H)

# Sort eigenvalues and eigenvectors in ascending order
Eigval_ordered = np.argsort(Eigval)

# Limit the solution to the specified number of states (States)
z = Eigval_ordered[0:States]

Energies = (Eigval[z] / Eigval[z][0])

#print(Energies)
for i, E_i in enumerate(Energies):
  print(f"Energies[{i}] = {E_i:5.3f}")

# Plot 3 wavefunctions for the States - Not all states plotted

# Define the size of the plot
plt.figure(figsize=(12, 6))

# Loop to Plot each state in States
for i in range(States):  # for i in range(len(z))
  y = []
  y = np.append(y, Eigvec[:,z[i]])
  y = np.append(y,0)
  y = np.insert(y,0,0)
  plt.plot(x, y, lw=3, label="State: {} ".format(i))
  plt.xlabel('x', size=14)
  plt.ylabel('$\psi$(x)', size=16)

plt.legend()
plt.grid(alpha=0.2) # Schrödinger Equation
plt.title("Normalized Wavefunctions for the 1D Schrödinger Equation
\n", size=14)
plt.show()
```

Figure 2.1 displays the results of the Python code example for the Schrödinger's equation.

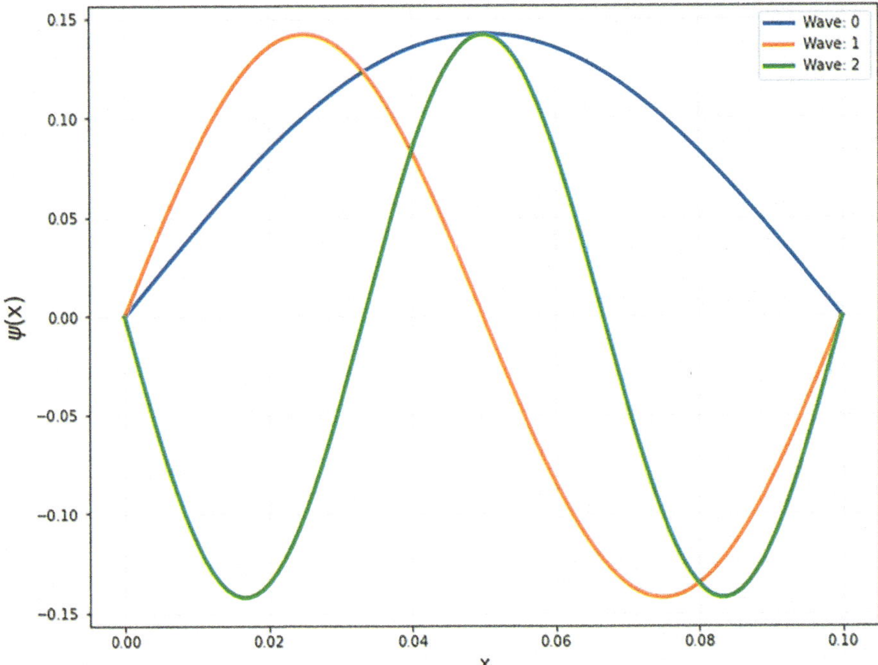

Fig. 2.1 Normalized wave functions for the 1D Schrödinger equation (by authors)

2.4 Introduction to Hamiltonians in Quantum Computing

In quantum mechanics, the Hamiltonian is one of the most important operators that dictate the evolution of a quantum system. It encompasses all the energy contributions within the system, including kinetic and potential energies. Understanding Hamiltonians is essential for designing and analyzing quantum algorithms, simulating physical systems, and optimizing quantum circuits. At the core of quantum mechanics lies the concept of the Hamiltonian, a fundamental operator that encapsulates the total energy of a quantum system. Figure 2.2 summarizes the importance of the Hamiltonian in quantum computing.

This subsection explains what Hamiltonians are, their significance in quantum computing, and provides illustrative Python code examples to demonstrate their applications.

2.4.1 Mathematical Definition of Hamiltonians

Mathematically, the Hamiltonian H is a Hermitian operator (i.e., $H = H^{\dagger}$) acting on the Hilbert space of a quantum system. Its primary role is to govern the time evolution of quantum states through the Schrödinger equation:

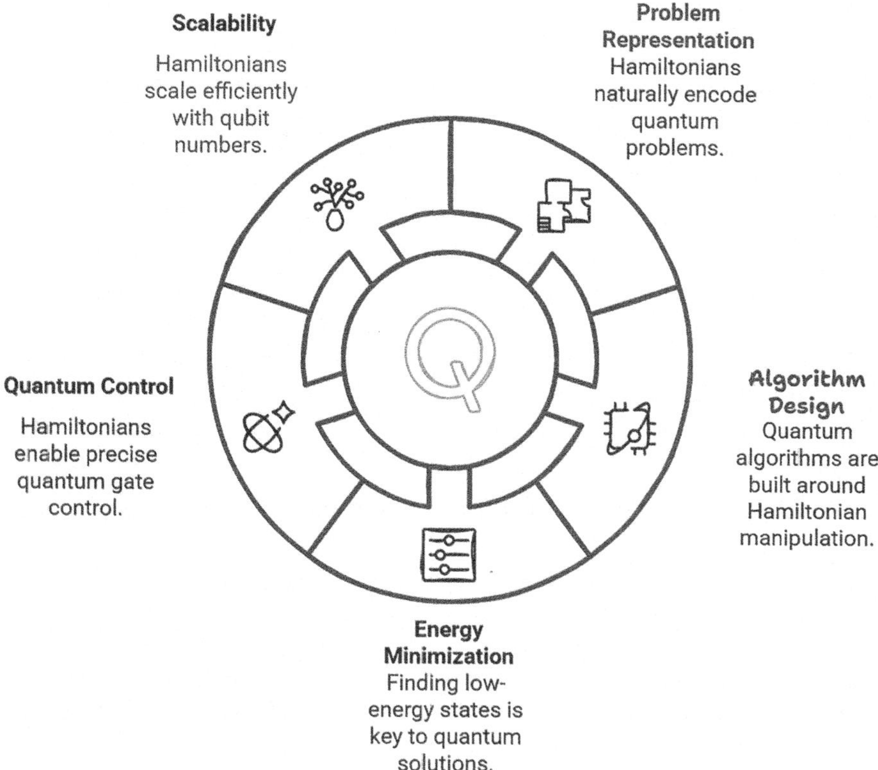

Fig. 2.2 Importance of Hamiltonians in quantum computing (by authors using Napkin.ai)

$$i\hbar \frac{\partial}{\partial t} \,|\psi(t)\rangle = H\,|\psi(t)\rangle$$

where

- i is the imaginary unit.
- $\hbar$ is the reduced Planck constant.
- $|\psi(t)\rangle$ is the state vector of the system at time t.

Hamiltonians must satisfy the following properties:

1. It must be Hermitian to ensure that all eigenvalues (energy levels) are real, a necessity for physical observables.
2. It must hold true when expressed as a sum of tensor products of Pauli matrices when dealing with multi-qubit systems.

2.4.2 Example: Two-Qubit Hamiltonian

Consider a simple two-qubit system with the Hamiltonian:

$$H = c_0 I \otimes I + c_1 Z \otimes I + c_2 I \otimes Z + c_3 Z \otimes Z + c_4 X \otimes X$$

where

- I is the Identity matrix.
- X, Y, Z are the Pauli matrices.
- c_0, c_1, c_2, c_3, c_4 are real coefficients representing the strength of each term.
- This Hamiltonian encapsulates both individual qubit terms and interactions between qubits.

2.4.3 Hamiltonians in Quantum Computing

Hamiltonians are integral to various areas in quantum computing, from representing computational problems to guiding the execution of quantum algorithms. In quantum simulation, Hamiltonians model the physical systems being simulated. For instance, simulating molecular structures, material properties, or spin systems requires accurate representations of their Hamiltonians. Table 2.1 summarizes the topic of quantum states and Hamiltonians.

The following subsections explore Hamiltonian applications in detail.

2.4.4 Example of Quantum Algorithms Using Hamiltonians

Several quantum algorithms leverage Hamiltonians to solve optimization problems and perform simulations. Two prominent examples, which we will cover in this book, are:

2.4.4.1 Quantum Approximate Optimization Algorithm (QAOA)

QAOA is designed to find approximate solutions to combinatorial optimization problems. It utilizes a parameterized quantum circuit constructed from two Hamiltonians:

Table 2.1 Quantum states and Hamiltonians

Phenomenon	Description
Quantum states	Represented as vectors in a Hilbert space, often denoted $|\psi\rangle$
Hamiltonian action	The Hamiltonian acts on these states to determine their evolution and energy properties

(a) Cost Hamiltonian (H_C): Encodes the optimization problem's objective function.
(b) Mixing Hamiltonian (H_M): Drives the system's evolution to explore the solution space.

The algorithm alternates between applying these Hamiltonians and adjusting parameters to minimize the expectation value of the cost Hamiltonian.

2.4.4.2   Variational Quantum Eigensolver (VQE)

VQE aims to find the ground-state energy of a Hamiltonian, making it valuable for quantum chemistry and material science. It employs a hybrid quantum-classical approach:

(a) Parameterized quantum circuit to prepare trial states.
(b) Classical optimization (in digital computers) to adjust parameters to minimize the energy expectation value.

Python Code Example: Hamiltonian

The Hamiltonian we want to demonstrate in the Python code represents two qubits with individual Z terms and interaction terms $Z \otimes Z$ and $X \otimes X$. Understanding the energy spectrum of a Hamiltonian is crucial. For this objective, the code computes Hamiltonian eigenvalues and eigenvectors to identify the system's energy levels and ground state. The Hamiltonian in this code example has four eigenvalues corresponding to the four possible two-qubit states. The lowest eigenvalue (-2.0) represents the ground-state energy. The code starts by defining a Hamiltonian using Qiskit's `PauliSumOp`. This class allows for the representation of Hamiltonians as sums of tensor products of Pauli operators (Fig. 2.3).

```python
#------------------------------------------------------------------
# Hamiltonian Introduction
# Chapter 2 in the QUANTUM COMPUTING AND QUANTUM MACHINE LEARNING BOOK
#------------------------------------------------------------------
# Version 1.0
# Qiskit changes frequently.
# We recommend using the latest version from the book code repository at:
# https://aqtinitiative.org/quantum-computing-for-engineers

# (c) 2025 Jesse Van Griensven, Roydon Fraser, and Jose Rosas
# License: MIT - Citation required
#------------------------------------------------------------------
import numpy as np
import matplotlib.pyplot as plt
from qiskit.opflow import I, X, Y, Z, PauliSumOp
```

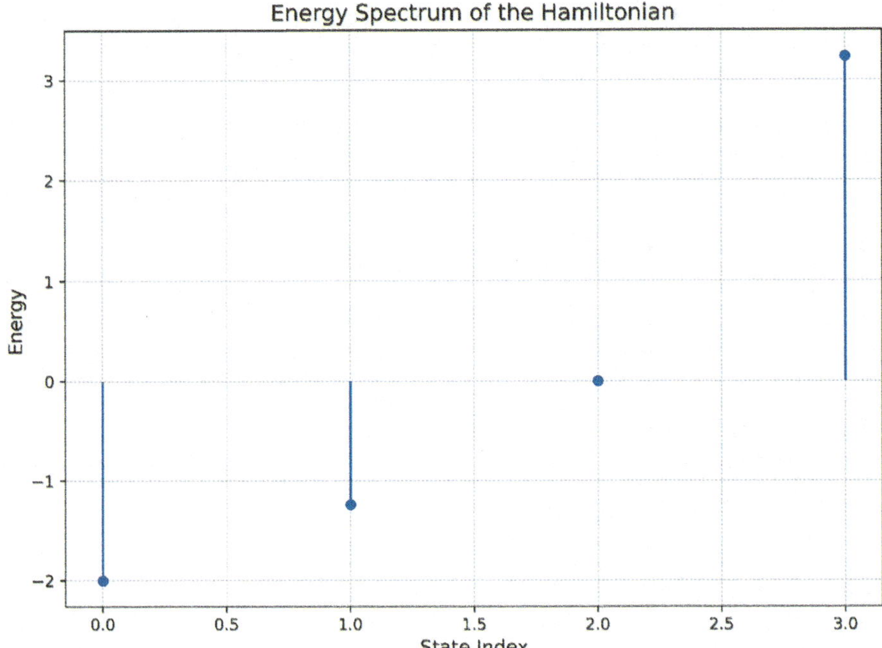

Fig. 2.3 Hamilton eigenvalue results: Energy Levels

```
# Define the Hamiltonian for a 2-qubit system.
# Example: H = Z0 + Z1 + Z0 Z1 + X0 X1
H = (1.0 * (Z ^ I)) + (1.0 * (I ^ Z)) + (1.0 * (Z ^ Z)) + (1.0 * (X ^ X))

print("Hamiltonian H:")
print(H)
# Output:
# Hamiltonian H:
# 1.0 * Z ⊗ I + 1.0 * I ⊗ Z + 1.0 * Z ⊗ Z + 1.0 * X ⊗ X

# Convert Hamiltonian to matrix form
H_matrix = H.to_matrix()

# Compute eigenvalues and eigenvectors
eigenvalues, eigenvectors = np.linalg.eigh(H_matrix)

print("Eigenvalues of H:")
print(eigenvalues)

print("\nEigenvectors of H:")
print(eigenvectors)
```

```python
# Expected Output:
# Eigenvalues of H:
# [-2.0 -2.0 2.0 2.0]
#
# Eigenvectors of H:
# [[ 0.70710678  0.70710678  0.       0.     ]
# [ 0.70710678 -0.70710678  0.       0.     ]
# [ 0.        0.        0.70710678  0.70710678]
# [ 0.        0.        0.70710678 -0.70710678]]


# Plotting Energy Spectra
# Visualizing the energy spectrum gives insight into the system's
properties.
#The code below creates a stem plot of the eigenvalues.
# X-Axis (State Index): Represents the different quantum states.
# Y-Axis (Energy): Shows the corresponding energy levels.
# Stem Plot: Visualizes discrete energy levels, highlighting ground and
excited states.


# Plot the energy spectrum with increased resolution
plt.figure(figsize=(8, 6), dpi=600)


# Create a stem plot of the eigenvalues
plt.stem(eigenvalues, use_line_collection=True, basefmt=" ")


# Labeling the axes and adding a title
plt.xlabel('State Index', fontsize=12)
plt.ylabel('Energy', fontsize=12)
plt.title('Energy Spectrum of the Hamiltonian', fontsize=14)


# Adding a grid for better readability
plt.grid(True, which='both', linestyle='--', linewidth=0.5)


# Adjust layout to prevent clipping of labels/titles
plt.tight_layout()


# Save the plot as a high-resolution PNG file
plt.savefig("Energy_Spectrum_High_Res.png", dpi=600)


# Display the plot
plt.show()
```

2.5 Qubits and Quantum States

Quantum computing is built upon the concept of qubits, which serve as the fundamental units of quantum information. Unlike classical bits, qubits leverage quantum mechanical properties such as superposition and entanglement to perform computations. This section explores the distinctions between classical bits and qubits, the mathematical framework for multi-qubit systems, and the role of entanglement in quantum computation.

Qubits and quantum states form the foundation of quantum computing, enabling powerful computational capabilities through superposition and entanglement. The distinctions between classical bits and qubits highlight the novel principles underlying quantum information. Multi-qubit systems and entanglement expand these principles, providing the tools for quantum teleportation, cryptography, and advanced computation. Understanding these concepts is crucial for harnessing the full potential of quantum technologies.

2.5.1 Classical Bits Vs. Qubits

Classical bits and qubits differ fundamentally in how they represent and process information. Understanding these differences is crucial for comprehending the power of quantum computing.

2.5.1.1 Classical Bits

Classical bits are the basic units of information in conventional computing. They exist in one of two states, 0 or 1, and their values are manipulated through logical operations performed by gates, such as AND, OR, and NOT. For example, a classical register of n bits can store one of 2^n possible configurations, but only one at a time.

The deterministic nature of classical bits makes them suitable for tasks that require well-defined and sequential computations. However, this characteristic also limits their ability to handle problems requiring exploration of vast solution spaces.

2.5.1.2 Qubits

Qubits extend the capabilities of classical bits by exploiting the superposition principle. A qubit can exist in a linear combination of states $|0\rangle$ and $|1\rangle$:

$$|\psi\rangle = \alpha\,|0\rangle + \beta\,|1\rangle$$

where α and β are complex probability amplitudes satisfying $|\alpha|^2 + |\beta|^2 = 1$.
In this representation:

- $|0\rangle$ and $|1\rangle$ form the computational basis.
- $|\alpha|^2$ represents the probability of measuring the qubit in the state $|0\rangle$.
- $|\beta|^2$ represents the probability of measuring the qubit in the state $|1\rangle$.

Superposition enables qubits to process multiple states simultaneously, exponentially increasing computational power as the number of qubits grows. For example, a system of n qubits can represent 2^n states concurrently.

2.5.1.3   Visualization: The Bloch Sphere Representation

The Bloch sphere provides a geometric representation of a qubit's state, enabling visualization of its quantum properties. A qubit state can be expressed as:

$$|\psi\rangle = \cos\frac{\theta}{2}\,|0\rangle + e^{i\phi}\sin\frac{\theta}{2}\,|1\rangle$$

where θ and ϕ are spherical coordinates.
On the Bloch sphere:

1. The north pole represents $|0\rangle$, and the south pole represents $|1\rangle$.
2. Points on the sphere's surface correspond to superpositions of $|0\rangle$ and $|1\rangle$.
3. The azimuthal angle ϕ determines the relative phase between the states (Fig. 2.4).

The Bloch sphere highlights the continuous nature of quantum states, which contrasts with the discrete states of classical bits. This representation is instrumental in understanding quantum gates, which correspond to rotations on the sphere.

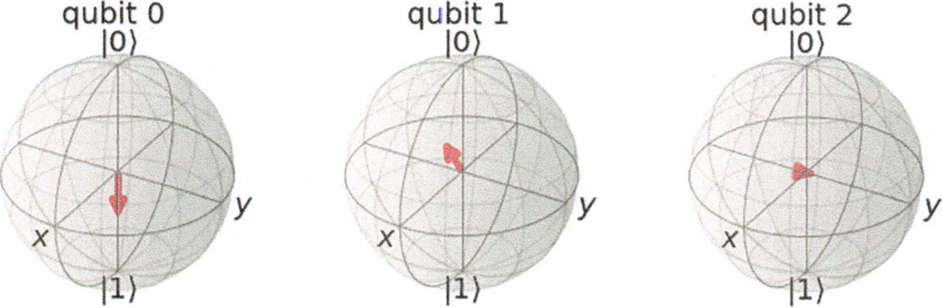

Fig. 2.4 Three Bloch spheres with different azimuthal angles

2.5.2 Multi-Qubit Systems and Entanglement

When qubits interact, they form multi-qubit systems whose properties and computational capabilities far exceed those of individual qubits. The mathematical foundation for describing these systems lies in the tensor product, which combines individual qubit states into a joint state.

2.5.2.1 Tensor Product of Quantum States

For a system with two qubits, each in a state $|\psi_1\rangle = \alpha_1 \, | \, 0\rangle + \beta_1 \, | \, 1\rangle$ and $|\psi_2\rangle = \alpha_2 \, | \, 0\rangle + \beta_2 \, | \, 1\rangle$, the combined state is represented as:

$$|\psi\rangle = |\psi_1\rangle \otimes |\psi_2\rangle = \alpha_1\alpha_2 \, |00\rangle + \alpha_1\beta_2 \, |01\rangle + \beta_1\alpha_2 \, |10\rangle + \beta_1\beta_2 \, |11\rangle$$

The tensor product operation ensures that the combined state accounts for all possible configurations of the two qubits. For an n-qubit system, the state is expressed as a vector in a 2^n-dimensional Hilbert space.

2.5.2.2 Entangled States

Entanglement is a unique quantum phenomenon that arises in multi-qubit systems. Unlike separable states, where each qubit's state can be independently described, entangled states exhibit correlations that persist regardless of the distance between the qubits.

A well-known example of an entangled state is the Bell state:

$$|\phi^+\rangle = \frac{1}{\sqrt{2}}(\, |00\rangle + \, |11\rangle)$$

In this state, measuring one qubit immediately determines the state of the other. For instance, if the first qubit is found in $|0\rangle$, the second qubit must also be in $|0\rangle$.

Entanglement is mathematically characterized by the inability to express the joint state as a product of individual states:

$$|\psi\rangle \neq |\psi_1\rangle \otimes |\psi_2\rangle$$

2.5.2.3 Importance of Entanglement

Entanglement is a critical resource for quantum computation and communication. Its applications include:

(a) **Quantum Teleportation**

Teleportation uses entanglement to transfer quantum states between distant locations without physically transmitting particles. The protocol involves sharing an entangled pair and performing local measurements and classical communication.

(b) **Quantum Cryptography**

Protocols like E91 rely on entanglement to detect eavesdropping in quantum key distribution. Any attempt to intercept the entangled particles introduces measurable disturbances, ensuring secure communication.

(c) **Quantum Computing**

Entanglement enables quantum computers to perform operations that involve global correlations among qubits, which are essential for algorithms like Shor's algorithm and Grover's search.

2.5.2.4   Case Study: Entanglement in Quantum Algorithms

In Grover's search algorithm, entanglement and superposition are used to amplify the probability of the correct solution. The algorithm starts with an equal superposition of all states and applies a sequence of quantum gates to entangle the qubits and perform amplitude amplification. This process reduces the search complexity from $O(N)$ in classical systems to $O\left(\sqrt{N}\right)$ in quantum systems.

Entanglement is also leveraged in error correction schemes, where logical qubits are encoded in entangled states of multiple physical qubits to protect against errors caused by decoherence.

2.5.2.5   Visualization and Experimental Realizations

Entanglement can be visualized using the density matrix formalism, which provides a statistical representation of quantum states. Experimentally, entanglement is realized using techniques such as:

(a) **Spontaneous Parametric Down-Conversion (SPDC)**

A nonlinear optical process is used to generate entangled photon pairs.

(b) **Ion Traps**

Here, entanglement is achieved by manipulating the shared vibrational modes of trapped ions.

2.6   Quantum Operators and Gates

Quantum operators and gates form the foundation of quantum computing by defining the transformations that can be applied to quantum states. Operators in quantum mechanics generalize classical functions and matrices to act on quantum states, while quantum gates implement these operators in physical systems to perform computations. This section delves into the mathematical framework of quantum operators and their application through quantum gates, highlighting key examples and their significance in quantum computing.

Quantum operators and gates are the building blocks of quantum computation, enabling the manipulation and transformation of quantum states. Single-qubit gates like the Hadamard and phase shift gates introduce superposition and relative phase, while multi-qubit gates like CNOT create entanglement and enable conditional operations. These tools form the foundation for implementing quantum algorithms and developing robust quantum systems.

2.6.1 Operators in Quantum Mechanics

Operators in quantum mechanics are mathematical entities that act on quantum states to describe physical transformations or measurements. These operators are typically linear transformations represented as matrices, which operate on vectors in a Hilbert space.

2.6.1.1 Definition

An operator $\widehat{O}$ is a linear transformation such that for any two vectors $|\psi\rangle$ and $|\phi\rangle$ and any scalars a and b:

$$\widehat{O}(a\,|\,\psi\rangle + b\,|\,\phi\rangle) = a\widehat{O}\,|\,\psi\rangle + b\widehat{O}\,|\,\phi\rangle$$

Operators are used to represent observables in quantum mechanics, such as position, momentum, and spin. The application of an operator to a quantum state often yields another quantum state or a scalar value corresponding to a measurable quantity.

2.6.1.2 Hermitian Operators

Hermitian operators play a crucial role in quantum mechanics because they represent observable quantities. An operator $\widehat{O}$ is Hermitian if it satisfies:

$$\widehat{O} = \widehat{O}^{\dagger}$$

where $\widehat{O}^{\dagger}$ is the conjugate transpose of $\widehat{O}$. The eigenvalues of a Hermitian operator are real, ensuring that measurements yield real values. For example, the Hamiltonian $\widehat{H}$, which represents the total energy of a quantum system, is a Hermitian operator.

2.6.1.3 Key Example: Pauli Matrices

The Pauli matrices are a set of Hermitian and unitary operators that are fundamental in describing single-qubit operations. They are:

$$\sigma_x = \begin{bmatrix} 0 & 1 \\ 1 & 0 \end{bmatrix}, \quad \sigma_y = \begin{bmatrix} 0 & -i \\ i & 0 \end{bmatrix}, \quad \sigma_z = \begin{bmatrix} 1 & 0 \\ 0 & -1 \end{bmatrix}$$

These matrices represent rotations about the x, y, and z axes of the Bloch sphere. For example:

- σ_x: Flip the qubit state $|0\rangle \leftrightarrow |1\rangle$.
- σ_y: Apply a $\pi/2$ phase shift while flipping the state.
- σ_z: Apply a π phase shift to the $|1\rangle$ component of the state.

These operators are used extensively in quantum gates and error correction codes.

2.7 Quantum Gates

Quantum gates implement unitary transformations on qubits, analogous to classical logic gates in digital computing. A quantum gate operates on one or more qubits and is represented by a unitary matrix U, which satisfies:

$$UU^\dagger = I$$

where I is the identity matrix. The unitarity of U ensures that quantum gates preserve the total probability of the quantum state.

2.7.1 Quantum Circuit Diagrams and Their Notation

Quantum circuits are commonly represented using circuit diagrams. These diagrams use lines for qubits and symbols for gates, making it easier to visualize the computation. The main key notations are summarized below:

1. **Qubits**
   Represented as horizontal lines.
2. **Gates**
   Represented as symbols placed on the qubit lines.

   (a) Single-qubit gates (e.g., H, X, R_z) are drawn as labeled boxes.
   (b) Multi-qubit gates (e.g., CNOT) are drawn with connections between qubits.

3. **Measurements**

Represented as a meter or M symbol, indicating the measurement operation.

Example: Circuit Diagram for Entanglement

A circuit that creates a Bell state:

1. Applies a Hadamard gate to qubit 0.
2. Applies a CNOT gate with qubit 0 as the control and qubit 1 as the target.

Example of a Circuit Diagram:

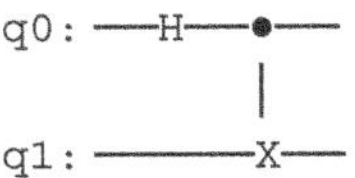

2.7.2 Single-Qubit Gates

Single-qubit gates act on individual qubits and perform rotations or phase shifts.

2.7.3 Hadamard Gate

The Hadamard gate creates a superposition of the basis states $|0\rangle$ and $|1\rangle$. Its matrix representation is:

$$H = \frac{1}{\sqrt{2}} \begin{bmatrix} 1 & 1 \\ 1 & -1 \end{bmatrix}$$

When applied to $|0\rangle$, the Hadamard gate produces the state:

$$H|0\rangle = \frac{1}{\sqrt{2}}(|0\rangle + |1\rangle)$$

Similarly, when applied to $|1\rangle$:

$$H|1\rangle = \frac{1}{\sqrt{2}}(|0\rangle - |1\rangle)$$

The Hadamard gate is widely used in quantum algorithms, such as Grover's search and Shor's algorithm, to initialize superposition states.

2.7.4 *Phase Shift Gate*

Phase shift gates are essential for implementing controlled phase rotations in multi-qubit circuits. The phase shift gate adds a phase ϕ to the $|1\rangle$ component of a qubit. Its matrix representation is:

$$R_\phi = \begin{bmatrix} 1 & 0 \\ 0 & e^{i\phi} \end{bmatrix}$$

For example, a $\pi/2$ phase shift gate ($\phi = \pi/2$) modifies the quantum state $|\psi\rangle$

$$|\psi\rangle = \alpha |0\rangle + \beta |1\rangle$$

to:

$$|\psi'\rangle = \alpha |0\rangle + \beta e^{i\pi/2} |1\rangle$$

2.7.5 *Multi-Qubit Gates*

Multi-qubit gates operate on two or more qubits, enabling interactions and entanglement.

2.7.5.1 CNOT Gate (Controlled NOT)

The CNOT gate flips the state of a target qubit $|t\rangle$ if the control qubit $|c\rangle$ is in the $|1\rangle$ state. Its matrix representation is:

$$\text{CNOT} = \begin{bmatrix} 1 & 0 & 0 & 0 \\ 0 & 1 & 0 & 0 \\ 0 & 0 & 0 & 1 \\ 0 & 0 & 1 & 0 \end{bmatrix}$$

The CNOT gate transforms the basis states as follows:

- $|00\rangle \rightarrow |00\rangle$
- $|01\rangle \rightarrow |01\rangle$
- $|10\rangle \rightarrow |11\rangle$
- $|11\rangle \rightarrow |10\rangle$

The CNOT gate is fundamental for creating entangled states. For example, applying a Hadamard gate to the first qubit of $|00\rangle$, followed by a CNOT gate, produces the Bell state:

$$|\phi^+\rangle = \frac{1}{\sqrt{2}}(|00\rangle + |11\rangle)$$

2.7.5.2 Applications of Multi-Qubit Gates

The following is a sample of multi-qubit gate applications:

1. **Entanglement Generation**
   Multi-qubit gates like CNOT enable the creation of entangled states, which is essential for quantum communication protocols and error correction.
2. **Quantum Algorithms**
   Controlled gates implement conditional operations in algorithms like Grover's search and the quantum Fourier transform.
3. **Error Correction**
   Multi-qubit gates are used to encode logical qubits into entangled states of physical qubits, protecting against decoherence and gate errors.

2.8 Expanding the Dirac Notation

Dirac notation, also known as bra-ket notation, is a compact and elegant mathematical framework used extensively in quantum mechanics and quantum computing. This notation simplifies the representation of quantum states, inner products, and operators, providing a consistent language for describing quantum systems. This section introduces the core components of Dirac notation, including ket and bra representations, inner and outer products, and their corresponding matrix forms.

Dirac notation provides a powerful framework for describing quantum states, inner and outer products, and operators. Its matrix representations facilitate explicit calculations and form the basis for implementing quantum algorithms and simulating quantum systems. Understanding these concepts is fundamental to mastering quantum mechanics and quantum computing.

2.8.1 Ket and Bra Notation

Dirac notation employs two fundamental components: kets ($|\psi\rangle$) and bras ($\langle\psi|$), which represent quantum states and their duals, respectively.

A **ket,** denoted $|\psi\rangle$, represents a column vector in a complex vector space. For example, a single-qubit quantum state can be expressed as:

$$|\psi\rangle = \begin{bmatrix} \alpha \\ \beta \end{bmatrix}$$

where α and β are complex numbers satisfying $|\alpha|^2 + |\beta|^2 = 1$.

A bra, denoted $\langle\psi|$, is the Hermitian conjugate (complex conjugate transpose) of a ket. It is a row vector given by:

$$\langle\psi| = [\alpha^* \;\; \beta^*]$$

where α^* and β^* are the complex conjugates of α and β.

2.8.2 Inner Product

The inner product of two quantum states $|\psi\rangle$ and $|\phi\rangle$ is represented as $\langle\phi \mid \psi\rangle$. This operation produces a scalar value that measures the overlap between the states:

$$\langle\phi\rangle = \alpha_1^* \alpha_2 + \beta_1^* \beta_2$$

where $|\phi\rangle = \begin{bmatrix} \alpha_1 \\ \beta_1 \end{bmatrix}$ and $|\psi\rangle = \begin{bmatrix} \alpha_2 \\ \beta_2 \end{bmatrix}$.

The inner product satisfies the following properties:

1. **Linearity**

$$\langle\phi| \, (a\,|\psi_1\rangle + b\,|\psi_2\rangle) = a\langle\phi\,|\,\psi_1\rangle + b\langle\phi\,|\,\psi_2\rangle$$

2. **Hermitian Symmetry**

$$\langle\phi\,|\,\psi\rangle = (\langle\psi\,|\,\phi\rangle)^*$$

The result of the inner product is a complex number whose magnitude determines the probability amplitude of one state transitioning to another.

2.8.3 Outer Product

The outer product of two states, $|\psi\rangle\langle\phi|$, constructs an operator that maps one state to another. For example:

$$|\psi\rangle = \begin{bmatrix} \alpha \\ \beta \end{bmatrix}, \quad \langle\phi| = [\alpha_1^* \;\; \beta_1^*]$$

then:

$$|\psi\rangle\langle\phi| = \begin{bmatrix} \alpha \\ \beta \end{bmatrix} \begin{bmatrix} \alpha_1^* & \beta_1^* \end{bmatrix} = \begin{bmatrix} \alpha\alpha_1^* & \alpha\beta_1^* \\ \beta\alpha_1^* & \beta\beta_1^* \end{bmatrix}$$

The outer product is frequently used to define projectors and quantum operators. For example, the projector onto a state $|\psi\rangle$ is given by:

$$P = |\psi\rangle\langle\psi|$$

Projectors are essential in measurement theory, where they represent the probability of collapsing a quantum state into a specific basis state upon observation.

2.8.4 Matrix Representation

While Dirac notation provides a compact abstract representation of quantum states and operators, it is often useful to express these entities in matrix form for explicit calculations.

2.8.4.1 Representing Kets and Bras in Matrix Form

A single-qubit state $|\psi\rangle = \alpha\,|\,0\rangle + \beta\,|\,1\rangle$ can be represented as a column vector:

$$|\psi\rangle = \begin{bmatrix} \alpha \\ \beta \end{bmatrix}$$

The corresponding bra is:

$$\langle\psi| = \begin{bmatrix} \alpha^* & \beta^* \end{bmatrix}$$

For a multi-qubit system, the state $|\psi\rangle = \alpha\,|\,00\rangle + \beta\,|\,01\rangle + \gamma\,|\,10\rangle + \delta\,|\,11\rangle$ is represented as:

$$|\psi\rangle = \begin{bmatrix} \alpha \\ \beta \\ \gamma \\ \delta \end{bmatrix}$$

2.8.4.2 Matrix Representation of Operators

Quantum operators, such as the Pauli matrices, can be expressed in matrix form. For example:

- The identity operator I is:

$$I = \begin{bmatrix} 1 & 0 \\ 0 & 1 \end{bmatrix}$$

- The Pauli-X (NOT gate) operator σ_x is:

$$\sigma_x = \begin{bmatrix} 0 & 1 \\ 1 & 0 \end{bmatrix}$$

- The Pauli-Z operator σ_z is:

$$\sigma_z = \begin{bmatrix} 1 & 0 \\ 0 & -1 \end{bmatrix}$$

2.8.4.3 Operator Action on Kets

The action of an operator $\widehat{O}$ on a ket $|\psi\rangle$ is computed as matrix multiplication. For example, applying σ_x to $|\psi\rangle = \begin{bmatrix} \alpha \\ \beta \end{bmatrix}$:

$$\sigma_x |\psi\rangle = \begin{bmatrix} 0 & 1 \\ 1 & 0 \end{bmatrix} \begin{bmatrix} \alpha \\ \beta \end{bmatrix} = \begin{bmatrix} \beta \\ \alpha \end{bmatrix}$$

2.8.4.4 Inner Product in Matrix Form

The inner product $\langle \phi | \psi \rangle$ is the matrix product of a row vector and a column vector. For $|\phi\rangle = \begin{bmatrix} \alpha_1 \\ \beta_1 \end{bmatrix}$ and $|\psi\rangle = \begin{bmatrix} \alpha_2 \\ \beta_2 \end{bmatrix}$, the inner product is:

$$\langle \phi | \psi \rangle = \begin{bmatrix} \alpha_1^* & \beta_1^* \end{bmatrix} \begin{bmatrix} \alpha_2 \\ \beta_2 \end{bmatrix} = \alpha_1^* \alpha_2 + \beta_1^* \beta_2$$

2.8.4.5 Outer Product in Matrix Form

The outer product $|\psi\rangle\langle\phi|$ results in a matrix. For $|\psi\rangle = \begin{bmatrix} \alpha \\ \beta \end{bmatrix}$ and $\langle\phi| = \begin{bmatrix} \alpha_1^* & \beta_1^* \end{bmatrix}$:

$$|\psi\rangle\langle\phi| = \begin{bmatrix} \alpha \\ \beta \end{bmatrix} \begin{bmatrix} \alpha_1^* & \beta_1^* \end{bmatrix} = \begin{bmatrix} \alpha\alpha_1^* & \alpha\beta_1^* \\ \beta\alpha_1^* & \beta\beta_1^* \end{bmatrix}$$

2.8.4.6 Applications

(a) **Quantum Measurement**

Projectors represented as $|\psi\rangle\langle\psi|$ are used to calculate the probability of measuring a quantum state in a particular basis.

(b) **Quantum Operations**

Unitary operators, such as quantum gates, are represented in matrix form and applied to kets to simulate quantum computations.

(c) **Tensor Products for Multi-Qubit States**

Multi-qubit states and operations are represented as tensor products of individual qubits and operators, enabling analysis of entangled systems.

2.9 Quantum Measurement and Its Implications

Measurement in quantum mechanics is a fundamental process that differentiates quantum systems from classical systems. Unlike classical measurements, which do not alter the state of a system, quantum measurements inherently disturb the state being observed. This phenomenon is governed by the measurement postulates of quantum mechanics, which define how quantum states collapse and how measurement probabilities are calculated. The implications of quantum measurement extend across probabilistic outcomes, system coherence, and practical applications in engineering fields.

Quantum measurement is a defining feature of quantum mechanics, characterized by the collapse of the wave function and probabilistic outcomes. Its implications extend to the behavior of quantum systems, introducing challenges such as loss of coherence and non-determinism. From an engineering perspective, quantum measurement underpins critical applications in cryptography, sensing, and computation, necessitating careful design to optimize performance and preserve quantum states.

2.9.1 Measurement Postulates

The measurement postulates of quantum mechanics describe how observations influence the state of a quantum system and determine the likelihood of specific outcomes.

2.9.1.1 Collapse of the Wave Function

When a quantum system is measured, its wave function $|\psi\rangle$ collapses into one of the eigenstates of the observable being measured. An observable is represented by a

Hermitian operator $\widehat{O}$, whose eigenstates $|a\rangle$ correspond to possible measurement outcomes.

If the system is initially in the state $|\psi\rangle$, and the observable $\widehat{O}$ is measured, the wave function collapses to the eigenstate $|a\rangle$ associated with the measured eigenvalue a:

$$|\psi\rangle \to |a\rangle$$

This collapse is instantaneous and irreversible, reflecting the transition from a superposition of states to a definite state.

2.9.1.2 Probability of Measurement

The probability of measuring a specific outcome a, associated with the eigenstate $|a\rangle$, is given by the squared magnitude of the inner product between the initial state $|\psi\rangle$ and the eigenstate $|a\rangle$:

$$P(a) = |\langle a\rangle|^2$$

where $|\psi\rangle$ is normalized such that $\langle\psi\,|\,\psi\rangle = 1$.

For a multi-dimensional system, the completeness relation ensures that the sum of probabilities over all possible outcomes equals 1:

$$\sum_a P(a) = \sum_a |\langle a\,|\,\psi\rangle|^2 = 1$$

This probabilistic nature of quantum measurement reflects the inherent uncertainty in quantum mechanics and contrasts with the deterministic predictions of classical physics.

2.9.1.3 Spin Measurement in a Two-Level System

Consider a qubit in the state:

$$|\psi\rangle = \alpha\,|0\rangle + \beta\,|1\rangle$$

where $|\alpha|^2 + |\beta|^2 = 1$.

Measuring the observable $\widehat{\sigma}_z$, whose eigenstates are $|0\rangle$ (eigenvalue $+1$) and $|1\rangle$ (eigenvalue -1), yields:

- Probability of $+1$: $P(+1) = |\alpha|^2$
- Probability of -1: $P(-1) = |\beta|^2$

After measurement, the qubit collapses to either $|0\rangle$ or $|1\rangle$, depending on the outcome.

2.9.2 *Implications of Measurement*

Quantum measurement has profound implications for the behavior of quantum systems, introducing non-deterministic outcomes and altering system states.

2.9.2.1 Non-deterministic Outcomes

Unlike classical systems, where measurements reveal pre-existing values of properties, quantum systems exhibit probabilistic behavior. The exact outcome of a measurement cannot be predicted with certainty but is instead governed by the probability distribution derived from the system's wave function.

This non-determinism is not due to experimental limitations but is intrinsic to the nature of quantum mechanics. It introduces fundamental uncertainty, as highlighted by Heisenberg's uncertainty principle.

2.9.2.2 Effects on System States

Measurement collapses the wave function, destroying any superposition or coherence in the system. This loss of coherence is critical in quantum computing and quantum information processing, as it limits the ability to maintain and manipulate quantum states over time.

2.9.2.3 Applications

Decoherence

In a multi-qubit system, measurements on one qubit can entangle or disentangle the state of other qubits, affecting the overall computation.

No-Cloning Theorem

The collapse of the wave function implies that quantum states cannot be perfectly copied, a principle that underpins the security of quantum cryptography.

2.9.3   *Engineering Perspective*

Quantum measurement plays a pivotal role in practical applications, including quantum cryptography, sensors, and quantum computing. Engineers must navigate the trade-offs between precision and state preservation to optimize these technologies.

2.9.3.1   Role of Quantum Measurements in Quantum Cryptography and Sensors

Quantum Cryptography

Quantum key distribution (QKD) protocols, such as BB84, rely on the measurement-induced collapse of quantum states to ensure security. Any eavesdropping attempt introduces detectable disturbances in the quantum states being transmitted.

Quantum Sensors

Quantum measurements enhance the sensitivity of sensors used in applications, such as gravitational wave detection and magnetic field imaging. For instance, quantum entanglement and superposition improve the precision of atomic clocks and interferometers.

2.9.3.2   Trade-Offs in Precision and State Preservation

Quantum systems face inherent trade-offs between measurement accuracy and the preservation of quantum states. Engineers must design measurement protocols that balance these competing demands.

Weak Measurements

Weak measurements extract partial information from a quantum system without causing complete collapse. This approach is used in feedback control systems to stabilize quantum states.

Projective Measurements

Strong, projective measurements provide precise information about a quantum state but result in its immediate collapse. These are employed in tasks like state preparation and error correction.

2.9.3.3 Applications in Quantum Computing

Readout Mechanisms

Quantum computers use measurement to extract computational results from qubits. High-fidelity readout mechanisms are essential for minimizing errors in the final output.

Error Detection and Correction

Quantum error correction codes rely on repeated measurements of ancillary qubits to detect and correct errors without collapsing the logical qubits.

2.10 Classical Circuits

This subsection provides a comprehensive introduction to classical circuits, bits, logic gates, and their mathematical representation. Classical circuits are the foundation of traditional computing, using classical bits to perform logical operations. This section explores the concept of classical bits, their manipulation using logic gates, and how matrix operations can represent these computations. Each topic is expanded with detailed explanations, equations, and Python codes to further clarify the subject matter. By leveraging matrix operations and Python examples, readers can bridge the gap between classical and quantum circuits, setting the stage for understanding quantum computation. The most commonly used classical logic gates are presented in Fig. 2.5.

2.10.1 Classical Computing Basics

Classical computing forms the backbone of modern information technology, utilizing binary states and deterministic logic for processing data.

2.10.1.1 Information Representation

Classical computers represent information as bits, which can be either 0 or 1. Bits are typically implemented as voltage levels in digital circuits, where a high voltage represents 1 and a low voltage represents 0.

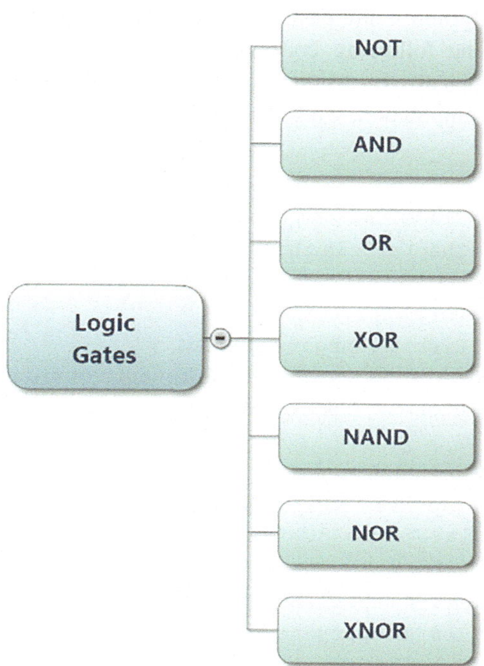

Fig. 2.5 Commonly used classical gates

Table 2.2 Central processing unit components

Elements in CPU architecture	Function
Arithmetic logic unit (ALU)	Performs arithmetic and logical operations
Control unit	Decodes instructions and manages data flow
Registers	Temporarily store data for processing

2.10.1.2   Central Processing Unit (CPU)

The CPU is the primary computational engine of a classical computer. It processes instructions sequentially using Boolean logic gates, such as AND, OR, and NOT, to manipulate binary data. The architecture of a CPU includes items listed in Table 2.2.

2.10.1.3   Boolean Logic and Sequential Processing

The CPU operates based on Boolean algebra, evaluating logical expressions such as:

$$f(x, y) = x \wedge y$$

where $\wedge$ represents the AND operation. This deterministic approach is ideal for tasks requiring exact computations and linear workflows.

Python Code Example: Classical Computation
The following code demonstrates a simple classical computation: evaluating a logical expression. This example illustrates the deterministic nature of classical computation, where the output is precisely determined by the input.

```python
#--------------------------------------------------------------------
# Classical computation of a Boolean expression
# Chapter 2 in the QUANTUM COMPUTING AND QUANTUM MACHINE LEARNING BOOK
#--------------------------------------------------------------------
# Version 1.0
# Qiskit changes frequently.
# We recommend using the latest version from the book code repository at:
# https://aqtinitiative.org/quantum-computing-for-engineers

# (c) 2025 Jesse Van Griensven, Roydon Fraser, and Jose Rosas
# License: MIT - Citation required
#--------------------------------------------------------------------

x = 1 # Representing True
y = 0 # Representing False

# Logical AND operation
result = x & y
print("AND result:", result)

# Logical OR operation
result = x | y
print("OR result:", result)
```

2.10.2 Classical Bit

A classical bit is the fundamental unit of classical information. A bit can exist in one of the two distinct states:

$$\text{Bit} = 0 \quad \text{or} \quad \text{Bit} = 1.$$

In digital systems, a bit can represent binary states, such as:

- Voltage levels: 0 (low voltage) and 1 (high voltage).
- Logical conditions: 0 (false) and 1 (true).

Bits are combined to form multi-bit representations, enabling the storage and manipulation of complex data.

2.10.3 Classical Circuits

A classical circuit processes bits through a sequence of logical operations. These circuits consist of the operations indicated in Table 2.3.

The output of a classical circuit depends on the configuration of gates and the values of the inputs.

Example: Simple Classical Circuit
A circuit with two inputs, A and B, and an AND gate produces:

$$\text{Output} = A \wedge B$$

2.10.4 Logic Gates

Logic gates are the building blocks of classical circuits with well-defined operations. They perform operations based on Boolean algebra. Each gate has a truth table defining its behavior for all possible inputs. These are explained in the following subsections.

There are many types of classical logic gates. Table 2.4 summarizes the most commonly used ones.

2.10.5 Matrix Operations on Circuits

Logic gates can be represented mathematically using matrices. For example:

Table 2.3 Classical circuits

Operation	Action
Inputs	Binary values representing the starting state of the system
Logic gates	Perform logical operations on the inputs
Outputs	The result of the operations

Table 2.4 Common types of logic gates

Logic gate type	Description
AND	Outputs 1 if both inputs are 1
OR	Outputs 1 if at least one input is 1
NOT	Inverts the input ($0 \rightarrow 1$, $1 \rightarrow 0$)
XOR	Outputs 1 if exactly one input is 1

- Inputs are represented as vectors:

$$|x\rangle = \begin{bmatrix} x_0 \\ x_1 \end{bmatrix}, \quad x_0, x_1 \in \{0, 1\}$$

- Gates are represented as matrices:

$$M = \begin{bmatrix} m_{00} & m_{01} \\ m_{10} & m_{11} \end{bmatrix}$$

The output of a gate is computed as:

$$|y\rangle = M\,|x\rangle$$

Example: Matrix Representation of the NOT Gate
The NOT gate flips the input:

$$M_{\mathsf{NOT}} = \begin{bmatrix} 0 & 1 \\ 1 & 0 \end{bmatrix}$$

Applying the NOT gate:

$$M_{\mathsf{NOT}} \begin{bmatrix} 1 \\ 0 \end{bmatrix} = \begin{bmatrix} 0 \\ 1 \end{bmatrix}$$

2.10.6 The AND Gate

The AND gate outputs 1 only if both inputs are 1. Its truth table is:

A	B	Output
0	0	0
0	1	0
1	0	0
1	1	1

2.10.6.1 Matrix Representation

The AND gate can be represented using a matrix that maps two input bits to a single output bit.

2.10.7   The NOT Gate

The NOT gate inverts the input:

- If the input is 0, the output is 1.
- If the input is 1, the output is 0.

  Truth Table:

Input	Output
0	1
1	0

Matrix Representation

$$M_{\mathsf{NOT}} = \begin{bmatrix} 0 & 1 \\ 1 & 0 \end{bmatrix}$$

2.10.7.1   XOR Gate

The XOR gate outputs 1 if the inputs are different. Its truth table is:

A	B	Output
0	0	0
0	1	1
1	0	1
1	1	0

Matrix Representation

$$M_{\mathsf{XOR}} = \begin{bmatrix} 1 & 0 & 0 & 0 \\ 0 & 0 & 0 & 1 \\ 0 & 0 & 0 & 1 \\ 0 & 1 & 1 & 0 \end{bmatrix}$$

Python Code Example: Matrix Operations on Classical Circuits
Below is a Python example that demonstrates matrix operations for a classical
circuit:

```
#---------------------------------------------------------------
# Matrix Operations on Classical Circuits
# Chapter 2 in the QUANTUM COMPUTING AND QUANTUM MACHINE LEARNING BOOK
#---------------------------------------------------------------
# Version 1.0
```

```python
# Qiskit changes frequently.
# We recommend using the latest version from the book code repository at:
# https://aqtinitiative.org/quantum-computing-for-engineers

# (c) 2025 Jesse Van Griensven, Roydon Fraser, and Jose Rosas
# License:  MIT - Citation of this work required
#------------------------------------------------------------------
import numpy as np


#------------------------------------------------------------------
def sprint(Matrix, decimals=4):
    """ Prints a numpy Matrix in a nice format with sympy and specified
precision """
    # Define the SYMPY routines we need
    import sympy as sp
    # The input Matrix can be a Numpy or a Sympy Matrix
    SMatrix = sp.Matrix(Matrix)
    # Apply the rounding to each element
    SMatrix = SMatrix.applyfunc(lambda x: round(float(x), decimals) if
isinstance(x, (float, int, sp.Float)) else x)
    display(SMatrix)
    return
#------------------------------------------------------------------

# NOT logic gates
M_NOT = np.array([[0, 1],
         [1, 0]])

# AND logic gate
M_AND = np.array([[1, 0, 0, 0],
         [0, 0, 0, 0],
         [0, 0, 0, 0],
         [0, 0, 0, 1]])

# Define input states
input_NOT = np.array([1, 0])      # |0>
input_AND = np.array([1, 0, 0, 0]) # |00>

# Apply NOT gate
output_NOT = np.dot(M_NOT, input_NOT)

# Apply AND gate
output_AND = np.dot(M_AND, input_AND)
```

```
# Print results
print ("NOT Gate Output:", output_NOT)
sprint(output_NOT)

print ("AND Gate Output:", output_AND)
sprint(output_AND)
```

2.11 Using Matrices to Compute Results in a Classical Circuit

Classical digital circuits can be analyzed in several ways, including Boolean algebra, truth tables, and Karnaugh maps. Another interesting (though less common) method is to represent classical gates or entire circuits as matrices that transform input bit patterns into output bit patterns. While matrix representations are more naturally associated with quantum gates (which are typically unitary operators), we can still encode classical logic operations in a matrix form that acts on vectors indicating the state of the bits.

This section contains explanations for a simple two-bit classical circuit composed of a NOT gate and an AND gate, it demonstrates how to build its truth table, and then it explains how one can construct a matrix that reproduces the same logical behavior. This section also contains a Python example illustrating how these matrices can be used to compute circuit outputs.

2.11.1 Simple Classical Circuit: NOT Followed by AND

Consider a circuit with two input bits, x_1 and x_2. The first bit x_1 is left unchanged, while the second bit x_2 goes through a NOT gate (inverting its value). The resulting bits are then combined in an AND gate to produce the final output y. A diagram of this circuit is shown in Fig. 2.6.

Hence, the logic is:

1. $x_2' = \mathrm{NOT}(x_2)$
2. $y = x_1 \, \mathrm{AND} \, x_2'$

2.11.2 Truth Table

First, it would be more evident to construct the truth table. Each input bit (x_1, x_2) can take the following values:

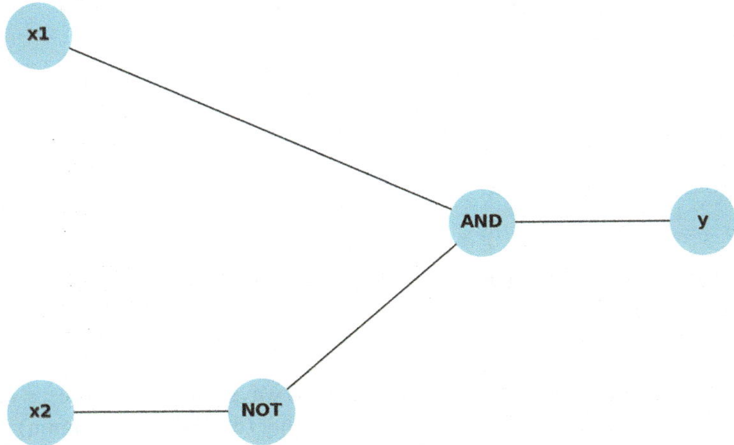

Fig. 2.6 Classical circuit NOT gate followed by an AND gate (by the authors)

$(0,0),\ (0,1),\ (1,0),\ (1,1)$. One computes $x_2' = \mathsf{NOT}(x_2)$ and then $y = x_1\,\mathsf{AND}\,x_2'$.

x_1	x_2	$\mathrm{NOT}(x_2) = x_2'$	$y = x_1\,\mathrm{AND}\,x_2'$
0	0	1	0
0	1	0	0
1	0	1	1
1	1	0	0

Thus, the final output y is 0, 0, 1, 0 for each of the four input combinations $(00, 01, 10, 11)$.

2.11.3 Vector and Matrix Representations

To use a matrix to compute the results of a classical circuit, one first represents the input bit patterns as basis vectors. For two bits, one can form a four-dimensional vector space with the basis states:

$$|00\rangle = \begin{pmatrix} 1 \\ 0 \\ 0 \\ 0 \end{pmatrix}, \quad |01\rangle = \begin{pmatrix} 0 \\ 1 \\ 0 \\ 0 \end{pmatrix}, \quad |10\rangle = \begin{pmatrix} 0 \\ 0 \\ 1 \\ 0 \end{pmatrix}, \quad |11\rangle = \begin{pmatrix} 0 \\ 0 \\ 0 \\ 1 \end{pmatrix}$$

Each four-dimensional vector can encode which two-bit pattern is present. For example, $|10\rangle$ corresponds to the input bits $(x_1, x_2) = (1,0)$.

2.11.4   Matrix for the NOT Gate (Single-Bit)

If one looks at a single-bit NOT operation, it flips 0 to 1 and 1 to 0. In a one-bit vector space (two-dimensional), the matrix for NOT is:

$$\text{NOT} = \begin{pmatrix} 0 & 1 \\ 1 & 0 \end{pmatrix}$$

Acting on a single-bit vector $\begin{pmatrix} 1 \\ 0 \end{pmatrix}$ (which represents bit $= 0$) produces $\begin{pmatrix} 0 \\ 1 \end{pmatrix}$ (which represents bit $= 1$) and vice versa.

2.11.5   Matrix for a Two-Bit Gate

For two-bit input vectors, gates can be represented by 4×4 matrices. Here, one wants a matrix M that directly gives us the single classical output y in some encoded form. Since the final circuit produces only one output bit, there are different ways of embedding that into a 2-bit or 4-bit representation.

One approach is to have the matrix map each four-dimensional input vector to another four-dimensional vector, where the output bit is stored in a designated part of the resulting vector. For instance, one might define the transformed state as $|0y\rangle$ if the output bit is y. This means:

For input $|00\rangle$, the output is $|00\rangle$ if $y = 0$.
For input $|01\rangle$, the output is $|00\rangle$ if $y = 0$
For input $|10\rangle$, the output is $|01\rangle$ if $y = 1$
For input $|11\rangle$, the output is $|00\rangle$ if $y = 0$

labeling the basis $|00\rangle, |01\rangle, |10\rangle, |11\rangle$ as rows 1–4, respectively, and likewise for the columns. In that convention, the transformation matrix M can be written so that $M \cdot |10\rangle = |01\rangle$ and all other basis vectors map to $|00\rangle$. One can express M as:

$$M = \begin{pmatrix} 1 & 0 & 0 & 0 \\ 0 & 1 & 0 & 0 \\ 0 & 0 & 0 & 0 \\ 0 & 0 & 1 & 0 \end{pmatrix}$$

Analyzing this matrix row-by-row:

1. The first row is $(1,0,0,0)$, sending $|00\rangle$ and $|01\rangle$ to $|00\rangle$ (and ignoring other inputs in that row).
2. The second row is $(0,1,0,0)$, so $|01\rangle$ also maps to $|01\rangle$ if one interprets it that way, or one can adjust to unify them.
3. The third and fourth rows handle the mapping from $|10\rangle$ to $|01\rangle$ specifically and $|11\rangle$ to $|00\rangle$.

In practice, one must carefully define how the single-bit output is embedded into the final 2-bit (or four-dimensional) vector. The key idea remains that each of the four input basis states gets mapped to the correct output encoding via a row in the matrix.

Python Code Example: Computing Circuit Outputs with Matrices

The Python code example below shows how one might use matrix multiplication to compute the output of our simple circuit. The code steps are:

1. Define the basis vectors for $|00\rangle, |01\rangle, |10\rangle, |11\rangle$.
2. Define our transformation matrix M.
3. Multiply each basis vector by M and observe the result.

```python
#----------------------------------------------------------------------
# Computing Circuit Outputs with Matrices
# Chapter 2 in the QUANTUM COMPUTING AND QUANTUM MACHINE LEARNING BOOK
#----------------------------------------------------------------------
# Version 1.0
# Qiskit changes frequently.
# We recommend using the latest version from the book code repository at:
# https://aqtinitiative.org/quantum-computing-for-engineers

# (c) 2025 Jesse Van Griensven, Roydon Fraser, and Jose Rosas
# License: MIT - Citation required
#----------------------------------------------------------------------
import numpy as np

# Define basis states for 2-bit inputs: |00>, |01>, |10>, |11>
state_00 = np.array([1, 0, 0, 0])
state_01 = np.array([0, 1, 0, 0])
state_10 = np.array([0, 0, 1, 0])
state_11 = np.array([0, 0, 0, 1])

# Define the matrix M that represents the circuit:
# We'll adopt a particular embedding so that y=0 maps to |00>, y=1 maps to
|01>.
M = np.array([
    [1, 0, 1, 0],  # This row says: if input is |00> or |10>, map to output |00>
    [0, 1, 0, 0],  # We'll refine to ensure unique mapping
    [0, 0, 0, 0],
    [0, 0, 0, 1]
], dtype=int)

# Adjust the matrix to ensure correct mapping:
# We want:
```

```
# - |00> -> |00> (y=0)
# - |01> -> |00> (y=0)
# - |10> -> |01> (y=1)
# - |11> -> |00> (y=0)
M = np.array([
   [1, 1, 0, 0], # row for final |00>, captures both input |00> and |01>
   [0, 0, 1, 0], # row for final |01>, captures input |10>
   [0, 0, 0, 0],
   [0, 0, 0, 0]
], dtype=int)

# Let's see how each basis state transforms
inputs = [state_00, state_01, state_10, state_11]
labels = ["|00>", "|01>", "|10>", "|11>"]

for inp, lab in zip(inputs, labels):
   out_vec = M.dot(inp)
   print(f"Input {lab} -> Output vector = {out_vec}")
```

2.11.5.1   Explanation of the Matrix Construction

1. Row 1 in M (index 0) is [1, 1, 0, 0], meaning that if the input is either $|00\rangle$ or $|01\rangle$, the resulting output will be mapped to the final state $|00\rangle$.
2. Row 2 in M (index 1) is [0, 0, 1, 0], meaning that if the input is $|10\rangle$, the resulting output is $|01\rangle$.
3. Rows 3 and 4 are [0, 0, 0, 0], effectively ignoring other final states or collisions in mapping. One could refine these to ensure $|11\rangle$ also goes to $|00\rangle$ by editing the first row or another row accordingly.

When running the code, the printed output vector reveals how each of the four input bit patterns is mapped to a distinct or shared final vector, consistent with the logical behavior. The code output is:

```
Input |00> -> Output vector = [1 0 0 0]
Input |01> -> Output vector = [1 0 0 0]
Input |10> -> Output vector = [0 1 0 0]
Input |11> -> Output vector = [0 0 0 0]
```

If one interprets $|00\rangle$ as output bit $y = 0$ and $|01\rangle$ as output bit $y = 1$, one gets the correct truth table for $y = x_1$ AND NOT(x_2).

2.12   Classical Logic Gates and Quantum Gates

Table 2.5 summarizes the key differences between classical and quantum gates. In classical computing:

1. A logic gate takes binary inputs (0 s and 1 s) and produces binary outputs.
2. Operations like AND, OR, and NOT are irreversible, except for the NOT gate.

Example: The classical NOT gate flips the input:

$$\text{Input} : 0 \rightarrow \text{Output} : 1, \quad \text{Input} : 1 \rightarrow \text{Output} : 0.$$

Example: The quantum equivalent of the classical NOT gate is the X (Pauli-X) gate. It flips the $|0\rangle$ state to $|1\rangle$ and vice versa:

$$X|0\rangle = |1\rangle, \quad X|1\rangle = |0\rangle$$

Mathematically, X is represented as:

$$X = \begin{bmatrix} 0 & 1 \\ 1 & 0 \end{bmatrix}$$

2.13   Practical Implementation of Quantum Gates

This section explores the definition of quantum gates, their comparison with classical logic gates, and their significance in quantum circuits.

Quantum gates are operations applied to one or more qubits, represented mathematically by unitary matrices. A unitary matrix U satisfies the condition:

$$U^\dagger U = I$$

where

Table 2.5 Key differences between classical and quantum gates

Feature	Classical gates	Quantum gates
Input/output	Binary (0 or 1)	Quantum states
Reversibility	Typically irreversible	Always reversible due to unitarity
Operation type	Deterministic	Deterministic (unitary), but measurements are probabilistic (due to superposition)
Multiple states	Single state at a time	Operate on superposition of states

- $U^\dagger$ is the conjugate transpose (Hermitian adjoint) of U.
- I is the identity matrix.

This condition ensures that quantum gates are reversible and preserve the total probability (i.e., normalization) of the quantum state.

2.14 Mathematical Representation of Quantum Gates

Quantum gates are represented mathematically by unitary matrices, which define how they manipulate quantum states. The mathematical framework underlying quantum gates is essential for understanding their behavior and implementing quantum algorithms. This section delves into the matrix representation of gates, fundamental linear algebra concepts, and the conditions that ensure unitarity.

2.15 Single-Qubit Gates

Single-qubit gates are the simplest quantum gates, operating on individual qubits to manipulate their states. These gates are represented by 2×2 unitary matrices, ensuring that quantum operations are reversible and preserve probability amplitudes. The identity gate maintains the state, Pauli gates manipulate qubit states through flips and phase changes, the Hadamard gate introduces superposition, and phase shift gates allow for precise phase control. Together, these gates provide the essential tools for constructing quantum circuits and implementing quantum algorithms.

Quantum gates are modeled as unitary matrices, which act on qubits to transform their quantum states. For a single qubit, the state is represented as a vector:

$$|\psi\rangle = \begin{bmatrix} \alpha \\ \beta \end{bmatrix}$$

where α and β are complex amplitudes, and $|\alpha|^2 + |\beta|^2 = 1$.

When a quantum gate U is applied, it transforms the state vector as:

$$|\psi'\rangle = U|\psi\rangle$$

where U is a unitary matrix that preserves the norm of the state.

In this section, we explore the identity gate, the Pauli gates, the Hadamard gate, and the phase shift gates, with their mathematical representations, physical interpretations, and practical applications.

2.15.1   Identity Gate (**I**)

The identity gate is the quantum equivalent of the classical no-operation (NOP) gate. It leaves the state of the qubit unchanged, effectively acting as a "do-nothing" operation. The matrix representation of the identity gate is represented by the 2×2 identity matrix:

$$I = \begin{bmatrix} 1 & 0 \\ 0 & 1 \end{bmatrix}$$

For any qubit state:

$$|\psi\rangle = \alpha\,|0\rangle + \beta\,|1\rangle$$

applying I results in:

$$I\,|\psi\rangle = \begin{bmatrix} 1 & 0 \\ 0 & 1 \end{bmatrix} \begin{bmatrix} \alpha \\ \beta \end{bmatrix} = \begin{bmatrix} \alpha \\ \beta \end{bmatrix}$$

The state remains unchanged.

2.15.2   Pauli Gates (**X, Y, Z**)

The Pauli gates are fundamental single-qubit gates that correspond to rotations of π radians (180°) about the x, y, and z axes of the Bloch sphere. They are named after Wolfgang Pauli and play a central role in quantum computing.

2.15.3   Pauli-X Gate (**X**)

The X gate is the quantum analog of the classical NOT gate, flipping the $|0\rangle$ and $|1\rangle$ states. The matrix representation of the Pauli-X gate is:

$$X = \begin{bmatrix} 0 & 1 \\ 1 & 0 \end{bmatrix}$$

The effect of the NOT gate on a qubit is demonstrated below:

$$X\,|0\rangle = |1\rangle, \quad X\,|1\rangle = |0\rangle$$

Python Code Example: Pauli-X Gate Using Numpy

```python
#-------------------------------------------------------------------
# Checking X Pauli Gate
# Chapter 2 in the QUANTUM COMPUTING AND QUANTUM MACHINE LEARNING BOOK
#-------------------------------------------------------------------
# Version 1.0
# Qiskit changes frequently.
# We recommend using the latest version from the book code repository at:
# https://aqtinitiative.org/quantum-computing-for-engineers


# (c) 2025 Jesse Van Griensven, Roydon Fraser, and Jose Rosas
# License: MIT - Citation of this work required
#-------------------------------------------------------------------
import numpy as np


#-------------------------------------------------------------------
def sprint(Matrix, decimals=4):
  """ Prints a Matrix with real and imaginary parts rounded to 'decimals'
"""

  import sympy as sp
  SMatrix = sp.Matrix(Matrix)  # Convert to Sympy Matrix if it's not
already

  def round_complex(x):
    """Round real and imaginary parts of x to the given number of
decimals."""
    c = complex(x)  # handle any real or complex Sympy expression
    r = round(c.real, decimals)
    i = round(c.imag, decimals)

    # If imaginary part is negligible, treat as purely real
    if abs(i) < 10**(-decimals): return sp.Float(r)
    else: return sp.Float(r) + sp.Float(i)*sp.I

  # Display the rounded Sympy Matrix
  display(SMatrix.applyfunc(round_complex))

  return
#-------------------------------------------------------------------


# Note: This code is very verbose in order to clearly explain the topics


# Define the Pauli Gate
X_gate = np.array([[0, 1], [1, 0]])
```

```
print("Pauli X Gate:")
sprint(X_gate)

# Define the matrix dagger
X_dagger = X_gate.conj().T
print("X_dagger:")
sprint(X_dagger)

# Dot Produt X with dagger of X
Result = np.dot(X_gate, X_dagger)

print("Pauli-X unitary test should result in the Identity matrix")
sprint(Result)
```

2.15.4   Pauli-Y Gate (Y)

The Y gate combines bit-flip and phase-flip operations. The matrix representation of the Pauli-Y gate is:

$$Y = \begin{bmatrix} 0 & -i \\ i & 0 \end{bmatrix}$$

The effect of the Pauli-Y gate on a qubit is demonstrated below:

$$Y|0\rangle = i|1\rangle, \quad Y|1\rangle = -i|0\rangle$$

2.15.5   Pauli-Z Gate (Z)

The Z gate performs a phase flip, leaving $|0\rangle$ unchanged while flipping the sign of $|1\rangle$. The matrix representation of the Pauli-Z gate is:

$$Z = \begin{bmatrix} 1 & 0 \\ 0 & -1 \end{bmatrix}$$

The effect of the Pauli-Z gate on a qubit is demonstrated below:

$$Z|0\rangle = |0\rangle, \quad Z|1\rangle = -|1\rangle$$

2.15.6   Hadamard Gate (H)

The Hadamard gate creates superposition states, making it one of the most important gates in quantum computing. It transforms the basis states $|0\rangle$ and $|1\rangle$ into an equal superposition. The matrix representation of the Hadamard gate is:

$$H = \frac{1}{\sqrt{2}} \begin{bmatrix} 1 & 1 \\ 1 & -1 \end{bmatrix}$$

The effect of the Hadamard gate on a qubit is demonstrated below:
For $|0\rangle$:

$$H|0\rangle = \frac{1}{\sqrt{2}}(|0\rangle + |1\rangle)$$

For $|1\rangle$:

$$H|1\rangle = \frac{1}{\sqrt{2}}(|0\rangle - |1\rangle)$$

Applications of the Hadamard gate include:

- Preparing superposition states.
- Generating entangled states when combined with CNOT gates.

Python Code Example: Applying the Hadamard Gate to the Zero State

```python
#-------------------------------------------------------------------
# Applying the Hadamard Gate to the Zero state
# Chapter 2 in the QUANTUM COMPUTING AND QUANTUM MACHINE LEARNING BOOK
#-------------------------------------------------------------------
# Version 1.0
# Qiskit changes frequently.
# We recommend using the latest version from the book code repository at:
# https://aqtinitiative.org/quantum-computing-for-engineers

# (c) 2025 Jesse Van Griensven, Roydon Fraser, and Jose Rosas
# License: MIT - Citation of this work required
#-------------------------------------------------------------------
import numpy as np
#-------------------------------------------------------------------
def sprint(Matrix, decimals=4):
    """ Prints a numpy Matrix in a nice format with sympy and specified
precision """
    # Define the SYMPY routines we need
    import sympy as sp
```

```python
  # The input Matrix can be a Numpy or a Sympy Matrix
  SMatrix = sp.Matrix(Matrix)
  # Apply the rounding to each element
  SMatrix = SMatrix.applyfunc(lambda x: round(float(x), decimals) if
isinstance(x, (float, int, sp.Float)) else x)
  display(SMatrix)
  return
#-----------------------------------------------------------------------


# Define State \0>
state_0 = np.array([[1], [0]])


# Define Hadamard gate
H = (1 / np.sqrt(2)) * np.array([[1, 1], [1, -1]])


# Apply H to |0>
new_state = np.dot(H, state_0)
print("New State in Superposition:\n")
sprint(new_state)
```

Python Code Example: Applying the Hadamard Gate to the Pauli-X Gate

```python
#-----------------------------------------------------------------------
# Hadamard and Pauli Gates
# Chapter 2 in the QUANTUM COMPUTING AND QUANTUM MACHINE LEARNING BOOK
#-----------------------------------------------------------------------
# Version 1.0
# Qiskit changes frequently.
# We recommend using the latest version from the book code repository at:
# https://aqtinitiative.org/quantum-computing-for-engineers


# (c) 2025 Jesse Van Griensven, Roydon Fraser, and Jose Rosas
# License: MIT - Citation of this work required
#-----------------------------------------------------------------------


import numpy as np


#-----------------------------------------------------------------------
def sprint(Matrix, decimals=4):
  """ Prints a formatted Matrix with specified precision """
  # Define the SYMPY routines we need
  import sympy as sp
  # The input Matrix can be a Numpy or a Sympy Matrix
  SMatrix = sp.Matrix(Matrix)
  # Apply the rounding to each element
  SMatrix = SMatrix.applyfunc(lambda x: round(float(x), decimals) if
```

```
isinstance(x, (float, int, sp.Float)) else x)
  display(SMatrix)
  return
#--------------------------------------------------------------------

# Identity Gate
I = np.array([[1, 0], [0, 1]])

# Pauli-X Gate - the NOT Gate
X = np.array([[0, 1], [1, 0]])

# Hadamard Gate
H = (1. / np.sqrt(2.)) * np.array([[1, 1], [1, -1]])

# Apply gates to a state |0>
state    = np.array([[1], [0]])  # |0>
new_state = np.dot(H, state)
print("New State:\n")
sprint(new_state)
```

2.15.7   *Phase Shift Gates* ($\mathbf{R_\phi}$)

Phase shift gates apply a phase rotation to the $|1\rangle$ component of a qubit's state. These gates are parametrized by an angle ϕ. The matrix representation of the phase shift gates is:

$$R_\phi = \begin{bmatrix} 1 & 0 \\ 0 & e^{i\phi} \end{bmatrix}$$

The effect of the phase shift gates on a qubit is demonstrated below. For a general qubit state $|\psi\rangle = \alpha \, | \, 0\rangle + \beta \, | \, 1\rangle$:

$$R_\phi \, |\,\psi\rangle = \alpha\,|0\rangle + \beta e^{i\phi}\,|\,1\rangle$$

Special Cases of Phase Shift Are the *S* and *T* Gates

1. ***S* Gate**
     The *S* gate is a phase shift gate with $\phi = \frac{\pi}{2}$:

$$S = \begin{bmatrix} 1 & 0 \\ 0 & i \end{bmatrix}$$

2. ***T* Gate**

The **T** gate is a phase shift gate with $\phi = \frac{\pi}{4}$:

$$T = \begin{bmatrix} 1 & 0 \\ 0 & e^{i\pi/4} \end{bmatrix}$$

Python Code Example: The Phase Gate

```python
#------------------------------------------------------------------
# Phase Shift gate R_phi
# Chapter 2 in the QUANTUM COMPUTING AND QUANTUM MACHINE LEARNING BOOK
#------------------------------------------------------------------
# Version 1.0
# Qiskit changes frequently.
# We recommend using the latest version from the book code repository at:
# https://aqtinitiative.org/quantum-computing-for-engineers

# (c) 2025 Jesse Van Griensven, Roydon Fraser, and Jose Rosas
# License: MIT - Citation of this work required
#------------------------------------------------------------------
import numpy as np


#------------------------------------------------------------------
def sprint(Matrix, decimals=4):
  """ Prints a Matrix with real and imaginary parts rounded to 'decimals'
"""
  import sympy as sp
  SMatrix = sp.Matrix(Matrix)  # Convert to Sympy Matrix if it's not
already

  def round_complex(x):
    """Round real and imaginary parts of x to the given number of
decimals."""
    c = complex(x)  # handle any real or complex Sympy expression
    r = round(c.real, decimals)
    i = round(c.imag, decimals)

    # If imaginary part is negligible, treat as purely real
    if abs(i) < 10**(-decimals): return sp.Float(r)
    else: return sp.Float(r) + sp.Float(i)*sp.I

  # Display the rounded Sympy Matrix
  display(SMatrix.applyfunc(round_complex))
```

```python
    return
#---------------------------------------------------------------

def phase_shift_gate(phi):
    # Function to create the Phase Shift gate R_phi
    """Returns the matrix representation of the Phase Shift Gate R_phi."""
    return np.array([[1, 0],
            [0, np.exp(1j * phi)]])
#---------------------------------------------------------------

# Define a qubit state |ψ> = α|0> + β|1>
qubit_state = np.array([[1], # Amplitude of |0>
            [1]]) # Amplitude of  1>

# Define the phase angle (in radians)
phi = np.pi / 4  # 45 degrees

# Create the phase shift gate R_phi
R_phi = phase_shift_gate(phi)

# Apply the phase shift gate to the qubit state
new_state = np.dot(R_phi, qubit_state)

# Output the results
print("Phase Shift Gate (R_phi):")
sprint(R_phi)

print("\nOriginal Qubit State |ψ>:")
sprint(qubit_state)

print("\nNew Qubit State after applying R_phi:")
sprint(new_state)
```

2.16 Basic Linear Algebra Concepts Relevant to Quantum Gates

The following is a review of the linear algebra concepts we need to understand in order to operate with quantum gates.

2.16.1   Tensor Products

The tensor product is used to represent the combined state of multiple qubits. If $|\psi_1\rangle$ and $|\psi_2\rangle$ are single-qubit states, their joint state is:

$$|\psi\rangle = |\psi_1\rangle \otimes |\psi_2\rangle$$

For two qubits:

$$\begin{bmatrix} a \\ b \end{bmatrix} \otimes \begin{bmatrix} c \\ d \end{bmatrix} = \begin{bmatrix} ac \\ ad \\ bc \\ bd \end{bmatrix}$$

For example, the tensor product of $|0\rangle = \begin{bmatrix} 1 \\ 0 \end{bmatrix}$ and $|1\rangle = \begin{bmatrix} 0 \\ 1 \end{bmatrix}$ is:

$$|0\rangle \otimes |1\rangle = \begin{bmatrix} 1 \\ 0 \\ 0 \\ 0 \end{bmatrix}$$

Python Code Example: Tensor Product

```python
#-------------------------------------------------------------------
# Tensor Product
# Chapter 2 in the QUANTUM COMPUTING AND QUANTUM MACHINE LEARNING BOOK
#-------------------------------------------------------------------
# Version 1.0
# Qiskit changes frequently.
# We recommend using the latest version from the book code repository at:
# https://aqtinitiative.org/quantum-computing-for-engineers

# (c) 2025 Jesse Van Griensven, Roydon Fraser, and Jose Rosas
# License:  MIT - Citation of this work required
#-------------------------------------------------------------------
import numpy as np
from numpy import kron
#-------------------------------------------------------------------
def sprint(Matrix, decimals=4):
    """ Prints a Matrix with real and imaginary parts rounded to 'decimals'
"""
    import sympy as sp
    SMatrix = sp.Matrix(Matrix)  # Convert to Sympy Matrix if it's not
already
```

```python
def round_complex(x):
    """Round real and imaginary parts of x to the given number of
decimals."""
    c = complex(x)  # handle any real or complex Sympy expression
    r = round(c.real, decimals)
    i = round(c.imag, decimals)
    # If imaginary part is negligible, treat as purely real
    if abs(i) < 10**(-decimals): return sp.Float(r)
    else: return sp.Float(r) + sp.Float(i)*sp.I

# Display the rounded Sympy Matrix
display(SMatrix.applyfunc(round_complex))
return
#---------------------------------------------------------------------

# Define single-qubit states
q0 = np.array([[1], [0]]) # |0>
q1 = np.array([[0], [1]]) # |1>

# Compute the tensor product
joint_state = kron(q0, q1)

# Output results
print("q0: |0>:")
sprint(q0)
print("q1: |1>:")
sprint(q1)
print("Tensor Product:")
sprint(joint_state)
```

2.16.2 Kronecker Products

The Kronecker product is the mathematical operation used to compute the tensor
product for matrices. For matrices A and B:

$$A \otimes B = \begin{bmatrix} a_{11}B & a_{12}B \\ a_{21}B & a_{22}B \end{bmatrix}$$

For example, if:

$$A = \begin{bmatrix} 1 & 2 \\ 3 & 4 \end{bmatrix}, \quad B = \begin{bmatrix} 0 & 5 \\ 6 & 7 \end{bmatrix}$$

then:

$$A \otimes B = \begin{bmatrix} 0 & 5 & 0 & 10 \\ 6 & 7 & 12 & 14 \\ 0 & 15 & 0 & 20 \\ 18 & 21 & 24 & 28 \end{bmatrix}$$

Python Code Example: Kronecker Product—The Tensor Product for Matrices

```python
#------------------------------------------------------------------
# Kronecker Product - The Tensor Product for Matrices
# Chapter 2 in the QUANTUM COMPUTING AND QUANTUM MACHINE LEARNING BOOK
#------------------------------------------------------------------
# Version 1.0
# Qiskit changes frequently.
# We recommend using the latest version from the book code repository at:
# https://aqtinitiative.org/quantum-computing-for-engineers

# (c) 2025 Jesse Van Griensven, Roydon Fraser, and Jose Rosas
# License: MIT - Citation of this work required
#------------------------------------------------------------------
import numpy as np
from numpy import kron
#------------------------------------------------------------------
def sprint(Matrix, decimals=4):
  """ Prints a Matrix with real and imaginary parts rounded to 'decimals'
"""
  import sympy as sp
  SMatrix = sp.Matrix(Matrix)  # Convert to Sympy Matrix if it's not
already

  def round_complex(x):
    """Round real and imaginary parts of x to the given number of
decimals."""
    c = complex(x)  # handle any real or complex Sympy expression
    r = round(c.real, decimals)
    i = round(c.imag, decimals)
    # If imaginary part is negligible, treat as purely real
    if abs(i) < 10**(-decimals): return sp.Float(r)
    else: return sp.Float(r) + sp.Float(i)*sp.I

  # Display the rounded Sympy Matrix
  display(SMatrix.applyfunc(round_complex))
  return
#------------------------------------------------------------------
```

```
# Define matrices
A = np.array([[1, 2], [3, 4]])
B = np.array([[0, 5], [6, 7]])

# Compute Kronecker product
Kronecker = np.kron(A, B)

# Output results
print("Matrix A: ")
sprint(A)
print("Matrix B: ")
sprint(B)
print("Kronecker Product:")
sprint(Kronecker)
```

2.16.3   Matrix Multiplication

Matrix multiplication is used to apply quantum gates and combine operations. If A and B are matrices:

$$C = A \cdot B$$

where the element at position (i, j) in C is:

$$C_{ij} = \sum_k A_{ik} B_{kj}$$

For example, applying a CNOT gate:

$$\text{CNOT} = \begin{bmatrix} 1 & 0 & 0 & 0 \\ 0 & 1 & 0 & 0 \\ 0 & 0 & 0 & 1 \\ 0 & 0 & 1 & 0 \end{bmatrix}$$

2.16.4   Conditions for Unitarity

A matrix U is unitary if:

$$U^\dagger U = I$$

where $U^\dagger$ is the conjugate transpose of U, and I is the identity matrix.

Properties of Unitary Matrices
1. Norm Preservation: $|\psi'\rangle = U \, |\psi\rangle$, and $\langle \psi' | \psi'\rangle = \langle \psi | \psi\rangle$.
2. Reversibility: Unitary operations can be undone by applying $U^\dagger$.

For example, the Pauli-X gate is unitary:

$$X^\dagger X = \begin{bmatrix} 0 & 1 \\ 1 & 0 \end{bmatrix} \begin{bmatrix} 0 & 1 \\ 1 & 0 \end{bmatrix} = \begin{bmatrix} 1 & 0 \\ 0 & 1 \end{bmatrix}$$

Python Code Example: Verification Unitary Gate

```python
#------------------------------------------------------------------------
# Checking X Pauli Gate Unitary
# Chapter 2 in the QUANTUM COMPUTING AND QUANTUM MACHINE LEARNING BOOK
#------------------------------------------------------------------------
# Version 1.0
# Qiskit changes frequently.
# We recommend using the latest version from the book code repository at:
# https://aqtinitiative.org/quantum-computing-for-engineers

# (c) 2025 Jesse Van Griensven, Roydon Fraser, and Jose Rosas
# License: MIT - Citation of this work required
#------------------------------------------------------------------------
import numpy as np


#------------------------------------------------------------------------
def sprint(Matrix, decimals=4):
    """ Prints a numpy Matrix in a nice format with sympy and specified
precision """
    # Define the SYMPY routines we need
    import sympy as sp
    # The input Matrix can be a Numpy or a Sympy Matrix
    SMatrix = sp.Matrix(Matrix)
    # Apply the rounding to each element
    SMatrix = SMatrix.applyfunc(lambda x: round(float(x), decimals) if
isinstance(x, (float, int, sp.Float)) else x)
    display(SMatrix)
    return
#------------------------------------------------------------------------

# Note: This code is very verbose in order to clearly explain the topics

# Define the Pauli Gate
X_gate = np.array([[0, 1], [1, 0]])
print("Pauli X Gate:")
sprint(X_gate)
```

```
# Define the matrix dagger
X_dagger = X_gate.conj().T
print("X_dagger:")
sprint(X_dagger)

# Dot Produt X with dagger of X
Result = np.dot(X_gate, X_dagger)

print("Pauli-X unitary test should result in the Identity matrix")
sprint(Result)
```

2.17   Quantum Gates in Quantum Circuits

Quantum gates are fundamental to implementing quantum algorithms and performing computations. They manipulate qubits to create superposition, entanglement, and interference, which are the three key phenomena enabling quantum advantage.

2.17.1   Enabling Quantum Superposition

Quantum gates allow qubits to exist in a superposition of $|0\rangle$ and $|1\rangle$, expanding the computational space exponentially.

Example: The Hadamard gate creates a superposition:

$$H\,|0\rangle = \frac{1}{\sqrt{2}}\left(|0\rangle + |1\rangle\right)$$

The Hadamard gate is represented as:

$$H = \frac{1}{\sqrt{2}}\begin{bmatrix} 1 & 1 \\ 1 & -1 \end{bmatrix}$$

Python Code Example: The Hadamard Gate Using Qiskit
```
#--------------------------------------------------------------------
# The Hadamard Gate using Qiskit
# Chapter 2 in the QUANTUM COMPUTING AND QUANTUM MACHINE LEARNING BOOK
#--------------------------------------------------------------------
# Version 1.0
# Qiskit changes frequently.
```

```
# We recommend using the latest version from the book code repository at:
# https://aqtinitiative.org/quantum-computing-for-engineers

# (c) 2025 Jesse Van Griensven, Roydon Fraser, and Jose Rosas
# License: MIT - Citation of this work required
#--------------------------------------------------------------------
import warnings
warnings.filterwarnings('ignore')

from qiskit import QuantumCircuit
from qiskit.visualization import import circuit_drawer
#--------------------------------------------------------------------

# Create a single-qubit circuit
qc = QuantumCircuit(1)
qc.h(0)  # Apply the Hadamard gate
qc.measure_all()

# Visualize the circuit
print("Quantum Circuit:")
print(qc.draw())
display(circuit_drawer(qc, output='mpl', style="iqp"))
```

The circuit developed in the code above is shown in Fig. 2.7.

2.17.2   *Enabling Quantum Entanglement*

Quantum gates, such as the Hadamard gate (see Sect. 2.7.3) along with the Controlled-NOT (CNOT) gate (see Sect. 2.7.5.1), enable entanglement, a unique quantum phenomenon where the states of qubits become interdependent. The following is an example that entangles two qubits using Hadamard and CNOT gates:

1. Apply H to qubit 1 to create superposition.
2. Apply CNOT to entangle qubit 0 and qubit 1.
3. Apply H to qubit 1 a second time.

Fig. 2.7 Creating entangled qubits using Hadamard gates and CNOT gate

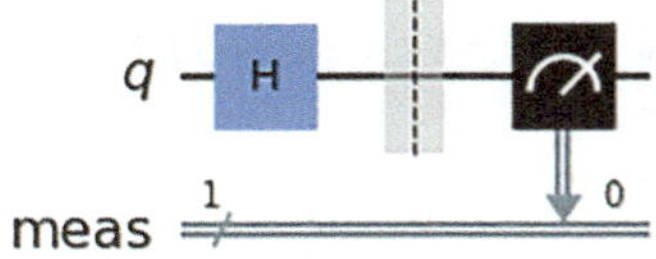

Python Code Example: CNOT Gate Using Qiskit

```python
#-------------------------------------------------------------------
# CNOT Gate using Qiskit
# Chapter 2 in the QUANTUM COMPUTING AND QUANTUM MACHINE LEARNING BOOK
#-------------------------------------------------------------------
# Version 1.0
# Qiskit changes frequently.
# We recommend using the latest version from the book code repository at:
# https://aqtinitiative.org/quantum-computing-for-engineers


# (c) 2025 Jesse Van Griensven, Roydon Fraser, and Jose Rosas
# License: MIT - Citation of this work required
#-------------------------------------------------------------------
import warnings
warnings.filterwarnings('ignore')

import numpy as np
from qiskit import QuantumCircuit
from qiskit.visualization import import circuit_drawer
#-------------------------------------------------------------------


qc = QuantumCircuit(2)

# Hadamard on target qubit
qc.h(1)

# Apply the CNOT gate
qc.cx(0, 1)

# Apply the Hadamard on target qubit again
qc.h(1)

# Visualize the circuit
print("Quantum Circuit:")
print(qc.draw())
display(circuit_drawer(qc, output='mpl', style="iqp"))
```

The CNOT gate circuit developed in the code above is displayed in Fig. 2.8.

Fig. 2.8 CNOT gate circuit

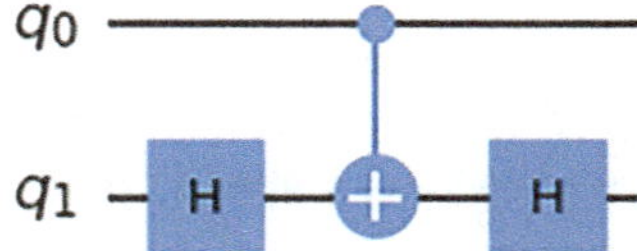

2.17.3 Universal Quantum Computation

Quantum gates form a universal gate set, allowing any quantum operation to be decomposed into a sequence of these gates. The Clifford + T set is one example of a universal gate set:

1. Clifford Gates: H, S, $CNOT$.
2. T: A non-Clifford gate enabling universality.

2.17.4 Implementing Quantum Algorithms

Quantum gates enable the construction of circuits for quantum algorithms like the following, which will be explained in future chapters:

1. Shor's Algorithm: For factoring large integers.
2. Grover's Algorithm: For searching unsorted databases.

Python Code Example: Grover's Oracle
A simple quantum circuit implementing Grover's Oracle

```
#-----------------------------------------------------------------
# Grover's Oracle
# Chapter 2 in the QUANTUM COMPUTING AND QUANTUM MACHINE LEARNING BOOK
#-----------------------------------------------------------------
# Version 1.0
# Qiskit changes frequently.
# We recommend using the latest version from the book code repository at:
# https://aqtinitiative.org/quantum-computing-for-engineers

# (c) 2025 Jesse Van Griensven, Roydon Fraser, and Jose Rosas
# License: MIT - Citation required
#-----------------------------------------------------------------
import warnings
warnings.filterwarnings('ignore')

from qiskit import QuantumCircuit
from qiskit.visualization import circuit_drawer
#-----------------------------------------------------------------

qc = QuantumCircuit(2)
qc.h([0, 1])  # Apply Hadamard to create superposition
qc.cz(0, 1)   # Apply Controlled-Z
```

Fig. 2.9 A simple Grover's
Oracle circuit

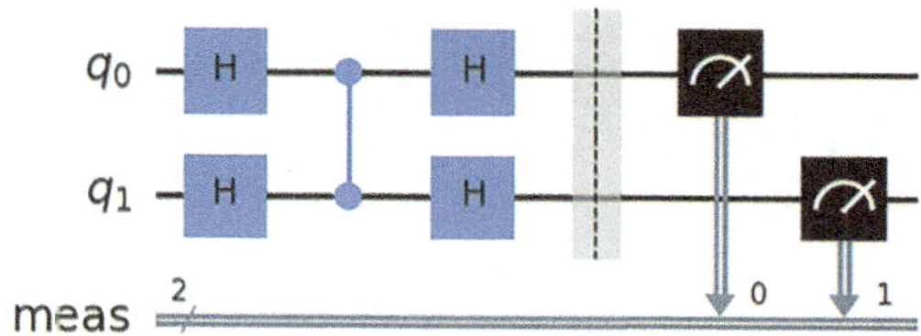

```
qc.h([0, 1]) # Apply Hadamard again
qc.measure_all()

# Visualize the circuit
print("Grover's Oracle Simple Quantum Circuit:")
print(qc.draw())
display(circuit_drawer(qc, output='mpl', style="iqp"))
```

The CNOT gate circuit developed in the code above is displayed in Fig. 2.9.

2.18 Multi-Qubit Gates

Multi-qubit gates are essential for creating interactions between qubits, enabling the construction of complex quantum circuits. They play a pivotal role in implementing quantum algorithms, entanglement, and other operations requiring qubit correlations. Multi-qubit gates like the controlled gates, SWAP gate, and Toffoli gate are critical for implementing quantum algorithms. Controlled gates enable conditional operations and entanglement, while the SWAP gate is indispensable for qubit routing in large systems. The Toffoli gate bridges quantum and classical reversible computing, highlighting its versatility.

This section explores key multi-qubit gates, including controlled gates, the SWAP gate, and the Toffoli gate, with their mathematical representations, applications, and relevance in quantum computing.

Controlled gates act on multiple qubits, where one or more qubits (control qubits) determine whether an operation is applied to the target qubit(s). These gates are fundamental for introducing entanglement and performing conditional operations.

2.18.1 Controlled-NOT (CNOT) Gate

The Controlled-NOT gate, or CNOT, flips the state of the target qubit if the control qubit is in the $|1\rangle$ state.

2.18.1.1 Matrix Representation

The CNOT gate for a 2-qubit system is represented by the 4×4 matrix:

$$\text{CNOT} = \begin{bmatrix} 1 & 0 & 0 & 0 \\ 0 & 1 & 0 & 0 \\ 0 & 0 & 0 & 1 \\ 0 & 0 & 1 & 0 \end{bmatrix}$$

2.18.1.2 Effect on Basis States

$$\text{CNOT}\,|00\rangle = |00\rangle, \quad \text{CNOT}\,|01\rangle = |01\rangle$$
$$\text{CNOT}\,|10\rangle = |11\rangle, \quad \text{CNOT}\,|11\rangle = |10\rangle$$

Python Code Example: CNOT Gate—Two Qubit Using Numpy

```python
#------------------------------------------------------------------
# CNOT Gate - Two Qubit using numpy
# Chapter 2 in the QUANTUM COMPUTING AND QUANTUM MACHINE LEARNING BOOK
#------------------------------------------------------------------
# Version 1.0
# Qiskit changes frequently.
# We recommend using the latest version from the book code repository at:
# https://aqtinitiative.org/quantum-computing-for-engineers

# (c) 2025 Jesse Van Griensven, Roydon Fraser, and Jose Rosas
# License: MIT - Citation of this work required
#------------------------------------------------------------------
import numpy as np
#------------------------------------------------------------------
def sprint(Matrix, decimals=4):
    """ Prints a Matrix with real and imaginary parts rounded to 'decimals'
"""
    import sympy as sp
    SMatrix = sp.Matrix(Matrix)  # Convert to Sympy Matrix if it's not
already

    def round_complex(x):
        """Round real and imaginary parts of x to the given number of
decimals."""
        c = complex(x)  # handle any real or complex Sympy expression
        r = round(c.real, decimals)
```

```
    i = round(c.imag, decimals)
    # If imaginary part is negligible, treat as purely real
    if abs(i) < 10**(-decimals): return sp.Float(r)
    else: return sp.Float(r) + sp.Float(i)*sp.I

  # Display the rounded Sympy Matrix
  display(SMatrix.applyfunc(round_complex))
  return
#-----------------------------------------------------------------------

# Define the CNOT gate
CNOT = np.array([
   [1, 0, 0, 0],
   [0, 1, 0, 0],
   [0, 0, 0, 1],
   [0, 0, 1, 0]
])

# Define a 2-qubit state
state = np.array([1, 0, 0, 0]) # |00>
new_state = np.dot(CNOT, state)
print("New State:")
sprint(new_state)
```

2.18.1.3   Applications

(a) Generating entanglement, such as in Bell state preparation:

$$|\psi\rangle = \frac{1}{\sqrt{2}}(|00\rangle + |11\rangle)$$

(b) Implementing quantum algorithms like Grover's and Shor's.

2.18.2   Controlled-Z (CZ) Gate

The Controlled-Z gate flips the phase of the target qubit if the control qubit is in the $|1\rangle$ state. The matrix representation of the CZ gate is:

$$CZ = \begin{bmatrix} 1 & 0 & 0 & 0 \\ 0 & 1 & 0 & 0 \\ 0 & 0 & 1 & 0 \\ 0 & 0 & 0 & -1 \end{bmatrix}$$

The effect on basis states is

$$CZ\,|00\rangle = |00\rangle, \quad CZ\,|01\rangle = |01\rangle$$
$$CZ\,|10\rangle = |10\rangle, \quad CZ\,|11\rangle = -\,|11\rangle$$

2.18.2.1 Applications

(a) Used in quantum algorithms that require phase control.
(b) Facilitates certain forms of entanglement in cluster-state quantum computing.

2.19 General Controlled Operations

A general controlled operation applies a unitary gate U to the target qubit
(s) conditioned on the control qubit(s).

2.19.1 Matrix Representation

For a single control qubit and a 1-qubit gate U:

$$\text{Controlled} - U = \begin{bmatrix} I & 0 \\ 0 & U \end{bmatrix}$$

where I is the identity matrix for the control qubit in the $|0\rangle$ state.

2.19.2 Applications

- Implementing conditional logic in quantum circuits.
- Building universal gate sets for arbitrary quantum operations.

2.19.3 SWAP Gate

The SWAP gate exchanges the states of two qubits, effectively routing information
between them.

2.19.3.1 Matrix Representation

The SWAP gate is represented by:

$$SWAP = \begin{bmatrix} 1 & 0 & 0 & 0 \\ 0 & 0 & 1 & 0 \\ 0 & 1 & 0 & 0 \\ 0 & 0 & 0 & 1 \end{bmatrix}$$

2.19.3.2 Effect on Basis States

$$SWAP\,|01\rangle = |10\rangle, \quad SWAP\,|10\rangle = |01\rangle$$

Python Code Example: SWAP Gate Using Numpy

```python
#---------------------------------------------------------------------
# SWAP Gate using numpy
# Chapter 2 in the QUANTUM COMPUTING AND QUANTUM MACHINE LEARNING BOOK
#---------------------------------------------------------------------
# Version 1.0
# Qiskit changes frequently.
# We recommend using the latest version from the book code repository at:
# https://aqtinitiative.org/quantum-computing-for-engineers

# (c) 2025 Jesse Van Griensven, Roydon Fraser, and Jose Rosas
# License: MIT - Citation of this work required
#---------------------------------------------------------------------
import numpy as np
#---------------------------------------------------------------------
def sprint(Matrix, decimals=4):
    """Prints a Matrix with real and imaginary parts rounded to decimals
"""
    import sympy as sp
    SMatrix = sp.Matrix(Matrix)  # Convert to Sympy Matrix if it's not
already

    def round_complex(x):
        """Round real and imaginary parts a given number of decimals"""
        c = complex(x)  # handle any real or complex Sympy expression
        r = round(c.real, decimals)
        i = round(c.imag, decimals)
        # If imaginary part is negligible, treat as purely real
```

```python
    if abs(i) < 10**(-decimals): return sp.Float(r)
    else: return sp.Float(r) + sp.Float(i)*sp.I

  # Display the rounded Sympy Matrix
  display(SMatrix.applyfunc(round_complex))
  return
#------------------------------------------------------------------------

# Define the SWAP gate
SWAP = np.array([
  [1, 0, 0, 0],
  [0, 0, 1, 0],
  [0, 1, 0, 0],
  [0, 0, 0, 1]
])

# State |01>
state = np.array([0, 1, 0, 0]) #

# Print the initial state |01>
print("Initial State:")
sprint(state)
print()

# Apply SWAP to a 2-qubit state
new_state = np.dot(SWAP, state)

# Output Results
print("New State after SWAP:")
sprint(new_state)
```

2.19.3.3 Applications in Qubit Routing

(a) **Quantum Networks**: Moving qubits across a quantum processor.
(b) **Error Correction**: Rearranging qubits in multi-qubit systems for error correction codes.

2.19.4 Toffoli Gate

The Toffoli gate, also known as the Controlled-Controlled-NOT (CCNOT) gate, flips the state of the target qubit if both control qubits are in the $|1\rangle$ state.

1. Matrix Representation

$$
\text{Toffoli} =
\begin{bmatrix}
1 & 0 & 0 & 0 & 0 & 0 & 0 & 0 \\
0 & 1 & 0 & 0 & 0 & 0 & 0 & 0 \\
0 & 0 & 1 & 0 & 0 & 0 & 0 & 0 \\
0 & 0 & 0 & 1 & 0 & 0 & 0 & 0 \\
0 & 0 & 0 & 0 & 1 & 0 & 0 & 0 \\
0 & 0 & 0 & 0 & 0 & 1 & 0 & 0 \\
0 & 0 & 0 & 0 & 0 & 0 & 0 & 1 \\
0 & 0 & 0 & 0 & 0 & 0 & 1 & 0
\end{bmatrix}
$$

2. Effect on Basis States

$$
\text{Toffoli} \,|\,110\rangle = |\,111\rangle, \quad \text{Toffoli}\,|\,111\rangle = |\,110\rangle
$$

3. Relevance for Reversible Computing

(a) The Toffoli gate is a universal gate for classical reversible computing.

(b) It plays a key role in quantum circuits for arithmetic operations, such as addition and multiplication.

2.20 Parametric Quantum Gates

Parametric quantum gates are a class of gates that depend on a continuous parameter, typically representing an angle of rotation. These gates are fundamental for precise control over qubit states and are widely used in quantum algorithms. Rotation gates allow for transformations around the Bloch sphere's axes, while universal gate sets, including Clifford and T gates, ensure that any quantum operation can be implemented. These tools are indispensable for quantum computing, enabling both fundamental operations and advanced quantum algorithm design.

This section covers rotation gates, their matrix representations, physical realizations, and the importance of universal gate sets.

2.20.1 Rotation Gates (R_x, R_y, R_z)

Rotation gates perform rotations of a qubit's state around the x, y, or z axis of the Bloch sphere. They provide a way to manipulate qubit states continuously and are crucial for creating arbitrary single-qubit operations.

2.20.2   *Rotation Around the* x-*Axis (*$\mathbf{R_x}$*)*

The R_x gate rotates the qubit state by an angle θ about the x-axis. Its matrix representation is:

$$R_x(\theta) = \left[\cos \frac{\theta}{2} - i \sin \frac{\theta}{2} - i \sin \frac{\theta}{2} \cos \frac{\theta}{2} \right]$$

When applied to a qubit state:

$$R_x(\theta)\,|\psi\rangle = \cos \frac{\theta}{2}\,|\psi\rangle - i \sin \frac{\theta}{2} X\,|\psi\rangle$$

2.20.3   *Rotation Around the* y-*Axis (*$\mathbf{R_y}$*)*

The R_y gate rotates the qubit state by an angle θ about the y-axis. Its matrix representation is:

$$R_y(\theta) = \left[\cos \frac{\theta}{2} - \sin \frac{\theta}{2} \sin \frac{\theta}{2} \cos \frac{\theta}{2} \right]$$

When applied to a qubit state:

$$R_y(\theta)\,|\psi\rangle = \cos \frac{\theta}{2}\,|\psi\rangle - \sin \frac{\theta}{2} Y\,|\psi\rangle$$

2.20.4   *Rotation Around the* z-*Axis (*$\mathbf{R_z}$*)*

The R_z gate rotates the qubit state by an angle θ about the z-axis. Its matrix representation is:

$$R_z(\theta) = \begin{bmatrix} e^{-i\theta/2} & 0 \\ 0 & e^{i\theta/2} \end{bmatrix}$$

When applied to a qubit state:

$$R_z(\theta)\,|\psi\rangle = e^{-i\theta/2}\,|0\rangle + e^{i\theta/2}\,|1\rangle$$

Physical Realizations

In physical quantum systems, rotation gates are implemented using various techniques:

1. In superconducting qubits, microwave pulses control the amplitude and phase to induce rotations.
2. In trapped ion systems, laser pulses adjust the internal states of ions to achieve desired rotations.
3. Photonic qubits use beam splitters and phase shifters to simulate rotations.

Python Code Example: Rotation Gates Using Numpy

```python
#------------------------------------------------------------------
# Rotation Gates using numpy
# Chapter 2 in the QUANTUM COMPUTING AND QUANTUM MACHINE LEARNING BOOK
#------------------------------------------------------------------
# Version 1.0
# Qiskit changes frequently.
# We recommend using the latest version from the book code repository at:
# https://aqtinitiative.org/quantum-computing-for-engineers

# (c) 2025 Jesse Van Griensven, Roydon Fraser, and Jose Rosas
# License: MIT - Citation of this work required
#------------------------------------------------------------------
import numpy as np
#------------------------------------------------------------------
def sprint(Matrix, decimals=4):
  """ Prints a Matrix with real and imaginary parts rounded to decimals
"""
  import sympy as sp
  SMatrix = sp.Matrix(Matrix)  # Convert to Sympy Matrix if it's not
already

  def round_complex(x):
    """Round real and imaginary of the given number of decimals """
    c = complex(x)  # handle any real or complex Sympy expression
    r = round(c.real, decimals)
    i = round(c.imag, decimals)
    # If imaginary part is negligible, treat as purely real
    if abs(i) < 10**(-decimals): return sp.Float(r)
    else: return sp.Float(r) + sp.Float(i)*sp.I

  # Display the rounded Sympy Matrix
  display(SMatrix.applyfunc(round_complex))
  return
#------------------------------------------------------------------
```

```python
def R_x(theta):
  theta2 = theta/2.
  return np.array([
    [np.cos(theta2), -1j * np.sin(theta2)],
    [-1j * np.sin(theta2), np.cos(theta2)]
  ])

def R_y(theta):
  theta2 = theta/2.
  return np.array([
    [np.cos(theta2), -np.sin(theta2)],
    [np.sin(theta2), np.cos(theta2)]
  ])

def R_z(theta):
  theta2 = theta/2.
  return np.array([
    [np.exp(-1j  * theta2), 0],
    [0, np.exp(1j * theta2)]
  ])

theta = np.pi / 4. # Example rotation angle
print("R_x(theta):")
sprint(R_x(theta))
print("R_y(theta):")
sprint(R_y(theta))
print("R_z(theta):")
sprint(R_z(theta))
```

2.21 Universal Gate Sets

A universal gate set is a small collection of gates capable of approximating any quantum operation to arbitrary precision. Universal gates are fundamental for quantum computing because they allow the implementation of any unitary transformation on a quantum system.

2.21.1 The Concept of Universality

Universality ensures that quantum computers can execute any quantum algorithm by combining a finite set of gates. For single-qubit operations, rotation gates (R_x, R_y, R_z) are sufficient to construct arbitrary transformations. For multi-qubit systems, a universal set must include gates that generate entanglement, such as the Controlled-NOT (CNOT) gate.

2.21.2   Minimal Universal Gate Sets

Two commonly used universal gates are:

1. **Clifford Gates**
   Hadamard (H), phase (S), and CNOT gates.
2. ***T* Gate**
   A non-Clifford gate is essential for universality.

The T gate is a phase shift gate with $\phi = \pi/4$:

$$T = \begin{bmatrix} 1 & 0 \\ 0 & e^{i\pi/4} \end{bmatrix}$$

Examples of applications of universal gate sets are:

1. Implementing arbitrary quantum algorithms, such as Grover's search or Shor's factoring algorithm.
2. Decomposing complex unitary operations into simpler gates for efficient hardware implementation.
3. Fault-tolerant quantum computation using logical gates in error-corrected systems.

Python Code Example: Decomposing a Gate Using Universal Set

```python
#------------------------------------------------------------------
# Decomposing a Gate Using Universal Set
# Chapter 2 in the QUANTUM COMPUTING AND QUANTUM MACHINE LEARNING BOOK
#------------------------------------------------------------------
# Version 1.0
# Qiskit changes frequently.
# We recommend using the latest version from the book code repository at:
# https://aqtinitiative.org/quantum-computing-for-engineers


# (c) 2025 Jesse Van Griensven, Roydon Fraser, and Jose Rosas
# License: MIT - Citation of this work required
#------------------------------------------------------------------
import warnings
warnings.filterwarnings('ignore')
import numpy as np
from qiskit import QuantumCircuit
```

Fig. 2.10 Example circuit to decompose a gate using a universal set (two types of drawing)

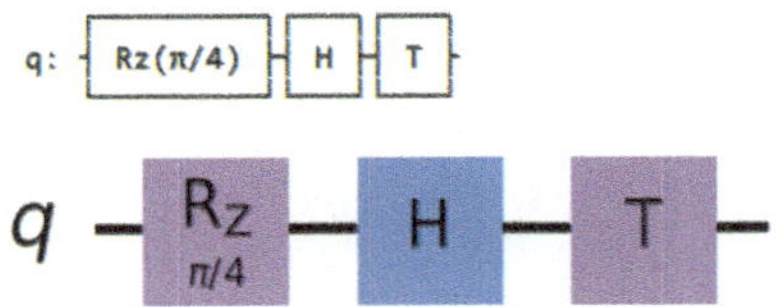


```
from qiskit.visualization import circuit_drawer
#-------------------------------------------------------------------

qc = QuantumCircuit(1)

# Apply R_z using a universal gate set
qc.rz(np.pi/4., 0)

# Hadamard gate
qc.h(0)

# T gate
qc.t(0)

# Visualize the circuit
print("Quantum Circuit:")
print(qc.draw())
display(circuit_drawer(qc, output='mpl', style="iqp"))
```

The circuit created in the code above is presented in Fig. 2.10.

2.21.3 Quantum Gate Sequences

Quantum gate sequences are the foundation of quantum circuit design, allowing the construction of complex quantum operations from simpler building blocks. Composite gates simplify the implementation of higher-level operations, while gate decomposition ensures compatibility with physical hardware. Quantum gate synthesis further enhances the flexibility of quantum circuits, allowing for the approximation and implementation of any unitary operation. Understanding these techniques is crucial for designing efficient quantum algorithms and circuits.

Note that in a quantum circuit, the matrix operations are as indicated in Fig. 2.11.

This section explores composite gates, gate decomposition, and quantum gate synthesis.

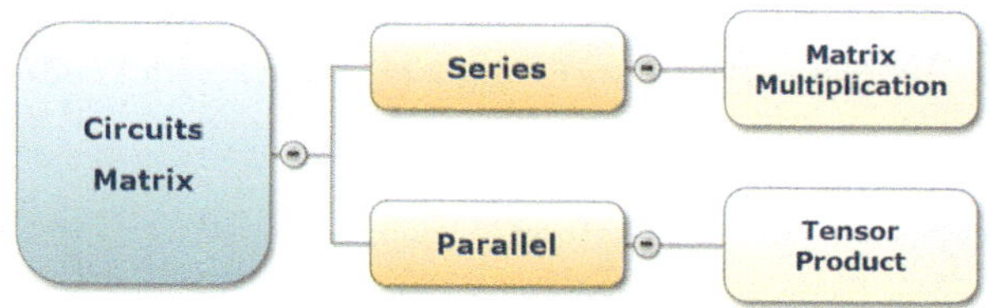

Fig. 2.11 Sequential quantum gate operations (by the authors)

2.21.4   Composite Gates and Gate Decomposition

Composite gates are sequences of basic quantum gates that achieve a specific operation when applied together. These gates combine the effects of simpler gates to implement more complex transformations on qubit states.

2.21.5   Composite Gates

A composite gate is formed by applying a sequence of unitary gates to qubits. If U_1, U_2, ..., U_n are unitary operations, the composite operation U is given by:

$$U = U_n U_{n-1} \cdots U_1$$

This sequence is applied from right to left due to the way matrix multiplication is performed in quantum mechanics.

Example: Building a Controlled-Hadamard (CH) Gate

A Controlled-Hadamard (CH) gate applies a Hadamard gate to the target qubit only when the control qubit is in the $|1\rangle$ state. It can be decomposed into a sequence of basic gates using the Controlled-NOT (CNOT) and Hadamard gates:

1. Apply a Hadamard gate to the target qubit.
2. Apply a CNOT gate with the control qubit as the control and the target qubit as the target.
3. Apply another Hadamard gate to the target qubit.

Python Code Example: CNOT Gate Using Qiskit

```
#----------------------------------------------------------------
# CNOT Gate using Qiskit
# Chapter 2 in the QUANTUM COMPUTING AND QUANTUM MACHINE LEARNING BOOK
#----------------------------------------------------------------
# Version 1.0
# Qiskit changes frequently.
# We recommend using the latest version from the book code repository at:
# https://aqtinitiative.org/quantum-computing-for-engineers
```

```
# (c) 2025 Jesse Van Griensven, Roydon Fraser, and Jose Rosas
# License: MIT - Citation of this work required
#--------------------------------------------------------------------
import warnings
warnings.filterwarnings('ignore')

import numpy as np
from qiskit import QuantumCircuit
from qiskit.visualization import circuit_drawer
#--------------------------------------------------------------------


qc = QuantumCircuit(2)

# Hadamard on target qubit
qc.h(1)

# Apply the CNOT gate
qc.cx(0, 1)

# Apply the Hadamard on target qubit again
qc.h(1)

# Visualize the circuit
print("Quantum Circuit:")
print(qc.draw())
display(circuit_drawer(qc, output='mpl', style="iqp"))
```

2.21.6 Gate Decomposition

Gate decomposition refers to expressing a quantum operation as a sequence of simpler gates from a universal gate set. Decomposition is essential for implementing complex gates on hardware, as physical quantum devices typically support only a limited set of basic gates.

Python Code Example: Decomposing a Controlled-Z (CZ) Gate
The CZ gate applies a phase flip to the target qubit when the control qubit is in the $|1\rangle$ state. It can be decomposed as:

$$CZ = H \cdot CNOT \cdot H$$

where H is the Hadamard gate and CNOT is the Controlled-NOT gate.

```python
#------------------------------------------------------------------
# Decomposing a Controlled-Z (CZ) Gate
# Chapter 2 in the QUANTUM COMPUTING AND QUANTUM MACHINE LEARNING BOOK
#------------------------------------------------------------------
# Version 1.0
# Qiskit changes frequently.
# We recommend using the latest version from the book code repository at:
# https://aqtinitiative.org/quantum-computing-for-engineers

# (c) 2025 Jesse Van Griensven, Roydon Fraser, and Jose Rosas
# License:  MIT - Citation of this work required
#------------------------------------------------------------------
import warnings
warnings.filterwarnings('ignore')

import numpy as np
from qiskit import QuantumCircuit
from qiskit.visualization import circuit_drawer
#------------------------------------------------------------------

# Create a quantum circuit with 2 qubits
qc = QuantumCircuit(2)

# Apply the Hadamard gate on target qubit
qc.h(1)

# print circuit at this point
print(qc.draw())

# Apply CNOT gate
qc.cx(0, 1)

# print circuit at this point
print(qc.draw())

# Hadamard on target qubit again
qc.h(1)

# print the final circuit
print(qc.draw())
display(circuit_drawer(qc, output='mpl', style="iqp"))
```

2.22   Quantum Gate Synthesis

Quantum gate synthesis is the process of constructing a desired unitary operation by combining gates from a universal gate set. Gate synthesis ensures that any quantum operation can be implemented with finite precision on physical hardware. There are many applications of quantum gate synthesis. These are described below.

1. **Quantum Algorithm Implementation**
   Algorithms like Shor's and Grover's require specific unitary transformations that are synthesized from basic gates.
2. **Error Correction**
   Logical gates in error-corrected systems are synthesized from physical gates to ensure fault tolerance.
3. **Hardware Optimization**
   Synthesis optimizes gate sequences to minimize resource usage on quantum hardware, such as reducing the number of CNOT gates.

2.22.1   Decomposing Complex Unitary Operations

Any unitary matrix U can be decomposed into a sequence of basic gates using mathematical algorithms. Common approaches for gate decomposition include:

2.22.1.1   QR Decomposition

This method decomposes a unitary matrix into a product of a unitary matrix Q and an upper triangular matrix R. In quantum computing, R can be implemented as a series of phase gates.

2.22.1.2   Cosine-Sine Decomposition (CSD)

CSD decomposes U into smaller unitary blocks that can be implemented with standard gates like R_z, R_y, and CNOT gates.

2.22.2   Synthesis Using Universal Gate Sets

For universal gate sets like Clifford + T, synthesis involves approximating the target operation to a desired precision using gates from the set.

Python Code Example: Decomposing a Rotation Gate

A rotation gate $R_x(\theta)$ can be decomposed into H, $R_z(\phi)$, and H:

$$R_x(\theta) = H \cdot R_z(\theta) \cdot H$$

```python
#-------------------------------------------------------------------
# Decomposing a Rotation Gate
# Chapter 2 in the QUANTUM COMPUTING AND QUANTUM MACHINE LEARNING BOOK
#-------------------------------------------------------------------
# Version 1.0
# Qiskit changes frequently.
# We recommend using the latest version from the book code repository at:
# https://aqtinitiative.org/quantum-computing-for-engineers

# (c) 2025 Jesse Van Griensven, Rcydon Fraser, and Jose Rosas
# License: MIT - Citation of this work required
#-------------------------------------------------------------------
import warnings
warnings.filterwarnings('ignore')

import numpy as np
from qiskit import QuantumCircuit
from qiskit.visualization import circuit_drawer
#-------------------------------------------------------------------
# Rotation angle
theta = np.pi / 3.

# Create the Circuit with 1 qubit
qc = QuantumCircuit(1)

# Apply Hadamard
qc.h(0)

# Apply Rz - Z Rotation gate
qc.rz(theta, 0)  # Apply Rz

# Draw the circuit at this point
print(qc.draw())
display(circuit_drawer(qc, output='mpl', style="iqp"))

# Apply Hadamard again
qc.h(0)
```

```
# Draw the final circuit
print(qc.draw())
display(circuit_drawer(qc, output='mpl', style="iqp"))
```

2.23   Quantum Circuits and Gate Layout

Quantum circuits are the computational structures in quantum computing analogous
to classical logic circuits. They consist of qubits and quantum gates arranged in
sequences to perform specific tasks or algorithms. Circuit diagrams provide a clear
way to represent and analyze quantum computations. Simple circuits, like those for
Bell state preparation and quantum teleportation, demonstrate how gates combine to
achieve fundamental tasks in quantum information processing. Understanding how
these quantum gates combine into circuits, their diagrammatic representation, and
their implementation in real-world examples is essential for designing and analyzing
quantum algorithms.

2.23.1   *How Quantum Gates Are Combined Into Circuits*

In quantum circuits, gates are applied sequentially or in parallel to manipulate the
quantum states of qubits. Each operation is time-ordered, meaning the outcome of
one operation can serve as the input for the next. The gates collectively perform a
unitary transformation on the initial quantum state.

2.23.2   *General Structure of a Quantum Circuit*

Table 2.6 summarizes the elements that comprise a quantum circuit.

For n qubits, the overall unitary operation U performed by the circuit is the
product of individual gate operations:

$$U = U_k \cdot U_{k-1} \cdots U_1$$

Table 2.6 Elements of quantum circuits

Element	Description
Input qubits	Initialized in a known state, usually $|0\rangle$
Quantum gates	Applied to transform the qubits' states
Measurements	Performed at the end to extract classical information from the quantum states

2.23.3   Parallel Gates

When gates act on different qubits simultaneously, they are said to act in parallel. This is equivalent to applying the tensor product of their respective unitary matrices.

Example

A quantum circuit that applies a Hadamard gate to the first qubit and a Pauli-X gate to the second qubit in parallel is represented as:

$$U = H \otimes X$$

2.23.4   Bell State Preparation

The Bell state $|\Phi^+\rangle = \frac{1}{\sqrt{2}}(|00\rangle + |11\rangle)$ is an entangled two-qubit state. It serves as the foundation for many quantum protocols, including quantum teleportation and superdense coding.

Steps to Prepare a Bell State

1. Initialize both qubits in $|0\rangle$.
2. Apply a Hadamard gate (H) to the first qubit, creating a superposition:

$$H|0\rangle = \frac{1}{\sqrt{2}}(|0\rangle + |1\rangle)$$

3. Apply a CNOT gate with the first qubit as the control and the second qubit as the target. This entangles the qubits:

$$\text{CNOT}\left(\frac{1}{\sqrt{2}}(|0\rangle + |1\rangle)|0\rangle\right) = \frac{1}{\sqrt{2}}(|00\rangle + |11\rangle)$$

Python Code Example: Bell State

```
#------------------------------------------------------------------
# Bell State Preparation
# Chapter 2 in the QUANTUM COMPUTING AND QUANTUM MACHINE LEARNING BOOK
#------------------------------------------------------------------
# Version 1.0
# Qiskit changes frequently.
# We recommend using the latest version from the book code repository at:
# https://aqtinitiative.org/quantum-computing-for-engineers
```

```
# (c) 2025 Jesse Van Griensven, Roydon Fraser, and Jose Rosas
# License: MIT - Citation of this work required
#---------------------------------------------------------------------
import warnings
warnings.filterwarnings('ignore')

import numpy as np
from qiskit import QuantumCircuit
from qiskit.visualization import circuit_drawer
#---------------------------------------------------------------------


# Create a quantum circuit with 2 qubits
qc = QuantumCircuit(2)

# Apply Hadamard gate to the first qubit
qc.h(0)

# Apply CNOT gate (control: qubit 0, target: qubit 1)
qc.cx(0, 1)

# Visualize the circuit
print("Quantum Circuit:")
print(qc.draw())
display(circuit_drawer(qc, output='mpl', style="iqp"))
```

2.23.5 Basic Quantum Teleportation Circuit

Quantum teleportation is a protocol for transferring the state of a qubit from one
location to another without physically moving the qubit. This process relies on
entanglement, measurement, and classical communication

The following are the steps for quantum teleportation:

1. **Initialize the Qubits**

    (a) Qubit 0 (q_0): Contains the state to be teleported ($|\psi\rangle$).
    (b) Qubits 1 and 2 (q_1, q_2): Start in $|0\rangle$.

2. **Entangle Qubits 1 and 2**

    (a) Apply a Hadamard gate to q_1.
    (b) Apply a CNOT gate (control: q_1, target: q_2).

3. **Bell Measurement**

    (a) Apply a CNOT gate (control: q_0, target: q_1).
    (b) Apply a Hadamard gate to q_0.
    (c) Measure q_0 and q_1.

4. Apply Conditional Gates

Use the classical measurement outcomes to apply corrections to q_2, restoring $|\psi\rangle$ on q_2.

Python Code Example: Quantum Teleportation Circuit

```python
#-------------------------------------------------------------------
# Quantum Teleportation Circuit
# Chapter 2 in the QUANTUM COMPUTING AND QUANTUM MACHINE LEARNING BOOK
#-------------------------------------------------------------------
# Version 1.0
# Qiskit changes frequently.
# We recommend using the latest version from the book code repository at:
# https://aqtinitiative.org/quantum-computing-for-engineers

# (c) 2025 Jesse Van Griensven, Roydon Fraser, and Jose Rosas
# License: MIT - Citation of this work required
#-------------------------------------------------------------------
import warnings
warnings.filterwarnings('ignore')

import numpy as np
from qiskit import QuantumCircuit
from qiskit.visualization import circuit_drawer
#-------------------------------------------------------------------

# Create a quantum circuit with 3 qubits
qc = QuantumCircuit(3, 2)
print(qc.draw())

# Entangle qubits 1 and 2
qc.h(1)
qc.cx(1, 2)
print(qc.draw())

# Perform Bell measurement on qubits 0 and 1
qc.cx(0, 1)
qc.h(0)
qc.measure([0, 1], [0, 1])
print(qc.draw())

# Apply conditional corrections to qubit 2
qc.cx(1, 2)
qc.cz(0, 2)
```

```
# Visualize the final circuit
print("Final Quantum Teleportation Circuit:")
print(qc.draw())
display(circuit_drawer(qc, output='mpl', style="iqp"))
```

2.23.6   Measurement in Quantum Circuits

Measurement in quantum circuits is a critical process in quantum computing, serving as the bridge between quantum and classical worlds. It allows the extraction of classical information from a quantum system by collapsing a qubit's state into one of its basis states. Post-measurement states determine the classical information obtained from the quantum system. The integration of measurement with gate operations enables critical functionalities, including feedback control, error correction, and algorithmic decision-making.

This section explores how measurement operations differ from quantum gates, the concept of post-measurement states, and the integration of measurement with gate operations.

2.23.7   How Measurement Operations Differ from Gates

Quantum gates are unitary operations, meaning they preserve the overall quantum state and probability distribution of the system. In contrast, measurement is a non-unitary operation that irreversibly collapses a quantum state into one of its eigenstates. The main key differences are listed below:

1. **Reversibility**

    (a) Gates are reversible because they are unitary transformations. Applying a gate's inverse undoes its operation.

    (b) Measurements are irreversible; once a quantum state collapses, the original state cannot be restored.

2. **Outcome**

    (a) Gates produce a new quantum state.

    (b) Measurements yield a classical outcome (e.g., 0 or 1) based on the probability amplitudes of the quantum state.

3. **Effect on State**

    (a) Gates transform a quantum state without collapsing it.

    (b) Measurements collapse the state to the eigenstate corresponding to the observed outcome.

2.23.8   Mathematical Representation of Measurement

Measurement is represented by using projection operators. For a single qubit, measuring in the computational basis ($|0\rangle$ and $|1\rangle$):

(a) The projection operators are:

$$P_0 = |0\rangle\langle 0|, \quad P_1 = |1\rangle\langle 1|$$

(b) If the state of the qubit before measurement is:

$$|\psi\rangle = \alpha|0\rangle + \beta|1\rangle$$

the probability of measuring $|0\rangle$ is:

$$P(0) = |\alpha|^2$$

and the probability of measuring $|1\rangle$ is:

$$P(1) = |\beta|^2$$

Measurements can be generalized to other bases using measurement operators. For example, measuring in the Hadamard basis ($|+\rangle, |-\rangle$) involves the states:

$$|+\rangle = \frac{1}{\sqrt{2}}(|0\rangle + |1\rangle), \quad |-\rangle = \frac{1}{\sqrt{2}}(|0\rangle - |1\rangle)$$

Python Code Example: Probability Measurement

```
#-------------------------------------------------------------------
# Probability Measurement
# Chapter 2 in the QUANTUM COMPUTING AND QUANTUM MACHINE LEARNING BOOK
#-------------------------------------------------------------------
# Version 1.0
# Qiskit changes frequently.
# We recommend using the latest version from the book code repository at:
# https://aqtinitiative.org/quantum-computing-for-engineers


# (c) 2025 Jesse Van Griensven, Roydon Fraser, and Jose Rosas
# License: MIT - Citation of this work required
#-------------------------------------------------------------------
```

```
import warnings
warnings.filterwarnings('ignore')

import numpy as np
#-------------------------------------------------------------------------

# Define computational basis states
state = np.array([np.sqrt(0.6), np.sqrt(0.4)]) # |ψ> = √0.6|0> + √0.4|
1>
prob_0 = np.abs(state[0])**2 # Probability of measuring |0>
prob_1 = np.abs(state[1])**2 # Probability of measuring |1>

# Print probabilities with 4 decimal places
print(f"Probability of |0>: {prob_0:.4f}")
print(f"Probability of |1>: {prob_1:.4f}")
```

2.23.9   Post-Measurement States

Post-measurement states refer to the quantum states of a system immediately after a measurement has been performed. In quantum mechanics, measurement is not just a passive readout; it fundamentally affects the state of the system by "collapsing" it onto an eigenstate (or a subspace) corresponding to the measurement outcome. These post-measurement states determine the subsequent evolution and possible future measurements of the system. In quantum circuits, for example, once a qubit is measured in the computational basis, it effectively collapses into either $|0\rangle$ or $|1\rangle$, losing all the superposition or entanglement it might have had with other qubits (unless special types of measurements are used).

2.23.9.1   State Collapse

After a measurement, the quantum state collapses into the eigenstate corresponding to the observed outcome. For a measurement result of $|0\rangle$, the state collapses to:

$$|\psi_{\text{post}}\rangle = |0\rangle$$

If $|1\rangle$ is measured, the state becomes:

$$|\psi_{\text{post}}\rangle = |1\rangle$$

2.23.9.2   Mixed States After Repeated Measurements

Repeated measurements on an ensemble of identically prepared states yield a probability distribution over the eigenstates. This distribution represents a mixed state, described by a density matrix:

$$\rho = P(0)\,|0\rangle\langle 0| + P(1)\,|1\rangle\langle 1|$$

Consider the state:

$$|\psi\rangle = \frac{1}{\sqrt{2}}\left(|0\rangle + |1\rangle\right)$$

If measured in the computational basis:

- There is a 50% chance of collapsing to $|0\rangle$.
- There is a 50% chance of collapsing to $|1\rangle$.

Python Code Example: Quantum Measurement after Simulation

```python
#-----------------------------------------------------------------
# Quantum Measurement After Simulation
# Chapter 2 in the QUANTUM COMPUTING AND QUANTUM MACHINE LEARNING BOOK
#-----------------------------------------------------------------
# Version 1.0
# Qiskit changes frequently.
# We recommend using the latest version from the book code repository at:
# https://aqtinitiative.org/quantum-computing-for-engineers

# (c) 2025 Jesse Van Griensven, Roydon Fraser, and Jose Rosas
# License: MIT - Citation of this work required
#-----------------------------------------------------------------
import warnings
warnings.filterwarnings('ignore')

import numpy as np
from qiskit import Aer, execute
from qiskit import QuantumCircuit
from qiskit.visualization import circuit_drawer
#-----------------------------------------------------------------

# Create a single-qubit circuit
qc = QuantumCircuit(1, 1)

# Apply Hadamard gate to create superposition
qc.h(0)
```

```python
# Measure qubit
qc.measure(0, 0)

# Simulate the circuit
backend = Aer.get_backend('qasm_simulator')
result = execute(qc, backend, shots=1000).result()
counts = result.get_counts()

print("Measurement Results:", counts)
print()

# Visualize the final circuit
print("The Final Measurement Circuit")
print(qc.draw())
display(circuit_drawer(qc, output='mpl', style="iqp"))
```

2.23.10   Integration of Measurement with Gate Operations

In quantum circuits, measurements are often combined with gate operations to achieve desired computational outcomes. Measurement provides classical feedback that can influence subsequent operations.

2.23.11   Measurement-Driven Operations

Classical measurement results can control further quantum gates. This feedback is essential in algorithms like quantum error correction and quantum teleportation.

Python Code Example: Quantum Teleportation in the Quantum Teleportation Protocol

1. The state of a qubit is measured.
2. The measurement outcomes are used to determine the corrective gates to apply to another qubit.

```python
#--------------------------------------------------------------------
# Quantum Teleportation Circuit
# Chapter 2 in the QUANTUM COMPUTING AND QUANTUM MACHINE LEARNING BOOK
#--------------------------------------------------------------------
# Version 1.0
# Qiskit changes frequently.
# We recommend using the latest version from the book code repository at:
# https://aqtinitiative.org/quantum-computing-for-engineers
```

```python
# (c) 2025 Jesse Van Griensven, Roydon Fraser, and Jose Rosas
# License: MIT - Citation of this work required
#-------------------------------------------------------------------------
import numpy as np
from qiskit import QuantumCircuit
#-------------------------------------------------------------------------

# Create a quantum circuit with 3 qubits and 2 classical bits
qc = QuantumCircuit(3, 2)

# Entangle qubits 1 and 2
qc.h(1)
qc.cx(1, 2)

# Bell measurement on qubits 0 and 1
qc.cx(0, 1)
qc.h(0)
qc.measure([0, 1], [0, 1])

# Conditional gates based on measurement outcomes
qc.cx(1, 2)
qc.cz(0, 2)

print(qc.draw())
```

2.23.12   Measurement for State Verification

Measurement is used to verify the correctness of a state during circuit execution. This is critical in experiments to confirm that the circuit produces the intended quantum state.

2.23.13   Applications in Quantum Algorithms

The following contains example of quantum algorithm applications:

1. **Grover's Algorithm**
   Measurement identifies the marked item in the search space.
2. **Shor's Algorithm**
   Measurement extracts the periodicity of a function.

3. **Gate Fidelity and Error Rates**

   (a) Metrics for gate quality:

       - Gate fidelity.
       - Depolarizing noise.

   (b) Impact of noise on gates.
   (c) Strategies for mitigating gate errors.

Exercise Questions: Fundamentals of Quantum Mechanics
1. **Wave-Particle Duality**

   (a) Explain the concept of wave-particle duality using the results of the double-slit experiment.
   (b) Derive the expression for the probability density $|\Psi(x, t)|^2$ from the wave function

$$\Psi(x, t) = A e^{i(kx - \omega t)}$$

2. **Applications of Wave-Particle Duality**

   (a) How does the principle of wave-particle duality influence the design of photonic devices such as lasers?
   (b) Discuss the role of wave interference in nanoscale transistor operation.

3. **Heisenberg's Uncertainty Principle**

   (a) Provide the mathematical statement of Heisenberg's uncertainty principle and explain its implications for position and momentum measurements.
   (b) For a particle confined to a one-dimensional box of length L, calculate the minimum uncertainty in its momentum.

4. **Practical Applications of the Uncertainty Principle**

   (a) Describe how the uncertainty principle supports the security of quantum key distribution (QKD).
   (b) Explain the effect of quantum confinement on the energy levels in quantum dots, using the uncertainty principle.

5. **Qubits and Superposition**

   (a) Compare and contrast classical bits and qubits, emphasizing the role of superposition in quantum computation.
   (b) Write the general form of a qubit state in terms of $|0\rangle$ and $|1\rangle$, and explain the significance of the coefficients.

6. **The Bloch Sphere**

   (a) Using the Bloch sphere representation, describe the qubit state $|\psi\rangle = \cos\frac{\theta}{2}|0\rangle + e^{i\phi}\sin\frac{\theta}{2}|1\rangle$.

   (b) What do the angles θ and ϕ represent in the Bloch sphere model?

7. **Multi-Qubit Systems**

   (a) For a two-qubit system, compute the tensor product of the states $|\psi_1\rangle = |0\rangle + |1\rangle$ and $|\psi_2\rangle = |0\rangle - |1\rangle$.

   (b) Explain the significance of the tensor product in representing multi-qubit systems.

8. **Entanglement**

   (a) Define quantum entanglement and explain why it is a non-classical phenomenon.

   (b) Derive the Bell state $|\Phi^+\rangle = \frac{1}{\sqrt{2}}(|00\rangle + |11\rangle)$ using a Hadamard gate and a CNOT gate.

9. **Applications of Entanglement**

   (a) How is entanglement used in quantum teleportation? Provide a brief outline of the protocol.

   (b) Describe how entanglement enhances the security of quantum communication protocols.

10. **Operators in Quantum Mechanics**

    (a) Define a Hermitian operator and explain its significance in quantum mechanics.

    (b) Compute the action of the Pauli-X operator on the state $|\psi\rangle = \frac{1}{\sqrt{2}}(|0\rangle + |1\rangle)$.

11. **Quantum Gates**

    (a) Write the matrix representation of the Hadamard gate and explain its role in creating superposition.

    (b) Derive the effect of a phase shift gate $R_\phi = \begin{bmatrix} 1 & 0 \\ 0 & e^{i\phi} \end{bmatrix}$ on a general qubit state $|\psi\rangle$.

12. **Quantum Measurement**

    (a) Explain the collapse of the wave function during measurement and provide the mathematical expression for the probability of measuring a specific state.

    (b) Discuss how measurement affects the coherence of a quantum system and its implications for quantum computing.

Additional Bibliography

1. S. Axler, *Linear Algebra Done Right* (Springer, 2015)
2. S. Gasiorowicz, *Quantum Physics* (Wiley, 2003)
3. G.H. Golub, C.F. Van Loan, *Matrix Computations* (Johns Hopkins University Press, 2013)
4. D.J. Griffiths, *Introduction to Quantum Mechanics* (Pearson, 2016)
5. R.A. Horn, C.R. Johnson, *Matrix Analysis* (Cambridge University Press, 2012)
6. S. Majidy, C. Wilson, R. Laflamme, *Building Quantum Computers—A Practical Introduction* (Cambridge Press, 2025)
7. M.A. Nielsen, I.L. Chuang, *Quantum Computation and Quantum Information* (Cambridge University Press, 2010)
8. P.J. Olver, C. Shakiban, *Applied Linear Algebra* (Prentice Hall, 2006)
9. D. Poole, *Linear Algebra: A Modern Introduction* (Cengage Learning, 2014)
10. R. Shankar, *Principles of Quantum Mechanics* (Springer, 1994)
11. G. Strang, *Linear Algebra and its Applications* (Brooks Cole, 2005)
12. L. Susskind, A. Friedman, *Quantum Mechanics: The Theoretical Minimum* (Basic Books, 2014)
13. L.N. Trefethen, D. Bau, *Numerical Linear Algebra* (SIAM, 1997)
14. N.S. Yanofsky, M.A. Mannucci, *Quantum Computing for Computer Scientists* (Cambridge University Press, 2008)
15. N. Zettili, *Quantum Mechanics: Concepts and Applications* (Wiley, 2009)

Chapter 3
Quantum and Classical Circuits

This chapter explores quantum and classical circuits, the major types of quantum computers, compares classical and quantum hardware architectures, highlights the need for error correction, and discusses engineering challenges. Future advancements in material science, fabrication, and interconnectivity are crucial to the scalability of quantum computing hardware.

3.1 Overview of Quantum Programming Languages

Quantum programming languages like Qiskit, Cirq, and Q# provide the tools engineers need to design, simulate, and execute quantum algorithms. While Qiskit offers comprehensive modules for quantum applications, Cirq focuses on NISQ devices, and Q# emphasizes high-level abstraction and seamless debugging. Together, these tools enable engineers to explore and harness the potential of quantum computing in diverse fields.

Quantum programming languages provide the critical interface between developers and quantum hardware. These languages allow engineers to design, test, and execute quantum algorithms, abstracting the complexity of hardware interactions while enabling fine-tuned control over quantum circuits. The most widely adopted quantum programming platforms include Qiskit, Cirq, and Microsoft's Q#, each with unique strengths tailored to specific use cases.

3.2 Qiskit

Qiskit, developed by IBM, is an open-source quantum computing framework that supports the complete development lifecycle. Its modular architecture caters to a wide range of quantum computing tasks, including circuit design, simulation, error

© The Author(s), under exclusive license to Springer Nature Switzerland AG 2025

J. Van Griensven Thé et al., *Quantum Computing and Quantum Machine Learning for Engineers and Developers*, https://doi.org/10.1007/978-3-031-98245-3_3

correction, and advanced quantum applications like quantum chemistry and machine learning.

Unfortunately, IBM may not be taking Qiskit users into consideration, with frequent changes that completely break the codes. For example, Qiskit functions are moved from one library to another, with no apparent benefit to the user, causing problems with dependencies and a tremendous waste of developers time. This way, this book's authors keep various versions of Qiskit on the book website.

3.2.1 Qiskit Modules

Qiskit, an open-source framework for quantum computing, is designed to provide researchers, developers, and engineers with the tools needed to experiment and develop quantum algorithms and applications. It comprises several distinct modules, each addressing specific aspects of quantum computing workflows. Table 3.1 is a detailed discussion of the core Qiskit modules, which are further discussed in the following subsections.

For clarity, these modules are graphically represented in Fig. 3.1.

3.3 Terra Module

The Terra module is the backbone of Qiskit, providing the software tools for developers to move from theoretical designs to practical implementations on quantum devices. The Terra module serves as the foundational layer of the Qiskit framework. It provides the essential tools for designing, building, and optimizing quantum circuits. With Terra, users can:

Table 3.1 Qiskit modules

Qiskit module name	Objective
Terra	The foundation for designing, building, and optimizing quantum circuits
Aer	A simulator for testing quantum circuits on classical hardware with realistic noise models
Ignis	Provides tools for error mitigation and error correction
Domain modules	High-level algorithms for quantum applications in chemistry, optimization, and artificial intelligence

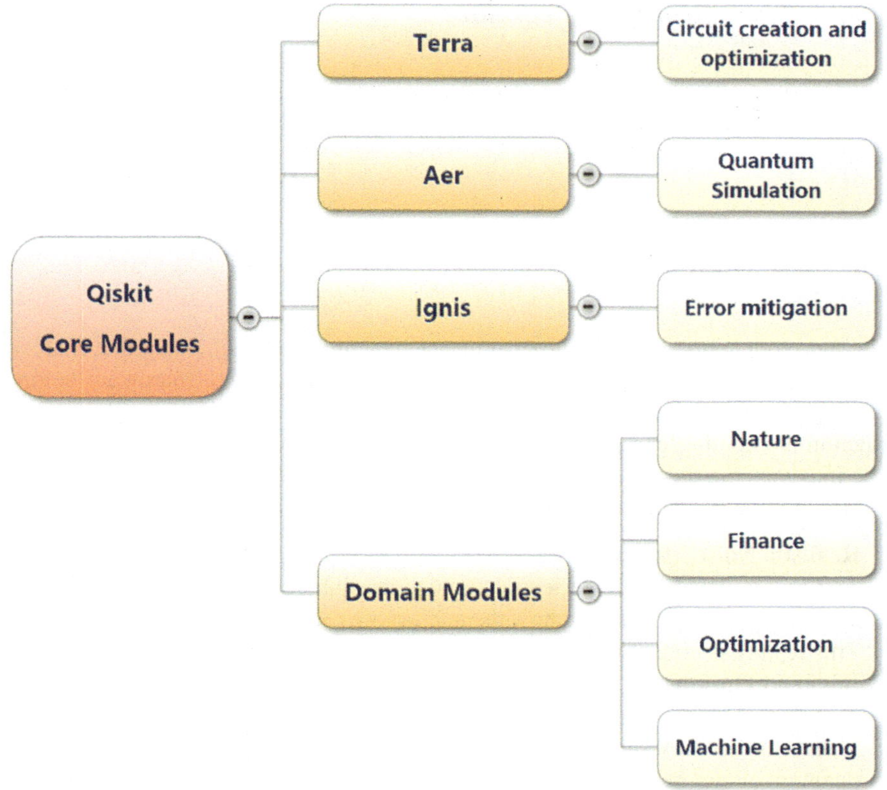

Fig. 3.1 Qiskit core modules (by the authors)

3.3.1   Design Quantum Circuits

Terra offers a flexible and intuitive interface for creating quantum circuits, which are the fundamental building blocks of quantum algorithms. Users can define circuits using standard quantum gates, control flows, and parameterized gates.

3.3.2   Transpilation

One of Terra's standout features is its ability to transpile quantum circuits. Transpilation involves translating a high-level circuit into a version that can be efficiently executed on a specific quantum hardware backend. This includes mapping logical qubits to physical qubits and optimizing gate sequences to minimize errors and resource usage.

3.3.3 Terra Backend Integration

Terra facilitates seamless integration with various quantum hardware and simulators. By defining circuits in Terra, users can execute them on IBM Quantum's devices or simulators.

3.4 Aer Module

The Aer module provides a high-performance quantum circuit simulator that enables users to test and validate their quantum algorithms on classical hardware before running them on actual quantum devices. By bridging the gap between theoretical quantum computing and physical implementation, the Aer module plays a critical role in advancing the development of quantum technologies.

Aer module key features include:

1. **Realistic Noise Models**

   Aer simulates the imperfections and noise inherent in quantum hardware, allowing users to understand how their algorithms behave in real-world scenarios. This feature is crucial for preparing robust algorithms against quantum errors.

2. **Backend Options**

   Aer includes multiple simulation backends, such as state vector, density matrix, and unitary simulation. These allow users to explore different aspects of quantum circuit behavior, from idealized operations to complex error dynamics.

3. **Performance**

   Aer is optimized for classical computing platforms, providing fast and efficient simulation of quantum circuits. This makes it a vital tool for prototyping and debugging quantum algorithms.

Qiskit Aer supports noisy simulations, making it ideal for testing algorithms for noisy intermediate-scale quantum (NISQ) devices. The main Aer module features are detailed in the table below and graphically in Fig. 3.2 (Table 3.2).

Python Code Example: Qiskit Aer Simulation

This example shows how Qiskit Aer simulates circuits with custom noise models. This code's final circuit is displayed in Fig. 3.3.

```
#---------------------------------------------------------------------
# Qiskit-Aer Simulator with custom Noise
# Chapter 3 in the QUANTUM COMPUTING AND QUANTUM MACHINE LEARNING BOOK
#---------------------------------------------------------------------
# Version 1.0
# Qiskit changes frequently.
# We recommend using the latest version from the book code repository at:
# https://aqtinitiative.org/quantum-computing-for-engineers
```

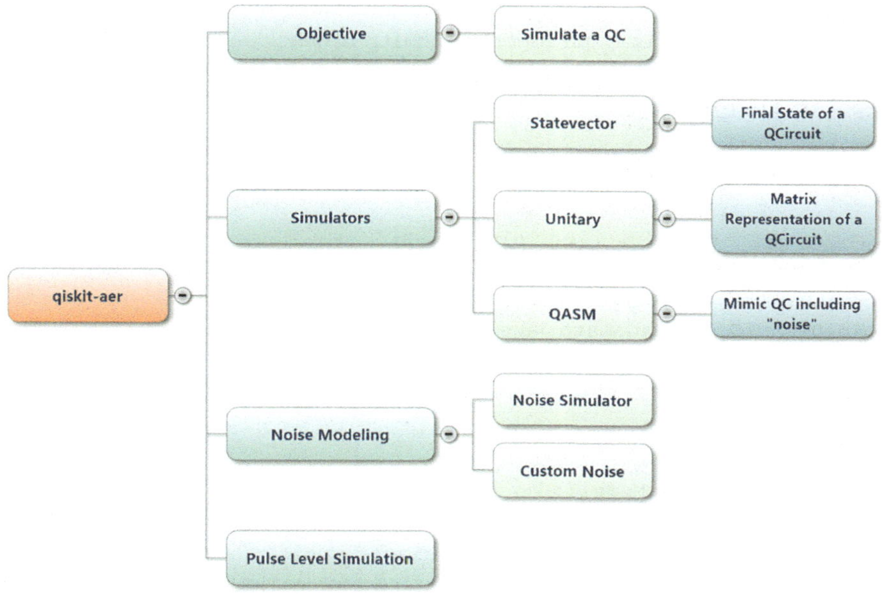

Fig. 3.2 Qiskit Aer sub-modules (by the authors)

Table 3.2 Main Aer features

Aer module feature	Description
Noise models	Allows the simulation of realistic quantum errors, such as decoherence and gate errors
Custom backends	Supports state vector, unitary, and density matrix simulations
High scalability	Handles circuits with thousands of qubits, depending on available computational resources

Fig. 3.3 Qiskit-Aer simulator with custom noise (by the authors)

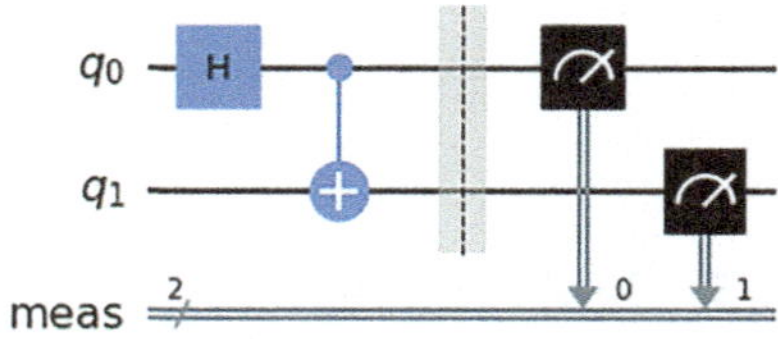


```
# (c) 2025 Jesse Van Griensven, Roydon Fraser, and Jose Rosas
# License: MIT - Citation required
#-----------------------------------------------------------------
import numpy as np
import matplotlib.pyplot as plt
```

```python
from qiskit import QuantumCircuit, Aer, execute
from qiskit.visualization import circuit_drawer
from qiskit.providers.aer.noise import NoiseModel,
depolarizing_error

print("Completed importing libraries")
#-----------------------------------------------------------------

# Create a simple quantum circuit
qc = QuantumCircuit(2)
qc.h(0)    # Apply Hadamard gate
qc.cx(0, 1)  # Apply CNOT gate
qc.measure_all()

# Define a noise model
noise_model = NoiseModel()

# Add 1-qubit depolarizing error for single-qubit gates
error_1q = depolarizing_error(0.01, 1)  # 1% depolarizing error
noise_model.add_all_qubit_quantum_error(error_1q, 'h')  # Apply to
'h' gate

# Add 2-qubit depolarizing error for two-qubit gates
error_2q = depolarizing_error(0.02, 2)  # 2% depolarizing error for
2-qubit gates
noise_model.add_all_qubit_quantum_error(error_2q, 'cx')  # Apply to
'cx' gate

# Simulate the circuit
simulator = Aer.get_backend('qasm_simulator')
result = execute(qc, simulator, noise_model=noise_model,
shots=1000).result()
print("Simulated Results with Noise:", result.get_counts())
```

3.5 Ignis Module

Quantum systems are intrinsically susceptible to errors arising from decoherence, environmental noise, and other fundamental quantum effects. The Ignis module addresses this challenge by providing tools for error characterization, mitigation, and correction. The Ignis module is indispensable for pushing quantum computing toward fault-tolerant systems, making it a cornerstone for future advancements in quantum technology.

Some of Ignis' key functionalities include:

1. **Error Characterization**

   Ignis offers protocols to characterize quantum errors, such as gate fidelities, coherence times, and cross talk between qubits. This information is vital for improving the performance of quantum systems.

2. **Error Mitigation**

   Ignis provides techniques to reduce the impact of errors on quantum computations. For instance, error mitigation methods can be applied during post-processing to obtain more accurate results from noisy quantum devices.

3. **Error Correction Frameworks**

   While quantum error correction remains a challenging frontier, Ignis supports research and experimentation with error-correcting codes. It provides tools to implement and test small-scale error-correcting schemes.

3.6   Qiskit Domain Modules

Qiskit underwent a significant restructuring when the Aqua module was deprecated. This reorganization divided Aqua's broad functionalities into distinct, domain-specific modules. Each module is designed to integrate seamlessly with Qiskit's core components, such as Terra, Aer, and Ignis, facilitating the development of robust and scalable quantum applications.

This subsection delves into each of these modules, detailing their functionalities, key features, and practical applications.

3.6.1   Qiskit Nature

Qiskit Nature focuses on applications in the natural sciences, particularly quantum chemistry and materials science. It provides tools and algorithms for simulating molecular structures, chemical reactions, and material properties using quantum computing techniques.

Table 3.3 summarizes the Qiskit Nature module's key features.

Figure 3.4 graphically depicts Qiskit's Nature module functionalities.

3.6.2   Qiskit Nature Use Case

A chemist seeking to explore new catalyst materials can use Qiskit Nature to simulate and analyze the electronic structure of various compounds, identifying those with optimal catalytic properties through precise quantum simulations.

Table 3.4 summarizes the practical applications of the Qiskit Nature module.

Table 3.3 Main Qiskit Nature modules

Nature module's key features	Functionality
Quantum chemistry simulations	Implements algorithms like the variational quantum eigensolver (VQE) and quantum phase estimation (QPE) to compute molecular energies, electronic structures, and reaction pathways with high precision
Molecular property calculations	Facilitates the calculation of various molecular properties, including dipole moments, excitation energies, and vibrational frequencies, which are essential for understanding chemical behavior
Material science applications	Enables the simulation of material properties at the quantum level, aiding in the design of new materials with desired characteristics, such as superconductivity, strength, and flexibility
Integration with classical tools	Seamlessly integrates with classical computational chemistry tools, allowing for hybrid simulations that combine quantum and classical methods for enhanced accuracy and efficiency

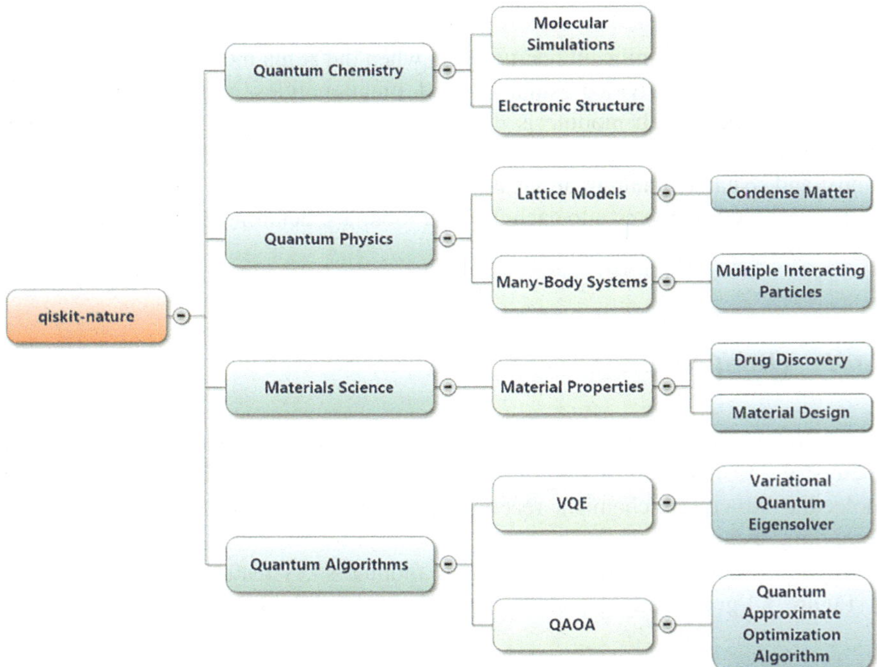

Fig. 3.4 Qiskit's Nature module functionalities (by the authors)

3.6.3 Qiskit Finance

Qiskit Finance applies quantum computing techniques to financial modeling and analysis. It offers algorithms and tools designed to tackle financial problems, enabling more efficient and accurate decision-making processes. Table 3.5 summarizes the Qiskit Finance module's key features.

Table 3.4 Applications that would benefit from using Qiskit's Nature module

Nature Module applications	Description
Drug discovery	Simulating molecular interactions to identify potential drug candidates and understand their binding mechanisms with target proteins
Catalysis research	Modeling catalytic processes to develop more efficient catalysts for industrial chemical reactions
Energy materials	Designing materials for energy storage and conversion, such as batteries and photovoltaic cells, by understanding their quantum properties

Table 3.5 Main Qiskit Finance modules

Finance module key features	Functionality
Portfolio optimization	Implements quantum algorithms to optimize the selection of assets in a portfolio, balancing expected returns against risk measures, such as volatility and value at risk (VaR)
Risk analysis	Utilizes quantum-enhanced methods to assess and mitigate financial risks, including market risk, credit risk, and operational risk, by analyzing large datasets and complex dependencies
Option pricing	Applies quantum algorithms to compute the fair value of financial derivatives, such as options and futures, by simulating underlying asset dynamics with high precision
Monte Carlo simulations	Enhances the efficiency of Monte Carlo simulations for pricing complex financial instruments and assessing risk by leveraging quantum parallelism

3.6.4   Qiskit Finance Use Case

An investment firm seeking to optimize its asset allocation strategy can employ Qiskit Finance to leverage quantum algorithms for portfolio optimization, enabling the firm to achieve better risk-adjusted returns by efficiently navigating the complex trade-offs inherent in financial markets. Table 3.6 summarizes the practical applications of the Finance module.

3.6.5   Qiskit Optimization

Qiskit Optimization is tailored to solve complex optimization problems using quantum algorithms. It provides tools for formulating, solving, and analyzing optimization tasks, leveraging both quantum and classical computational techniques.

Table 3.7 summarizes the key features of Qiskit's Optimization module.

Figure 3.5 graphically depicts Qiskit's Optimization module functionalities.

Table 3.6 Applications that would benefit from using Qiskit's Finance module

Finance module applications	Description
Asset management	Assisting asset managers in constructing optimized portfolios that align with client objectives and risk tolerances
Algorithmic trading	Developing quantum-enhanced trading algorithms that can analyze market data and execute trades with greater speed and accuracy
Insurance	Improving the assessment and pricing of insurance products by modeling and simulating various risk scenarios more effectively

Table 3.7 Main Qiskit Optimization modules

Optimization module key features	Functionality
Quantum optimization algorithms	Implements algorithms such as the quantum approximate optimization algorithm (QAOA) and the variational quantum eigensolver (VQE) tailored for optimization tasks, offering potential advantages in finding global minima
Quadratic unconstrained binary optimization (QUBO)	Facilitates the formulation of optimization problems in the QUBO framework, which is compatible with both quantum annealers and gate-based quantum computers
Hybrid quantum-classical approaches	Supports the development of hybrid algorithms that combine quantum optimization with classical heuristics, enhancing solution quality and convergence rates
Problem-specific encodings	Provides methods for encoding various optimization problems, such as scheduling, routing, and resource allocation, into quantum circuits suitable for quantum processing

3.6.6   Qiskit Optimization Use Case

A logistics company aiming to optimize delivery routes can utilize Qiskit Optimization to model the problem as a QUBO, applying QAOA to find near-optimal routing solutions that reduce travel time and fuel consumption more effectively than classical methods. Table 3.8 summarizes the practical applications of Qiskit's Optimization module.

3.6.7   Qiskit Machine Learning Module

Qiskit Machine Learning module is dedicated to integrating quantum computing with classical machine learning techniques. It aims to develop and implement quantum-enhanced machine learning algorithms that can potentially surpass classical methods in specific tasks.

Figure 3.6 graphically depicts Qiskit's Machine Learning module functionalities. Table 3.9 summarizes the key features of the QML module.

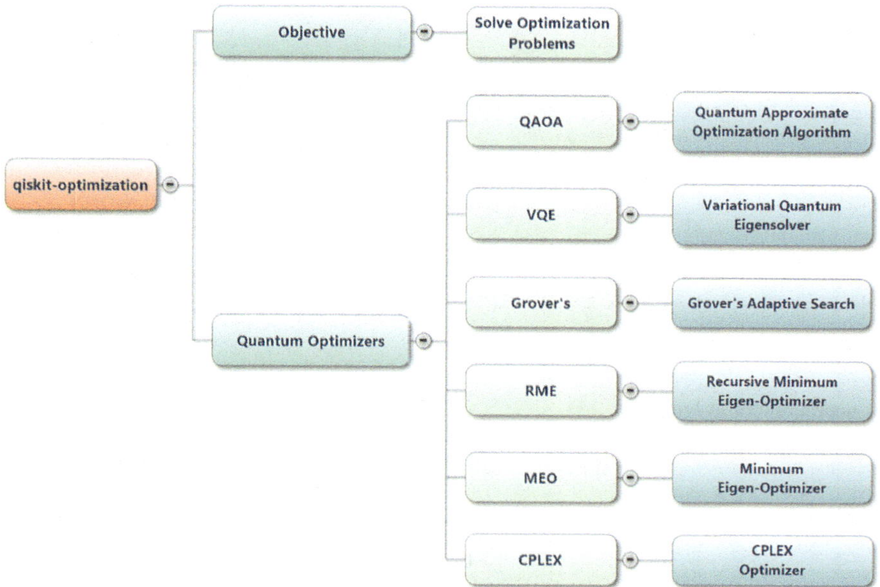

Fig. 3.5 Qiskit Optimization sub-modules (by the authors)

Table 3.8 Applications that would benefit from using Qiskit's Optimization module

Optimization practical application	Description
Logistics and supply chain management	Optimizing routing and scheduling to minimize costs and improve efficiency in transportation and distribution networks
Finance	Solving portfolio optimization problems to maximize returns while minimizing risk, taking into account various financial constraints and objectives
Manufacturing	Enhancing production processes by optimizing resource allocation, inventory management, and workflow scheduling

3.6.8   Qiskit ML: Use Case

A researcher aiming to improve image recognition systems may employ QSVM to classify images more accurately by exploiting quantum parallelism, potentially achieving better performance with fewer computational resources compared to classical SVMs. Table 3.10 summarizes the practical applications of the Machine Learning module.

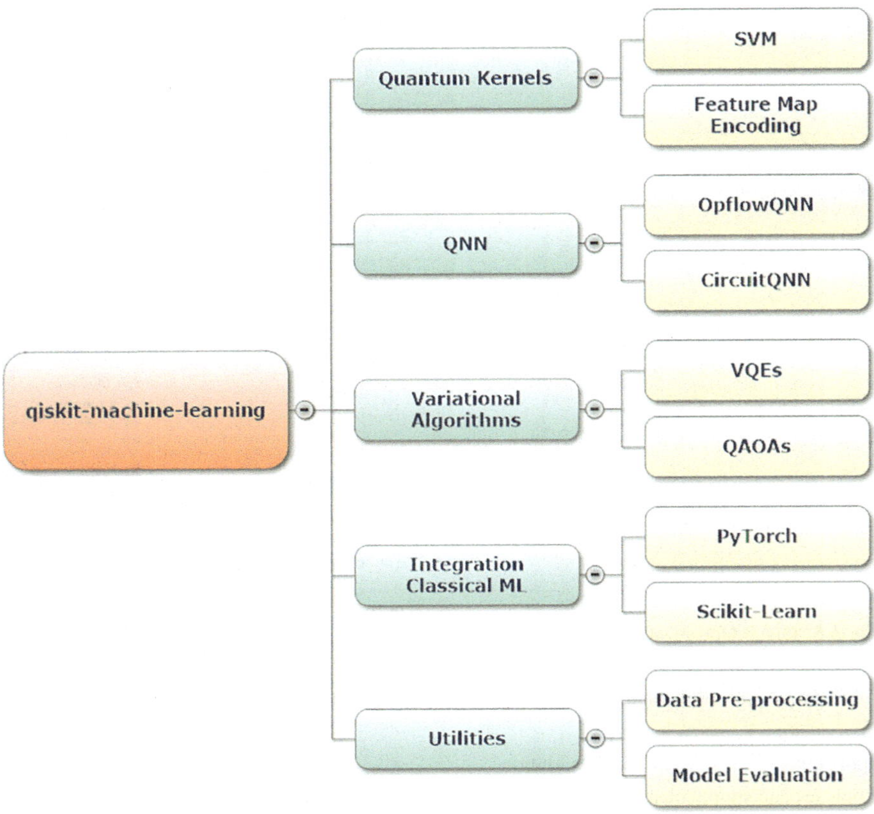

Fig. 3.6 Qiskit Machine Learning sub-module (by the authors)

3.6.9   Building a Quantum Circuit Example

The following example of a simple quantum circuit creates a Bell state. Note that the explanations on how to build quantum circuits will be conducted in future chapters.

Python Code Example: 2-Qubit Entanglement Circuit
This circuit creates entanglement between the two qubits and measures their states. The circuit generated by the code is presented in Fig. 3.7 and an example histogram of the frequency of the qubits in Fig. 3.8.

```
#-----------------------------------------------------------------
# 2-qubit entanglement circuit
# Chapter 3 in the QUANTUM COMPUTING AND QUANTUM MACHINE LEARNING BOOK
#-----------------------------------------------------------------
# Version 1.0
# Qiskit changes frequently.
```

Table 3.9 Qiskit Machine Learning module

Key QML feature	Functionality
Algorithms for machine learning	Provides implementations of quantum algorithms such as quantum support vector machines (QSVM), quantum neural networks (QNN), and quantum principal component analysis (QPCA). These algorithms leverage quantum properties like superposition and entanglement to perform computations that may offer speedups over their classical counterparts
Hybrid quantum-classical models	Facilitates the creation of hybrid models that combine quantum circuits with classical neural networks. This integration allows for the development of algorithms that can harness the strengths of both quantum and classical computing
Data encoding techniques	Offers various methods for encoding classical data into quantum states, a crucial step for quantum machine learning. Techniques include amplitude encoding, angle encoding, and basis encoding, each suited to different types of data and applications
Optimization tools	Includes tools for optimizing quantum circuits used in machine learning models, ensuring efficient use of quantum resources and minimizing errors during computation

Table 3.10 Qiskit Quantum Machine Learning module

QML module applications	Description
Classification and regression	Developing quantum classifiers and regression models that can handle high-dimensional data more efficiently
Clustering	Implementing quantum clustering algorithms to identify patterns and group similar data points in large datasets
Dimensionality reduction	Utilizing quantum algorithms to reduce the dimensionality of data, making it more manageable for analysis and visualization

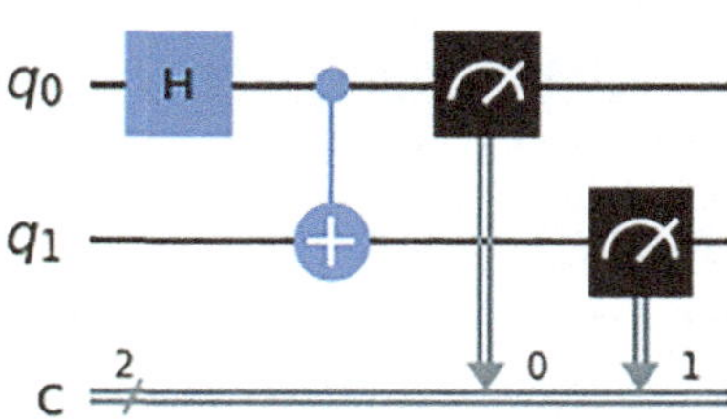

Fig. 3.7 Two-qubit entanglement circuit (by the authors)

```
# We recommend using the latest version from the book code repository at:
# https://aqtinitiative.org/quantum-computing-for-engineers

# (c) 2025 Jesse Van Griensven, Roydon Fraser, and Jose Rosas
# License: MIT - Citation required
#---------------------------------------------------------------------
import numpy as np
import matplotlib.pyplot as plt
```

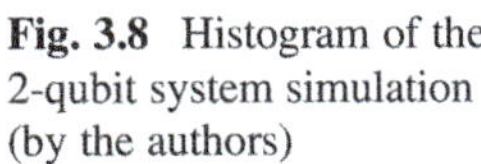

Fig. 3.8 Histogram of the 2-qubit system simulation (by the authors)

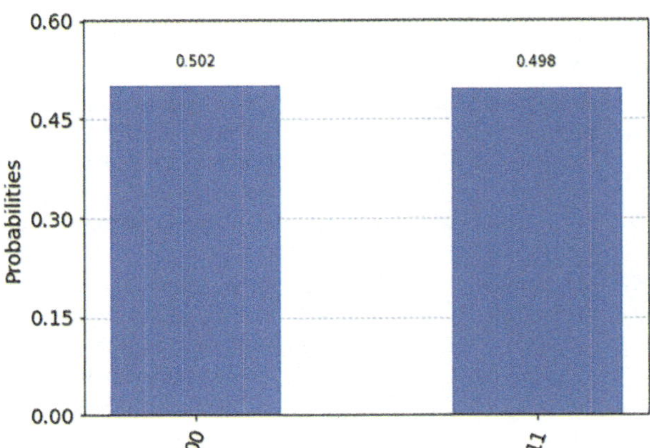

```python
import warnings
warnings.filterwarnings('ignore')

#-----------------------------------------------------------------
from qiskit import QuantumCircuit  as QC
from qiskit import QuantumRegister as QR

from qiskit import ClassicalRegister as CR

from qiskit import Aer
from qiskit import execute

from qiskit.visualization import plot_histogram
from qiskit.visualization import circuit_drawer
from qiskit.visualization import plot_state_qsphere
from qiskit.visualization import plot_bloch_multivector

#-----------------------------------------------------------------
def get_state_vector(qc):
  """ Execute the circuit and get the state vector """
  backend = Aer.get_backend('statevector_simulator')
  result  = execute(qc, backend).result()
  state_vector = result.get_statevector()

  return state_vector

#-----------------------------------------------------------------
def display_bloch(qc):
  """ Plots Bloch Sphere for the quantum circuit qc """
  # by Jesse Thé - Jan 2024
```

```python
    # Execute the circuit and get the state vector
    state_vector = get_state_vector(qc)

    # Draw the circuit
    display(circuit_drawer(qc, output='mpl', style="iqp"))

    # Plot Bloch sphere for the entire system
    display(plot_bloch_multivector(state_vector) )

    return

#------------------------------------------------------------------

# Create a quantum circuit with 2 qubits and 2 classical bits
qc = QC(2, 2)

# Apply a Hadamard gate on the first qubit
qc.h(0)

# Apply a CNOT gate with qubit 0 as control and qubit 1 as target
qc.cx(0, 1)

# Measure the qubits
qc.measure([0, 1], [0, 1])

# Draw the Circuit and the Bloch Sphere
display_bloch(qc)

# Simulate the circuit
simulator = Aer.get_backend('qasm_simulator')
result = execute(qc, simulator).result()

# Now print the measurements for qubit "00" and qubit "11"
# Note that the simulator ran the noperations 1024 times.
# Therefore, the probability is the count / 1024
print()
print("Measurement Results:", result.get_counts())
print()

# Now plot the probabilities for qubit "00" and qubit "11"
plot_histogram(result.get_counts())
```

3.6.10   *Advanced Applications: Quantum Machine Learning*

Using Qiskit's modules, engineers can build quantum machine learning models, such as variational classifiers.

Python Code: Quantum Classifier

```python
#------------------------------------------------------------------------
# Quantum Classifier circuit
# Chapter 3 in the QUANTUM COMPUTING AND QUANTUM MACHINE LEARNING BOOK
# Code demonstrates the use of Qiskit for quantum-enhanced machine
learning.
#------------------------------------------------------------------------
# Version 1.0
# (c) 2025 Jesse Van Griensven, Roydon Fraser, and Jose Rosas
# Licence: MIT - Citation required
#------------------------------------------------------------------------
import numpy as np
import matplotlib.pyplot as plt
import warnings
warnings.filterwarnings('ignore')


#------------------------------------------------------------------------
from qiskit import QuantumCircuit  as QC
from qiskit import QuantumRegister as QR

from qiskit import ClassicalRegister as CR

from qiskit import Aer
from qiskit import execute

from qiskit.visualization import plot_histogram
from qiskit.visualization import circuit_drawer
from qiskit.visualization import plot_state_qsphere
from qiskit.visualization import plot_bloch_multivector


#------------------------------------------------------------------------
def get_state_vector(qc):
    """ Execute the circuit and get the state vector """
    backend = Aer.get_backend('statevector_simulator')
    result  = execute(qc, backend).result()
    state_vector = result.get_statevector()

    return state_vector
```

```python
#-----------------------------------------------------------------------
def display_bloch(qc):
    """ Plots Bloch Sphere for the quantum circuit qc """
    # by Jesse Thé - Jan 2024

    # Execute the circuit and get the state vector
    state_vector = get_state_vector(qc)

    # Draw the circuit
    display(circuit_drawer(qc, output='mpl', style="iqp"))

    # Plot Bloch sphere for the entire system
    display(plot_bloch_multivector(state_vector) )

    return

#-----------------------------------------------------------------------

# Create a quantum circuit with 2 qubits and 2 classical bits
qc = QC(2, 2)

# Apply a Hadamard gate on the first qubit
qc.h(0)

# Apply a CNOT gate with qubit 0 as control and qubit 1 as target
qc.cx(0, 1)

# Measure the qubits
qc.measure([0, 1], [0, 1])

# Draw the Circuit and the Bloch Sphere
display_bloch(qc)

# Simulate the circuit
simulator = Aer.get_backend('qasm_simulator')
result = execute(qc, simulator).result()

print()
print("Measurement Results:", result.get_counts())
```

3.7 Google's Cirq

Cirq, developed by Google, is a Python-based framework designed specifically for noisy intermediate-scale quantum (NISQ) devices. Its focus on gate-level programming and hardware-specific optimizations makes it ideal for engineers working with near-term quantum hardware.

Cirq is a Python-based quantum computing framework developed by Google. Cirq's purpose is for noisy intermediate-scale quantum (NISQ) devices. These devices represent the current stage of quantum hardware, characterized by limited qubits and susceptibility to noise. Cirq distinguishes itself by providing an efficient platform tailored to the needs of near-term quantum computing applications. Its focus on gate-level programming and hardware-specific optimizations makes it particularly appealing to engineers and researchers striving to harness the potential of NISQ devices.

Cirq is designed to bridge the gap between theoretical quantum computing and practical applications, offering a suite of tools that allow developers to design, simulate, and execute quantum circuits with exceptional control. By leveraging its robust feature set, Cirq enables experimentation with quantum algorithms and facilitates advancements in various fields, from optimization problems to quantum machine learning.

3.7.1 Cirq Features

Cirq's features are summarized in Table 3.11 and further discussed in the following subsections.

3.7.1.1 Gate-Level Control

One of the standout features of Cirq is its emphasis on gate-level control. This allows developers to directly manipulate individual quantum gates within a circuit, providing an unparalleled level of precision in circuit design. This granularity is essential for researchers seeking to optimize their quantum algorithms, as it enables them to fine-tune operations and reduce unnecessary gate overhead.

Table 3.11 Google's Cirq features

Cirq feature	Description
Gate-level control	Direct manipulation of quantum gates for precise circuit design
Hardware optimization	Tailored for execution on Google's quantum processors
Integration with classical systems	Allows seamless combination of classical and quantum computations

Gate-level control also serves as a powerful educational tool. It enables users to gain a deeper understanding of how quantum circuits operate at a fundamental level. This feature is especially beneficial for those new to quantum programming, as it fosters an appreciation for the complexities of quantum computation while offering hands-on experience.

3.7.1.2 Hardware Optimization

Cirq is tailored for execution on Google's quantum processors, such as the Sycamore processor, which has achieved significant milestones in quantum computing. The framework is optimized to account for the specific characteristics of Google's hardware, including gate fidelity, qubit connectivity, and noise profiles.

This hardware-specific focus ensures that circuits designed in Cirq are highly efficient and well-suited for real-world quantum experiments. Researchers can simulate their circuits using Cirq's tools and then execute them on actual quantum devices with minimal adjustments. The integration of hardware-aware features allows for a seamless transition from simulation to implementation, maximizing the utility of available quantum resources. Moreover, Cirq's compatibility with Google's quantum processors positions it as a leading platform for exploring the capabilities of NISQ devices. This optimization not only enhances performance but also accelerates the development of practical quantum applications.

3.7.1.3 Cirq Simulator

Cirq's simulator is designed for gate-level circuits, focusing on efficient simulation of NISQ-era devices. It supports the simulation of large quantum circuits and integration with Google's hardware.

Python Code Example: Cirq Simulation
This code demonstrates how to simulate a Bell state creation and measurement using Cirq.

```python
import cirq

# Define qubits
q0, q1 = cirq.LineQubit.range(2)

# Create a quantum circuit
circuit = cirq.Circuit(
    cirq.H(q0),  # Apply Hadamard gate
    cirq.CNOT(q0, q1),  # Apply CNOT gate
    cirq.measure(q0, key='m0'),
```

```
  cirq.measure(q1, key='m1')
)
```

```
# Simulate the circuit
simulator = cirq.Simulator()
result = simulator.run(circuit, repetitions=1000)
print("Cirq Simulation Results:", result)
```

3.7.2 Cirq Integration with Classical Systems

Quantum computing often works in tandem with classical systems to solve complex problems. Cirq facilitates this integration by providing tools for seamless communication between classical and quantum components. This hybrid approach is crucial for many applications, as classical systems are used to preprocess data, control quantum operations, and analyze results.

For example, a classical machine learning model can preprocess data before passing it to a quantum algorithm implemented in Cirq. After the quantum computation, the results can be post-processed using classical tools. This synergy between classical and quantum computing enables researchers to leverage the strengths of both paradigms.

Integration with classical systems also simplifies the development of workflows where quantum computing is just one part of a larger computational pipeline. By allowing classical and quantum components to interact smoothly, Cirq ensures that developers can focus on the problem at hand rather than the intricacies of system compatibility.

Python Code Example: Cirq

```python
import cirq

# Define two qubits
qubit_0 = cirq.GridQubit(0, 0)
qubit_1 = cirq.GridQubit(0, 1)

# Create a quantum circuit
circuit = cirq.Circuit(
  cirq.H(qubit_0), # Apply Hadamard gate on qubit 0
  cirq.CNOT(qubit_0, qubit_1), # Apply CNOT with qubit 0 as control
  cirq.measure(qubit_0, key='m0'),
  cirq.measure(qubit_1, key='m1')
)

# Simulate the circuit
simulator = cirq.Simulator()
```

```
result = simulator.run(circuit, repetitions=10)
print("Measurement Results:")
print(result)
```

Cirq provides low-level control for engineers to customize quantum circuits for specific tasks.

Cirq Error Mitigation Applications

Cirq can implement error mitigation techniques to improve the reliability of NISQ devices. Engineers can design circuits that measure and compensate for noise-induced errors, enhancing the fidelity of quantum computations.

3.8 Microsoft's Q#

Q#, developed by Microsoft, is a quantum programming language tightly integrated with the Quantum Development Kit (QDK). It is designed for high-level abstraction and seamless debugging in a familiar development environment like Visual Studio.

Since the authors are uncertain about the future of Q#, we will only summarize key features and provide a simple code example.

3.8.1 Q# Features

Table 3.12 summarizes the key Q# features.

Quantum Teleportation in Q# Code

```
operation TeleportationProtocol() : Unit {
  using (qubits = Qubit[3]) {
    // Prepare an entangled pair
    H(qubits[1]);
    CNOT(qubits[1], qubits[2]);

    // Prepare the message state
    H(qubits[0]);
```

Table 3.12 Q# main features

Q# feature	Description
Abstractions for quantum operators	Simplifies the development of complex quantum algorithms
Interoperability	Allows integration with classical programming languages like python and C#
Debugging and simulation	Provides tools for debugging quantum programs and simulating their behavior on classical hardware

```
    // Teleportation operations
    CNOT(qubits[0], qubits[1]);
    H(qubits[0]);
    let (m0, m1) = (M(qubits[0]), M(qubits[1]));

    // Apply corrections based on measurements
    if (m1 == One) { Z(qubits[2]); }
    if (m0 == One) { X(qubits[2]); }

    // Verify the state of the teleported qubit
    let finalState = M(qubits[2]);
    Message($"Teleported qubit state: {finalState}");
    }
}
```

This code demonstrates how Q# simplifies the implementation of complex quantum protocols like teleportation.

3.9   Quantum Simulators and Cloud-Based Platforms

Quantum simulators and cloud-based platforms have revolutionized access to quantum computing resources, enabling researchers, engineers, and students to explore quantum systems without the need for dedicated hardware. Simulators emulate quantum behavior on classical hardware, offering a cost-effective and flexible way to test algorithms. Cloud platforms provide scalable, on-demand access to actual quantum devices and high-performance simulators.

Simulators such as Qiskit Aer, Cirq, and QuTiP enable cost-effective testing and development of quantum algorithms. Cloud platforms like IBM Quantum Platform, Google Quantum AI, and Amazon Braket provide access to cutting-edge quantum hardware, democratizing the field and accelerating innovation.

3.10   Quantum Simulators

Quantum simulators replicate the behavior of quantum systems using classical computation. They are indispensable for:

(a) Testing quantum algorithms under noise-free and noisy conditions.
(b) Understanding the impact of errors and optimizing circuits.
(c) Exploring quantum phenomena without requiring physical quantum devices.

QuTiP (Quantum Toolbox in Python)

Qiskit's Aer module is IBM's quantum simulator. The quantum simulation example below will employ another tool, QuTip. QuTiP specializes in simulating open quantum systems, making it ideal for studying decoherence and system–environment interactions.

Python Code: QuTiP Simulation

This example uses QuTiP to simulate the dynamics of a single qubit under a Hamiltonian. Results obtained from this code are displayed in Fig. 3.9.

```
#-------------------------------------------------------------------
# Simple Qutip run
# Chapter 3 in the QUANTUM COMPUTING AND QUANTUM MACHINE LEARNING BOOK
#-------------------------------------------------------------------
# Version 1.0
#-------------------------------------------------------------------
# Simple Qutip Run
# Chapter 3 in the QUANTUM COMPUTING AND QUANTUM MACHINE LEARNING BOOK
#-------------------------------------------------------------------
# Version 1.0
# (c) 2025 Jesse Van Griensven, Roydon Fraser, and Jose Rosas
```

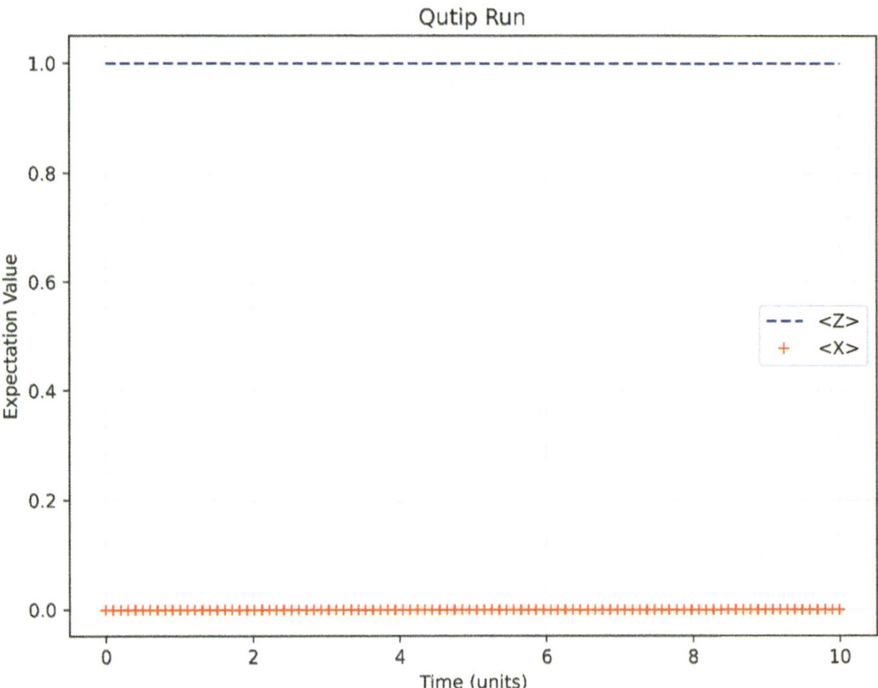

Fig. 3.9 Qutip run results (by the authors)

```python
# Licence: MIT - Citation of this work required
#------------------------------------------------------------------

from qutip import *  # Import QuTiP, a Python library for simulating
quantum systems
import numpy as np  # Import NumPy for numerical computations
import matplotlib.pyplot as plt  # Import Matplotlib for plotting
#------------------------------------------------------------------

# Define a qubit Hamiltonian
# The Hamiltonian defines the energy of the system. Here, it is 0.5 * sigmaz
().
# sigmaz() is the Pauli-Z operator, which corresponds to a measurement
along the Z-axis on the Bloch sphere.
# Physically, this describes a system with energy levels separated by 0.5
units, oriented along the Z-axis.
H = 0.5 * sigmaz()

# Define the initial state of the qubit
# basis(2, 0) represents the ground state |0> in a two-dimensional
Hilbert space.
# The qubit is initialized in the |0> state (aligned with the Z-axis).
psi0 = basis(2, 0)

# Define the time evolution parameters
# np.linspace(0, 10, 100) generates 100 evenly spaced time points
between 0 and 10.
times = np.linspace(0, 10, 100)

# Simulate the Schrödinger equation for time evolution
# sesolve is the Schrödinger equation solver in QuTiP.
# - H: The Hamiltonian (governs the system's dynamics).
# - psi0: The initial state of the system.
# - times: The time points at which to solve the Schrödinger equation.
# - [sigmaz(), sigmax()]: Expectation operators to compute observables
during evolution.
# - sigmaz(): Pauli-Z operator measures the Z-component of the qubit's
state (aligned or anti-aligned with Z-axis).
# - sigmax(): Pauli-X operator measures the X-component of the qubit's
state (probing coherence in the X-basis).
result = sesolve(H, psi0, times, [sigmaz(), sigmax()])

# Plot the results
```

```python
# Set up the plot
plt.figure(figsize=(8, 6), dpi=600)
plt.title("Qutip Run")

# Plot the expectation value of <Z>
plt.plot(times, result.expect[0], "b--", label='<Z>') # Expectation
value of sigmaz()

# Plot the expectation value of <X>
plt.plot(times, result.expect[1], "r+", label='<X>') # Expectation
value of sigmax()

# Add labels for the axes
plt.xlabel("Time (units)") # Time axis
plt.ylabel("Expectation Value") # Expectation value axis

# Add gridlines for readability
plt.grid(alpha=0.2)

# Add a legend to differentiate <Z> and <X> plots
plt.legend()

# Show the plot
plt.show()
```

3.11   Cloud-Based Platforms

Cloud-based platforms have democratized quantum computing by providing access
to quantum processors and simulators over the internet. These platforms cater to
researchers and engineers who need access to state-of-the-art quantum hardware and
large-scale computational resources.

3.11.1   IBM Quantum Experience

IBM Quantum Experience was IBM's original cloud-based interface for accessing
quantum computing resources. It has since been rebranded as the IBM Quantum
Platform, encompassing IBM Quantum Composer, IBM Quantum Lab, and Qiskit
Runtime. The platform provides both free and subscription-based access to real
quantum processors and simulators. IBM Platform key features are summarized in
Table 3.13.

Table 3.13 IBM quantum simulators and quantum processors

IBM quantum platform	Description
Simulators	Access to ideal and noisy quantum simulators via Qiskit Aer, supporting circuit testing and error modeling
Quantum hardware	Cloud-based access to superconducting quantum processors, such as *ibmq_quito* and the 127-qubit **eagle** processor
Developer tools	Qiskit SDK, quantum composer, and Jupyter-based quantum lab for circuit design, simulation, and execution

The IBM Quantum Platform is a pioneering environment that democratizes access to quantum computing by providing researchers, developers, and students with tools to explore, prototype, and deploy quantum algorithms. It supports both free and premium access to cloud-based simulators and physical quantum processors, making it a valuable resource for individuals and institutions developing, testing, and validating quantum applications.

At the heart of the IBM Quantum Platform is the seamless integration of cutting-edge quantum hardware, cloud-based tools, and a user-friendly interface. The platform has become a cornerstone for researchers, educators, and developers seeking to engage with and contribute to the rapidly evolving field of quantum computing. By providing streamlined access to both simulators and real quantum processors, the IBM Quantum Platform effectively bridges the gap between theoretical models and practical implementation.

3.11.2 Simulators in the IBM Quantum Platform

One of the most valuable features of the IBM Quantum Platform is its access to high-performance quantum simulators. These simulators, including the Qiskit Aer library, enable users to test, refine, and optimize quantum algorithms without needing immediate access to physical quantum processors. They are particularly effective for exploring algorithmic behavior, performing circuit-level debugging, and understanding quantum circuit dynamics in a controlled environment.

IBM's cloud-based noisy simulators are especially noteworthy. These simulators model the behavior of real quantum devices by incorporating realistic noise sources, such as gate errors, decoherence, and readout errors. By simulating these imperfections, users can assess how their algorithms will perform on actual hardware, making these tools essential for developing noise-resilient quantum circuits and implementing error mitigation techniques.

Simulators also serve a critical role in education and training. They allow students, researchers, and developers to experiment with quantum computing concepts in a reproducible, low-risk environment. This accessibility supports learning, fosters innovation, and enables broader participation in quantum research without the need for specialized or expensive hardware.

3.11.3 Quantum Hardware in the IBM Quantum Platform

The IBM Quantum Platform provides unparalleled access to a wide range of superconducting quantum processors, from entry-level devices to state-of-the-art systems. Users can interact with real quantum hardware, such as the 5-qubit ibmq_quito processor (IBM Q Vigo replacement), which is ideal for educational purposes and early experimentation, to advanced devices like the 127-qubit Eagle processor, which marks a significant milestone in scalable quantum hardware development.

The platform's flexibility supports users across the spectrum of expertise. Beginners can gain hands-on experience with smaller systems to build foundational knowledge, while experienced researchers can explore more sophisticated algorithms using high-qubit-count processors. This versatility makes the IBM Quantum Platform a valuable resource for both academic and industrial users.

Access to physical quantum processors also accelerates progress in fields, such as quantum optimization, post-quantum cryptography, quantum chemistry, and materials science. By enabling the execution of quantum algorithms on real hardware, the IBM Quantum Platform transforms theoretical models into executable applications. This capability not only advances quantum technology but also fosters meaningful collaboration across scientific and engineering disciplines.

3.11.4 Developer Tools in the IBM Quantum Experience

The IBM Quantum Platform provides a comprehensive suite of developer tools that streamline the process of creating, testing, and deploying quantum algorithms. At the core of these tools is Qiskit, an open-source quantum computing framework that integrates seamlessly with Python, the leading language for scientific computing and machine learning.

Qiskit's modular architecture allows developers to design quantum circuits programmatically, enabling the construction of complex algorithms. Key components include Qiskit Terra for circuit creation and optimization and Qiskit Aer for high-performance simulation. This modularity provides the flexibility required to address a wide range of quantum computing challenges.

The platform's native integration with Python enhances its usability, allowing developers to combine quantum routines with classical programming paradigms. For example, quantum subroutines can be embedded in classical workflows to support hybrid algorithms—a key capability in variational quantum computing and quantum machine learning. This synergy between classical and quantum code execution is a hallmark of the IBM Quantum Platform, making it a practical resource for real-world applications.

In addition to Qiskit, the IBM Quantum Platform offers extensive documentation, interactive tutorials, and a globally active community of users. These resources empower developers to learn, share insights, and collaborate across research and industry. The combination of accessible tools, educational materials, and community support makes the IBM Quantum Platform an ideal environment for building and advancing quantum development skills.

Python Code Example: Cloud Execution on IBM Quantum

```python
#-------------------------------------------------------------------
# Cloud Execution on IBM Quantum
# Chapter 3 in the QUANTUM COMPUTING AND QUANTUM MACHINE LEARNING BOOK
#-------------------------------------------------------------------
# Version 1.0
# Qiskit changes frequently.
# We recommend using the latest version from the book code repository at:
# https://aqtinitiative.org/quantum-computing-for-engineers

# (c) 2025 Jesse Van Griensven, Roydon Fraser, and Jose Rosas
# License: MIT - Citation required
#-------------------------------------------------------------------
import numpy as np
import matplotlib.pyplot as plt
import warnings
warnings.filterwarnings('ignore')


#-------------------------------------------------------------------
from qiskit import IBMQ, QuantumCircuit, execute

# Load IBMQ account
IBMQ.load_account()
provider = IBMQ.get_provider(hub='ibm-q')

# Select a quantum backend
backend = provider.get_backend('ibmq_quito')

# Define a quantum circuit
qc = QuantumCircuit(2, 2)
qc.h(0)
qc.cx(0, 1)
qc.measure([0, 1], [0, 1])

# Execute on real quantum hardware
job  = execute(qc, backend, shots=1024)
result = job.result()
print("Hardware Results:", result.get_counts())
```

This code demonstrates running a quantum circuit on IBM's cloud-based quantum processor.

3.11.5 Google Quantum AI Cloud

Google's Quantum AI platform provides access to Sycamore, a state-of-the-art quantum processor. It supports gate-level programming and integration with Cirq for algorithm development and testing.

3.11.6 Amazon Braket

Amazon Braket is a fully managed quantum computing service by AWS that provides access to quantum hardware from multiple providers, including IonQ, Rigetti, QuEra, and Oxford Quantum Circuits (OQC). It offers flexibility for researchers and developers to choose the most suitable quantum devices for their specific workloads.

A key feature of Amazon Braket is its cross-vendor compatibility, enabling access to diverse quantum technologies, such as gate-based, ion-trap, and neutral atom devices. Although direct access to D-Wave hardware via Braket was discontinued in late 2022, users can still interface with D-Wave's Leap™ system through the AWS Marketplace.

Amazon Braket supports hybrid quantum-classical workflows through its Braket Hybrid Jobs feature, making it well-suited for solving optimization and machine learning tasks. This integration of classical and quantum computing resources enables users to experiment with real hardware, simulate quantum circuits, and benchmark performance across technologies, all within a unified development environment.

3.12 Superconducting Qubits

Superconducting qubits are among the most advanced and widely researched technologies for building quantum computers. Leveraging the principles of superconductivity and quantum mechanics, they offer a highly controllable and scalable platform for implementing qubits. These qubits rely on superconducting circuits that use the quantum properties of Josephson junctions to encode and manipulate quantum states.

Superconducting qubits are the foundation of many leading quantum computing efforts, including those by IBM, Google, and Rigetti. Their relatively fast gate speeds and compatibility with established microfabrication techniques make them a frontrunner in the race to build practical quantum computers. Superconducting qubits operate at cryogenic temperatures, typically below 10 millikelvin, achieved using dilution refrigerators. At these extremely low temperatures, superconductivity emerges, allowing electrons to flow through materials without resistance.

3.12.1 The Role of the Josephson Junction

The Josephson junction is the cornerstone of superconducting qubits. It consists of a thin insulating layer sandwiched between two superconducting materials. While the insulating layer blocks the flow of normal electrical current, it permits the tunneling of superconducting electron pairs (cooper pairs). This quantum tunneling introduces a nonlinear inductance into the circuit, which is critical for creating and manipulating quantum states.

This nonlinearity breaks the harmonic nature of the potential energy well, allowing the system to have discrete energy levels suitable for defining quantum states. The two lowest energy levels are used to encode the qubit states, $|0\rangle$ and $|1\rangle$.

3.12.2 Mathematical Representation

The behavior of a superconducting qubit is described by its Hamiltonian, which accounts for both the kinetic and potential energy of the system. The Hamiltonian for a superconducting qubit can be expressed as:

$$H = \frac{p^2}{2m} + \frac{1}{2}m\omega^2 x^2 + \frac{\beta}{4}x^4$$

where

- p is the momentum operator.
- x is the position operator.
- m is the effective mass.
- ω is the natural frequency of the circuit.
- β accounts for the nonlinearity introduced by the Josephson junction.

The term $\frac{1}{2}m\omega^2 x^2$ represents a harmonic oscillator, while the $\frac{\beta}{4}x^4$ term introduces nonlinearity. This nonlinearity allows the energy levels to become anharmonic, ensuring that the $|0\rangle$ and $|1\rangle$ states are distinguishable and can be manipulated independently.

3.12.3 Qubit Initialization and Control

To initialize and control qubits, one requires the following:

1. **Initialization**

   Superconducting qubits are initialized into their ground state $|0\rangle$ by cooling the system to cryogenic temperatures. Thermal energy at such low temperatures is insufficient to excite the qubit to higher energy states.

2. **Control**

   Microwave pulses are applied to the circuit to transition the qubit between the $|0\rangle$ and $|1\rangle$ states or create superposition states. These pulses are tuned to the resonance frequency of the qubit, which is determined by the spacing between its energy levels.

3.12.4 Key Features of Superconducting Qubits

The following is a list of the key features of superconducting qubits:

1. **Fast Gate Speeds**

   Superconducting qubits enable gate operations on the timescale of nanoseconds. This speed is critical for performing quantum computations before decoherence effects degrade the quantum states.

2. **Scalability**

   The scalability of superconducting qubits is a major advantage. Fabrication techniques derived from the semiconductor industry can be adapted to produce large arrays of qubits. Techniques like photolithography and etching are used to create superconducting circuits on silicon or sapphire substrates.

3. **High Controllability**

   Superconducting qubits offer precise control over quantum states. This is achieved using carefully designed microwave pulses and advanced cryogenic infrastructure.

4. **Integration with Classical Systems**

   Superconducting qubits are compatible with classical control systems, enabling seamless integration with readout electronics, error correction circuits, and quantum-classical hybrid algorithms.

5. **Error Rates and Decoherence**

   While superconducting qubits exhibit relatively low error rates, decoherence remains a challenge. Decoherence times (T1 and T2) are typically on the order of microseconds, limiting the number of operations that can be performed. Ongoing research focuses on improving coherence times through better materials and circuit designs.

3.13 Examples of Superconducting Qubit Architectures

There are dozens of different quantum computer architectures. Superconducting qubit architectures, in the early 2020s, are the dominant ones employed by IBM and others. These architectures are described in the following subsections.

3.13.1 Transmon Qubits

Transmon qubits are a widely used variant of superconducting qubits. They improve upon earlier designs by enhancing resistance to charge noise. This is achieved by shunting the Josephson junction with a large capacitance, which reduces the sensitivity of the qubit to small charge fluctuations. An image of a transmon qubit is shown in Fig. 3.10.

3.13.2 Flux Qubits

Flux qubits use the magnetic flux through a superconducting loop to encode quantum information. They are particularly useful for coupling multiple qubits (Fig. 3.11).

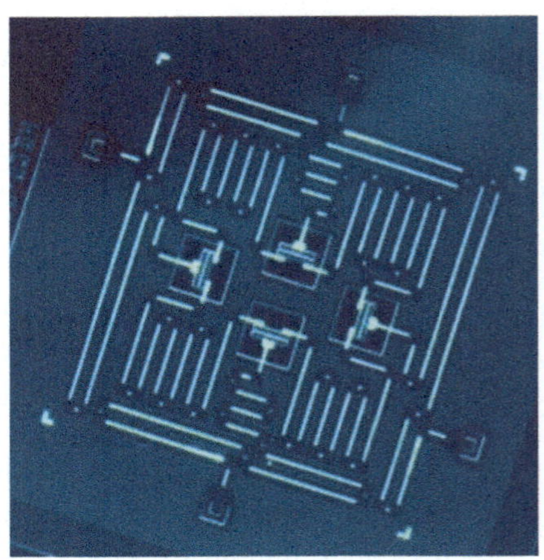

Fig. 3.10 A transmon qubit (from Wikipedia)

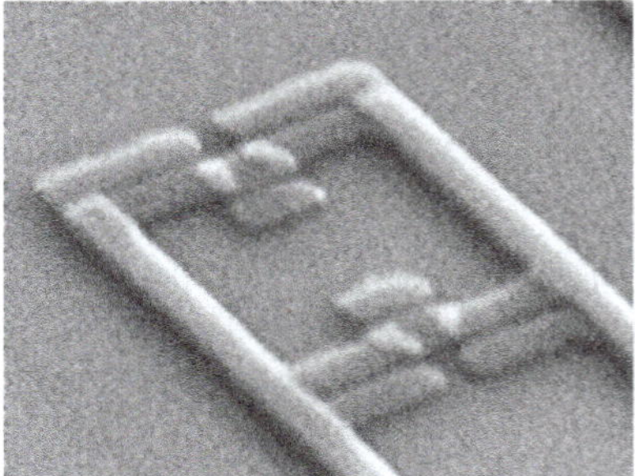

Fig. 3.11 Four-junction flux qubit by the Royal Holloway University of London (Wikipedia)

3.13.3 *Phase Qubits*

Phase qubits encode information in the phase difference across a Josephson junction. While they are less common than transmon qubits, they provide a complementary approach to qubit design.

3.13.4 *Applications of Superconducting Qubits*

Superconducting qubits form the backbone of many quantum computing platforms. Companies like IBM, Google, and Rigetti use these qubits to perform quantum computations, from simulating quantum systems to running optimization algorithms.

3.13.4.1 Superconducting Qubits Error Correction

The scalability and precision of superconducting qubits make them ideal for implementing quantum error correction. Error correction codes like the surface code can be realized using large arrays of superconducting qubits.

3.13.4.2 Quantum Simulation

Superconducting qubits are used to simulate quantum systems that are difficult to study experimentally, such as complex molecules or high-energy physics phenomena.

3.13.4.3 Superconducting Qubits Examples

(a) **IBM Quantum Platform**
    A cloud-based platform providing access to superconducting qubits for research and development.
(b) **Google's Sycamore Processor**
    Achieved quantum supremacy by performing a computation that was infeasible for classical supercomputers.

Python Code Example
The following Python code demonstrates the creation of a quantum circuit for a superconducting qubit system using IBM's Qiskit framework. This circuit applies a

Hadamard gate to prepare the qubit in a superposition state $|0\rangle + |1\rangle$ and measures the result.

```
#----------------------------------------------------------------------
# Quantum Circuit for a Superconducting Qubit System
# Demonstrates the preparation of a qubit in a superposition state using a
# Hadamard gate.
#----------------------------------------------------------------------
# Version 1.0
# Qiskit changes frequently.
# We recommend using the latest version from the book code repository at:
# https://aqtinitiative.org/quantum-computing-for-engineers

# (c) 2025 Jesse Van Griensven, Roydon Fraser, and Jose Rosas
# License: MIT - Citation required
#----------------------------------------------------------------------

from qiskit import QuantumCircuit, Aer, execute
from qiskit.visualization import plot_histogram
from qiskit.visualization import circuit_drawer

import numpy as np
import matplotlib.pyplot as plt
import warnings
warnings.filterwarnings('ignore')
#----------------------------------------------------------------------

# Step 1: Create a Quantum Circuit with 1 Qubit and 1 Classical Bit
qc = QuantumCircuit(1, 1)  # 1 qubit, 1 classical bit

# Step 2: Apply the Hadamard Gate to Create Superposition
qc.h(0)  # Applies a Hadamard gate to qubit 0, creating (|0⟩ + |1⟩)/√2

# Step 3: Measure the Qubit
qc.measure(0, 0)  # Measure the qubit and store the result in classical bit 0

# Step 4: Simulate the Circuit
simulator = Aer.get_backend('qasm_simulator')  # Use the QASM simulator
result  = execute(qc, simulator, shots=1024).result()  # Execute the circuit with 1024 shots
counts  = result.get_counts()  # Get the measurement results
```

```
# Step 5: Display the Results
print("Measurement Results:", counts)  # Print the measurement results

# Plot the results as a histogram
plot_histogram(counts, title="Superposition Measurement Results")
plt.show()

# Draw the quantum circuit
display(circuit_drawer(qc, output='mpl', style="iqp"))
```

3.14 Trapped Ions

Trapped ion quantum computers are a leading platform for implementing quantum computation due to their precision, reliability, and scalability. In these systems, individual ions confined within electromagnetic traps serve as qubits, with quantum operations performed using laser pulses. The technology is widely recognized for its high fidelity and long coherence times, making it a promising candidate for building fault-tolerant quantum computers.

1. **Mechanism of Trapped Ion Quantum Computing**

   Trapped ions leverage the quantum mechanical properties of isolated ions held in electromagnetic traps. These systems encode qubits in the internal electronic states of the ions, typically using two stable energy levels, represented as $|0\rangle$ and $|1\rangle$. Quantum operations, such as single-qubit rotations and multi-qubit entanglement, are achieved by applying carefully controlled laser pulses.

2. **Trapping Ions**

   The ions are confined using a combination of static and oscillating electric fields within a device called a Paul trap or a Penning trap:

   (a) **Paul Trap**

   Utilizes time-dependent electric fields to create a potential well that traps ions in a localized region.

   (b) **Penning Trap**

   Combines static electric fields with a magnetic field to confine ions.

   These traps stabilize the ions in space, creating a linear chain or lattice of qubits. The strong Coulomb repulsion between the ions ensures that they remain well-separated while allowing collective motion, which is crucial for quantum operations.

3. **Encoding Qubits**

   The qubit states $|0\rangle$ and $|1\rangle$ correspond to different electronic energy levels of the ion. Common choices for encoding include:

(a) **Hyperfine States**

Two energy levels are separated by microwave frequencies.

(b) **Optical States**

Energy levels are separated by optical frequencies, manipulated directly by lasers.

Transitions between these states are induced by laser pulses, which can also create superposition states:

$$|\psi\rangle = \alpha\,|\,0\rangle + \beta\,|\,1\rangle$$

where α and β are complex amplitudes satisfying $|\alpha|^2 + |\beta|^2 = 1$.

4. **Quantum Operations with Lasers**

Laser pulses are used to manipulate the qubit states and entangle ions:

(a) **Single-Qubit Gates.**

Achieved by driving transitions between $|0\rangle$ and $|1\rangle$ using resonant laser pulses.

(b) **Two-Qubit Gates.**

Entanglement is achieved by coupling the internal states of the ions to their collective vibrational modes, enabling controlled quantum interactions.

The interaction is described by the Hamiltonian:

$$H = \hbar\Omega\left(a^\dagger + a\right)\left(\sigma_+ + \sigma_-\right)$$

where

- $a^\dagger$ and a are the creation and annihilation operators for vibrational modes.
- σ_+ and σ_- are the spin-raising and lowering operators.
- Ω is the Rabi frequency, which controls the interaction strength.

This Hamiltonian governs how the ions interact with each other and with the applied laser fields, enabling precise control over their quantum states.

3.14.1 *Advantages of Trapped Ion Quantum Computers*

The following is a list of the key features of trapped ion systems:

3.14.1.1 High Fidelity

Trapped ion systems exhibit some of the lowest error rates in quantum gate operations. This is attributed to the high precision of laser control and the isolation of the ions from environmental noise.

(a) **Error Rates**: Single-qubit gates can achieve fidelities exceeding 99.9%, and two-qubit gates often exceed 99%.
(b) **Practical Impact**: High fidelity reduces the need for extensive error correction, making the systems more efficient for practical computations.

3.14.1.2   Long Coherence Times

The isolated nature of trapped ions ensures minimal interaction with the environment, leading to long coherence times.

(a) **Typical Coherence Times**: This can range from milliseconds to minutes, depending on the system.
(b) **Impact on Computation**: Long coherence times allow for complex quantum algorithms to be executed before decoherence significantly affects the qubits.

3.14.2   Applications of Trapped Ion Quantum Computers

Trapped ions are well-suited for a variety of quantum computing tasks due to their precision and flexibility. The following list addresses the potential for its applications.

3.14.2.1   Quantum Simulation

Trapped ions are used to simulate quantum systems that are challenging to model classically. Applications include:

(a) Modeling complex molecules.
(b) Studying quantum phase transitions in condensed matter physics.

3.14.2.2   Quantum Algorithms

The high fidelity of trapped ion gates makes them ideal for implementing algorithms such as:

(a) Quantum Fourier transform (QFT), the quantum equivalent of the Fourier transform.
(b) Shor's algorithm for integer factorization.
(c) Grover's algorithm for searching unsorted databases.

3.14.2.3 Quantum Networking

Trapped ions can be used as nodes in quantum networks, with their states entangled over long distances using photons. This enables secure quantum communication and distributed quantum computing.

3.14.3 Trapped Ion Challenges

The trapped ion technology showed initial potential, but it still has many limitations. These challenges are listed below:

3.14.3.1 Scalability

While trapped ion systems are scalable in principle, practical implementation requires significant engineering efforts. As the number of ions increases, controlling individual qubits and managing cross talk between ions becomes challenging.

3.14.3.2 Speed of Operations

The gate speeds in trapped ion systems are slower compared to other technologies, such as superconducting qubits. Efforts are underway to optimize laser control and hardware to enhance operational speed.

3.14.3.3 Integration with Classical Systems

Integrating trapped ion processors with classical control systems is critical for scaling up computations. This includes developing advanced control electronics and error correction protocols.

3.14.3.4 Hybrid Architectures

Future quantum systems may combine trapped ions with other qubit technologies, such as superconducting qubits or photonic qubits, to leverage the strengths of each approach.

Python Code Example: Entanglement
This circuit code demonstrates entanglement, where a Hadamard gate and a controlled-NOT gate create a Bell state. The results are presented in a histogram contained in Fig. 3.12 and the circuit for the simulation in Fig. 3.13.

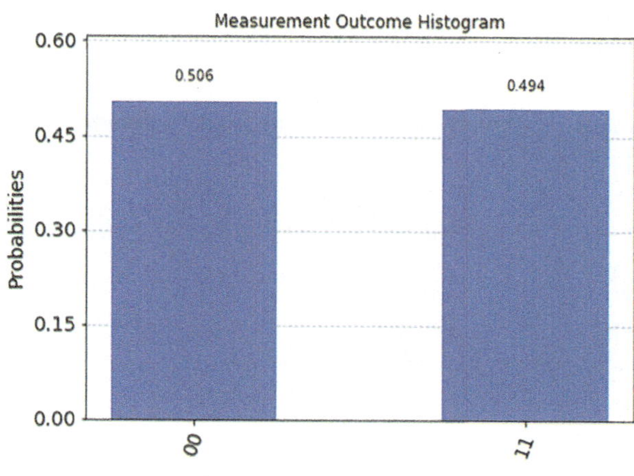

Fig. 3.12 Results of the quantum simulation of a trapped ion system (by the authors)

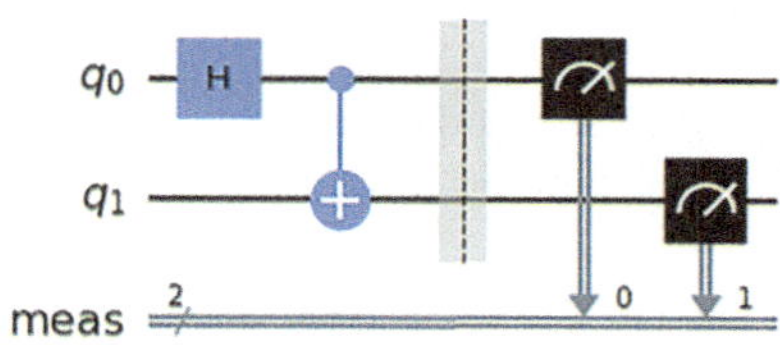

Fig. 3.13 Simulation circuit for a trapped ion system (by the authors)

```
#------------------------------------------------------------------
# Simulation of a trapped ion system
# Chapter 3 in the QUANTUM COMPUTING AND QUANTUM MACHINE LEARNING BOOK
#------------------------------------------------------------------
# Version 1.0
# Qiskit changes frequently.
# We recommend using the latest version from the book code repository at:
# https://aqtinitiative.org/quantum-computing-for-engineers

# (c) 2025 Jesse Van Griensven, Roydon Fraser, and Jose Rosas
# License: MIT - Citation of this work required
#------------------------------------------------------------------
import numpy as np
import matplotlib.pyplot as plt

from qiskit import QuantumCircuit, Aer, execute
from qiskit.visualization import circuit_drawer
from qiskit.visualization import plot_histogram
#------------------------------------------------------------------

# If you're in a Jupyter notebook, uncomment the following line:
# %matplotlib inline
```

```python
# Create a two-qubit quantum circuit
qc = QuantumCircuit(2)

# Apply a Hadamard gate to the first qubit
qc.h(0)

# Apply a controlled-NOT gate between the first and second qubit
qc.cx(0, 1)

# Measure the qubits
qc.measure_all()

# Simulate the circuit
simulator = Aer.get_backend('aer_simulator')
result = execute(qc, simulator, shots=1024).result()

# Retrieve measurement results
counts = result.get_counts()

# Print the measurement results
print("Measurement counts:", counts)

# Create the histogram figure
hist_figure = plot_histogram(counts, title="Measurement Outcome
Histogram")

# Display the plot in an interactive window or inline (in a notebook)
plt.show()

# Save the figure to a file (uncomment the line below if desired)
hist_figure.savefig("histogram.png")

# Draw the circuits
display(circuit_drawer(qc, output='mpl', style="iqp"))
```

3.15   Photonic Systems

Photonic quantum computers use light particles, or photons, as the fundamental units of quantum information (qubits). By encoding information into the photon's properties, such as polarization, phase, or path, these systems harness the unique advantages of light to process and transmit quantum information. Unlike other quantum technologies that require extreme conditions like cryogenic temperatures, photonic systems operate at room temperature, offering a practical approach to building scalable and robust quantum computers.

3.15.1 Approach to Photonic Quantum Computing

Photonic systems rely on optical circuits to manipulate and measure quantum states. These circuits consist of components like beam splitters, phase shifters, and mirrors, which enable precise control of photon behavior. The quantum states of photons are typically represented in a two-dimensional Hilbert space, where $|0\rangle$ and $|1\rangle$ denote the basis states.

3.15.1.1 Quantum State Representation

The quantum state of a single photon can be written as:

$$|\psi\rangle = \alpha |0\rangle + \beta |1\rangle.$$

where

- α and β are the complex probability amplitudes.
- $|\alpha|^2 + |\beta|^2 = 1$, ensuring the total probability is conserved.

In a polarization-based encoding:

- $|0\rangle$ might represent horizontal polarization.
- $|1\rangle$ might represent vertical polarization.

3.15.1.2 Manipulating Photon States

Photonic systems use optical components to manipulate photon states:

Beam Splitters

Split or combine light beams, enabling superposition and interference.
The output states of a beam splitter are related to the input states by:

$$|\psi_{\text{out}}\rangle = T |\psi_{\text{in}}\rangle_1 + R |\psi_{\text{in}}\rangle_2$$

where T and R are the transmission and reflection coefficients.

Phase Shifters

Adjust the phase of photons to control interference patterns. A phase shift ϕ transforms the state as:

$$|\psi_{\text{out}}\rangle = e^{i\phi}|\psi_{\text{in}}\rangle$$

Mirrors

Reflect photons, redirecting their path.

3.15.1.3   Measurement

Measurement in photonic systems is performed using photon detectors, such as avalanche photodiodes or superconducting nanowire single-photon detectors. These devices identify the presence and state of photons, collapsing the quantum state into one of its basis states ($|0\rangle$ or $|1\rangle$).

Python Code Example: Simulating Photon Manipulation

```python
#------------------------------------------------------------------------
# Simulating Photon Manipulation
# Chapter 3 in the QUANTUM COMPUTING AND QUANTUM MACHINE LEARNING BOOK
#------------------------------------------------------------------------
# Version 1.0
# Qiskit changes frequently.
# We recommend using the latest version from the book code repository at:
# https://aqtinitiative.org/quantum-computing-for-engineers

# (c) 2025 Jesse Van Griensven, Roydon Fraser, and Jose Rosas
# License: MIT - Citation of this work required
#------------------------------------------------------------------------
import numpy as np
import matplotlib.pyplot as plt

from qiskit import QuantumCircuit
from qiskit.visualization import circuit_drawer
#------------------------------------------------------------------------

# Initialize a photonic qubit in |0>
qc = QuantumCircuit(1)
qc.h(0)  # Apply Hadamard gate (similar to beam splitter effect)
qc.p(np.pi/4, 0)  # Phase shift by pi/4
```

Fig. 3.14 Quantum circuit
to simulate a photon
manipulation (by the
authors)

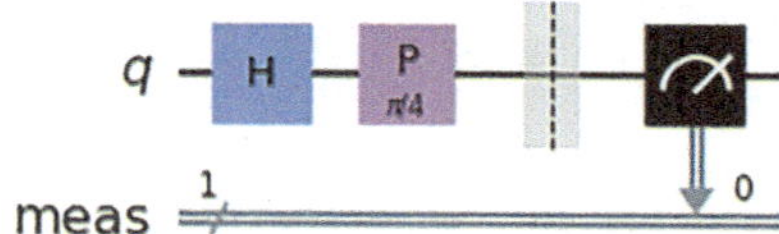

```
# Measure the state
qc.measure_all()
print(qc)


# Draw the circuits
display(circuit_drawer(qc, output='mpl', style="iqp"))
```

The circuit created in the Python code example above is shown below (Fig. 3.14).

3.15.2   Strengths of Photonic Systems

Photonic systems have many advantages over other quantum computing architectures. These are described below.

3.15.2.1   Room Temperature Operation

Unlike superconducting or trapped ion systems that require cryogenic cooling, photonic quantum computers operate at room temperature. This significantly reduces the complexity and cost of their implementation, making them an attractive option for practical deployment.

3.15.2.2   Decoherence Immunity

Photons are inherently less susceptible to environmental noise, such as thermal or magnetic fluctuations. This immunity makes them robust carriers of quantum information, particularly for tasks like quantum communication and networking.

3.15.2.3   Long-Distance Communication

Photons can travel long distances with minimal loss when transmitted through optical fibers or free space, making them ideal for quantum networking and secure communication protocols, such as quantum key distribution (QKD).

3.15.3 Challenges in Photonic Systems

Photonic quantum computers have an immense potential, which is not yet fully explored. The current challenges facing photonic systems are described below.

3.15.3.1 Scalability

Scaling up photonic quantum systems involves precise control over a large number of optical components. As the number of qubits increases, the complexity of aligning and maintaining beam splitters, phase shifters, and mirrors grows exponentially. Demonstrating scalability often involves problems like boson sampling, where the behavior of multiple photons through an optical network becomes computationally intensive.

3.15.3.2 Loss Minimization

Photon loss is a major challenge in photonic systems. It can occur due to imperfections in optical components or absorption in fibers, leading to a reduction in fidelity.

(a) Using high-quality mirrors and anti-reflective coatings.
(b) Employing advanced photon sources, such as quantum dot emitters or parametric down-conversion systems, to ensure reliable photon generation.

3.15.4 Applications of Photonic Quantum Systems

This subsection describes the main potential applications of photonic quantum systems.

3.15.4.1 Quantum Communication

Photonic systems form the backbone of quantum communication technologies, enabling the secure transfer of information through quantum key distribution (QKD). The BB84 Protocol is a widely used QKD protocol that relies on photon polarization to encode secure keys.

Python Code Example: Simulating QKD

```
# ------------------------------------------------------------------
# Quantum Basis Encoding
# Chapter 3 in the QUANTUM COMPUTING AND QUANTUM MACHINE LEARNING BOOK
# ------------------------------------------------------------------
# Version 1.0
```

```
# Qiskit changes frequently.
# We recommend using the latest version from the book code repository at:
# https://aqtinitiative.org/quantum-computing-for-engineers

# (c) 2025 Jesse Van Griensven, Roydon Fraser, and Jose Rosas
# License: MIT - Citation of this work required
#-----------------------------------------------------------------
from qiskit import QuantumCircuit
from qiskit.visualization import circuit_drawer

# BB84 Protocol Simulation
qc = QuantumCircuit(1)
qc.h(0)  # Prepare photon in superposition
qc.measure_all()
print(qc)

# Draw the circuits
display(circuit_drawer(qc, output='mpl', style="iqp"))
```

3.15.4.2   Quantum Networking

Photonic systems enable entanglement distribution across large distances, laying the groundwork for quantum internet infrastructure.

3.15.4.3   Boson Sampling

Boson sampling uses photonic systems to solve specific problems that are computationally hard for classical systems. Although not universal, boson sampling showcases quantum advantage in niche areas.

3.15.4.4   Quantum Computation

Advances in photonic systems are pushing them toward universal quantum computation. Techniques like cluster-state quantum computing use entangled photons to perform quantum gates.

3.15.5   *Photonic Systems Challenges*

Photonic quantum systems represent a promising approach to quantum computing, with their unique strengths in room temperature operation, decoherence immunity, and long-distance communication. While challenges like scalability and photon loss remain, advancements in integrated photonics and error correction techniques continue to push the boundaries of what these systems can achieve. With their potential for secure communication, networking, and computation, photonic systems are poised to play a pivotal role in the future of quantum technology.

3.15.5.1   Integrated Photonic Circuits

Developing compact photonic chips that integrate beam splitters, phase shifters, and detectors can dramatically improve scalability.

3.15.5.2   Loss-Tolerant Architectures

Research into error correction and loss mitigation will help overcome photon loss, enabling high-fidelity operations in larger systems.

3.15.5.3   Hybrid Quantum Systems

Combining photonic qubits with other qubit types, such as trapped ions or superconducting qubits, could leverage the strengths of multiple platforms for better overall performance.

Python Code Example: Phase Shift and Superposition Similar to Photonic QC
This example code simulates a phase shift and superposition analogous to photonic quantum operations. The circuit is displayed in Fig. 3.15 before the measurement and again in Fig. 3.16 after it is performed. Note the difference.

```
#-------------------------------------------------------------------------
# Simulates a phase shift and superposition
# Analogous to photonic quantum operations.
# Simulating a basic photonic system
# Chapter 3 in the QUANTUM COMPUTING AND QUANTUM MACHINE LEARNING BOOK
```

Fig. 3.15 Quantum circuit before applying measurement (by the authors)

Fig. 3.16 Quantum circuit after applying measurement (by the authors)

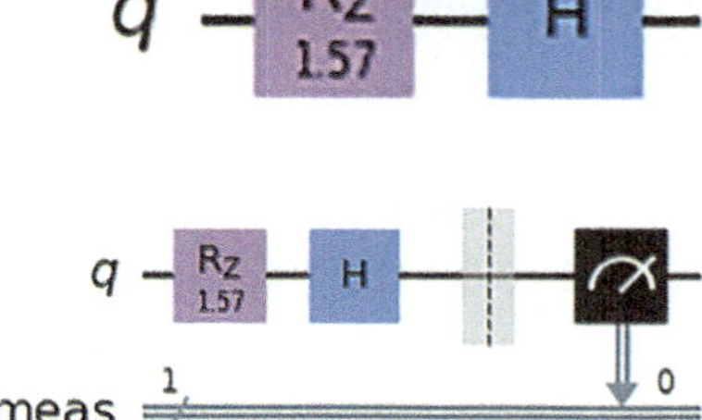

```python
#------------------------------------------------------------------
# Version 1.0
# Qiskit changes frequently.
# We recommend using the latest version from the book code repository at:
# https://aqtinitiative.org/quantum-computing-for-engineers

# (c) 2025 Jesse Van Griensven, Roydon Fraser, and Jose Rosas
# License: MIT - Citation of this work required
#------------------------------------------------------------------
from qiskit import QuantumCircuit, Aer, execute
from qiskit.visualization import circuit_drawer

import warnings
warnings.filterwarnings('ignore')
# Create a new figure with specified size and DPI
import matplotlib.pyplot as plt
plt.figure(figsize=(12,8), dpi=300)
#------------------------------------------------------------------

# Create a single-qubit quantum circuit
qc = QuantumCircuit(1)

# Apply a phase gate (simulating a photonic phase shift)
qc.rz(1.57, 0)  # 90-degree phase shift = Pi / 2 ~ 1.571

# Apply a Hadamard gate
qc.h(0)

# Draw the circuits
print("Quantum Circuit Before Applying Measurement")
display(circuit_drawer(qc, output='mpl', style="iqp"))

# Measure the qubit
qc.measure_all()

# Draw the circuits
print("Quantum Circuit After Applying Measurement")
display(circuit_drawer(qc, output='mpl', style="iqp"))

# Simulate the circuit
simulator = Aer.get_backend('aer_simulator')
result = execute(qc, simulator).result()
```

```
# Print the measurement results
print("Measurements Results:")
print(result.get_counts())
```

3.15.6   Comparison of Classical and Quantum Hardware Architecture

The architectures of classical and quantum computing systems differ fundamentally due to their underlying principles of operation. Classical computers operate based on binary logic, manipulating bits through well-established semiconductor technologies. Quantum computers, however, leverage the unique properties of quantum mechanics, such as superposition and entanglement, to process qubits. This section explores the structure, components, and comparative metrics of classical and quantum hardware architectures.

The architectural differences between classical and quantum systems highlight their respective strengths and limitations. Classical computing remains indispensable for deterministic tasks and large-scale data processing, while quantum computing offers transformative potential in solving complex, high-dimensional problems. Understanding these architectures is essential for leveraging their complementary capabilities in engineering and computational sciences.

3.15.7   Quantum Hardware Architecture

Quantum computing represents a paradigm shift from classical computing, utilizing qubits and quantum mechanical phenomena to process information.

3.15.7.1   Qubits

Unlike classical bits, qubits can exist in superposition states, simultaneously representing $|0\rangle$ and $|1\rangle$. A single qubit is described as:

$$|\psi\rangle = \alpha\,|0\rangle + \beta\,|1\rangle$$

where α and β are complex probability amplitudes, and $|\alpha|^2 + |\beta|^2 = 1$.

3.15.7.2   Quantum Registers

Quantum registers are collections of qubits used for computations. For an n-qubit register, the state space consists of 2^n possible states, enabling quantum parallelism.

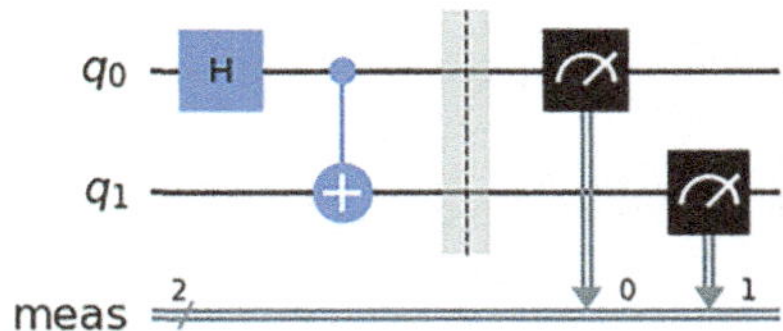

Fig. 3.17 Two qubit with a CNOT gate and measurement (by the authors)

3.15.7.3 Control Systems

Quantum control systems generate precise pulses, such as microwave signals, to manipulate qubits. These pulses implement quantum gates like the Hadamard (H) and Controlled-NOT (CNOT) gates, which form the building blocks of quantum circuits.

Python Code Example: Creating a 2-Qubit Quantum Circuit with a CNOT Gate

The following code demonstrates a quantum circuit with two qubits and a CNOT gate, as shown in Fig. 3.17. This circuit creates an entangled state $\frac{1}{\sqrt{2}}(|00\rangle + |11\rangle)$, demonstrating quantum superposition and entanglement.

```
#-------------------------------------------------------------------
# Creating a Quantum Circuit with two qubits and a CNOT gate
# Chapter 3 in the QUANTUM COMPUTING AND QUANTUM MACHINE LEARNING BOOK
#-------------------------------------------------------------------
# Version 1.0
# Qiskit changes frequently.
# We recommend using the latest version from the book code repository at:
# https://aqtinitiative.org/quantum-computing-for-engineers

# (c) 2025 Jesse Van Griensven, Roydon Fraser, and Jose Rosas
# License: MIT - Citation of this work required
#-------------------------------------------------------------------
from qiskit import QuantumCircuit, Aer, execute
from qiskit.visualization import circuit_drawer

import warnings
warnings.filterwarnings('ignore')
#-------------------------------------------------------------------

# Create a quantum circuit with 2 qubits
qc = QuantumCircuit(2)

# Apply a Hadamard gate to the first qubit
qc.h(0)
```

```
# Apply a CNOT gate (entanglement)
qc.cx(0, 1)

# Measure the qubits
qc.measure_all()

# Simulate the circuit
simulator = Aer.get_backend('aer_simulator')
result = execute(qc, simulator).result()

# Print the measurement results
print(result.get_counts())

# Draw the circuits
display(circuit_drawer(qc, output='mpl', style="iqp"))
```

3.15.8   Classical Vs. Quantum Comparative Metrics

The differences between classical and quantum hardware are evident across various performance and architectural metrics. Table 3.14 summarizes these differences.

3.15.8.1   Speed

Quantum computers excel in parallel computation due to superposition, processing multiple states simultaneously. For example, Shor's algorithm for integer factorization achieves exponential order speedup when compared to classical methods:

$$\text{Classical Runtime} : O\left(e^{c(\log N)^{1/3}(\log\log N)^{2/3}}\right)$$

$$\text{Quantum Runtime} : O\left((\log N)^2(\log\log N)(\log\log\log N)\right)$$

Table 3.14 Classical vs. quantum computing

Metric	Classical computing	Quantum computing
Information unit	Bit (0 or 1)	Qubit
Processing speed	Sequential	Parallelized (e.g., Shor's algorithm)
Error correction	Error-correcting codes (ECC) in memory	Quantum error correction (QEC)
Architecture	Silicon transistors	Qubits with control systems

3.15.8.2 Energy Consumption

Classical computers are energy-efficient and operate at room temperature. Quantum systems require cryogenic cooling to maintain qubit coherence, significantly increasing energy consumption.

3.15.8.3 Error Rates

Quantum systems experience higher error rates due to decoherence and noise. Quantum error correction (QEC) techniques, such as surface codes, mitigate these issues by encoding logical qubits in entangled states of multiple physical qubits:

$$|0_L\rangle = |000\rangle, \quad |1_L\rangle = |111\rangle$$

3.15.9 Applications in Engineering

Quantum computers have a critical role in future engineering applications. These are described below:

1. **Optimization Problems**
   Quantum systems outperform classical solvers in optimization tasks, such as portfolio optimization and route planning, by leveraging algorithms like QAOA.
2. **Material Science**
   Quantum computers simulate molecular interactions more efficiently, enabling the discovery of new materials.
3. **Signal Processing**
   Quantum Fourier transform (QFT) accelerates frequency analysis in engineering applications.

3.16 Error Correction and Fault Tolerance in Quantum Systems

Quantum computing systems are inherently prone to errors due to their fragile quantum states and susceptibility to environmental disturbances. Achieving reliable quantum computation requires robust error correction and fault-tolerant techniques to mitigate the effects of decoherence, noise, and gate imperfections. This section explores the sources of errors in quantum systems, the principles of quantum error correction (QEC), the threshold theorem, and the implementation of surface codes.

Error correction and fault tolerance are critical for the scalability of quantum computing systems. Techniques such as Shor's code, the Steane code, and the surface code address the challenges posed by decoherence and noise. The threshold theorem provides a theoretical foundation for achieving fault-tolerant quantum computation, while practical implementations like the surface code enable robust error correction in real-world systems. Understanding and applying these methods are essential for advancing quantum computing technology.

3.16.1 Sources of Error in Quantum Systems

Quantum systems face significant challenges from various sources of error, which can degrade the performance and reliability of quantum computations.

3.16.1.1 Decoherence

Decoherence arises from the interaction of a quantum system with its surrounding environment, causing the loss of quantum coherence. The state of a qubit $|\psi\rangle = \alpha \, | \, 0 \rangle + \beta \, | \, 1 \rangle$ decays over time as it becomes entangled with environmental states, leading to a mixed state:

$$\rho = | \, \psi \rangle \langle \psi \, | \rightarrow p \, | \, 0 \rangle \langle 0 \, | + (1 - p) \, | \, 1 \rangle \langle 1 \, |$$

where ρ is the density matrix, and p represents the decoherence probability.

3.16.1.2 Gate Errors

Imperfections in the implementation of quantum gates introduce errors in quantum computations. For instance, an intended unitary operation U may be executed as $U' = U + \epsilon$, where ϵ represents a deviation due to noise or control inaccuracies. These errors accumulate over the course of a computation, necessitating error correction mechanisms.

3.16.2 Quantum Error Correction (QEC)

Quantum error correction (QEC) protects quantum information by encoding logical qubits into multiple physical qubits, allowing the detection and correction of errors.

3.16.2.1 Shor's Error Correction Code

Shor's code is a pioneering QEC scheme that encodes one logical qubit into nine physical qubits. It corrects both bit-flip and phase-flip errors. The logical states are encoded as:

$$|0_L\rangle = \frac{1}{\sqrt{8}}(|000\rangle + |111\rangle) \otimes (|000\rangle + |111\rangle) \otimes (|000\rangle + |111\rangle)$$

$$|1_L\rangle = \frac{1}{\sqrt{8}}(|000\rangle - |111\rangle) \otimes (|000\rangle - |111\rangle) \otimes (|000\rangle - |111\rangle).$$

These states are designed to detect and correct errors by measuring parity checks and applying corrective operations.

3.16.2.2 Steane Error-Correcting Code

The Steane code encodes one logical qubit into seven physical qubits. It is based on the classical [7,4] Hamming code and protects against single-qubit errors. The logical states are:

$$|0_L\rangle = \frac{1}{\sqrt{8}}(|0000000\rangle + |1010101\rangle + |0110011\rangle + |1100110\rangle$$

$$+ |0001111\rangle + |1011010\rangle + |0111100\rangle + |1101001\rangle)$$

$$|1_L\rangle = X|0_L\rangle$$

where X is the Pauli-X operator.

Python Code Example: Error-Detection Circuit
The following code demonstrates the implementation of a simple error-detection circuit. This example introduces a bit-flip error and detects it using parity checks.

```
#------------------------------------------------------------------
# Quantum Error Detection Code
# Chapter 3 in the QUANTUM COMPUTING AND QUANTUM MACHINE LEARNING BOOK
#------------------------------------------------------------------
# Version 1.0
# Qiskit changes frequently.
# We recommend using the latest version from the book code repository at:
# https://aqtinitiative.org/quantum-computing-for-engineers

# (c) 2025 Jesse Van Griensven, Roydon Fraser, and Jose Rosas
# License: MIT - Citation of this work required
```

```python
#-----------------------------------------------------------------
from qiskit import QuantumCircuit, Aer, execute
from qiskit.visualization import circuit_drawer

import warnings
warnings.filterwarnings('ignore')
#-----------------------------------------------------------------

# Create a quantum circuit for Shor's code
qc = QuantumCircuit(9, 3)

# Encode a logical |0> state
for i in range(0, 9, 3):
    qc.h(i)
    qc.cx(i, i+1)
    qc.cx(i, i+2)

# Introduce a bit-flip error on one qubit
qc.x(4)

# Measure parity checks
qc.cx(0, 3)
qc.cx(1, 4)
qc.cx(2, 5)
qc.measure([3, 4, 5], [0, 1, 2])

# Simulate the circuit
simulator = Aer.get_backend('aer_simulator')
result = execute(qc, simulator).result()

# Print the measurement results
print(result.get_counts())

# Draw the circuits
display(circuit_drawer(qc, output='mpl', style="iqp"))
```

Figure 3.18 displays the quantum error detection code.

3.16.3　Threshold Theorem

The threshold theorem establishes that fault-tolerant quantum computation is achievable if the error rates in quantum operations are below a critical value, known as the threshold:

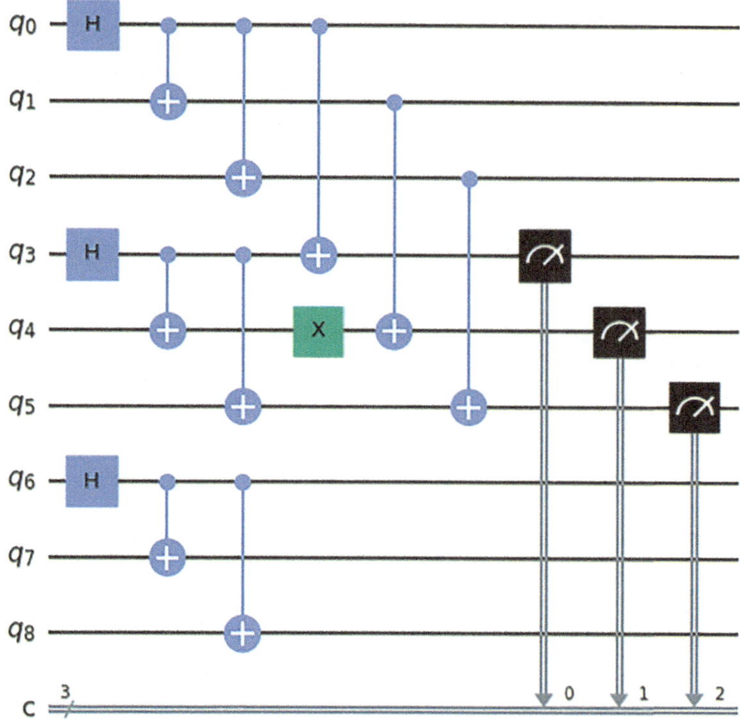

Fig. 3.18 Quantum circuit for error detection (by the authors)

$$p < p_{\text{threshold}}$$

where $p_{\text{threshold}}$ depends on the error correction code and the underlying physical system. The theorem ensures that logical error rates can be reduced exponentially by increasing the number of physical qubits and applying QEC repeatedly.

3.16.3.1 Implications

1. Fault tolerance enables scalable quantum computation even with noisy hardware.
2. The value of $p_{\text{threshold}}$ typically ranges between 10^{-3} and 10^{-2} for most QEC codes.

3.16.4 *Surface Code*

The surface code is a leading QEC scheme due to its simplicity and robustness against physical defects. It encodes logical qubits in a two-dimensional lattice of physical qubits and detects errors by measuring stabilizers.

3.16.4.1 Structure

Each qubit in the lattice is classified as either a data qubit or an ancillary (syndrome) qubit. Stabilizers, represented as parity checks, are measured using syndrome qubits to identify errors without collapsing the logical state.

3.16.4.2 Error Correction Process

(a) Measure stabilizers to detect bit-flip and phase-flip errors.
(b) Use a classical algorithm (e.g., minimum-weight perfect matching) to infer the error locations.
(c) Apply corrective operations to restore the logical state.

3.16.4.3 Advantages

(a) The surface code has a $p_{threshold}$ of approximately 1%, making it highly tolerant of physical errors.
(b) It requires only local interactions between qubits, simplifying hardware implementation.

Example Python Code: Surface Code Simulation
The following code simulates a simple surface code stabilizer measurement:

```python
#-----------------------------------------------------------------
# Quantum Surface Code Simulation
# Chapter 3 in the QUANTUM COMPUTING AND QUANTUM MACHINE LEARNING BOOK
#-----------------------------------------------------------------
# Version 1.0
# (c) 2025 Jesse Van Griensven, Roydon Fraser, and Jose Rosas
# Licence: MIT - Citation of this work required
#-----------------------------------------------------------------
from qiskit import QuantumCircuit, Aer, execute
from qiskit.visualization import import circuit_drawer

# Create a quantum circuit for a 3x3 surface code
qc = QuantumCircuit(9)

# Initialize data qubits in |0> state
for i in range(9):
    qc.initialize([1, 0], i)
```

```python
# Apply stabilizer measurement (e.g., parity checks)
qc.cx(0, 1)
qc.cx(1, 2)
qc.cx(3, 4)
qc.cx(4, 5)
qc.cx(6, 7)
qc.cx(7, 8)

# Measure the ancillary qubits
qc.measure_all()

# Simulate the circuit
simulator = Aer.get_backend('aer_simulator')
result = execute(qc, simulator).result()

# Print the measurement results
print(result.get_counts())
```

3.16.5   Engineering Challenges in Quantum Hardware Development

Quantum hardware development faces numerous engineering challenges due to the complex and delicate nature of quantum systems. Scaling up qubit counts, maintaining coherence, and minimizing noise is critical to achieving practical and scalable quantum computers. This section addresses key challenges in quantum hardware development and explores the strategies employed to overcome them.

The development of quantum hardware involves overcoming significant engineering challenges, including scaling qubit systems, fabricating high-precision qubits, managing decoherence and noise, establishing interconnectivity, and addressing energy requirements. Advanced techniques such as dynamic decoupling, photonic interconnects, and efficient cryogenic cooling are essential for building scalable and robust quantum systems. These efforts pave the way for practical quantum computing and its transformative applications.

3.16.6   Scaling Quantum Systems

Scaling quantum systems involves increasing the number of qubits while maintaining their coherence and fidelity. This challenge grows exponentially with the size of the quantum system.

As the number of qubits increases, the system becomes more susceptible to environmental interactions, leading to decoherence and noise. Coupling qubits for large-scale quantum circuits introduces additional complexity, requiring precise control over each qubit and its interactions. These are potential solutions to this problem:

3.16.6.1   Advanced Cryogenics

Maintaining qubits at millikelvin temperatures using dilution refrigerators reduces thermal noise and enhances coherence times. These refrigerators provide an ultra-low-temperature environment essential for superconducting qubits.

3.16.6.2   Noise Isolation Techniques

Shielding quantum systems from electromagnetic interference and vibrational noise is critical. Techniques include:

(a) Magnetic shielding with mu-metal enclosures.
(b) Vibration isolation using suspended platforms and damping systems (Fig. 3.19).

Python Code Example: Simulating Qubit Scaling
The following code demonstrates a multi-qubit quantum circuit. This circuit highlights the increased complexity when scaling to multiple qubits, including the challenge of managing entanglement and coherence. In this case, the code example employs a "Linear Entanglement," where a CNOT gate is applied between the control qubit and the qubit right below.

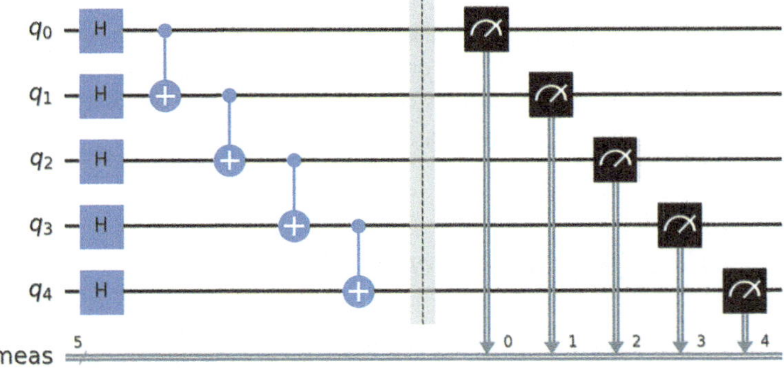

Fig. 3.19 Simulating qubit scaling (by the authors)

```python
#----------------------------------------------------------------------
# Simulating Qubit Scaling
# Chapter 3 in the QUANTUM COMPUTING AND QUANTUM MACHINE LEARNING BOOK
#----------------------------------------------------------------------
# Version 1.0
# Qiskit changes frequently.
# We recommend using the latest version from the book code repository at:
# https://aqtinitiative.org/quantum-computing-for-engineers

# (c) 2025 Jesse Van Griensven, Roydon Fraser, and Jose Rosas
# License: MIT - Citation of this work required
#----------------------------------------------------------------------
from qiskit import QuantumCircuit, Aer, execute
from qiskit.visualization import circuit_drawer

# Create a 5-qubit quantum circuit
qc = QuantumCircuit(5)

# Apply Hadamard gates to create superpositions
for i in range(5):
  qc.h(i)

# Entangle the qubits
# Using "Linear Entanglement", where a CNOT is applied on the qubit below
for i in range(4):
  qc.cx(i, i + 1)

# Measure all qubits
qc.measure_all()

# Simulate the circuit
simulator = Aer.get_backend('aer_simulator')
result  = execute(qc, simulator).result()

# Print the measurement results
print(result.get_counts())

# Visualize the circuit
print("Quantum Circuit:")
qc.draw(output='mpl', style={'backgroundcolor': '#FFFFFF', 'dpi':
900}, scale=2)
```

3.17　Fabrication of Qubits

The fabrication of qubits requires precise engineering techniques tailored to the specific physical implementation, such as superconducting circuits or ion traps.

3.17.1　Superconducting Qubits

Superconducting qubits rely on Josephson junctions, which are fabricated using nanofabrication techniques. These involve:

(a) Patterning superconducting materials with electron beam lithography.
(b) Depositing insulating layers with atomic layer deposition to achieve precise thicknesses.

3.17.2　Ion Traps

Trapped-ion systems require accurate alignment of laser beams to manipulate the electronic states of ions. Fabrication involves:

(a) Microfabrication of electrodes for ion confinement.
(b) Development of high-power, narrow-linewidth lasers for precision control.

3.18　Decoherence and Noise Management

Decoherence and noise are significant challenges in quantum hardware, limiting the coherence time T_2, which represents the duration at which a qubit can retain quantum information.

3.18.1　Decoherence Time (T_2)

The coherence time is inversely proportional to the noise power density:

$$T_2 \propto \frac{1}{\text{Noise Power Density}}$$

3.18.2   Noise Minimization Strategies

(a) **Dynamic Decoupling.**
Applying sequences of control pulses cancels out noise effects and extends coherence times. Examples include Carr–Purcell–Meiboom–Gill (CPMG) sequences.

(b) **Material Purity Improvement.**
Enhancing the purity of materials used in qubit fabrication reduces defects and noise. For instance, high-purity silicon is used in superconducting qubit substrates.

Example Python Code: Simulating Noise Effects
The following code uses Qiskit's noise model to simulate the impact of decoherence. This simulation demonstrates how decoherence affects the outcome of quantum operations.

```python
#-----------------------------------------------------------------
# Simulating Noise Effects
# Chapter 3 in the QUANTUM COMPUTING AND QUANTUM MACHINE LEARNING BOOK
#-----------------------------------------------------------------
# Version 1.0
# (c) 2025 Jesse Van Griensven, Rcydon Fraser, and Jose Rosas
# Licence: MIT - Citation of this work required
#-----------------------------------------------------------------
from qiskit import QuantumCircuit, Aer, execute
from qiskit.visualization import circuit_drawer
# Import the noise functions from aer.noise
from qiskit.providers.aer.noise import NoiseModel
from qiskit.providers.aer.noise import thermal_relaxation_error
#-----------------------------------------------------------------

# Create a quantum circuit
qc = QuantumCircuit(1)
qc.h(0)  # Apply a Hadamard gate
qc.measure_all()

# Define a noise model with thermal relaxation
noise_model = NoiseModel()
t1 = 50e-6  # Relaxation time (T1)
t2 = 70e-6  # Dephasing time (T2)
gate_time = 1e-6
error = thermal_relaxation_error(t1, t2, gate_time)
noise_model.add_all_qubit_quantum_error(error, ['id', 'h'])
```

```
# Simulate the circuit with noise
simulator = Aer.get_backend('aer_simulator')
result = execute(qc, simulator, noise_model=noise_model).result()

# Print the measurement results
print(result.get_counts())
```

3.19 Interconnectivity and Quantum Networks

Establishing robust interconnectivity between quantum processors is essential for building scalable quantum networks. There are two main challenges to the implementation of quantum networks. The first is to link multiple quantum processors, while maintaining coherence. The second is transmitting quantum information over long distances without significant loss.

Potential solutions include the following:

1. **Photonic Interconnects**

    Using photons as carriers of quantum information enables the transmission of qubits between nodes. Photonic systems utilize entangled photon pairs to establish secure communication channels.

2. **Quantum Repeaters**

    Quantum repeaters extend the range of quantum communication by entangling intermediate nodes in a network. These repeaters correct photon loss and decoherence, enabling scalable quantum networks (Fig. 3.20).

Python Code Example: Simulating Quantum Networks
The following code simulates a simple quantum network with entanglement. This circuit represents the entanglement between two nodes in a quantum network.

```
#----------------------------------------------------------------------
# Simulating Quantum Networks with 2 nodes (Alice and Bob)
# Chapter 3 in the QUANTUM COMPUTING AND QUANTUM MACHINE LEARNING BOOK
#----------------------------------------------------------------------
# Version 1.0
# Qiskit changes frequently.
```

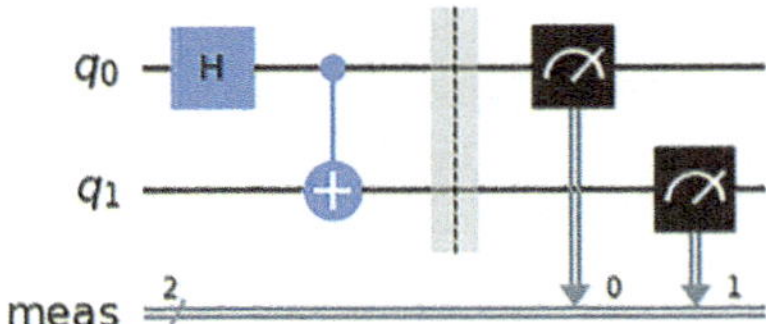

Fig. 3.20 Simulation of a quantum network with two nodes (by the authors)

```python
# We recommend using the latest version from the book code repository at:
# https://aqtinitiative.org/quantum-computing-for-engineers

# (c) 2025 Jesse Van Griensven, Rcydon Fraser, and Jose Rosas
# License: MIT - Citation of this work required
#----------------------------------------------------------------
from qiskit import QuantumCircuit, Aer, execute
from qiskit.visualization import import circuit_drawer

import warnings
warnings.filterwarnings('ignore')
#----------------------------------------------------------------

# Create a quantum circuit for two nodes
# Qubit 0: Alice
# Qubit 1: Bob
qc = QuantumCircuit(2)

# Generate entanglement
qc.h(0)
qc.cx(0, 1)

# Measure the qubits
qc.measure_all()

# Simulate the circuit
simulator = Aer.get_backend('aer_simulator')
result   = execute(qc, simulator).result()

# Print the measurement results
print(result.get_counts())

# Draw the circuits
display(circuit_drawer(qc, output='mpl', style="iqp"))
```

3.20   Quantum Computers Energy and Cryogenic Requirements

Quantum systems, particularly those using superconducting qubits, require operation at extremely low temperatures to maintain coherence.

3.20.1 Cryogenic Environments

Dilution refrigerators cool qubits to millikelvin temperatures, reducing thermal noise and enabling quantum coherence.

3.20.2 Energy Dissipation Equation

The power dissipation in a cryogenic environment is given by:

$$P = I^2 R$$

where

- P is the power dissipated.
- I is the current.
- R is the resistance.

Minimizing energy dissipation is essential for scaling quantum hardware, as large-scale systems require efficient cryogenic cooling.

3.21 Quantum Annealing

Quantum annealing is an advanced quantum optimization technique designed to solve complex combinatorial optimization problems by leveraging the principles of quantum mechanics. Unlike classical methods that rely on iterative and often exhaustive searches, quantum annealing explores solution spaces more efficiently using quantum superposition, tunneling, and entanglement to estimate the adiabatic evolution. These quantum phenomena allow a quantum system to explore multiple pathways simultaneously and escape local minima more effectively, enabling faster convergence to the global minimum of an objective function. Quantum annealing reliance on physical Hamiltonian dynamics makes it a promising alternative to classical optimization techniques for NP-hard and combinatorial problems.

3.21.1 Optimization with Quantum Annealing

Optimization problems arise across various domains, such as logistics, scheduling, graph theory, and machine learning. Many of these problems, especially NP-hard problems, are computationally intensive due to the exponentially large solution spaces. Classical optimization techniques, such as brute-force search or heuristic algorithms, struggle to find solutions efficiently as the problem size increases.

Quantum annealing presents a potential breakthrough by harnessing quantum mechanics to navigate through the solution space in parallel. By encoding the problem into a quantum system and allowing the system to evolve under quantum dynamics, quantum annealing aims to find the optimal solution much faster than classical methods.

3.21.2 Difference from Gate-Based Quantum Computing

While gate-based quantum computing executes algorithms through programmable sequences of quantum gates, quantum annealing leverages a specialized, non-universal gate-based architecture to solve optimization problems. It exploits the Hamiltonian dynamics of a quantum system to evolve from an initial configuration toward a low-energy (optimal) solution. Unlike universal quantum processors, annealers are tailored for specific problem classes and do not support arbitrary gate-level programming.

Key distinctions between gate-based quantum computing and quantum annealing include:

1. **Gate-Based Quantum Computing.**

    (a) Relies on well-defined quantum circuits and a sequence of quantum gates.
    (b) Requires precise control of qubits and operations, making it more challenging to implement on noisy quantum hardware.

2. **Quantum Annealing.**

    (a) Evolves the system's energy landscape by gradually changing the Hamiltonian.
    (b) Focuses on minimizing the system's energy to find the optimal solution, which is easier to implement on current quantum hardware (e.g., D-Wave systems).

This distinction makes quantum annealing particularly suitable for solving specific classes of optimization problems without requiring the full control and error correction necessary in universal gate-based quantum computation.

3.21.3 Applications of Quantum Annealing

Quantum annealing has a wide range of applications, particularly in problems involving combinatorial optimization, where the goal is to minimize (or maximize) an objective function subject to constraints. Examples include:

3.21.3.1   Combinatorial Optimization

Problems like the traveling salesman problem (finding the shortest route to visit a set of cities) and graph partitioning are well-suited for quantum annealing.

3.21.3.2   Machine Learning

Quantum annealing can optimize machine learning algorithms, such as clustering, regression, and support vector machines, by solving cost functions efficiently.

3.21.3.3   Logistics

Applications include route optimization for delivery systems, supply chain management, and resource allocation.

3.21.3.4   Energy Minimization

Quantum annealing is used to solve problems in material science, where finding the ground state of a physical system minimizes energy consumption.

These applications demonstrate the potential of quantum annealing to solve real-world problems across industries where classical approaches face computational bottlenecks.

3.21.4   Principles of Quantum Annealing

Quantum annealing is based on the adiabatic theorem of quantum mechanics. The adiabatic theorem states that a quantum system starting in the ground state of an initial Hamiltonian will remain in the ground state as long as the Hamiltonian evolves slowly enough toward a final Hamiltonian. The final Hamiltonian encodes the solution to the optimization problem.

This principle forms the foundation of quantum annealing, where the system evolves from a simple quantum state to one that represents the solution to the problem.

3.21.5   Hamiltonian and Optimization for Quantum Annealing

In quantum mechanics, the Hamiltonian of a system represents its total energy. For quantum annealing, two Hamiltonians are defined:

3.21.5.1 Initial Hamiltonian H_0

(a) Represents a simple quantum system whose ground state is easy to prepare.
(b) A typical choice for H_0 is:

$$H_0 = -\sum_i \sigma_x^i$$

where σ_x^i are Pauli-X operators acting on the qubits. The ground state of H_0 is a superposition of all possible states.

3.21.5.2 Problem Hamiltonian H_P

(a) Encodes the optimization problem. Its ground state corresponds to the optimal solution of the objective function.
(b) For example, for a cost function $C(z)$ defined on binary variables z, the problem Hamiltonian is:

$$H_P = \sum_z C(z)\,|z\rangle\langle z|$$

The system's Hamiltonian evolves over time as a combination of these two Hamiltonians:

$$H(t) = A(t)H_0 + B(t)H_P$$

where

- $A(t)$ and $B(t)$ are control functions that vary with time.
- At the beginning $(t = 0)$, $A(0) = 1$ and $B(0) = 0$, so $H(t)$ starts as H_0.
- At the end $(t = T$, where T is the total evolution time), $A(T) = 0$ and $B(T) = 1$, so $H(t)$ becomes H_P.

This gradual evolution ensures that the system remains in its ground state throughout the process, eventually reaching the ground state of H_P, which corresponds to the optimal solution of the problem.

3.21.6 Adiabatic Evolution in Quantum Annealing

The success of quantum annealing relies on the system evolving adiabatically, meaning the Hamiltonian changes slowly enough that the system remains in its ground state at all times. The adiabatic condition can be formally expressed as:

$$\frac{\left\langle \psi_1(t) \mid \frac{dH(t)}{dt} \mid \psi_0(t) \right\rangle}{E_1(t) - E_0(t)} \ll 1$$

where

- $|\psi_0(t)\rangle$ and $|\psi_1(t)\rangle$ are the ground and first excited states of $H(t)$, respectively.
- $E_0(t)$ and $E_1(t)$ are the corresponding energy levels.
- This condition implies that the system must evolve slowly enough to avoid excitations to higher energy states. Faster evolution can cause transitions to excited states, leading to suboptimal solutions.

3.21.7 Role of Quantum Fluctuations in Quantum Annealing

Quantum annealing leverages quantum fluctuations to explore the solution space. Quantum fluctuations arise from the uncertainty in quantum states due to the superposition principle and quantum tunneling.

3.21.7.1 Quantum Superposition

At the start of the process, the system is in a superposition of all possible states, allowing it to explore multiple solutions simultaneously.

3.21.7.2 Quantum Tunneling

A significant advantage of quantum annealing is its ability to tunnel through energy barriers. In classical systems, escaping local minima requires thermal energy or random exploration. Quantum tunneling provides a more efficient route by allowing the system to move through potential barriers instead of climbing over them.

3.21.8 Quantum Tunneling: Escaping Local Minima

Local minima are a common challenge in optimization problems. In a classical system, the search for the global minimum can be trapped in a local minimum, as escaping requires significant thermal energy or random perturbations. Quantum annealing overcomes this limitation through quantum tunneling, where the system can traverse energy barriers without requiring extra energy.

For example:

1. A classical system must climb an energy barrier of height ΔE to move from one valley (local minimum) to another.
2. In quantum systems, tunneling allows the system to "pass-through" the barrier, making the search for the global minimum more efficient.

This property is particularly beneficial for optimization problems with rugged energy landscapes, where the solution space contains many local minima.

3.22 Mathematical Formulation of Quantum Annealing

The mathematical foundation of quantum annealing relies on encoding the optimization problem into a quantum system, evolving it adiabatically, and extracting the optimal solution through measurements. At its core, quantum annealing maps a classical cost function to a problem Hamiltonian whose ground state encodes the solution to the optimization problem.

3.22.1 Optimization Problem as a Cost Function

The starting point for quantum annealing is an optimization problem defined as the minimization of a classical cost function $C(z)$, where z represents a configuration of variables. The goal is to find the configuration z^* that minimizes $C(z)$:

$$z^* = \arg \min_z C(z)$$

This cost function can often be represented as a sum over binary variables, where $z_i \in \{-1, 1\}$ or equivalently $z_i \in \{0, 1\}$. Many combinatorial optimization problems, such as the traveling salesman problem or graph partitioning, can be mapped to such cost functions.

3.22.2 Problem Hamiltonian H_P in Quantum Annealing

In quantum annealing, the cost function $C(z)$ is encoded into a problem Hamiltonian H_P. The ground state of H_P (i.e., the state with the lowest energy) corresponds to the solution z^* of the optimization problem.

The problem Hamiltonian is defined as:

$$H_P = \sum_z C(z) \, |z\rangle\langle z|$$

where

- $|z\rangle$ are the basis states representing all possible solutions.
- $C(z)$ is the cost function evaluated for configuration z.

In a physical system, this Hamiltonian corresponds to the "energy landscape" of the problem, where each configuration $|z\rangle$ has an energy $C(z)$. The quantum system seeks the ground state, i.e., the configuration $|z^*\rangle$ with the minimum energy.

3.22.3 Initial Hamiltonian $\mathbf{H_0}$

The initial Hamiltonian H_0 is chosen to have a simple ground state that is easy to prepare. A common choice for H_0 is:

$$H_0 = -\sum_i \sigma_x^i$$

where

- σ_x^i are the Pauli-X operators acting on the i-th qubit.

The ground state of H_0 is the equal superposition of all computational basis states:

$$|\psi(0)\rangle = \frac{1}{\sqrt{2^N}} \sum_z |z\rangle$$

where N is the number of qubits, and the sum runs over all 2^N possible states. This state represents a quantum system that initially explores all possible solutions simultaneously in superposition.

3.22.4 Time-Dependent Hamiltonian Evolution in Quantum Annealing

Quantum annealing relies on the system evolving under a time-dependent Hamiltonian $H(t)$, which smoothly interpolates between the initial Hamiltonian H_0 and the problem Hamiltonian H_P. The total Hamiltonian at time t is given by:

$$H(t) = A(t)H_0 + B(t)H_P$$

where $A(t)$ and $B(t)$ are time-dependent control functions that satisfy:

- $A(0) = 1$ and $B(0) = 0$ at the start of the process.
- $A(T) = 0$ and $B(T) = 1$ at the end of the process ($t = T$).

Here T is the total annealing time, which determines how slowly the system evolves.

At $t = 0$, the system begins in the ground state of H_0. As t progresses, $H(t)$ gradually changes from H_0 to H_P. By the adiabatic theorem of quantum mechanics, if the evolution is sufficiently slow, the system will remain in the ground state throughout the process, ultimately reaching the ground state of H_P. The final ground state represents the optimal solution to the optimization problem.

3.22.5 Measurement and Final Solution

At the end of the annealing process ($t = T$), the Hamiltonian is purely the problem Hamiltonian:

$$H(T) = H_P$$

The system is then measured on a computational basis, and the resulting state corresponds to one of the possible solutions. Due to the nature of quantum mechanics, the ground state (optimal solution) is the most probable measurement outcome. Repeating the annealing process multiple times increases the likelihood of obtaining the correct solution.

3.23 Quantum Annealing Vs. Simulated Annealing

Quantum annealing differs fundamentally from its classical counterpart, simulated annealing, in terms of the mechanisms used to explore the solution space. The table below summarizes the key differences:

Feature	Quantum annealing	Simulated annealing
Nature	Quantum fluctuations and tunneling	Thermal fluctuations in classical systems
Escape from local minima	Tunneling through energy barriers	Thermal jumps over energy barriers
Parallelism	Explores multiple solutions simultaneously	Sequential exploration of solutions
Speed	Faster convergence for some problems	Slower for complex, high-dimensional problems

Quantum annealing leverages quantum tunneling to escape local minima, whereas simulated annealing relies on thermal energy to "jump" over barriers. This gives quantum annealing a potential advantage in rugged energy landscapes where local minima are prevalent.

3.24   Applications of Quantum Annealing

Quantum annealing has been successfully applied to various real-world optimization problems. Below are some key domains where quantum annealing demonstrates significant utility. Some of the quantum annealing optimization problems are described below:

1. **Combinatorial Optimization**

    (a) **Traveling Salesman Problem (TSP).**

    Quantum annealing optimizes the shortest route among a set of cities.

    (b) **Graph Partitioning.**

    Dividing a graph into subgraphs while minimizing edge cuts.
2. **Logistics and Scheduling**

    Quantum annealing efficiently solves complex scheduling problems, such as airline crew scheduling and job-shop scheduling.

3.24.1   Machine Learning with Quantum Annealing

Quantum annealing can be used in machine learning for many applications. Some of these are listed below.

3.24.1.1   Feature Selection

Quantum annealing identifies optimal subsets of features, enhancing the efficiency of machine learning algorithms.

3.24.1.2   Clustering

Quantum clustering algorithms leverage annealing to partition large datasets into clusters efficiently.

3.24.2   Drug Discovery and Molecular Modeling

Quantum annealing accelerates drug discovery by optimizing molecular configurations to minimize their energy states, facilitating the design of new pharmaceutical compounds.

3.24.3 *Finance Portfolio Optimization with Quantum Annealing*

Quantum annealing finds the optimal allocation of financial assets to maximize returns while minimizing risk.

3.24.4 *Quantum Annealing in Practice*

Commercial quantum annealers, such as those developed by D-Wave Systems, provide practical implementations of quantum annealing for solving real-world problems. These systems operate using the Ising model, which represents optimization problems as spin systems.

3.24.5 *Ising Model Formulation*

The Ising Hamiltonian is defined as:

$$H = -\sum_i h_i \sigma_z^i - \sum_{i<j} J_{ij} \sigma_z^i \sigma_z^j$$

where

- σ_z^i are Pauli-Z operators representing spins.
- h_i are local magnetic field strengths.
- J_{ij} are coupling coefficients representing interactions between spins.

This formulation allows optimization problems to be mapped directly onto the Ising model, enabling quantum annealers to solve them efficiently.

Exercise Questions: Quantum and Classical Circuits
1. **Classical Circuits**

    (a) Define a classical bit and explain how it differs from a qubit in quantum computing.
    (b) Using an AND gate, describe the logical operation that occurs when two inputs are $A = 1$ and $B = 0$.

2. **Logic Gates**

    (a) Write the truth table for the XOR gate and explain its functionality.
    (b) Derive the matrix representation of the NOT gate and show its effect on the input state $|1\rangle$.

3. **Matrix Operations on Classical Circuits**

   (a) How can matrix operations represent classical logic gates? Provide an example.
   (b) Given the input state $|01\rangle$, compute the output of an AND gate represented as a matrix.

4. **Superconducting Qubits**

   (a) Explain the role of Josephson junctions in the operation of superconducting qubits.
   (b) Using the Hamiltonian for a superconducting qubit, describe how the nonlinear term ensures distinguishable qubit states:

   $$H = \frac{p^2}{2m} + \frac{1}{2}m\omega^2 x^2 + \frac{\beta}{4}x^4$$

5. **Trapped Ion Qubits**

   (a) Describe how trapped ions are confined using Paul or Penning traps.
   (b) Using the interaction Hamiltonian $H = \hbar\Omega(a^\dagger + a)(\sigma_+ + \sigma_-)$, explain how entanglement is achieved in trapped ion systems.

6. **Photonic Systems**

   (a) Explain how beam splitters and phase shifters manipulate photon states in photonic quantum systems.
   (b) Provide an example of a photonic system's measurement process and its result on the state $|0\rangle$.

7. **Error Correction and Fault Tolerance**

   (a) What are the key differences between Shor's and Steane's error-correcting code?
   (b) Explain how the surface code measures stabilizers and corrects errors without collapsing the quantum state.

8. **Hardware Comparison**

   (a) Compare classical and quantum computing architectures in terms of scalability and energy efficiency.
   (b) Discuss the role of cryogenic systems in maintaining qubit coherence in superconducting quantum computers.

9. **Engineering Challenges**

   (a) Identify two significant challenges in scaling quantum systems and propose solutions to address them.
   (b) Explain how noise isolation techniques improve the performance of quantum hardware.

10. **Quantum Networking**

   (a) Discuss the use of photonic interconnects in quantum networks and their role in enabling long-distance quantum communication.

   (b) Explain the concept of quantum repeaters and their importance in quantum networking.

11. **Python Applications**

   (a) Write a Python program using Qiskit to simulate a Bell state using a Hadamard gate and a CNOT gate.

   (b) Modify the program to introduce a bit-flip error and simulate its detection using an error correction circuit.

12. **Quantum Circuit Applications**

   (a) Describe how quantum circuits are used to simulate molecular interactions in material science.

   (b) Explain how the quantum Fourier transform (QFT) is implemented in a quantum circuit and its application in signal processing.

Additional Bibliography

1. P. Benioff, *The Computer as a Physical System: A Microscopic Quantum Mechanical Hamiltonian Model of Computers as Represented by Turing Machines* (Springer, 1982)
2. C.H. Bennett, G. Brassard, *Quantum Cryptography: Public Key Distribution and Coin Tossing* (Springer, 1984)
3. I.L. Chuang, M.A. Nielsen, *Quantum Computation and Quantum Information* (Cambridge University Press, 2000)
4. R.P. Feynman, *Simulating Physics with Computers* (Addison-Wesley, 1982)
5. J. Gruska, *Quantum Computing* (McGraw-Hill, 1999)
6. N. Gisin et al., *Quantum Nonlocality and Reality: 50 Years of Bell's Theorem* (Cambridge University Press, 2014)
7. A. Kitaev, *Classical and Quantum Computation* (American Mathematical Society, 2002)
8. R. Laflamme, D. Cory, *Quantum Error Correction for Quantum Memories* (Springer, 1998)
9. D.A. Lidar, T.A. Brun (eds.), *Quantum Error Correction* (Cambridge University Press, 2013)
10. S. Lloyd, *Programming the Universe: A Quantum Computer Scientist Takes on the Cosmos* (Alfred A. Knopf, 2006)
11. S. Majidy, C. Wilson, R. Laflamme, *Building Quantum Computers—A Practical Introduction* (Cambridge Press, 2025)
12. C. Monroe, *Quantum Networks with Atoms and Photons* (Wiley, 2014)
13. J. Preskill, *Lecture Notes on Quantum Computation* (California Institute of Technology, 1998)
14. E.G. Rieffel, W. Polak, *Quantum Computing: A Gentle Introduction* (MIT Press, 2011)
15. P.W. Shor, *Algorithms for Quantum Computation: Discrete Logarithms and Factoring* (IEEE Press, 1994)
16. A.M. Steane, *The Theory of Quantum Error Correction* (Oxford University Press, 1997)

17. J. Stolze, D. Suter, *Quantum Computing: A Short Course from Theory to Experiment* (Wiley-VCH, 2004)
18. V. Vedral, *Decoding Reality: The Universe as Quantum Information* (Oxford University Press, 2010)
19. W.H. Zurek, *Decoherence and the Transition from Quantum to Classical* (Cambridge University Press, 2003)

Chapter 4
Quantum Data Structures and Encoding

Data encoding is fundamental to computation in both classical and quantum systems. In any computational framework, data must be represented in a format that the system can interpret and process. The encoding process transforms real-world information, such as numbers, text, images, or measurements, into a structured format that can be processed by algorithms.

In classical computing, encoding typically involves converting data into binary form (0s and 1s) because classical systems use transistors that operate in binary states (on/off). In contrast, quantum systems use quantum bits (qubits), which can represent data in more complex ways due to properties, such as superposition and entanglement. Efficient encoding methods are critical for the performance of both classical and quantum algorithms. Poor encoding can lead to unnecessary computational overhead, increased error rates, and longer processing times. In classical computing, this might manifest as increased time complexity or memory usage, while in quantum computing, inefficient encoding can degrade the performance of quantum circuits and increase the probability of decoherence.

This chapter integrates theoretical foundations, practical examples, and mathematical rigor, equipping engineers to effectively encode and utilize data in quantum systems.

4.1 Differences in Classical Vs. Quantum Encoding

The differences between classical and quantum encoding arise from the fundamental properties of classical bits versus quantum bits, as shown in Table 4.1.

In classical systems, data is encoded in bits, where each bit takes a value of either **0** or **1**. For example, the number **5** in binary is encoded as **101**:

Table 4.1 Classical vs. quantum encoding of data

Aspect	Classical encoding	Quantum encoding
Basic unit of data	Bit (0 or 1)	Qubit
Representation	Binary (e.g., 1101)	Quantum state
Data storage	Deterministic	Probabilistic
Error handling	Error correction (parity, ECC)	Quantum error correction (QEC)
Data transmission	Classical signals	Quantum states (photons, electrons)

$$5_{10} = 101_2$$

In quantum systems, data is encoded using qubits, which can exist in superposition states, meaning they represent both $|0\rangle$ and $|1\rangle$ simultaneously. A general qubit state can be expressed as:

$$|\psi\rangle = \alpha|0\rangle + \beta|1\rangle$$

where α and β are complex numbers that satisfy the normalization condition:

$$|\alpha|^2 + |\beta|^2 = 1$$

This allows quantum computers to encode both bit states simultaneously, enabling exponentially more data in fewer qubits than classical bits. For example, n qubits can represent 2^n states simultaneously.

4.2   Importance of Efficient Encoding for Quantum Algorithms

Efficient encoding is essential for both classical and quantum systems, but it plays an even more critical role in quantum computing due to the fragility of quantum states. The success of quantum algorithms, such as Shor's algorithm (for factoring large numbers) or Grover's algorithm (for database search), depends heavily on how data is encoded into qubits.

Consider an example of encoding a binary number into qubits:

(a) In classical computing, the number **6** would be encoded as **110** in binary.
(b) In quantum computing, the same number could be encoded as a superposition of qubit states:

$$|\psi_6\rangle = |110\rangle = |1\rangle \otimes |1\rangle \otimes |0\rangle$$

However, quantum algorithms often require more complex encoding schemes to optimize performance. One such scheme is "amplitude encoding," where data is encoded directly into the amplitudes of a quantum state:

$$|\psi\rangle = \frac{1}{\sqrt{n}} \sum_{i=0}^{n-1} x_i\,|i\rangle$$

In this encoding, a dataset with n entries is represented by a single quantum state. This is particularly useful for quantum machine learning applications where large datasets need to be processed efficiently.

Python Code Example: Classical Vs. Quantum Encoding
Below is a Python example demonstrating how to encode a classical binary number into both a classical and quantum system. The output of the code below displays a quantum circuit with the appropriate X-gates applied to encode the number "6" in a quantum register.

```
#-------------------------------------------------------------------
# Quantum Encoding
# Chapter 4 in the QUANTUM COMPUTING AND QUANTUM MACHINE LEARNING BOOK
#-------------------------------------------------------------------
# Version 1.0
# Qiskit changes frequently.
# We recommend using the latest version from the book code repository at:
# https://aqtinitiative.org/quantum-computing-for-engineers


# (c) 2025 Jesse Van Griensven, Roydon Fraser, and Jose Rosas
# License: MIT - Citation of this work required
#-------------------------------------------------------------------
# Quantum Encoding - Import Qiskit Libraries
from qiskit import QuantumCircuit, Aer, transpile, assemble
from qiskit.visualization import plot_histogram
from qiskit.visualization import circuit_drawer
#-------------------------------------------------------------------
def classical_encode(number):
# Classical Encoding
  return bin(number)[2:]
#-------------------------------------------------------------------
def quantum_encode(number):
  binary_representation = bin(number)[2:]
  num_qubits = len(binary_representation)

  # Create a quantum circuit
  qc = QuantumCircuit(num_qubits)
```

```
# Apply X gates to encode the binary number
for i, bit in enumerate(reversed(binary_representation)):
  if bit == '1':
    qc.x(i)

qc.measure_all()
return qc
#---------------------------------------------------------------------

# Example usage
print("Classical Encoding of 6:", classical_encode(6))

# Quantum Encoding Circuit for number 6
qc = quantum_encode(6)
qc.draw('mpl')
```

4.3 Quantum Data Encoding Concepts

Quantum data encoding fundamentally differs from classical encoding methods due to the unique properties of qubits. Unlike classical bits, which can take values of either 0 or 1, qubits can exist in a superposition of both states simultaneously. Additionally, qubits are represented using complex numbers and can be visualized geometrically on the Bloch sphere (shown in Fig. 4.2). These features allow quantum systems to process and encode data more efficiently for certain computational tasks, such as cryptography, optimization, and machine learning.

In this chapter, we explore the foundational concepts of quantum data encoding, including qubits, basis states, and the Bloch sphere representation.

4.3.1 Qubits and the Superposition Principle

A qubit (quantum bit) is the basic unit of quantum information. Unlike classical bits, which are binary and can only be in one of the two states (0 or 1), qubits can exist in a linear combination of the two basis states, known as superposition.

A general qubit state can be written as:

$$|\psi\rangle = \alpha |0\rangle + \beta |1\rangle$$

where

- $|0\rangle$ and $|1\rangle$ are the basis states (often called the computational basis states).
- α and β are complex numbers that represent the amplitudes of the respective basis states.
- The condition $|\alpha|^2 + |\beta|^2 = 1$ ensures that the qubit state is normalized.

For example, a qubit in a pure superposition state might look like this:

$$|\psi\rangle = \frac{1}{\sqrt{2}}\,|0\rangle + \frac{1}{\sqrt{2}}\,|1\rangle$$

This state implies that the qubit has an equal probability of collapsing to either $|0\rangle$ or $|1\rangle$ when measured. The probabilities are determined by the squared magnitudes of the amplitudes:

$$P(|0\rangle) = |\alpha|^2, \quad P(|1\rangle) = |\beta|^2$$

The superposition principle allows quantum computers to process many possible states simultaneously, enabling parallel computation.

4.4   Data Loading Problem in Quantum Computing

The data loading problem refers to the challenge of efficiently encoding large classical datasets into quantum states. While quantum computers offer significant advantages in processing data, the process of data input can be a major bottleneck, especially for large-scale datasets. Unlike classical systems, where data can be loaded and processed in a straightforward manner, quantum systems require complex operations to prepare quantum states that represent classical information.

Several challenges make data encoding in quantum computing difficult:

1. **Exponential Data Requirements**

   Quantum states are represented in a 2^n dimensional Hilbert space, where n is the number of qubits. Therefore, encoding a large dataset requires exponentially more qubits as the size of the dataset increases.

2. **Normalization**

   Quantum states must be normalized to ensure that the total probability amplitude sums to exactly **1**. This adds an additional layer of complexity to the encoding process.

3. **State Preparation Time**

   Efficient state preparation is critical for quantum algorithms to achieve speedups. If the time to prepare a quantum state exceeds the time to solve the problem classically, any quantum advantage is lost.

4. **Noise and Error Rates**

Quantum systems are prone to decoherence and errors, making it essential to design encoding methods that are resilient to noise.

5. **Scalability**

Many encoding techniques work well for small datasets but do not scale efficiently for larger datasets. Developing scalable encoding strategies is crucial for practical quantum applications.

4.5   Quantum Data Encoding Methods

Encoding classical data into quantum states is a critical step in quantum computing, enabling classical information to be processed using quantum algorithms. Effective data encoding ensures efficient utilization of quantum resources and improves the performance of computations. This section explores four major methods of quantum data encoding: basis encoding, amplitude encoding, angle encoding, and hybrid encoding.

Quantum data encoding methods are foundational to quantum computing and enable the transition from classical to quantum information processing. Basis encoding provides a direct representation of binary data, while amplitude encoding allows for a compact representation of large datasets. Angle encoding is ideal for variational quantum algorithms, and hybrid encoding combines methods to address complex datasets. Advancements in these encoding techniques will continue to enhance the applicability and performance of quantum algorithms across diverse fields.

4.5.1   Encoding Information in Quantum States

Quantum data encoding involves mapping classical data into qubit states. There are several methods to achieve this, depending on the application. The following is a small list that this book will use to encode data.

4.5.1.1   Basis Encoding

Classical data is mapped directly to the computational basis states. For example, a binary number **110** is encoded as the quantum state $|110\rangle$.

4.5.1.2   Amplitude Encoding

The values of a dataset are encoded into the amplitudes x_i of a quantum state. For example, for a dataset with n values, the quantum state can be represented as:

$$|\psi\rangle = \frac{1}{\sqrt{n}} \sum_{i=0}^{n-1} x_i \,|\, i\rangle$$

4.5.1.3   Angle Encoding

Data is encoded into the angles of qubit rotations. For instance, a classical value **x** can be encoded as a rotation angle θ on the Bloch sphere:

$$R_y(\theta)\,|\,0\rangle = \cos\left(\frac{\theta}{2}\right)\,|\,0\rangle + \sin\left(\frac{\theta}{2}\right)\,|\,1\rangle$$

4.5.1.4   Phase Encoding

Information is encoded into the phase of qubit states. For example, a classical value x can be represented as a phase shift ϕ applied to a qubit:

$$|\psi\rangle = \frac{1}{\sqrt{2}}\left(|\,0\rangle + e^{i\phi}\,|\,1\rangle\right)$$

In this method, the phase ϕ modulates the superposition state of the qubit, allowing the encoding of complex information through phase relationships (Table 4.2).

4.5.2   Basis Encoding

Basis encoding maps classical data directly to the computational basis states of qubits. Each data point is represented as a binary string, which corresponds to a specific quantum state.

Table 4.2 Summary of quantum encoding methods

Encoding method	Description
Basis encoding	Maps classical data directly to computational basis states
Amplitude encoding	Encodes data into the amplitudes of a quantum state, allowing the representation of large datasets efficiently
Angle encoding	Utilizes rotation angles on the Bloch sphere to map classical values to qubit states
Phase encoding	Incorporates phase shifts into qubit states to encode information through phase relationships

For an n-qubit system, a classical value x in the range $\{0, 1, ..., 2^n - 1\}$ is mapped to a quantum state:

$$x \in \{0, 1, ..., 2^n - 1\} \rightarrow |x\rangle$$

Basis encoding is often used in quantum algorithms requiring direct bit-level manipulation, such as Grover's search and quantum counting.

For example, the binary value $x = 5$ is represented as $|101\rangle$ in a 3-qubit system.

Python Code Example: Basis Encoding
This code encodes the binary value 5 into the computational basis state $|101\rangle$.

```python
#------------------------------------------------------------------
# Quantum Basis Encoding
# Chapter 4 in the QUANTUM COMPUTING AND QUANTUM MACHINE LEARNING BOOK
#------------------------------------------------------------------
# Version 1.0
# Qiskit changes frequently.
# We recommend using the latest version from the book code repository at:
# https://aqtinitiative.org/quantum-computing-for-engineers

# (c) 2025 Jesse Van Griensven, Roydon Fraser, and Jose Rosas
# License: MIT - Citation of this work required
#------------------------------------------------------------------
from qiskit import QuantumCircuit, Aer, execute
from qiskit.visualization import import plot_bloch_multivector
#------------------------------------------------------------------

# Create a quantum circuit with 3 qubits
qc = QuantumCircuit(3)
# Encode the binary value 5 (|101>)
qc.x(0)  # Set qubit 0 to |1>
qc.x(2)  # Set qubit 2 to |1>

# Simulate and visualize the Bloch sphere
simulator  = Aer.get_backend('statevector_simulator')
result    = execute(qc, simulator).result()
statevector = result.get_statevector()
print(result.get_counts())
plot_bloch_multivector(statevector)
```

Figure 4.1 displays the state of each qubit after the encoding.

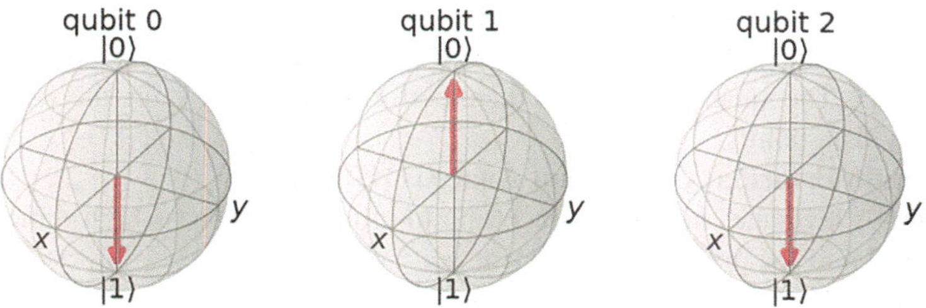

Fig. 4.1 Bloch sphere representation of qubit encoding of 5, which is binary 101. (By the authors)

4.5.3 *Amplitude Encoding*

Amplitude encoding is a more advanced and efficient method of encoding classical data into quantum states. In this approach, the values of a classical dataset are encoded into the amplitudes of a quantum state. This method leverages the superposition property of qubits, enabling a quantum system to represent exponentially large datasets using a small number of qubits.

Given a classical vector $x = [x_0, x_1, ..., x_{n-1}]$, amplitude encoding maps it to a quantum state as follows:

$$|\psi\rangle = \frac{1}{\sqrt{\sum_{i=0}^{n-1} |x_i|^2}} \sum_{i=0}^{n-1} x_i |i\rangle$$

where

- n is the length of the classical vector.
- The coefficients x_i are the amplitudes of the corresponding quantum states $|i\rangle$.

For example, a classical vector $x = [1, 2, 3, 4]$ is encoded as:

$$|\psi\rangle = \frac{1}{\sqrt{30}}(1\,|00\rangle + 2\,|01\rangle + 3\,|10\rangle + 4\,|11\rangle)$$

This approach is particularly useful in quantum machine learning, where large datasets must be processed efficiently. By encoding data into amplitudes, quantum systems can achieve exponential speedups for tasks, such as quantum support vector machines and quantum principal component analysis.

Amplitude encoding is commonly used in the following quantum machine-learning applications:

1. **Quantum Support Vector Machines (QSVMs)**
   In QSVMs, training data is encoded into quantum states using amplitude encoding. The quantum system then performs operations on these states to classify data points.
2. **Quantum Principal Component Analysis (QPCA)**
   QPCA leverages amplitude encoding to efficiently process high-dimensional data and find the principal components.
3. **Quantum Neural Networks (QNNs)**
   Amplitude encoding is used to input classical data into quantum neural networks for training and inference.

Python Code Example: Amplitude Encoding

```python
#---------------------------------------------------------------------
# Quantum Amplitude Encoding
# Encodes a 4-point dataset into the amplitudes of a 2-qubit quantum
state.

# Chapter 4 in the QUANTUM COMPUTING AND QUANTUM MACHINE LEARNING BOOK
#---------------------------------------------------------------------
# Version 1.0
# Qiskit changes frequently.
# We recommend using the latest version from the book code repository at:
# https://aqtinitiative.org/quantum-computing-for-engineers

# (c) 2025 Jesse Van Griensven, Roydon Fraser, and Jose Rosas
# License: MIT - Citation of this work required
#---------------------------------------------------------------------
import warnings
warnings.filterwarnings('ignore')

from qiskit import QuantumCircuit
from qiskit.visualization import circuit_drawer
import numpy as np
#---------------------------------------------------------------------

# Normalize the data
data = np.array([1, 2, 3, 4])
norm = np.linalg.norm(data)
normalized_data = data / norm
# Create a quantum circuit
qc = QuantumCircuit(2)  # 2 qubits for 4 data points
qc.initialize(normalized_data, [0, 1])
```

```
print("Quantum Circuit for Amplitude Encoding:")
print(qc)

# Draw the circuit
display(circuit_drawer(qc, output='mpl', style="iqp"))
```

4.5.4   Angle Encoding

Imagine if the measure of daily time would be just "the day" and "the night," a binary concept like "0" and "1." Now, try to visualize how much more information we can encode if we have a clock, and each 15° could represent an hour. This approach would allow 12 times more information than the binary "0" and "1." This is the idea behind angle encoding, which maps classical data to the rotation angles of quantum gates, typically on the Bloch sphere. Data is encoded using a rotation gate, such as $R_y(x)$, applied to a qubit initially in the $|0\rangle$ state:

$$R_y(x)\,|0\rangle = \cos\left(\frac{x}{2}\right)|0\rangle + \sin\left(\frac{x}{2}\right)|1\rangle$$

Example
For $x = \pi$, the rotation results in a state:

$$|1\rangle = R_y(\pi)\,|0\rangle$$

Angle encoding is widely used in variational quantum algorithms and quantum machine learning, where it parameterizes quantum circuits for optimization.

Python Code Example: Angle Encoding
This example applies an R_y rotation to encode a data point into the quantum state of a qubit.

```
#-----------------------------------------------------------------
# Quantum Amplitude-Encoded State
# Chapter 4 in the QUANTUM COMPUTING AND QUANTUM MACHINE LEARNING BOOK
#-----------------------------------------------------------------
# Version 1.0
# Qiskit changes frequently.
# We recommend using the latest version from the book code repository at:
# https://aqtinitiative.org/quantum-computing-for-engineers

# (c) 2025 Jesse Van Griensven, Roydon Fraser, and Jose Rosas
# License: MIT - Citation of this work required
```

```
#---------------------------------------------------------------------
import warnings
warnings.filterwarnings('ignore')

from qiskit import QuantumCircuit
from qiskit.visualization import circuit_drawer
import numpy as np
#---------------------------------------------------------------------


# Create a quantum circuit
qc = QuantumCircuit(1)

# Encode data using an Ry rotation gate
data_point = np.pi / 4
qc.ry(data_point, 0)

print("Quantum Circuit for Angle Encoding:")
print(qc)

# Draw the circuit
display(circuit_drawer(qc, output='mpl', style="iqp"))
```

4.5.5   *Phase Encoding*

Phase encoding is a method of encoding classical data into the phases of qubit states. Instead of representing data in the amplitudes of the quantum state, phase encoding modifies the relative phase between basis states to store information.

A general quantum state can be represented as:

$$|\psi\rangle = \alpha |0\rangle + e^{i\phi}\beta |1\rangle$$

In phase encoding, classical data is mapped to the phase φ of a qubit. For example, a classical value x can be encoded as a phase rotation $\varphi = 2\pi x$.

Consider the state:

$$|\psi\rangle = \frac{1}{\sqrt{2}}\left(|0\rangle + e^{i\phi}|1\rangle\right)$$

For $x = 1$, the phase $\varphi = 2\pi$ results in a full rotation, leaving the qubit state unchanged. For $x = 0.5$, the phase $\varphi = \pi$ represents a half rotation.

Phase encoding is heavily used in the quantum Fourier transform (QFT), a quantum analog of the classical Fourier transform. The QFT transforms quantum states from the computational basis to the frequency domain, where phase information is crucial.

The QFT is a key component in several quantum algorithms, including:

1. **Shor's Algorithm**
   Uses the QFT to factor large numbers efficiently.
2. **Quantum Phase Estimation (QPE)**
   Relies on phase encoding to estimate the eigenvalues of a unitary operator.

Python Code Example: Phase Encoding

Below is a Python example of how to apply phase encoding.

```python
#----------------------------------------------------------------
# Quantum Phase Encoding
# Chapter 4 in the QUANTUM COMPUTING AND QUANTUM MACHINE LEARNING BOOK
#----------------------------------------------------------------
# Version 1.0
# Qiskit changes frequently.
# We recommend using the latest version from the book code repository at:
# https://aqtinitiative.org/quantum-computing-for-engineers

# (c) 2025 Jesse Van Griensven, Roydon Fraser, and Jose Rosas
# License: MIT - Citation of this work required
#----------------------------------------------------------------
import warnings
warnings.filterwarnings('ignore')

from qiskit import QuantumCircuit
from qiskit.visualization import circuit_drawer
#----------------------------------------------------------------

def phase_encoding(circuit, qubit, phase):
# Function to apply phase encoding
  circuit.rz(phase, qubit)
  return circuit
#----------------------------------------------------------------

# Example: Encode phase π on qubit 0
qc = QuantumCircuit(1)
qc = phase_encoding(qc, 0, 3.14159)

print("Quantum Circuit for Phase Encoding:")
print(qc)
```

```python
# Draw the circuit
display(circuit_drawer(qc, output='mpl', style="iqp"))
```

4.5.6 Hybrid Encoding

Hybrid encoding combines multiple encoding methods, such as amplitude and angle encoding, to leverage their respective strengths for complex datasets. Some of the advantages of hybrid encoding are:

1. Increases flexibility by enabling tailored encoding strategies for specific problems.
2. Improved expressiveness: by capturing complex relationships in datasets.

A hybrid encoding scheme may use amplitude encoding for numerical data and angle encoding for categorical features.

Python Code Example: Hybrid Encoding
This code demonstrates how hybrid encoding combines amplitude and angle methods to encode different types of data into a quantum state.

```python
#------------------------------------------------------------------
# Quantum Hybrid Encoding
# Chapter 4 in the QUANTUM COMPUTING AND QUANTUM MACHINE LEARNING BOOK
#------------------------------------------------------------------
# Version 1.0
# Qiskit changes frequently.
# We recommend using the latest version from the book code repository at:
# https://aqtinitiative.org/quantum-computing-for-engineers

# (c) 2025 Jesse Van Griensven, Roydon Fraser, and Jose Rosas
# License: MIT - Citation of this work required
#------------------------------------------------------------------
from qiskit import QuantumCircuit
from qiskit.visualization import circuit_drawer
import numpy as np
#------------------------------------------------------------------

# Create a quantum circuit
qc = QuantumCircuit(2)

# Amplitude encode numerical data
numerical_data = np.array([1, 2])
norm = np.linalg.norm(numerical_data)
```

```
normalized_data = numerical_data / norm
qc.initialize(normalized_data, [0])

# Angle encode categorical data
categorical_data = np.pi / 3
qc.ry(categorical_data, 1)

print("Quantum Circuit for Hybrid Encoding:")
print(qc)

# Draw the circuit
display(circuit_drawer(qc, output='mpl', style="iqp"))
```

4.6   Visual Representations of Encoding Methods

To better understand quantum data encoding, it is useful to visualize how different encoding techniques translate classical data into quantum states. The Bloch sphere is a powerful visualization tool that represents the state of a single qubit. Encoding classical data involves manipulating a qubit's position on the Bloch sphere through rotations and superpositions.

4.6.1   Bloch Sphere Representation

The Bloch sphere is a geometric representation of a qubit state, where any pure qubit state can be visualized as a point on the surface of a unit sphere. The general qubit state is expressed as:

$$|\psi\rangle = \cos\left(\frac{\theta}{2}\right)|0\rangle + e^{i\phi}\sin\left(\frac{\theta}{2}\right)|1\rangle$$

This can be represented on the Bloch sphere using two angles:

- θ (polar angle) represents the angle between the qubit state and the Z-axis.
- φ (azimuthal angle) represents the rotation around the Z-axis.

The Bloch sphere allows for an intuitive visualization of quantum state transformations (Fig. 4.2):

- $|0\rangle$: Located at the north pole.
- $|1\rangle$: Located at the south pole.
- $|+\rangle$: Located on the equator along the X-axis.

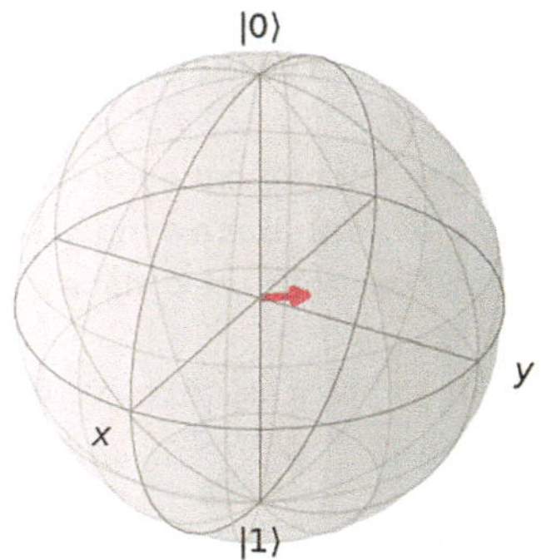

Fig. 4.2 Bloch sphere example, with $\theta = \frac{\pi}{3}$ and $\phi = \frac{\pi}{4}$. (By the authors)

Quantum gates can be visualized as rotations on the Bloch sphere. For example:

- **X Gate**: Rotates a qubit 180° around the X-axis.
- **Y Gate**: Rotates a qubit 180° around the Y-axis.
- **Z Gate**: Rotates a qubit 180° around the Z-axis.

These rotations are expressed mathematically as:

$$R_x(\theta) = e^{-i\frac{\theta}{2}X}, \quad R_y(\theta) = e^{-i\frac{\theta}{2}Y}, \quad R_z(\theta) = e^{-i\frac{\theta}{2}Z}$$

Python Code Example: Visualizing Qubits on the Bloch Sphere
We can visualize qubit states on the Bloch sphere:

```python
#------------------------------------------------------------------
# Quantum Visualize the Bloch Sphere
# Chapter 4 in the QUANTUM COMPUTING AND QUANTUM MACHINE LEARNING BOOK
#------------------------------------------------------------------
# Version 1.0
# Qiskit changes frequently.
# We recommend using the latest version from the book code repository at:
# https://aqtinitiative.org/quantum-computing-for-engineers

# (c) 2025 Jesse Van Griensven, Roydon Fraser, and Jose Rosas
# License: MIT - Citation of this work required
#------------------------------------------------------------------
import warnings
warnings.filterwarnings('ignore')

from qiskit import QuantumCircuit, Aer, transpile, assemble
from qiskit.visualization import plot_bloch_vector
from qiskit.visualization import circuit_drawer
import numpy as np
#------------------------------------------------------------------
```

```
# Define a qubit state using angles theta and phi
theta = np.pi / 3.
phi   = np.pi / 4.

# Plot the Bloch vector
bloch_vector = [np.sin(theta) * np.cos(phi), np.sin(theta) * np.sin
(phi), np.cos(theta)]
plot_bloch_vector(bloch_vector)
```

4.6.2   Basis Encoding Visualization

Basis encoding maps classical binary data directly to computational basis states. For example, a binary value of 101 maps to the quantum state $|101\rangle$. This encoding is straightforward but requires a number of qubits equal to the number of bits in the binary data.

Python Code Example: Basis Encoding
Qiskit is used to create code to visualize a basis-encoded state, which is shown in Fig. 4.3.

```
#---------------------------------------------------------------------
# Quantum Basis Encoding
# Chapter 4 in the QUANTUM COMPUTING AND QUANTUM MACHINE LEARNING BOOK
#---------------------------------------------------------------------
# Version 1.0
# Qiskit changes frequently.
# We recommend using the latest version from the book code repository at:
# https://aqtinitiative.org/quantum-computing-for-engineers

# (c) 2025 Jesse Van Griensven, Roydon Fraser, and Jose Rosas
# License:  MIT - Citation of this work required
#---------------------------------------------------------------------
```

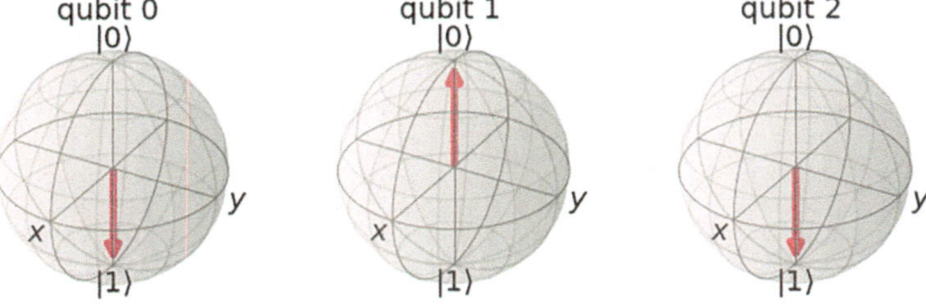

Fig. 4.3 Visualization of a basis-encoded state of binary value 101 to state $|101\rangle$. (By the authors)

```
from qiskit import QuantumCircuit, Aer, execute
from qiskit.visualization import plot_bloch_multivector
#----------------------------------------------------------------------

# Create a quantum circuit with 3 qubits
qc = QuantumCircuit(3)
# Encode the binary value 5 (|101>)
qc.x(0)  # Set qubit 0 to |1>
qc.x(2)  # Set qubit 2 to |1>

# Simulate and visualize the Bloch sphere
simulator  = Aer.get_backend('statevector_simulator')
result     = execute(qc, simulator).result()
statevector = result.get_statevector()
plot_bloch_multivector(statevector)
```

4.6.3   Amplitude Encoding Visualization

Amplitude encoding represents classical data as the amplitudes of a quantum state. The amplitudes of a qubit quantum state are the α and β in the superposition state $|\psi\rangle = \alpha\,|\,0\rangle + \beta\,|\,1\rangle$. It allows a dataset of size N to be encoded using $\log_2(N)$ qubits, making it more compact and efficient for large datasets.

Python Code Example: Amplitude Encoding
Qiskit was used to create the code below to visualize an amplitude-encoded stat, which is shown in Fig. 4.4.

```
#----------------------------------------------------------------------
# Quantum Amplitude-Encoded State Visualization
# Chapter 4 in the QUANTUM COMPUTING AND QUANTUM MACHINE LEARNING BOOK
```

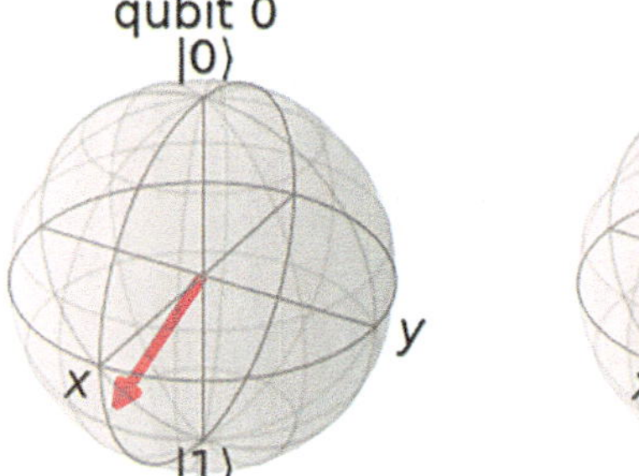
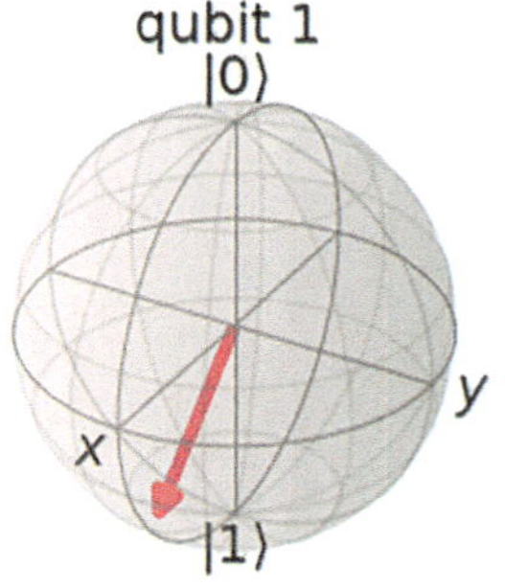

Fig. 4.4 Visualization of an amplitude-encoded state. (By the authors)

```python
#-----------------------------------------------------------------
# Version 1.0
# Qiskit changes frequently.
# We recommend using the latest version from the book code repository at:
# https://aqtinitiative.org/quantum-computing-for-engineers

# (c) 2025 Jesse Van Griensven, Roydon Fraser, and Jose Rosas
# License: MIT - Citation of this work required
#-----------------------------------------------------------------
import numpy as np
from qiskit import QuantumCircuit, Aer, execute
from qiskit.visualization import plot_bloch_multivector
from qiskit.visualization import circuit_drawer
#-----------------------------------------------------------------

# Normalize the dataset
data = np.array([1, 2, 3, 4])
norm = np.linalg.norm(data)
normalized_data = data / norm

# Create a quantum circuit for 2 qubits
qc = QuantumCircuit(2)
qc.initialize(normalized_data, [0, 1])

# Simulate and visualize
simulator  = Aer.get_backend('statevector_simulator')
result    = execute(qc, simulator).result()
statevector = result.get_statevector()
plot_bloch_multivector(statevector)
```

4.7   Advanced Quantum Encoding Techniques

In addition to the previous quantum encoding methods, several specialized techniques have been developed to efficiently map classical data into quantum states for specific applications. These techniques often leverage the unique properties of quantum systems, such as basis states, entanglement, and topological stability, to achieve higher levels of computational efficiency and robustness.

This section presents the details of four key specialized quantum encoding techniques. These methods will be presented in the following subsections:

1. Binary Phase Encoding.
2. Distributed Amplitude Encoding.
3. Entanglement-Based Encoding.
4. Topological Encoding.

4.7.1 Binary Phase Encoding

Binary phase encoding represents classical data as phases in quantum states. This method leverages the phase property of quantum states to encode information more efficiently.

Given a binary string $x = x_1 x_2 ... x_n$, the binary phase encoding maps it to a quantum state as:

$$|x\rangle = \frac{1}{\sqrt{2^n}} \sum_{i=0}^{2^n-1} (-1)^{x_i} |i\rangle$$

Python Code Example: Binary Phase Encoding (Fig. 4.5)

```
#-------------------------------------------------------------------
# Quantum Binary Phase Encoding
# Chapter 4 in the QUANTUM COMPUTING AND QUANTUM MACHINE LEARNING BOOK
#-------------------------------------------------------------------
# Version 1.0
# Qiskit changes frequently.
```

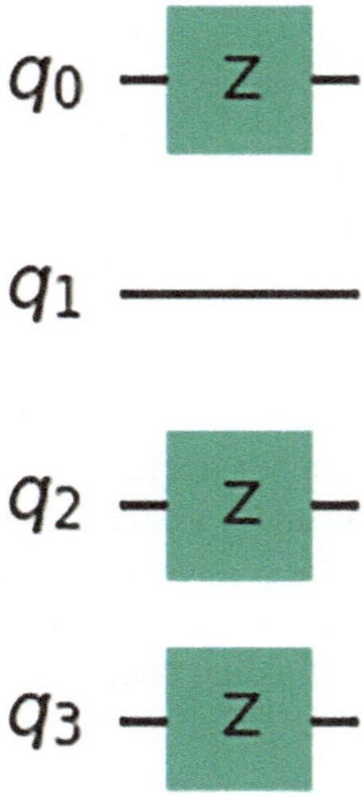

Fig. 4.5 Circuit for the encoding string 1011 using a phase shift gate. (By the authors)

```
# We recommend using the latest version from the book code repository at:
# https://aqtinitiative.org/quantum-computing-for-engineers

# (c) 2025 Jesse Van Griensven, Roydon Fraser, and Jose Rosas
# License: MIT - Citation of this work required
#----------------------------------------------------------------------
from qiskit import QuantumCircuit
from qiskit.visualization import circuit_drawer

# Example using binary data
binary_data = [1, 0, 1, 1]

# Create a quantum circuit with 4 qubits
qc = QuantumCircuit(len(binary_data))

# Apply phase shifts based on binary data
for i, bit in enumerate(binary_data):
    if bit == 1:
        qc.z(i)  # Apply Z gate to introduce a phase shift

# Print the circuit in a simplistic way
print(qc)
# Draw the circuit
display(circuit_drawer(qc, output='mpl', style="iqp"))
```

4.7.2 Distributed Amplitude Encoding

Distributed amplitude encoding allows a dataset to be split across multiple quantum states, reducing the required number of qubits and improving scalability.

For a dataset $\{x_1, x_2, \ldots, x_N\}$, distributed amplitude encoding maps the data into separate quantum states as:

$$|\psi_1\rangle = \sum_{i=1}^{\frac{N}{2}} x_i |i\rangle, \quad |\psi_2\rangle = \sum_{i=\frac{N}{2}+1}^{N} x_i |i\rangle$$

Python Code Example: Distributed Amplitude Encoding (Fig. 4.6)

```
#----------------------------------------------------------------------
# Distributed Amplitude Encoding
# Chapter 4 in the QUANTUM COMPUTING AND QUANTUM MACHINE LEARNING BOOK
```

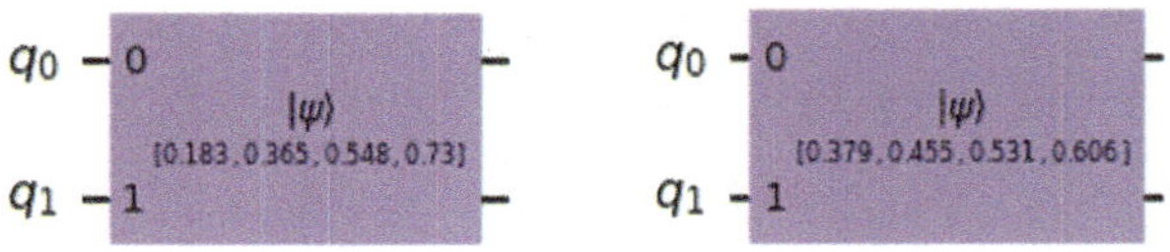

Fig. 4.6 Results for the distributed amplitude encoding. (By the authors)


```python
#------------------------------------------------------------------
# Version 1.0
# Qiskit changes frequently.
# We recommend using the latest version from the book code repository at:
# https://aqtinitiative.org/quantum-computing-for-engineers

# (c) 2025 Jesse Van Griensven, Roydon Fraser, and Jose Rosas
# License: MIT - Citation of this work required
#------------------------------------------------------------------
import numpy as np
from qiskit import QuantumCircuit
from qiskit.visualization import circuit_drawer
#------------------------------------------------------------------

# Split dataset into two parts
data = np.array([1, 2, 3, 4, 5, 6, 7, 8])
data1, data2 = np.split(data, 2)

# Normalize the data
norm1 = np.linalg.norm(data1)
norm2 = np.linalg.norm(data2)
normalized_data1 = data1 / norm1
normalized_data2 = data2 / norm2

# Create quantum circuits for both parts
qc1 = QuantumCircuit(2)  # Using 2
qc2 = QuantumCircuit(2)  # Using 2

# Initialize the circuits
qc1.initialize(normalized_data1, [0, 1])
qc2.initialize(normalized_data2, [0, 1])

print("Quantum Circuit 1:\n", qc1)
print("Quantum Circuit 2:\n", qc2)

# Draw the circuits
display(circuit_drawer(qc1, output='mpl', style="iqp"))
display(circuit_drawer(qc2, output='mpl', style="iqp"))
```

4.7.3 Entanglement-Based Encoding

Entanglement-based encoding leverages the phenomenon of quantum entanglement to represent and manipulate data across multiple qubits. In this method, the state of one qubit becomes correlated with the state of another, regardless of the physical distance between them.

Entanglement allows qubits to share information in such a way that the state of one qubit depends on the state of another. This unique property is used to encode data in a way that classical systems cannot replicate. For example, consider a two-qubit system in an entangled state:

$$|\psi\rangle = \frac{1}{\sqrt{2}}(|00\rangle + |11\rangle)$$

In this state:

- If one qubit is measured as $|0\rangle$, the other will also collapse to $|0\rangle$.
- If one qubit is measured as $|1\rangle$, the other will also collapse to $|1\rangle$.

This correlation between qubits can be used to encode distributed data across multiple qubits, enabling parallelism and non-local data representation. Entanglement-based encoding plays a center role in several quantum applications:

1. **Quantum Key Distribution (QKD)**
   In BB84 and E91 protocols, entanglement is used to securely distribute cryptographic keys between two parties. The security of QKD relies on the fact that any eavesdropping attempt on the entangled qubits will disrupt their correlated states, alerting the legitimate parties.
2. **Quantum Teleportation**
   Entanglement-based encoding is essential for quantum teleportation, a process by which the state of a qubit can be transmitted from one location to another without physically transferring the qubit itself. This is achieved by sharing an entangled pair of qubits between the sender and receiver.
3. **Quantum Networking**
   Entanglement-based encoding enables the creation of quantum networks, where entangled qubits are distributed across different nodes to enable secure communication and distributed quantum computing.

Python Code Example: Entangled-Based Encoding (Fig. 4.7)

```
#-------------------------------------------------------------------
# Quantum Entangled-Based Encoding
# Chapter 4 in the QUANTUM COMPUTING AND QUANTUM MACHINE LEARNING BOOK
#-------------------------------------------------------------------
# Version 1.0
# Qiskit changes frequently.
```

Fig. 4.7 Circuit for
quantum entangled-based
encoding. (By the authors)

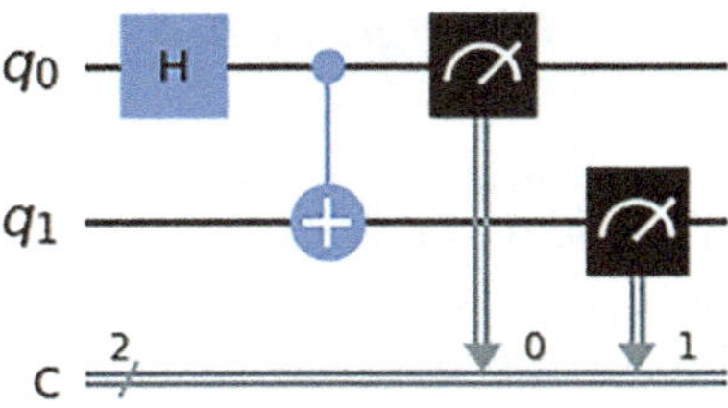

```
# We recommend using the latest version from the book code repository at:
# https://aqtinitiative.org/quantum-computing-for-engineers

# (c) 2025 Jesse Van Griensven, Roydon Fraser, and Jose Rosas
# License: MIT - Citation of this work required
#-----------------------------------------------------------------------
import warnings
warnings.filterwarnings('ignore')

# Import necessary libraries
from qiskit import QuantumCircuit, Aer, execute
from qiskit.visualization import plot_histogram
from qiskit.visualization import circuit_drawer
import matplotlib.pyplot as plt
#-----------------------------------------------------------------------

# Create a Quantum Circuit with 2 qubits and 2 classical bits
qc = QuantumCircuit(2, 2)

# Step 1: Create Entanglement
qc.h(0)      # Apply Hadamard gate to qubit 0 to create superposition
qc.cx(0, 1)   # Apply CNOT gate with qubit 0 as control and qubit 1 as
target

# Step 2: Measure the qubits
qc.measure([0,1], [0,1])

# Visualize the circuit
print("Quantum Circuit:")
display(circuit_drawer(qc, output='mpl', style="iqp"))

# Execute the circuit on the qasm simulator
simulator = Aer.get_backend('qasm_simulator')

# Execute the circuit with a specified number of shots (simulations)
shots = 1024
job  = execute(qc, simulator, shots=shots)
```

```python
# Get the results
result = job.result()
counts = result.get_counts(qc)

# Print the results
print("\nMeasurement Results: ')
print(counts)

# Plot the results
plot_histogram(counts)
plt.title("Entangled Qubits Measurement Results")
plt.show()
```

4.7.4 Topological Encoding

Topological encoding is a highly advanced quantum encoding technique that leverages the properties of topological qubits to achieve fault-tolerant quantum computation. In this method, data is encoded in the topological properties of quantum states, which are inherently protected from local errors.

Topological qubits are based on the concept of anyons, which are exotic particles that exist in two-dimensional spaces. These particles have unique properties that can be used to encode quantum information in a way that is resilient to errors. In topological encoding, data is represented by the braiding of anyons:

1. Braiding refers to the process of moving anyons around each other in specific patterns.
2. The resulting topological structure is used to encode data, which remains stable even if the system is subjected to noise or perturbations.

One of the most significant advantages of topological encoding is its inherent fault tolerance. Unlike traditional qubits, which are susceptible to decoherence and quantum errors, topological qubits are much more robust. Topological encoding is the basis for topological quantum computers, which are being developed by companies like Microsoft, using the surface code and Majorana fermions.

The surface code is a type of topological code that arranges qubits in a two-dimensional lattice, allowing for the correction of both bit-flip and phase-flip errors. The stability of topological qubits makes them ideal for long-term quantum storage and error correction, which are critical for the scalability of quantum computers.

Python Code Example: Topological Encoding

```python
#------------------------------------------------------------------
# Quantum Topological Encoding
# Chapter 4 in the QUANTUM COMPUTING AND QUANTUM MACHINE LEARNING BOOK
#------------------------------------------------------------------
# Version 1.0
# Qiskit changes frequently.
# We recommend using the latest version from the book code repository at:
# https://aqtinitiative.org/quantum-computing-for-engineers

# (c) 2025 Jesse Van Griensven, Roydon Fraser, and Jose Rosas
# License: MIT - Citation of this work required
#------------------------------------------------------------------


# Import necessary libraries
from qiskit import QuantumCircuit, Aer, execute
from qiskit.visualization import plot_histogram
from qiskit.ignis.verification import stabilizer
import matplotlib.pyplot as plt

# Define the code distance
d = 3  # Distance-3 Surface Code

# Calculate the number of data and ancilla qubits
num_data_qubits = d * d
num_ancilla_qubits = 2 * (d - 1) * (d - 1)

# Total qubits
total_qubits = num_data_qubits + num_ancilla_qubits

# Initialize the quantum circuit
qc = QuantumCircuit(total_qubits, num_ancilla_qubits)

# For demonstration, we'll focus on encoding a logical |0> state
# Initialize all data qubits to |0>, so no action is needed

# Define stabilizer measurements (simplified)
# In a full implementation, you'd define X and Z stabilizers across the
lattice

# Example: Measure a Z stabilizer on a plaquette
# Apply CNOT gates between data qubits and ancilla qubits

# Placeholder for stabilizer measurements
# This is a highly simplified version
```

```python
for anc in range(num_ancilla_qubits):
    # Assume ancilla qubits are at the end
    ancilla = num_data_qubits + anc
    # Apply Hadamard to ancilla for Z stabilizer
    qc.h(ancilla)
    # Apply CNOTs from data qubits to ancilla
    # This requires mapping data qubits to stabilizers
    # Here, we use a placeholder mapping
    data_qubits = [0, 1, 3, 4]  # Example data qubits for a stabilizer
    for dq in data_qubits:
        qc.cx(dq, ancilla)
    # Apply Hadamard again
    qc.h(ancilla)
    # Measure the ancilla
    qc.measure(ancilla, anc)

# Introduce a bit-flip error on data qubit 0
qc.x(0)

# Visualize the circuit
print("Surface Code Quantum Circuit:")
print(qc.draw())

# Execute the circuit on the qasm simulator
simulator = Aer.get_backend('qasm_simulator')

# Execute the circuit with a specified number of shots
shots = 1024
job = execute(qc, simulator, shots=shots)
result = job.result()
counts = result.get_counts(qc)

# Print the measurement results
print("\nMeasurement Results (Ancilla Qubits):")
print(counts)

# Plot the results
plot_histogram(counts)
plt.title("Stabilizer Measurement Results")
plt.show()
```

4.8 Compression Techniques for Quantum Data

As quantum hardware is limited in terms of qubit count and coherence time, efficient compression techniques are essential for quantum data encoding. These techniques help reduce the number of qubits required while preserving the integrity of the data.

4.8.1 Principal Component Analysis (PCA)

PCA is a dimensionality reduction technique that can be applied to quantum data. It identifies the most significant features in the dataset and discards less relevant information, reducing the number of qubits needed for encoding.

Python Code Example: Principal Component Analysis Compression

```python
#------------------------------------------------------------------
# Principal Component Analysis Compression
# Chapter 4 in the QUANTUM COMPUTING AND QUANTUM MACHINE LEARNING BOOK
#------------------------------------------------------------------
# Version 1.0
# Qiskit changes frequently.
# We recommend using the latest version from the book code repository at:
# https://aqtinitiative.org/quantum-computing-for-engineers

# (c) 2025 Jesse Van Griensven, Roydon Fraser, and Jose Rosas
# License: MIT - Citation of this work required
#------------------------------------------------------------------

from sklearn.decomposition import PCA
import numpy as np
#------------------------------------------------------------------

# Example dataset
data = np.random.rand(100, 10)  # 100 samples, 10 features
# Apply PCA to reduce to 2 dimensions
pca = PCA(n_components=2)
reduced_data = pca.fit_transform(data)
print("Reduced Data Shape:", reduced_data.shape)
```

4.8.2   Sparse Encoding

Sparse encoding focuses only on significant data points, reducing the need to encode zero or near-zero values. This technique is particularly useful for datasets with a large number of zeros.

Python Code Example: Sparse Encoding (Fig. 4.8)

```python
#-----------------------------------------------------------------
# Quantum Sparse Encoding
# Chapter 4 in the QUANTUM COMPUTING AND QUANTUM MACHINE LEARNING BOOK
#-----------------------------------------------------------------
# Version 1.0
# Qiskit changes frequently.
# We recommend using the latest version from the book code repository at:
# https://aqtinitiative.org/quantum-computing-for-engineers


# (c) 2025 Jesse Van Griensven, Roydon Fraser, and Jose Rosas
# License: MIT - Citation of this work required
#-----------------------------------------------------------------
import numpy as np
from qiskit import QuantumCircuit
from qiskit.visualization import circuit_drawer

# Example sparse dataset
sparse_data = np.array([0, 0, 1.5, 0, 0.8, 0])

# Normalize non-zero elements
indices = np.nonzero(sparse_data)[0]
values  = sparse_data[indices]
norm   = np.linalg.norm(values)
normalized_values = values / norm

# Create a quantum circuit
qc = QuantumCircuit(len(indices))
for i, value in enumerate(normalized_values):
    qc.ry(value, i)
```


Fig. 4.8 A quantum sparse encoding circuit. (By the authors)

```
print("Quantum Circuit for Sparse Encoding:")
print(qc)

# Draw the circuits
display(circuit_drawer(qc, output='mpl', style="iqp"))
```

4.9   Noise and Error Correction in Data Encoding

Quantum data encoding is vulnerable to various types of noise that can degrade the integrity of quantum states. Understanding quantum noise models and applying error correction codes is necessary to preserve the fidelity of encoded data. While full error correction remains challenging with current hardware, noise mitigation techniques provide a practical approach to improving the accuracy of quantum computations in the near term.

This section provides an in-depth analysis of quantum noise models, explores error correction techniques like surface codes, and includes Python examples to demonstrate the impact of noise on encoded states and how to mitigate it.

4.9.1   Quantum Noise Models

Quantum noise refers to unwanted interactions between a quantum system and its environment, which can cause errors in quantum computations. Common noise models include depolarizing noise, bit-flip noise, phase-flip noise, and amplitude damping. These noise models help us understand how quantum systems behave under real-world conditions (Fig. 4.9).

4.9.1.1   Depolarizing Noise

This type of noise randomly flips a qubit's state with a certain probability p. The depolarizing channel can be mathematically represented as:

$$\rho \rightarrow (1-p)\rho + \frac{p}{d}I$$

where ρ is the quantum state, p is the depolarization probability, d is the dimension of the state space, and I is the identity matrix.

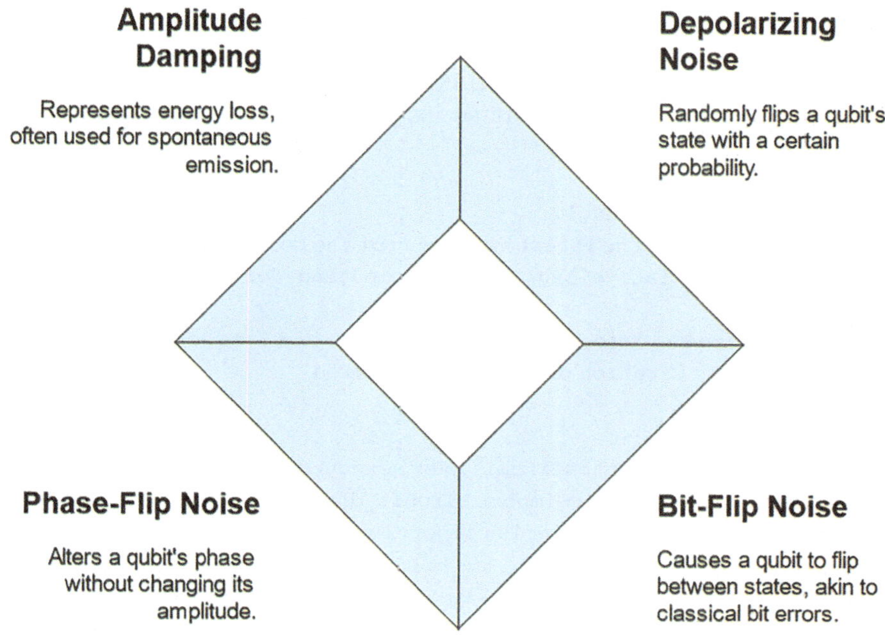

Fig. 4.9 Quantum encoding noise types. (By the authors)

4.9.1.2   Bit-Flip Noise

In this model, a qubit flips from $|0\rangle$ to $|1\rangle$ or vice versa with a probability p. This noise is analogous to classical bit errors.

4.9.1.3   Phase-Flip Noise

This noise alters the phase of a qubit without changing its amplitude. The phase-flip operation can be represented as a Pauli-Z gate applied to a qubit:

$$Z = \begin{bmatrix} 1 & 0 \\ 0 & -1 \end{bmatrix}$$

4.9.1.4   Amplitude Damping

This noise model describes energy loss from a quantum system, often used to represent spontaneous emission in quantum optics.

Python Code Example: Depolarizing Noise Simulation (Fig. 4.10)

```python
#----------------------------------------------------------------
# Quantum Depolarizing Noise Simulation
# Chapter 4 in the QUANTUM COMPUTING AND QUANTUM MACHINE LEARNING BOOK
#----------------------------------------------------------------
# Version 1.0
# Qiskit changes frequently.
# We recommend using the latest version from the book code repository at:
# https://aqtinitiative.org/quantum-computing-for-engineers

# (c) 2025 Jesse Van Griensven, Roydon Fraser, and Jose Rosas
# License: MIT - Citation of this work required
#----------------------------------------------------------------
import numpy as np
from qiskit import QuantumCircuit, Aer, execute
from qiskit.visualization import circuit_drawer
from qiskit.providers.aer.noise import NoiseModel,
depolarizing_error
#----------------------------------------------------------------

# Create a quantum circuit
qc = QuantumCircuit(1)

# Apply Hadamard gate
qc.h(0)

# Execute the measurement
qc.measure_all()

# Draw the circuits
display(circuit_drawer(qc, output='mpl', style="iqp"))

# Define a depolarizing noise model
noise_model = NoiseModel()
noise_model.add_all_qubit_quantum_error(depolarizing_error(0.1,
1), ['h'])

# Simulate the circuit with noise
simulator = Aer.get_backend('aer_simulator')
result  = execute(qc, simulator, noise_model=noise_model).result()
print(result.get_counts())
```

Fig. 4.10 Quantum depolarizing noise simulation circuit. (By the authors)

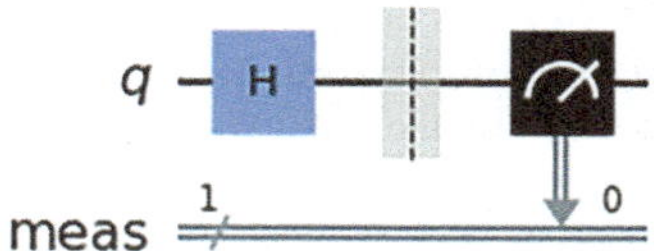

4.9.2 Error Correction Codes

To counteract the effects of quantum noise, error correction codes are essential. These codes allow quantum computers to detect and correct errors that occur during computations. One of the most widely studied error correction methods is the surface code, which operates continuously in parallel with quantum computations to protect logical qubits from errors.

4.9.2.1 Surface Codes

The surface code is a topological error correction code that encodes logical qubits into a grid of physical qubits. It is highly resilient to local noise and can correct both bit-flip and phase-flip errors. The surface code works by measuring stabilizers (parity checks) throughout the computation, detecting errors without collapsing the quantum state. These stabilizer measurements are performed repeatedly to ensure error correction is applied dynamically as errors occur.

For a surface code defined on an $n \times n$ grid of qubits, stabilizers are represented as:

$$S_X = \prod_{i \in X} X_i, \quad S_Z = \prod_{j \in Z} Z_j$$

where X_i and Z_j are Pauli operators applied to qubits in the grid.

Python Code Example: Surface Error Correction Code (Fig. 4.11)

```
#-------------------------------------------------------------------
# Quantum Surface Error Correction Code
# Chapter 4 in the QUANTUM COMPUTING AND QUANTUM MACHINE LEARNING BOOK
#-------------------------------------------------------------------
# Version 1.0
# Qiskit changes frequently.
# We recommend using the latest version from the book code repository at:
# https://aqtinitiative.org/quantum-computing-for-engineers

# (c) 2025 Jesse Van Griensven, Roydon Fraser, and Jose Rosas
# License: MIT - Citation of this work required
#-------------------------------------------------------------------
import numpy as np
```

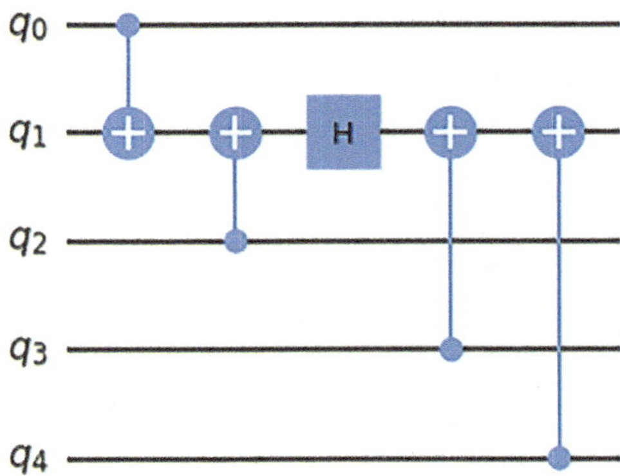

Fig. 4.11 A quantum surface error correction circuit. (By the authors)

```python
from qiskit import QuantumCircuit
from qiskit.visualization import circuit_drawer
#--------------------------------------------------------------------

# Create a simple surface code circuit
qc = QuantumCircuit(5)

# Apply stabilizer checks
qc.cx(0, 1)  # CNOT gate for X stabilizer
qc.cx(2, 1)
qc.h(1)     # Hadamard gate for Z stabilizer
qc.cx(3, 1)
qc.cx(4, 1)

print("Surface Code Quantum Circuit:")
print(qc)

# Draw the circuits
display(circuit_drawer(qc, output='mpl', style="iqp"))
```

4.9.3 Mitigating Noise in Encoded States

Noise mitigation techniques aim to reduce the impact of quantum noise without requiring full error correction. Table 4.3 summarizes common techniques for noise mitigation in encoded states.

Table 4.3 Quantum noise mitigating techniques

Noise mitigation technique	Description
Dynamical decoupling	This method applies sequences of control pulses to qubits to cancel out environmental noise
Error mitigation	Post-processing techniques that adjust the results of quantum computations to account for noise
Hybrid quantum-classical algorithms	These algorithms use classical computations to compensate for quantum noise

Fig. 4.12 A quantum error mitigation circuit. (By the authors)

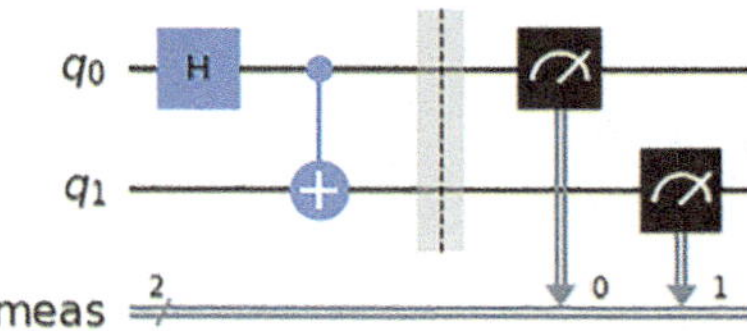

Python Code Example: Error Mitigation

The quantum error mitigation code is presented below, and the final circuit is shown in Fig. 4.12.

```python
#-------------------------------------------------------------------
# Quantum Error Mitigation
# Chapter 4 in the QUANTUM COMPUTING AND QUANTUM MACHINE LEARNING BOOK
#-------------------------------------------------------------------
# Version 1.0
# Qiskit changes frequently.
# We recommend using the latest version from the book code repository at:
# https://aqtinitiative.org/quantum-computing-for-engineers

# (c) 2025 Jesse Van Griensven, Roydon Fraser, and Jose Rosas
# License: MIT - Citation of this work required
#-------------------------------------------------------------------
import numpy as np
from qiskit.visualization import circuit_drawer

from qiskit import QuantumCircuit, Aer, execute
from qiskit.providers.aer.noise import NoiseModel, depolarizing_error
from qiskit.ignis.mitigation.measurement import complete_meas_cal
from qiskit.ignis.mitigation.measurement import CompleteMeasFitter

from qiskit.visualization import circuit_drawer
#-------------------------------------------------------------------
```

```python
# Create a quantum circuit
qc = QuantumCircuit(2)
qc.h(0)
qc.cx(0, 1)
qc.measure_all()

# Draw the circuits
display(circuit_drawer(qc, output='mpl', style="iqp"))

# Define noise model
noise_model = NoiseModel()

# Add 1-qubit depolarizing error to single-qubit gates
noise_model.add_all_qubit_quantum_error(depolarizing_error(0.05,
1), ['h'])

# Add 2-qubit depolarizing error to the 'cx' gate
noise_model.add_all_qubit_quantum_error(depolarizing_error(0.05,
2), ['cx'])

# Simulate with noise
simulator = Aer.get_backend('aer_simulator')
result = execute(qc, simulator, noise_model=noise_model).result()

# Apply error mitigation
# Pass a list of qubit indices instead of an integer
cal_circuits, state_labels = complete_meas_cal(qubit_list=[0, 1],
qr=qc.qregs[0])
meas_fitter = CompleteMeasFitter(result, state_labels)
mitigated_result = meas_fitter.filter.apply(result)

print("Mitigated Result:", mitigated_result.get_counts())
```

4.9.4   Basis States and Measurement

In quantum computing, a basis is a set of states that form a complete orthonormal set, allowing any quantum state to be expressed as a linear combination of these basis states.

4.9.4.1   Computational Basis Vs. Other Basis States

The computational basis is the most commonly used basis in quantum computing. It consists of the states $|0\rangle$ and $|1\rangle$ for a single qubit or $|00\rangle$, $|01\rangle$, $|10\rangle$, $|11\rangle$ for two qubits. However, other basis states can also be used for encoding and measurement.

4.9.4.2   Hadamard Basis (H-Basis)

The Hadamard basis states are $|+\rangle$ and $|-\rangle$, where:

$$|+\rangle = \frac{|0\rangle + |1\rangle}{\sqrt{2}}, \quad |-\rangle = \frac{|0\rangle - |1\rangle}{\sqrt{2}}$$

The Hadamard basis is often used in quantum algorithms such as Deutsch–Jozsa and Grover's search.

4.9.4.3   Pauli Bases

Quantum states can also be expressed in terms of the eigenvectors of the Pauli operators (X, Y, Z), which are used in various quantum error correction protocols.

4.9.5   *Measurement in Different Bases*

When measuring a qubit, the result depends on the basis in which the measurement is performed. If a qubit is measured on a computational basis, the result will be either $|0\rangle$ or $|1\rangle$, with probabilities determined by the state's amplitudes.

If the measurement is performed on a different basis (such as the Hadamard basis), the qubit will collapse to one of the basis states for that basis.

Example of measurement probabilities:

For the state:

$$|\psi\rangle = \frac{1}{\sqrt{3}}|0\rangle + \sqrt{\frac{2}{3}}|1\rangle$$

The probability of measuring $|0\rangle$ in the computational basis is:

$$P(|0\rangle) = |\frac{1}{\sqrt{3}}|^2 = \frac{1}{3}$$

The probability of measuring $|1\rangle$ is:

$$P(|1\rangle) = |\sqrt{\tfrac{2}{3}}|^2 = \tfrac{2}{3}$$

4.10   Quantum Circuits for Engineers: Design and Simulation

Quantum circuits are the fundamental building blocks for quantum computation. Designing and simulating quantum circuits involve a systematic approach to choosing gates, encoding data, and ensuring coherence throughout the computation. This section provides an overview of quantum circuit design principles, an example of encoding a vector into a quantum state, and tools for simulating quantum circuits.

Designing and simulating quantum circuits require a systematic approach to selecting gates and encoding data. Tools like Qiskit, Cirq, and PennyLane empower engineers to design efficient circuits, simulate their behavior, and optimize their performance. Understanding these design principles and tools is crucial for advancing quantum applications in engineering and computational sciences.

4.10.1   Review of Quantum Circuit Design Principles

The design of quantum circuits requires clarity of purpose, careful selection of quantum gates, and attention to operational fidelity.

4.10.1.1   Start with the Goal

The purpose of the quantum circuit determines its structure:

(a) Computation Circuits: Focused on solving specific problems, such as optimization or search.
(b) Data Preparation Circuits: Used for encoding classical data into quantum states.
(c) Hybrid Circuits: Combine data encoding with computation for tasks like quantum machine learning.

4.10.1.2   Choose Gates for Encoding and Computation

Quantum gates manipulate qubits to perform computations or encode data:

(a) **Basis Gates (X, Y, Z):** Fundamental operations that flip or rotate qubit states.
(b) **Hadamard Gate (H):** Creates superposition by applying an equal amplitude to $|0\rangle$ and $|1\rangle$:

$$H = \frac{1}{\sqrt{2}} \begin{bmatrix} 1 & 1 \\ 1 & -1 \end{bmatrix}$$

(c) **Rotation Gates (R_x, R_y, R_z):** Rotate qubits around specific axes on the Bloch sphere:

$$R_y(\theta) = \begin{bmatrix} \cos\left(\dfrac{\theta}{2}\right) & -\sin\left(\dfrac{\theta}{2}\right) \\ \sin\left(\dfrac{\theta}{2}\right) & \cos\left(\dfrac{\theta}{2}\right) \end{bmatrix}$$

4.10.1.3 Ensure Entanglement

For multi-qubit systems, entanglement is crucial for representing correlations in data. Gates like CNOT and CZ create entangled states that are key to quantum algorithms.

Python Code Example: Designing a Circuit
This example demonstrates the creation of a circuit that combines superposition, entanglement, and rotation gates (Fig. 4.13).

```
#----------------------------------------------------------------------
# Quantum Entangled-Based Encoding
# Chapter 4 in the QUANTUM COMPUTING AND QUANTUM MACHINE LEARNING BOOK
#----------------------------------------------------------------------
# Version 1.0
```

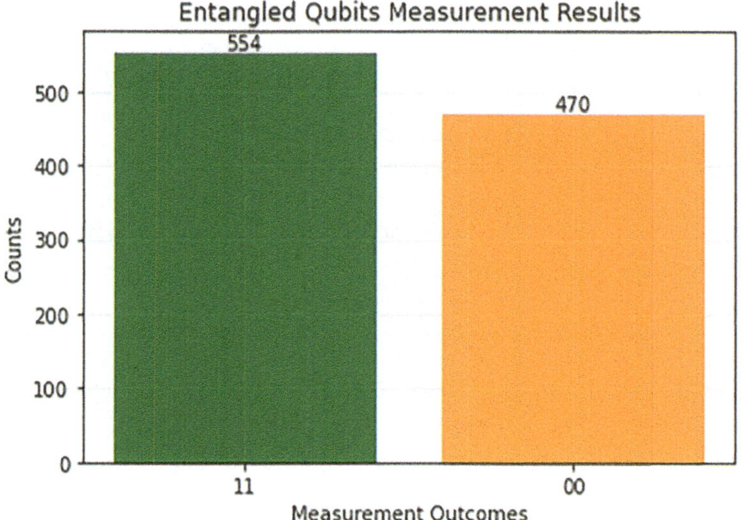

Fig. 4.13 Entangled qubits measurement results. (By the authors)

```python
# Qiskit changes frequently.
# We recommend using the latest version from the book code repository at:
# https://aqtinitiative.org/quantum-computing-for-engineers

# (c) 2025 Jesse Van Griensven, Roydon Fraser, and Jose Rosas
# License: MIT - Citation of this work required
#-------------------------------------------------------------------
import warnings
warnings.filterwarnings('ignore')

# Import necessary libraries
from qiskit import QuantumCircuit, Aer, execute
from qiskit.visualization import plot_histogram
from qiskit.visualization import circuit_drawer
import matplotlib.pyplot as plt
#-------------------------------------------------------------------

# Create a Quantum Circuit with 2 qubits and 2 classical bits
qc = QuantumCircuit(2, 2)

# Step 1: Create Entanglement
qc.h(0)      # Apply Hadamard gate to qubit 0 to create superposition
qc.cx(0, 1)   # Apply CNOT gate with qubit 0 as control and qubit 1 as
target

# Step 2: Measure the qubits
qc.measure([0,1], [0,1])

# Visualize the circuit
print("Quantum Circuit:")
display(circuit_drawer(qc, output='mpl', style="iqp"))

# Execute the circuit on the qasm simulator
simulator = Aer.get_backend('qasm_simulator')

# Execute the circuit with a specified number of shots (simulations)
shots = 1024
job  = execute(qc, simulator, shots=shots)

# Get the results
result = job.result()
counts = result.get_counts(qc)
```

```python
# Print the results
print("\nMeasurement Results:")
print(counts)

# Extract keys and values
labels = list(counts.keys())
values = list(counts.values())

# Create the histogram
# Note: using matplotlib directly - qiskit plot_histogram(counts) is
buggy
plt.bar(labels, values, color=['green', 'orange'])
plt.title("Entangled Qubits Measurement Results")
plt.xlabel("Measurement Outcomes")
plt.grid(alpha=0.2)
plt.ylabel("Counts")

# Add counts above the bars
for i, value in enumerate(values):
    plt.text(i, value + 5, str(value), ha='center', fontsize=10,
color='black')

plt.show()
```

4.10.2   Example: Encoding a Vector into a Quantum State

Encoding classical data into quantum states is an essential task in quantum computing. Consider a classical vector $v = [v_0, v_1, v_2, v_3]$.

4.10.2.1   Normalize the Vector

Normalization ensures the quantum state satisfies the condition $\langle \psi \mid \psi \rangle = 1$:

$$v' = \frac{v}{\sqrt{v_0^2 + v_1^2 + v_2^2 + v_3^2}}$$

4.10.2.2 Use Amplitude Encoding

The normalized vector is encoded into a 2-qubit quantum state:

$$|\psi\rangle = v_0' \,|00\rangle + v_1' \,|01\rangle + v_2' \,|10\rangle + v_3' \,|11\rangle$$

Python Code Example: Amplitude Encoding of a Four-Dimensional Vector
This code encodes a four-dimensional vector into a quantum state using amplitude
encoding.

```python
#------------------------------------------------------------------
# Amplitude Encoding of a 4-Dimensional Vector
# Chapter 4 in the QUANTUM COMPUTING AND QUANTUM MACHINE LEARNING BOOK
#------------------------------------------------------------------
# Version 1.0
# Qiskit changes frequently.
# We recommend using the latest version from the book code repository at:
# https://aqtinitiative.org/quantum-computing-for-engineers

# (c) 2025 Jesse Van Griensven, Roydon Fraser, and Jose Rosas
# License: MIT - Citation of this work required
#------------------------------------------------------------------
from qiskit import QuantumCircuit
from qiskit.visualization import circuit_drawer
import numpy as np

# Classical vector to encode
v = np.array([1, 2, 3, 4])

# Normalize the vector
norm = np.linalg.norm(v)
v_normalized = v / norm

# Create a quantum circuit for 2 qubits
qc = QuantumCircuit(2)

# Encode the normalized vector into the quantum state
qc.initialize(v_normalized, [0, 1])

print("Quantum Circuit for Vector Encoding:")
print(qc)
```

4.10.3   Tools for Simulation

Simulating quantum circuits is an essential step for validating designs and debugging algorithms. Several frameworks are available for circuit simulation:

4.10.3.1   Qiskit

Developed by IBM, Qiskit is a versatile software development kit (SDK) for designing, simulating, and running quantum circuits on both simulators and real quantum hardware. These are the two main features of Qiskit:

(a) Extensive support for circuit manipulation and visualization.
(b) Integration with IBM's quantum processors.

Python Code Example: Qiskit

```python
from qiskit import Aer, execute

# Simulate the previously defined quantum circuit
simulator = Aer.get_backend('aer_simulator')
result   = execute(qc, simulator).result()

# Get and display the result
counts = result.get_counts()
print("Simulation Result:", ccunts)
```

4.10.3.2   Cirq

Developed by Google, Cirq is a Python library for building and simulating quantum circuits, especially for hardware based on trapped ions and superconducting qubits.

Python Code Example: Cirq

```python
import cirq

# Create two qubits
q0, q1 = cirq.LineQubit.range(2)

# Define a circuit
circuit = cirq.Circuit(
  cirq.H(q0),
  cirq.CNOT(q0, q1),
  cirq.measure(q0, q1)
)
```

```
# Simulate the circuit
simulator = cirq.Simulator()
result = simulator.run(circuit)
print("Cirq Simulation Result:")
print(result)
```

4.10.3.3   PennyLane

PennyLane integrates quantum computation with machine learning, enabling hybrid quantum-classical workflows. It supports variational circuits and interfaces with frameworks like TensorFlow and PyTorch.

Python Code Example: PennyLane

```
#--------------------------------------------------------------------
# PennyLane Example
# Chapter 4 in the QUANTUM COMPUTING AND QUANTUM MACHINE LEARNING BOOK
#--------------------------------------------------------------------
# Version 1.0
# Qiskit changes frequently.
# We recommend using the latest version from the book code repository at:
# https://aqtinitiative.org/quantum-computing-for-engineers

# (c) 2025 Jesse Van Griensven, Roydon Fraser, and Jose Rosas
# License: MIT - Citation of this work required
#--------------------------------------------------------------------
import pennylane as qml

# Define a 2-qubit device
dev = qml.device("default.qubit", wires=2)

# Define a quantum function
@qml.qnode(dev)
def circuit(params):
  qml.RY(params[0], wires=0)
  qml.RY(params[1], wires=1)
  qml.CNOT(wires=[0, 1])
  return qml.probs(wires=[0, 1])

# Execute the circuit with sample parameters
params = [0.5, 1.0]
print("PennyLane Simulation Result:")
print(circuit(params))
```

4.11   Preparing Datasets for Quantum Machine Learning

Preparing datasets for QML is essential for leveraging quantum capabilities in data-driven applications. Data normalization ensures compatibility with quantum state requirements, feature mapping enables efficient representation of data in high-dimensional spaces, and hybrid encoding addresses the challenges of complex datasets. These preprocessing techniques are critical for the success of quantum machine learning algorithms across various fields.

Furthermore, preparing datasets for quantum machine learning (QML) involves tailoring classical data for quantum processing. Effective preprocessing ensures the compatibility of data with quantum encoding methods and optimizes the performance of QML algorithms. This section covers key aspects of dataset preparation, including data normalization, feature mapping, and encoding complex data structures.

4.11.1   Data Normalization

Data normalization is a critical preprocessing step for amplitude encoding, which requires that the dataset satisfies the quantum state normalization condition.

4.11.1.1   Normalization Condition

A quantum state $|\psi\rangle$ is valid only if the sum of the squares of its amplitudes equals 1:

$$\sum_i |x_i|^2 = 1$$

where x_i represents the components of the dataset.

4.11.1.2   Normalization Process

For a dataset $\mathbf{x} = \{x_1, x_2, \ldots, x_n\}$, normalization transforms the data as:

$$x_i' = \frac{x_i}{\sqrt{\sum_j |x_j|^2}}.$$

Python Code Example: Data Normalization
This code normalizes a dataset to satisfy the quantum state normalization condition, making it suitable for amplitude encoding.

```
#------------------------------------------------------------------
# Data Normalization
# Chapter 4 in the QUANTUM COMPUTING AND QUANTUM MACHINE LEARNING BOOK
#------------------------------------------------------------------
# Version 1.0
# Qiskit changes frequently.
# We recommend using the latest version from the book code repository at:
# https://aqtinitiative.org/quantum-computing-for-engineers

# (c) 2025 Jesse Van Griensven, Roydon Fraser, and Jose Rosas
# License: MIT - Citation of this work required
#------------------------------------------------------------------
import numpy as np

# Example dataset
data = np.array([2, 3, 5, 7])

# Normalize the data
norm_factor = np.linalg.norm(data)
normalized_data = data / norm_factor

print("Normalized Data:", normalized_data)
```

4.11.2   Feature Mapping

Feature mapping in QML transforms classical features into quantum states, enabling quantum algorithms to leverage high-dimensional spaces for better learning and classification. Quantum kernels use feature maps to transform classical data into quantum states:

$$\phi(x) = \exp\left(i\sum_{j} x_j Z_j \right) |0\rangle$$

where Z_j is a Pauli-Z gate applied to the j-th qubit, and x_j represents the j-th feature of the data point.

Two of the main advantages of feature mapping are:

1. Efficient representation of nonlinear relationships.
2. Enables quantum kernel-based methods, such as quantum support vector machines (QSVMs).

Python Code Example: Feature Mapping

This code encodes a data point into a quantum circuit using feature mapping with R_z gates.

```python
#-------------------------------------------------------------------
# Feature Mapping
# Chapter 4 in the QUANTUM COMPUTING AND QUANTUM MACHINE LEARNING BOOK
#-------------------------------------------------------------------
# Version 1.0
# Qiskit changes frequently.
# We recommend using the latest version from the book code repository at:
# https://aqtinitiative.org/quantum-computing-for-engineers

# (c) 2025 Jesse Van Griensven, Roydon Fraser, and Jose Rosas
# License: MIT - Citation of this work required
#-------------------------------------------------------------------
from qiskit import QuantumCircuit
from qiskit.visualization import circuit_drawer
import numpy as np


#-------------------------------------------------------------------
# Define a feature mapping function
def feature_map(data_point):
  n_qubits = len(data_point)
  qc = QuantumCircuit(n_qubits)
  for i, feature in enumerate(data_point):
   qc.rz(feature, i)  # Apply Z-rotation proportional to the feature
value
  return qc
#-------------------------------------------------------------------
# Example data point
data_point = np.array([0.5, 1.0, 1.5])

# Generate feature map
qc = feature_map(data_point)
print("Quantum Circuit for Feature Mapping:")
print(qc)

# Draw the circuit
display(circuit_drawer(qc, output='mpl', style="iqp"))
```

4.11.3   Encoding Complex Data Structures

QML applications often involve high-dimensional or structured data, such as categorical variables or time-series data. Preparing such datasets for quantum processing requires hybrid encoding techniques.

4.11.3.1   Hybrid Encoding for High-Dimensional Data

Hybrid encoding combines methods like amplitude encoding and angle encoding to manage large datasets efficiently. For example:

- Use amplitude encoding for numerical features.
- Use angle encoding for categorical features.

4.11.3.2   Mapping Categorical Features

Categorical variables can be mapped to discrete quantum states:

$$\text{Category } A \to |00\rangle, \quad \text{Category } B \to |01\rangle, \quad \text{Category } C \to |10\rangle$$

Python Code Example: Hybrid Encoding
This code demonstrates hybrid encoding for datasets containing both numerical and categorical features.

```python
#------------------------------------------------------------------
# Hybrid Encoding for numerical and categorical features
# Chapter 4 in the QUANTUM COMPUTING AND QUANTUM MACHINE LEARNING BOOK
#------------------------------------------------------------------
# Version 1.0
# Qiskit changes frequently.
# We recommend using the latest version from the book code repository at:
# https://aqtinitiative.org/quantum-computing-for-engineers

# (c) 2025 Jesse Van Griensven, Roydon Fraser, and Jose Rosas
# License: MIT - Citation of this work required
#------------------------------------------------------------------
from qiskit import QuantumCircuit
from qiskit.visualization import circuit_drawer
import numpy as np

# Numerical features
numerical_data = np.array([3.2, 5.7])
```

```
norm = np.linalg.norm(numerical_data)
normalized_numerical = numerical_data / norm

# Categorical feature (e.g., "Category B")
categorical_data = {"A": [0, 0], "B": [0, 1], "C": [1, 0]}
category_state  = categorical_data["B"]

# Create a quantum circuit
qc = QuantumCircuit(3)  # 2 qubits for numerical, 1 for categorical
qc.initialize(normalized_numerical, [0, 1])  # Amplitude encode
numerical data
qc.initialize(category_state, 2)  # Basis encode categorical data

print("Quantum Circuit for Hybrid Encoding:")
print(qc)

# Draw the circuit
display(circuit_drawer(qc, output='mpl', style="iqp"))
```

4.11.4   Applications of Preprocessed Quantum Datasets

Preprocessed quantum datasets, in addition to facilitating the development of quantum codes, are extensively employed in the following fields:

1. **Quantum Machine Learning Models**
   Preprocessed datasets are used in QSVMs, quantum neural networks, and variational quantum circuits for classification and regression tasks.
2. **Optimization Problems**
   Normalized data and feature maps enable efficient representation of optimization problems in quantum algorithms like QAOA.
3. **Quantum Feature Engineering**
   Hybrid encoding and kernel-based methods allow the extraction of meaningful patterns from complex datasets.

4.12   Real-World Data Encoding Examples in Quantum Computing

Quantum computing has the potential to revolutionize various industries by enabling efficient processing of large, complex datasets. However, effectively encoding real-world data into quantum states is a critical challenge. With continued advancements in quantum hardware and encoding techniques, the potential applications of quantum computing will only continue to grow.

This subsection presents practical examples of encoding different types of real-world datasets into quantum circuits, including image data for quantum image processing, financial data for portfolio optimization, and genomic data for bioinformatics applications. The following examples highlight the versatility of quantum data encoding techniques and their potential impact on quantum information processing.

4.12.1 Encoding Image Data for Quantum Image Processing

Image data is a key component in many applications, including computer vision, medical diagnostics, and security systems. Quantum image processing aims to leverage quantum algorithms to perform tasks such as image classification, enhancement, and compression more efficiently than classical methods. To achieve this, image data must first be encoded into quantum states.

Amplitude encoding is one of the most efficient methods for representing image data in quantum systems. It encodes pixel intensity values into the amplitudes of a quantum state, allowing for compact representation of large images. Consider a grayscale image with four pixels represented by the values [64, 128, 192, 255]. These values can be normalized and encoded into a two-qubit quantum state using amplitude encoding.

Python Code Example: Encoding Image Data

```
#------------------------------------------------------------------
# Encoding Image Data
# Chapter 4 in the QUANTUM COMPUTING AND QUANTUM MACHINE LEARNING BOOK
#------------------------------------------------------------------
# Version 1.0
# Qiskit changes frequently.
# We recommend using the latest version from the book code repository at:
# https://aqtinitiative.org/quantum-computing-for-engineers


# (c) 2025 Jesse Van Griensven, Roydon Fraser, and Jose Rosas
# License: MIT - Citation of this work required
#------------------------------------------------------------------


import numpy as np
from qiskit import QuantumCircuit
from qiskit.visualization import plot_bloch_multivector
from qiskit.visualization import circuit_drawer
#------------------------------------------------------------------
```

```python
# Image pixel values
pixels = np.array([64, 128, 192, 255])

# Normalize the pixel values
norm = np.linalg.norm(pixels)
normalized_pixels = pixels / norm

# Create a quantum circuit for 2 qubits
qc = QuantumCircuit(2)
qc.initialize(normalized_pixels, [0, 1])

print("Quantum Circuit for Image Data Encoding:")
print(qc)

# Draw the circuit
display(circuit_drawer(qc, output='mpl', style="iqp"))
```

4.12.2   Hybrid Encoding for Color Images

For color images, hybrid encoding techniques can be used to encode RGB (red, green, blue) values separately. Each color channel can be amplitude-encoded into a different set of qubits, allowing for parallel processing of color information.

Python Code Example: Encoding a 2x2 Color Image

```python
#------------------------------------------------------------------
# Encoding a 2x2 color image
# Chapter 4 in the QUANTUM COMPUTING AND QUANTUM MACHINE LEARNING BOOK
#------------------------------------------------------------------
# Version 1.0
# Qiskit changes frequently.
# We recommend using the latest version from the book code repository at:
# https://aqtinitiative.org/quantum-computing-for-engineers

# (c) 2025 Jesse Van Griensven, Roydon Fraser, and Jose Rosas
# License: MIT - Citation of this work required
#------------------------------------------------------------------

import numpy as np
from qiskit import QuantumCircuit
from qiskit.visualization import plot_bloch_multivector
from qiskit.visualization import circuit_drawer
#------------------------------------------------------------------
```

```python
# Example color image data (2x2 pixels)
red_channel   = np.array([255, 128, 64, 0])
green_channel = np.array([0, 64, 128, 255])
blue_channel  = np.array([64, 128, 192, 255])

# Normalize each channel
norm_red   = np.linalg.norm(red_channel)
norm_green = np.linalg.norm(green_channel)
norm_blue  = np.linalg.norm(blue_channel)
normalized_red   = red_channel   / norm_red
normalized_green = green_channel / norm_green
normalized_blue  = blue_channel  / norm_blue

# Create quantum circuits for each channel
qc_red   = QuantumCircuit(2)
qc_green = QuantumCircuit(2)
qc_blue  = QuantumCircuit(2)

qc_red.initialize( normalized_red,   [0, 1])
qc_green.initialize(normalized_green, [0, 1])
qc_blue.initialize( normalized_blue,  [0, 1])

print("Quantum Circuits for RGB Channels:")
print(qc_red)
print(qc_green)
print(qc_blue)

# Draw the circuits
display(circuit_drawer(qc_red,   output='mpl', style="iqp"))
display(circuit_drawer(qc_green, output='mpl', style="iqp"))
display(circuit_drawer(qc_blue,  output='mpl', style="iqp"))
```

4.12.3 Encoding Financial Data for Quantum Portfolio Optimization

Financial data, such as stock prices and risk metrics, plays a critical role in portfolio optimization. Quantum algorithms, like the quantum approximate optimization algorithm (QAOA), can optimize portfolios by finding the best asset allocation that minimizes risk and maximizes returns.

In portfolio optimization, decision variables can be encoded using basis encoding. For example, a binary decision variable x_i can represent whether a particular asset is included in the portfolio ($x_i = 1$) or not ($x_i = 0$).

Python Code Example: Encoding a Portfolio Selection Problem with Three Assets

```python
#-----------------------------------------------------------------
# Encoding a Portfolio Selection
# Chapter 4 in the QUANTUM COMPUTING AND QUANTUM MACHINE LEARNING BOOK
#-----------------------------------------------------------------
# Version 1.0
# Qiskit changes frequently.
# We recommend using the latest version from the book code repository at:
# https://aqtinitiative.org/quantum-computing-for-engineers

# (c) 2025 Jesse Van Griensven, Roydon Fraser, and Jose Rosas
# License: MIT - Citation of this work required
#-----------------------------------------------------------------

import numpy as np
from qiskit import QuantumCircuit
from qiskit.visualization import plot_bloch_multivector
from qiskit.visualization import circuit_drawer
#-----------------------------------------------------------------

# Create a quantum circuit with 3 qubits
qc = QuantumCircuit(3)

# Encode the portfolio selection
qc.x(0)  # Include asset 1
qc.x(2)  # Include asset 3

print("Quantum Circuit for Portfolio Selection:")
print(qc)

# Draw the circuit
display(circuit_drawer(qc, output='mpl', style="iqp"))
```

4.12.4   Amplitude Encoding for Asset Weights

In more advanced portfolio optimization models, asset weights can be amplitude-encoded into quantum states. This allows the representation of fractional asset allocations in a compact form.

Python Code Example: Encoding Asset Weights

```python
#-----------------------------------------------------------------
# Encoding Asset Weights
# Chapter 4 in the QUANTUM COMPUTING AND QUANTUM MACHINE LEARNING BOOK
#-----------------------------------------------------------------
# Version 1.0
# Qiskit changes frequently.
# We recommend using the latest version from the book code repository at:
# https://aqtinitiative.org/quantum-computing-for-engineers

# (c) 2025 Jesse Van Griensven, Roydon Fraser, and Jose Rosas
# License: MIT - Citation of this work required
#-----------------------------------------------------------------
import warnings
warnings.filterwarnings('ignore')

import numpy as np
from qiskit import QuantumCircuit
from qiskit.visualization import circuit_drawer
#-----------------------------------------------------------------

# Example asset weights
weights = np.array([0.3, 0.5, 0.2])

# Normalize the weights
norm_weights = np.linalg.norm(weights)
normalized_weights = weights / norm_weights

# Ensure the state vector has 2^n = 4 elements for a 2-qubit system
# Pad with zeros to make the length valid
state_vector = np.zeros(4)
state_vector[:len(normalized_weights)] = normalized_weights

# Create a quantum circuit
qc = QuantumCircuit(2)
qc.initialize(state_vector, [0, 1])

print("Quantum Circuit for Asset Weights:")
print(qc)

# Draw the circuit
display(circuit_drawer(qc, output='mpl', style="iqp"))
```

4.12.5 Encoding Genomic Data for Quantum Bioinformatics

In bioinformatics, genomic data is used to identify genetic variations and understand biological processes. Quantum computing can accelerate genomic analysis by encoding DNA sequences into quantum states and applying quantum algorithms to find patterns and mutations. DNA sequences consist of four nucleotides: adenine (A), cytosine (C), guanine (G), and thymine (T). Basis encoding can be used to represent each nucleotide as a unique binary string.

Python Code Example: Encoding a DNA Sequence "ACGT"

```python
#------------------------------------------------------------------
# Encoding a DNA sequence "ACGT"
# Chapter 4 in the QUANTUM COMPUTING AND QUANTUM MACHINE LEARNING BOOK
#------------------------------------------------------------------
# Version 1.0
# Qiskit changes frequently.
# We recommend using the latest version from the book code repository at:
# https://aqtinitiative.org/quantum-computing-for-engineers

# (c) 2025 Jesse Van Griensven, Roydon Fraser, and Jose Rosas
# License: MIT - Citation of this work required
#------------------------------------------------------------------

import numpy as np
from qiskit import QuantumCircuit
from qiskit.visualization import plot_bloch_multivector
from qiskit.visualization import circuit_drawer
#------------------------------------------------------------------

# DNA to binary mapping
dna_map = {"A": "00", "C": "01", "G": "10", "T": "11"}

# Example DNA sequence
dna_sequence = "ACGT"

# Create a quantum circuit
qc = QuantumCircuit(4)

# Encode the DNA sequence
for i, nucleotide in enumerate(dna_sequence):
    if dna_map[nucleotide][0] == "1":
        qc.x(2 * i)
    if dna_map[nucleotide][1] == "1":
        qc.x(2 * i + 1)
```

```
print("Quantum Circuit for DNA Sequence Encoding:")
print(qc)

# Draw the circuit
display(circuit_drawer(qc, output='mpl', style="iqp"))
```

In more complex genomic datasets, hybrid encoding can be used to represent numerical features (e.g., gene expression levels) alongside categorical features (e.g., nucleotide sequences).

Python Code Example: Encoding Gene Expression Levels and Nucleotide Sequences

```
#------------------------------------------------------------------
# Encoding a DNA sequence "ACGT"
# Chapter 4 in the QUANTUM COMPUTING AND QUANTUM MACHINE LEARNING BOOK
#------------------------------------------------------------------
# Version 1.0
# Qiskit changes frequently.
# We recommend using the latest version from the book code repository at:
# https://aqtinitiative.org/quantum-computing-for-engineers


# (c) 2025 Jesse Van Griensven, Roydon Fraser, and Jose Rosas
# License: MIT - Citation of this work required
#------------------------------------------------------------------


import numpy as np
from qiskit import QuantumCircuit
from qiskit.visualization import plot_bloch_multivector
from qiskit.visualization import circuit_drawer
#------------------------------------------------------------------


# Gene expression levels
gene_expression = np.array([1.5, 2.3, 0.8])
# Normalize the expression levels
norm_expression = np.linalg.norm(gene_expression)
normalized_expression = gene_expression / norm_expression
# Create a quantum circuit
qc = QuantumCircuit(3)
qc.initialize(normalized_expression, [0, 1, 2])
print("Quantum Circuit for Genomic Data Encoding:")
print(qc)

# Draw the circuit
display(circuit_drawer(qc, output='mpl', style="iqp"))
```

4.13 Challenges in Quantum Data Representation

Quantum data representation involves encoding classical data into quantum states for further processing. This process is fraught with challenges arising from the constraints of current quantum hardware, the complexities of quantum state preparation, and the limitations of scalability and resource allocation.

Quantum data representation faces challenges in high dimensionality, noise, scalability, and resource optimization. Addressing these issues involves techniques, such as dimensionality reduction, error mitigation, hybrid workflows, and resource-efficient encoding. As quantum hardware evolves, these solutions will play a crucial role in enabling effective quantum computation on real-world datasets.

4.13.1 High Dimensionality

Classical datasets often have high dimensionality, which poses significant challenges for quantum representation due to the limited number of qubits in existing quantum systems.

4.13.1.1 Dimensionality Reduction

To fit high-dimensional data into quantum hardware, dimensionality reduction techniques such as principal component analysis (PCA) are often employed. PCA reduces data by selecting the most significant eigenvectors of the covariance matrix:

$$\mathbf{X}_{reduced} = \mathbf{X}\mathbf{V}_{top-k}$$

where $\mathbf{V}_{top-k}$ represents the top k eigenvectors corresponding to the largest eigenvalues.

Python Code Example: Dimensionality Reduction with PCA
This code demonstrates the use of PCA to reduce a high-dimensional dataset to a lower-dimensional representation suitable for quantum encoding. Figure 4.14 demonstrates this dimensionality reduction from three-dimensional data into a two-dimensional space.

```
#-----------------------------------------------------------------
# Dimensionality Reduction with PCA
# Chapter 4 in the QUANTUM COMPUTING AND QUANTUM MACHINE LEARNING BOOK
#-----------------------------------------------------------------
# Version 1.0
# Qiskit changes frequently.
```

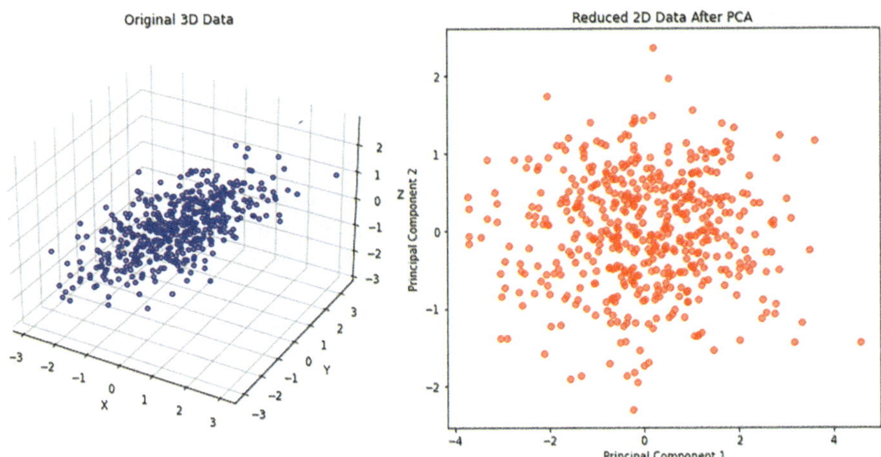

Fig. 4.14 PCA reduction of three-dimensional data to simpler 2D representation. (By the authors)

```python
# We recommend using the latest version from the book code repository at:
# https://aqtinitiative.org/quantum-computing-for-engineers

# (c) 2025 Jesse Van Griensven, Roydon Fraser, and Jose Rosas
# License: MIT - Citation of this work required
#-------------------------------------------------------------------------
# Import necessary libraries
import numpy as np
import matplotlib.pyplot as plt
from sklearn.decomposition import PCA
from mpl_toolkits.mplot3d import Axes3D
#-------------------------------------------------------------------------
# Step 1: Generate synthetic 3D data
# Fix the random seed to assure that all runs will have the same results
np.random.seed(42)

# Mean of the data
mean = [0, 0, 0]

# Covariance matrix
cov = [[1, 0.8, 0.6], [0.8, 1, 0.5], [0.6, 0.5, 1]]

# Generate 3D points
data_3d = np.random.multivariate_normal(mean, cov, 500)
#-------------------------------------------------------------------------
# Step 2: Visualize the original 3D data
fig = plt.figure(figsize=(12, 6))
```

```python
ax = fig.add_subplot(121, projection='3d')
ax.scatter(data_3d[:, 0], data_3d[:, 1], data_3d[:, 2], alpha=0.5,
c='blue')
ax.set_title("Original 3D Data")
ax.set_xlabel("X")
ax.set_ylabel("Y")
ax.set_zlabel("Z")
#-------------------------------------------------------------------
# Step 3: Perform PCA for dimensionality reduction to 2D
pca = PCA(n_components=2)  # Reduce to 2 dimensions
data_2d = pca.fit_transform(data_3d)
#-------------------------------------------------------------------
# Step 4: Explained variance ratio
explained_variance = pca.explained_variance_ratio_
print("Explained Variance Ratio by Principal Components:",
explained_variance)
#-------------------------------------------------------------------
# Step 5: Visualize the reduced 2D data
ax2 = fig.add_subplot(122)
ax2.scatter(data_2d[:, 0], data_2d[:, 1], alpha=0.5, c='red')
ax2.set_title("Reduced 2D Data After PCA")
ax2.set_xlabel("Principal Component 1")
ax2.set_ylabel("Principal Component 2")
ax2.grid(alpha=0.2)

plt.tight_layout()
plt.show()
```

4.13.2   Quantum PCA (QPCA)

QPCA can perform dimensionality reduction directly on quantum states using the quantum Fourier transform (QFT) to identify principal components. The covariance matrix is represented in quantum form, and eigenvalues are determined using phase estimation.

4.13.3   Quantum Noise and Quantum Error

Quantum gates introduce noise, which can lead to errors in the preparation of quantum states. Noise is often modeled as a depolarizing channel:

$$\rho \rightarrow (1-p)\rho + \frac{p}{d}I$$

where ρ is the density matrix of the quantum state, p is the depolarization probability, and d is the dimension of the state space.

Error Mitigation Techniques

1. Dynamical Decoupling: Applies sequences of control pulses to cancel out environmental noise.
2. Error Correction Codes: Techniques such as the surface code detect and correct errors during computation.

Python Code Example: Simulating Noise
This code simulates the effect of depolarizing noise on a quantum circuit.

```python
#------------------------------------------------------------------------
# Simulating Noise
# Chapter 4 in the QUANTUM COMPUTING AND QUANTUM MACHINE LEARNING BOOK
#------------------------------------------------------------------------
# Version 1.0
# Qiskit changes frequently.
# We recommend using the latest version from the book code repository at:
# https://aqtinitiative.org/quantum-computing-for-engineers

# (c) 2025 Jesse Van Griensven, Roydon Fraser, and Jose Rosas
# License: MIT - Citation of this work required
#------------------------------------------------------------------------

from qiskit import QuantumCircuit, Aer, execute
from qiskit.providers.aer.noise import NoiseModel,
depolarizing_error
from qiskit.visualization import circuit_drawer
#------------------------------------------------------------------------

# Define a quantum circuit
qc = QuantumCircuit(1)
qc.h(0)  # Apply a Hadamard gate
qc.measure_all()

# Add noise model
noise_model = NoiseModel()
noise_model.add_all_qubit_quantum_error(depolarizing_error(0.1,
1), ['h'])
```

```
# Simulate with noise
simulator = Aer.get_backend('aer_simulator')
result = execute(qc, simulator, noise_model=noise_model).result()
print("Noisy Simulation Result:", result.get_counts())

# Draw the circuit
display(circuit_drawer(qc, output='mpl', style="iqp"))
```

4.14   Qubit and Coherence Limitations

Current quantum systems have a limited number of qubits, and these qubits are prone to decoherence over time. The coherence time (T_2) is inversely proportional to the noise power density:

$$T_2 \propto \frac{1}{\text{Noise Power Density}}$$

4.14.1   Impact on Large Datasets

The limited qubit count constrains the ability to encode large datasets. For example, encoding N data points using amplitude encoding requires $\log_2(N)$ qubits.

Table 4.4 summarizes ways to mitigate existing quantum computer scalability problems.

Python Code Example: Hybrid Quantum-Classical Workflow
This example shows how classical preprocessing reduces the size of data before quantum encoding.

```
#------------------------------------------------------------------
# Hybrid Quantum-Classical Workflow
# Chapter 4 in the QUANTUM COMPUTING AND QUANTUM MACHINE LEARNING BOOK
#------------------------------------------------------------------
# Version 1.0
```

Table 4.4 Solutions for quantum computer scalability

Challenge	Solution
Quantum feature maps	Transform datasets into smaller, meaningful feature spaces before encoding
Hybrid quantum-classical systems	Use classical preprocessing to reduce dataset size while leveraging quantum algorithms for high-dimensional features

```
# Qiskit changes frequently.
# We recommend using the latest version from the book code repository at:
# https://aqtinitiative.org/quantum-computing-for-engineers


# (c) 2025 Jesse Van Griensven, Roydon Fraser, and Jose Rosas
# License: MIT - Citation of this work required
#------------------------------------------------------------------


from sklearn.preprocessing import StandardScaler
from qiskit import QuantumCircuit
from qiskit.visualization import circuit_drawer
#------------------------------------------------------------------


# Preprocess data with classical methods
data = [[1.2, 0.7], [0.9, 1.5], [1.8, 1.0]]
scaler = StandardScaler()
normalized_data = scaler.fit_transform(data)


# Create a quantum circuit to encode reduced features
qc = QuantumCircuit(2)
for i, feature in enumerate(normalized_data[0]):
    qc.ry(feature, i)  # Encode each feature as a rotation


print("Quantum Circuit for Scaled Data:")
print(qc)


# Draw the circuit
display(circuit_drawer(qc, output='mpl', style="iqp"))
```


4.15 Resource Optimization

Quantum algorithms often face trade-offs between the number of qubits used and the accuracy of results. For example, encoding a dataset with high fidelity requires more qubits and gates, increasing circuit depth and error rates (Table 4.5).

Table 4.5 Strategies for resource optimization

Optimization strategy	Description
Gate optimization	Minimize the number of gates used in the circuit. For instance, combine multiple operations into a single gate
Sparse encoding	Represent only significant features of the dataset to reduce qubit usage
Efficient error mitigation	Use error mitigation techniques that balance computational cost with fidelity improvement

Python Code Example: Sparse Encoding
This code encodes a sparse dataset, minimizing resource usage by focusing on nonzero elements.

```python
#------------------------------------------------------------------------
# Sparse Encoding
# Chapter 4 in the QUANTUM COMPUTING AND QUANTUM MACHINE LEARNING BOOK
#------------------------------------------------------------------------
# Version 1.0
# Qiskit changes frequently.
# We recommend using the latest version from the book code repository at:
# https://aqtinitiative.org/quantum-computing-for-engineers

# (c) 2025 Jesse Van Griensven, Roydon Fraser, and Jose Rosas
# License: MIT - Citation of this work required
#------------------------------------------------------------------------
import numpy as np
from qiskit import QuantumCircuit
from qiskit.visualization import circuit_drawer
#------------------------------------------------------------------------

# Example sparse dataset
sparse_data = np.array([0, 0, 1.5, 0, 0.8, 0])

# Normalize and encode only non-zero elements
indices = np.nonzero(sparse_data)[0]
values = sparse_data[indices]
norm = np.linalg.norm(values)
normalized_values = values / norm

# Create a quantum circuit
qc = QuantumCircuit(len(indices))
for i, value in enumerate(normalized_values):
  qc.ry(value, i)

print("Quantum Circuit for Sparse Encoding:")
print(qc)

# Draw the circuit
display(circuit_drawer(qc, output='mpl', style="iqp"))
```

4.16   Hardware-Specific Encoding Methods

Quantum encoding techniques must adapt to the physical characteristics of the quantum hardware platform on which they are implemented. Different types of quantum hardware, such as superconducting qubits, trapped ion qubits, photonic qubits, and spin qubits, have distinct methods for storing and processing quantum information. The choice of encoding method depends on the specific qubit technology and its underlying physical properties.

In this section, we explore how data is encoded in various hardware platforms, including superconducting qubits, trapped ion qubits, photonic qubits, and spin qubits. Each hardware-specific encoding method has unique advantages and challenges, and understanding these differences is essential for optimizing quantum algorithms for different quantum devices.

4.16.1   Superconducting Qubits

Superconducting qubits are among the most widely used qubit technologies in modern quantum computers. They rely on the principles of superconductivity and are implemented using devices such as transmon qubits. These qubits are formed by creating a Josephson junction, which is a nonlinear inductive element that allows for the precise control of quantum states.

The transmon qubit is a type of superconducting qubit that stores quantum information in the discrete energy levels of a superconducting circuit. The two lowest energy levels, typically labeled $|0\rangle$ and $|1\rangle$, form the computational basis states of the qubit.

The state of a transmon qubit is represented as a superposition of these two states:

$$|\psi\rangle = \alpha\,|0\rangle + \beta\,|1\rangle$$

Data encoding in transmon devices is achieved by applying microwave pulses to control the transition between the $|0\rangle$ and $|1\rangle$ states. The frequency, phase, and duration of these pulses determine the final quantum state.

Superconducting qubits can perform quantum gates at high speeds (nanosecond timescales) and are compatible with existing semiconductor fabrication technologies, making them easier to scale. However, superconducting qubits are prone to decoherence, limiting the time available for quantum computations, and they require extremely low temperatures (millikelvin range) to maintain superconductivity.

4.16.2 *Trapped Ion Qubits*

Trapped ion qubits are another leading quantum hardware platform. These qubits use charged atoms (ions) that are trapped and manipulated using electromagnetic fields. The quantum information is encoded in the internal energy levels of the ions, such as their electronic or hyperfine states.

In trapped ion systems, qubits are typically represented by two energy levels of an ion:

$$|0\rangle = \text{Ground state}, \quad |1\rangle = \text{Excited state}$$

Data encoding is performed using laser pulses to manipulate the ion's quantum state. The frequency and intensity of the laser pulses control transitions between the ground and excited states. The two main types of encodings in trapped ion qubits are described below:

1. **Hyperfine States Encoding**

   (a) Quantum information is stored in the hyperfine energy levels of the ion's electron.
   (b) This encoding is highly stable and less prone to decoherence.

2. **Optical States Encoding**

   (a) The qubit is encoded in the ion's optical transitions.
   (b) This method allows for faster gate operations but is more sensitive to noise.

Trapped ion qubits can maintain coherence for minutes or even hours and they have some of the highest gate fidelities among quantum platforms. However, trapped ion qubits have slower gate operations compared to superconducting qubits, and they require precise laser control and stable trapping mechanisms.

4.16.3 *Photonic Qubits*

Photonic qubits use particles of light (photons) to encode quantum information. Photonic qubits are particularly useful for quantum communication and quantum networking applications because photons can travel long distances with minimal decoherence. There are several methods to encode quantum information in photonic qubits:

1. **Polarization Encoding**

   (a) Quantum information is encoded in the polarization state of a photon.
   (b) The horizontal polarization $|H\rangle$ represents $|0\rangle$, and the vertical polarization $|V\rangle$ represents $|1\rangle$.

$$|\psi\rangle = \alpha |H\rangle + \beta |V\rangle$$

2. **Phase Encoding**

   (a) Quantum information is encoded in the relative phase between two paths in an interferometer.
   (b) This method is commonly used in quantum key distribution (QKD) protocols.

3. **Time Bin Encoding**

   (a) Quantum information is encoded in the arrival time of photons.
   (b) Early arrival represents $|0\rangle$, and late arrival represents $|1\rangle$.

Photonic qubits do not require cryogenic temperatures and they are less affected by environmental noise compared to other qubits. Additionally, photons can travel through fiber-optic cables or free space for long distances without significant loss. However, generating entangled photonic states is challenging.

4.16.4   Spin Qubits

Spin qubits use the spin states of electrons or nuclei to encode quantum information. These qubits are typically implemented using semiconductor quantum dots or diamond-based systems. In a spin qubit, the quantum information is encoded in the spin orientation of an electron or nucleus:

$$|0\rangle = \text{Spin down}, \quad |1\rangle = \text{Spin up}$$

Data encoding is achieved by applying magnetic fields or microwave pulses to manipulate the spin state. There are two main types of spin qubit. These are described below.

1. **Electron Spin Qubits**:

   (a) The qubit is encoded in the spin of an electron trapped in a quantum dot.
   (b) These qubits are compatible with semiconductor manufacturing technologies.

2. **Nuclear Spin Qubits**:

   (a) The qubit is encoded in the spin of a nucleus in a solid-state material.
   (b) These qubits have longer coherence times than electron spin qubits.

Spin qubits are compatible with existing semiconductor fabrication processes. They have long coherence times, particularly for nuclear spin qubits. However, manipulating spin qubits requires precise control over magnetic fields and microwave pulses. Additionally, spin qubits are sensitive to noise from surrounding nuclear spins in the material.

4.17   Encoding in Post-Quantum Cryptographic Schemes

Post-quantum cryptography (PQC) refers to cryptographic methods designed to remain secure even against attackers with access to powerful quantum computers. While quantum key distribution (QKD) offers information-theoretic security based on the laws of quantum physics, PQC relies on mathematical problems that are believed to be hard for both classical and quantum computers.

The encoding methods used in PQC are typically based on lattice-based cryptography, hash-based cryptography, code-based cryptography, and multivariate polynomial cryptography.

4.17.1   Lattice-Based Encoding

One of the most promising approaches in PQC is lattice-based cryptography, which encodes data as points in a high-dimensional lattice. The security of lattice-based schemes is based on problems such as the shortest vector problem (SVP) and the learning with errors (LWE) problem, which are believed to be hard for quantum computers to solve.

In lattice-based encoding:

1. Data is represented as vectors in a lattice.
2. The encryption process involves adding noise to the vector to make it difficult for an attacker to recover the original message.

   Given a lattice Λ and a vector v in the lattice, the encoded message is:

$$v' = v + e$$

where e is a small random error vector, the challenge for an attacker is to recover v from v', which is computationally hard.

4.17.2   Hash-Based Encoding

Hash-based cryptography uses one-way hash functions to encode data. These schemes are simple and secure against quantum attacks, as the security of hash functions is not affected by quantum algorithms like Shor's algorithm.

In hash-based encoding:

1. Data is encoded by applying a cryptographic hash function.
2. The encoded data is difficult to reverse, ensuring security against attackers.

4.17.3   Code-Based Encoding

Code-based cryptography uses error-correcting codes to encode data. The security of these schemes is based on the hardness of decoding random linear codes, a problem that remains hard even for quantum computers. In code-based encoding:

1. Data is encoded as a code word in an error-correcting code.
2. The encryption process involves introducing random errors into the code word.

Exercise Questions: Quantum Data Structures and Encoding
1. **Basis Encoding**

    (a) Define basis encoding and provide its mathematical representation for an n-qubit system.
    (b) Using basis encoding, represent the binary value $x = 6$ in a 3-qubit system.

2. **Amplitude Encoding**

    (a) Explain amplitude encoding and discuss its advantage in representing large datasets.
    (b) Derive the normalized quantum state for the dataset $x = [2, 4, 4]$ using amplitude encoding.

3. **Angle Encoding**

    (a) Write the mathematical expression for angle encoding using $R_y(x)$ and explain its significance.
    (b) Provide an example where angle encoding is applied to map the value $x = \pi/4$.

4. **Hybrid Encoding**

    (a) What is hybrid encoding, and why is it useful for complex datasets?
    (b) Design a hybrid encoding scheme for a dataset containing numerical and categorical features, such as age and gender.

5. **Quantum Circuit Design**

    (a) Describe the key principles for designing quantum circuits for data encoding.
    (b) Write a Python program using Qiskit to create a circuit that encodes a three-dimensional vector using amplitude encoding.

6. **Data Normalization**

    (a) Explain the role of data normalization in preparing datasets for amplitude encoding.
    (b) Normalize the dataset $[3, 4, 5]$ and provide the corresponding normalized quantum state.

7. **Feature Mapping**

   (a) What is quantum feature mapping, and how is it used in quantum machine learning?
   (b) Using rotation gates, map the feature vector $[\pi/2, \pi/3]$ into a quantum circuit.

8. **Encoding Complex Data Structures**

   (a) Discuss the challenges of encoding high-dimensional datasets into quantum systems.
   (b) Propose a hybrid encoding strategy for a dataset containing time-series and categorical data.

9. **Tools for Quantum Circuit Simulation**

   (a) Compare the simulation capabilities of Qiskit, Cirq, and PennyLane for quantum circuit design.
   (b) Using Qiskit, write a program to simulate a quantum circuit that encodes a 4-point dataset into a 2-qubit quantum state.

10. **Challenges in Quantum Data Representation**

    (a) Identify two major challenges in representing classical data as quantum states and discuss possible solutions.
    (b) Explain the concept of dimensionality reduction and its application in quantum principal component analysis (QPCA).

11. **Quantum Noise and Error Mitigation**

    (a) How does quantum noise affect the fidelity of quantum data encoding, and what techniques mitigate these errors?
    (b) Simulate the effect of depolarizing noise on a simple quantum circuit and discuss the results.

12. **Applications of Preprocessed Quantum Datasets**

    (a) Explain how normalized quantum datasets are used in quantum support vector machines (QSVMs).
    (b) Provide an example of how hybrid encoding enables better classification in quantum neural networks.

Appendix: Equations for Key Concepts

1. **Quantum Normalization Condition:**

$$\sum_{i=0}^{2^n - 1} |a_i|^2 = 1$$

2. **Amplitude Encoding for a Vector:**

$$|x\rangle = \frac{1}{\sqrt{\sum_i |x_i|^2}} \sum_{i=0}^{N-1} x_i |i\rangle$$

3. **Angle Encoding Example**:

$$R_y(\theta) = \begin{bmatrix} \cos(\theta/2) & -\sin(\theta/2) \\ \sin(\theta/2) & \cos(\theta/2) \end{bmatrix}$$

4. **Kernel Mapping in Quantum Feature Spaces**:

$$k(x,y) = \langle \phi(x) | \phi(y) \rangle$$

Additional Bibliography

1. S. Aaronson, *Quantum Computing Since Democritus* (Cambridge University Press, 2013)
2. C.H. Bennett, G. Brassard, Quantum cryptography: public key distribution and coin tossing, in *Proceedings of IEEE International Conference on Computers, Systems and Signal Processing*, (1984), pp. 175–179
3. D. Bouwmeester, A.K. Ekert, A. Zeilinger (eds.), *The Physics of Quantum Information: Quantum Cryptography, Quantum Teleportation, Quantum Computation* (Springer, Berlin, 2000)
4. G. Brassard, *Quantum Communication Complexity* (Cambridge University Press, 2002)
5. R. Cleve et al., Quantum algorithms revisited. Math. Programm. **97**(1), 17–25 (2003)
6. D. Deutsch, Quantum theory, the Church-Turing principle, and the universal quantum computer. Proc. R. Soc. Lond. A Math. Phys. Sci. **400**(1818), 97–117 (1985)
7. J. Gruska, *Quantum Computing* (McGraw-Hill, 1999)
8. L.K. Grover, A Fast Quantum Mechanical Algorithm for Database Search, in *Proceedings of the Twenty-Eighth Annual ACM Symposium on Theory of Computing (STOC)*, (1996), pp. 212–219
9. M. Hirvensalo, *Quantum Computing* (Springer, 2004)
10. A. Kitaev et al., *Classical and Quantum Computation* (American Mathematical Society, 2002)
11. S. Lloyd, Universal Quantum Simulators. Science **273**(5278), 1073–1078 (1996). https://doi.org/10.1126/science.273.5278.1073
12. S. Lloyd, *Programming the Universe: A Quantum Computer Scientist Takes on the Cosmos* (Alfred A. Knopf, 2006)
13. N.D. Mermin, *Quantum Computer Science: An Introduction* (Cambridge University Press, 2007)
14. A. Montanaro, Quantum algorithms: an overview. NPJ Quantum Inf. **2**, 15023 (2016)
15. M.A. Nielsen, I.L. Chuang, *Quantum Computation and Quantum Information: 10th Anniversary Edition* (Cambridge University Press, 2010)
16. J. Preskill, Quantum computing and the entanglement frontier. Proc. Natl. Acad. Sci. **111**(15), 5759–5763 (2014)
17. P.W. Shor, Polynomial-time algorithms for prime factorization and discrete logarithms on a quantum computer. SIAM J. Comput. **26**(5), 1484–1509 (1997)
18. A.M. Steane, Error correcting codes in quantum theory. Phys. Rev. Lett. **77**(5), 793–797 (1996)

19. V. Vedral, *Decoding Reality: The Universe as Quantum Information* (Oxford University Press, 2010)
20. W.H. Zurek, *Quantum Theory and Measurement* (Princeton University Press, 1983)
21. W.H. Zurek, Decoherence, Einselection, and the quantum origins of the classical. Rev. Mod. Phys. **75**(3), 715–775 (2003). https://dci.org/10.1103/RevModPhys.75.715

Chapter 5
Quantum Algorithms

Quantum gates are the building blocks of quantum circuits, analogous to classical logic gates in traditional computing. They operate on qubits and manipulate their quantum states through unitary transformations, enabling quantum computations. They are indispensable in quantum computing, serving as the fundamental operations for constructing quantum circuits. Unlike classical gates, they operate on quantum states, enabling superposition, entanglement, and interference. Quantum gates' mathematical foundation in unitary matrices ensures reversibility and fidelity, while their role in enabling universal computation underscores their importance in realizing practical quantum algorithms.

This chapter also introduces foundational quantum algorithms like QFT, Shor's, Grover's, and VQE, emphasizing their mathematical basis and applications in engineering. Engineers equipped with this knowledge can apply quantum algorithms to tackle complex challenges in optimization, secure communication, and simulation.

5.1 Physical Implementations of Quantum Gates

Quantum gates, as theoretical constructs, are realized in physical systems through carefully engineered interactions that manipulate qubits. The physical implementation of quantum gates varies significantly across different platforms, each offering unique strengths and challenges. Superconducting qubits excel in gate speed and scalability but face decoherence issues. Trapped ions offer high fidelity and precise control, albeit at slower speeds. Photonic systems enable room-temperature operation and robust quantum communication, while topological qubits promise fault tolerance with intricate implementation requirements.

This section explores the physical implementations of quantum gates in superconducting qubits, trapped ions, photonic systems, and topological qubits, highlighting the methods used, their performance, and the associated trade-offs.

J. Van Griensven Thé et al., *Quantum Computing and Quantum Machine Learning for Engineers and Developers*, https://doi.org/10.1007/978-3-031-98245-3_5

5.2 Superconducting Qubits

Superconducting qubits are among the most mature platforms for quantum computing. They are fabricated as electrical circuits using superconducting materials, where quantum states are encoded in the system's energy levels. Quantum gates are implemented using microwave pulses to manipulate these states.

5.2.1 Microwave Pulse-Based Gates

Superconducting qubits operate at cryogenic temperatures to maintain superconductivity. The two lowest energy levels of a superconducting circuit, typically a transmon qubit, are used as the qubit states $|0\rangle$ and $|1\rangle$. Quantum gates are applied by using resonant microwave pulses with precise amplitudes, frequencies, and phases.

1. **Single-Qubit Gates**

    These gates correspond to rotations around the Bloch sphere axes (R_x, R_y, R_z) and are implemented by tuning the frequency of the microwave pulses to the qubit's resonance frequency.

2. **Two-Qubit Gates**

    Gates such as the Controlled-NOT (CNOT) or Controlled-Z (CZ) gates are achieved by coupling two qubits through a tunable interaction mediated by a resonator or a shared superconducting element.

Python Code Example Single-Qubit Rotation

```python
#------------------------------------------------------------------
# Quantum Single-Qubit Rotation
# Chapter 5 in the QUANTUM COMPUTING AND QUANTUM MACHINE LEARNING BOOK
#------------------------------------------------------------------
# Version 1.0
# Qiskit changes frequently.
# We recommend using the latest version from the book code repository at:
# https://aqtinitiative.org/quantum-computing-for-engineers

# (c) 2025 Jesse Van Griensven, Roydon Fraser, and Jose Rosas
# License: MIT - Citation of this work required
#------------------------------------------------------------------
import numpy as np
from qiskit import QuantumCircuit
from qiskit.visualization import circuit_drawer
#------------------------------------------------------------------
```

```
# Create a single-qubit circuit
qc = QuantumCircuit(1)
qc.rx(np.pi / 2, 0)  # Rotate around X-axis by 90 degrees
qc.rz(np.pi, 0)    # Rotate around Z-axis by 180 degrees

# Draw the circuit
#print(qc.draw())
display(circuit_drawer(qc, output='mpl', style="iqp"))
```

5.2.2 Superconducting Qubit Limitations

Superconducting qubits have some limitations:

1. **Decoherence**
   Superconducting qubits are sensitive to environmental noise, leading to decoherence and errors during gate operations. Typical coherence times are on the order of microseconds, limiting the depth of quantum circuits.
2. **Cross Talk**
   Interaction between neighboring qubits can lead to unintended gate operations.
3. **Scalability**
   Fabricating large-scale superconducting qubit systems with high fidelity remains a significant engineering challenge.

5.3 Trapped Ions

Trapped ion quantum computing uses ions confined in electromagnetic traps as qubits. These qubits are manipulated using laser-induced interactions to implement quantum gates.

5.3.1 Laser-Induced Gate Operations

In a trapped ion system, qubits are encoded in the internal electronic states of the ions. Laser pulses are used to control the transitions between these states.

1. **Single-Qubit Gates**
   Implemented by shining a laser at an individual ion to induce transitions between $|0\rangle$ and $|1\rangle$. These gates correspond to rotations on the Bloch sphere.

2. **Two-Qubit Gates**

Achieved by coupling the internal states of two ions to their shared vibrational modes. A common two-qubit gate is the Mølmer–Sørensen gate, which creates entanglement between the ions.

For example, the Mølmer–Sørensen interaction is described by the Hamiltonian:

$$H = \hbar\Omega(a^\dagger + a)\left(\sigma_x^{(1)} + \sigma_x^{(2)}\right)$$

where $a^\dagger$ and a are the creation and annihilation operators for the vibrational modes, and $\sigma_x^{(i)}$ represents the Pauli-X operator for ion i.

Python Code Example: Simple Trapped Ion Circuit

```python
#------------------------------------------------------------------
# Quantum Simple Trapped Ion Circuit
# Chapter 5 in the QUANTUM COMPUTING AND QUANTUM MACHINE LEARNING BOOK
#------------------------------------------------------------------
# Version 1.0
# Qiskit changes frequently.
# We recommend using the latest version from the book code repository at:
# https://aqtinitiative.org/quantum-computing-for-engineers


# (c) 2025 Jesse Van Griensven, Roydon Fraser, and Jose Rosas
# License: MIT - Citation of this work required
#------------------------------------------------------------------
import numpy as np
from qiskit import QuantumCircuit
from qiskit.visualization import import circuit_drawer


#------------------------------------------------------------------

qc = QuantumCircuit(2)
qc.h(0)  # Apply Hadamard to ion 1
qc.cx(0, 1)  # Entangle ion 1 and ion 2
# Draw the circuit
print(qc.draw())
display(circuit_drawer(qc, output='mpl', style="iqp"))
```

5.3.2   Trapped Ion Speed Vs. Fidelity Trade-Offs

Table 5.1 summarizes the speed versus fidelity trade-off in trapped ion gates.

Table 5.1 Trapped ion speed and fidelity issues

Trade-off item	Description
Speed	Gate operations in trapped ion systems are slower compared to superconducting qubits, typically taking tens of microseconds
Fidelity	Trapped ions achieve high fidelity (greater than 99.9%) for both single- and two-qubit gates, making them reliable for error-corrected quantum computing
Scalability	While individual ions can be controlled with high precision, scaling up the number of ions in a trap or connecting multiple traps presents technical challenges

5.4 Photonic Systems

Photonic quantum computing uses photons as qubits, encoding quantum information in their polarization, path, or phase. Quantum gates are implemented using optical elements like beam splitters and phase shifters.

5.4.1 Beam Splitter and Phase Shifter Gates

Photons pass through linear optical elements that modify their properties to perform quantum gate operations.

1. **Beam Splitters**

    Split an incoming photon into superposition states. For example, a 50/50 beam splitter transforms the input states as:

$$|0\rangle \rightarrow \frac{1}{\sqrt{2}}(|0\rangle + |1\rangle), \quad |1\rangle \rightarrow \frac{1}{\sqrt{2}}(|0\rangle - |1\rangle)$$

2. **Phase Shifters**

    Add a phase factor to the photon's state, represented as:

$$R_\phi = \begin{bmatrix} 1 & 0 \\ 0 & e^{i\phi} \end{bmatrix}$$

3. Applications

    Photonic systems excel in quantum communication, such as quantum key distribution (QKD), and serve as a platform for quantum simulation and limited computational tasks like boson sampling.

4. Challenges

    Photon loss and the lack of deterministic two-photon gates make it challenging to scale photonic systems for universal quantum computing.

Python Code Example: Phase Shift

```python
#--------------------------------------------------------------------
# Quantum Phase Shift
# Chapter 5 in the QUANTUM COMPUTING AND QUANTUM MACHINE LEARNING BOOK
#--------------------------------------------------------------------
# Version 1.0
# Qiskit changes frequently.
# We recommend using the latest version from the book code repository at:
# https://aqtinitiative.org/quantum-computing-for-engineers

# (c) 2025 Jesse Van Griensven, Roydon Fraser, and Jose Rosas
# License: MIT - Citation of this work required
#--------------------------------------------------------------------
import numpy as np
import matplotlib.pyplot as plt
#--------------------------------------------------------------------
def sprint(Matrix, decimals=4):
  """ Prints a Matrix with real and imaginary parts rounded to 'decimals'
"""
  import sympy as sp
  SMatrix = sp.Matrix(Matrix)  # Convert to Sympy Matrix if it's not
already

  def round_complex(x):
    """Round real and imaginary parts of x to the given number of
decimals."""
    c = complex(x)  # handle any real or complex Sympy expression
    r = round(c.real, decimals)
    i = round(c.imag, decimals)
    # If imaginary part is negligible, treat as purely real
    if abs(i) < 10**(-decimals): return sp.Float(r)
    else: return sp.Float(r) + sp.Float(i)*sp.I

  # Display the rounded Sympy Matrix
  display(SMatrix.applyfunc(round_complex))
  return

#--------------------------------------------------------------------
def phase_shifter(phi):
  return np.array([
    [1, 0],
    [0, np.exp(1j * phi)]
  ])
#--------------------------------------------------------------------
```

```python
# Define the phase shift angle
phi = np.pi / 4.

# Apply phase shift to a sample quantum state (|1⟩)
initial_state = np.array([0, 1])  # |1⟩ state
shifted_state = phase_shifter(phi) @ initial_state

#--------------------------------------------------------------------
# Print input and output
print("Phase Shifter Matrix"); sprint(phase_shifter(phi))
print("Initial State:");    sprint(initial_state)
print("Shifted State:");    sprint(shifted_state)
#--------------------------------------------------------------------
# Plot the effect of the phase shift in the complex plane
fig, ax = plt.subplots()
ax.axhline(0, color='gray', lw=1)
ax.axvline(0, color='gray', lw=1)

# Plot initial and shifted states
ax.quiver(0, 0, initial_state[1].real, initial_state[1].imag,
     angles='xy', scale_units='xy', scale=1, color='blue',
label='Initial |1⟩')
ax.quiver(0, 0, shifted_state[1].real, shifted_state[1].imag,
     angles='xy', scale_units='xy', scale=1, color='red',
label='Shifted |1⟩')

# Formatting the plot
ax.set_xlim(-1.2, 1.2)
ax.set_ylim(-1.2, 1.2)
ax.set_xlabel("Real")
ax.set_ylabel("Imaginary")
ax.set_title("Effect of Phase Shift on |1⟩ in the Complex Plane")
ax.legend()
ax.grid(alpha=0.2)
plt.show()
```

Figure 5.1 displays the result of shifting state $|1\rangle$ by 45°.

5.5 Topological Qubits

Topological quantum computing encodes quantum information in the braiding of anyons, exotic particles that exist in two-dimensional spaces. Gates are implemented by braiding these anyons.

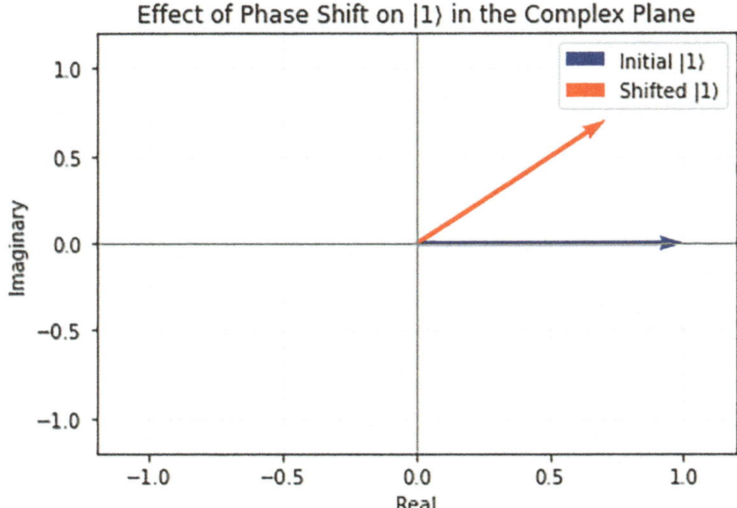

Fig. 5.1 State $|1\rangle$ Phase shifted by 45°. (By the authors)

5.5.1 Braiding Operations as Gates

The braiding of anyons corresponds to unitary operations on the quantum state. These operations are inherently fault-tolerant because the quantum information is stored in the global topology of the system, making it robust against local perturbations.

Exchanging two anyons applies a phase shift or rotation to the qubit state. These are two advantages:

(a) Topological qubits are immune to small local errors due to their global encoding of quantum information.
(b) The topological nature of the qubits ensures stability over long periods.

However, topological qubits have the following main challenges:

(a) Topological qubits require specific conditions, such as fractional quantum Hall states or Majorana zero modes, which are difficult to engineer in the lab.
(b) Precise control over braiding paths and operations is technically challenging.

5.6 Quantum Gate Models

The quantum gate model, also known as the circuit model, is one of the foundational paradigms for quantum computation. It is built upon the idea of sequentially applying quantum gates to qubits to perform computational tasks. While the gate-based model is the most widely implemented framework for quantum computing, alternative paradigms such as adiabatic quantum computing and measurement-based quantum computing provide complementary approaches. Each paradigm contributes to the broader landscape of quantum computing, expanding its capabilities and potential impact.

The gate-based quantum computing model is analogous to classical digital computing, where computations are carried out using a sequence of logical operations. In the quantum gate model, qubits are manipulated using quantum gates, which are represented as unitary operators. The sequence of gates forms a quantum circuit, which evolves the quantum state of the system to solve a computational problem.

5.6.1   Components of the Gate-Based Model

Quantum gates are operations that manipulate the states of qubits. These gates are analogous to classical logic gates but operate using unitary transformations. Unitary operators preserve the normalization condition and ensure the reversibility of quantum operations.

1. **Qubits**

   These are the fundamental units of quantum information, represented as $|0\rangle$ and $|1\rangle$ in a two-dimensional Hilbert space. Qubits can also exist in superposition states, enabling parallel computation.

2. **Quantum Gates**

   Unitary transformations applied to qubits. Common gates include:

   (a) Single-qubit gates: H, R_x, R_z, X, etc.
   (b) Multi-qubit gates: Controlled-NOT (CNOT), Controlled-Z (CZ), and Toffoli gates.

3. **Measurement**

   The process of extracting classical information from qubits, collapsing them into one of their basis states.

4. **Circuit**

   A sequence of gates and measurements representing a computation.

5.6.2   Mathematical Representation

In the gate-based model, the overall operation of a quantum circuit is a unitary transformation U applied to the initial state $|\psi_0\rangle$:

$$|\psi_{\text{final}}\rangle = U|\psi_0\rangle$$

The unitary transformation U is decomposed into a sequence of simpler gates:

$$U = U_k U_{k-1} \cdots U_1$$

5.6.3 *Advantages of the Gate-Based Model*

The following is a list of the main advantages of gate-based models:

1. **Universality**

   Any quantum computation can be expressed as a sequence of gates from a universal gate set
2. **Flexibility**

   The gate-based model supports a wide variety of algorithms, from Shor's factoring algorithm to Grover's search algorithm.
3. **Scalability**

   The model is well-suited for error correction and fault-tolerant quantum computation.

5.7 Adiabatic Quantum Computing

Adiabatic quantum computing (AQC) relies on the *adiabatic theorem*, which states that a quantum system remains in its ground state if the Hamiltonian governing it changes slowly enough. In AQC, the computation is encoded in the ground state of a time-dependent Hamiltonian.

1. **How It Works**

   (a) Begin with a simple Hamiltonian H_0, whose ground state is easy to prepare.
   (b) Slowly evolve H_0 into a final Hamiltonian H_f, whose ground state encodes the solution to the problem.
   (c) At the end of the evolution, the system is measured to extract the solution.

2. **Advantages**

   (a) Naturally robust to certain types of noise.
   (b) Well-suited for optimization problems like those in combinatorial optimization.

Table 5.2 Summary of quantum computing paradigms

Feature	Gate-based model	Adiabatic quantum computing	Measurement-based quantum computing
Key mechanism	Sequential gate application	Slow evolution of Hamiltonian	Measurements on an entangled state
General purpose	Yes	No (specialized for optimization problems)	Yes
Error tolerance	High (with fault tolerance)	Moderate	Dependent on measurement fidelity
Hardware requirements	Qubits and gates	Analog Hamiltonian controls	Entanglement generation and precise measurement

3. **Limitations**

   (a) Adiabatic evolution must be slow to prevent transitions to excited states, leading to longer computation times.
   (b) Not all algorithms can be efficiently implemented in AQC.

4. **Comparison with Gate-Based Model**

   (a) Gate-Based: Discrete operations suitable for general-purpose quantum computing.
   (b) AQC: Continuous evolution; specialized for optimization and simulation problems.

5.8   Measurement-Based Quantum Computing

Measurement-based quantum computing (MBQC), also known as the cluster-state model, performs computations through a series of measurements on an entangled multi-qubit state. The computation proceeds by progressively "measuring out" qubits, with the results determining subsequent operations. Table 5.2 summarizes the quantum computing paradigms.

1. **How It Works**

   (a) Prepare a highly entangled state, such as a cluster state.
   (b) Perform single-qubit measurements in specific bases.
   (c) Use the measurement outcomes to guide the computation.

2. **Advantages**

   (a) Reduces the need for continuous application of quantum gates during computation.
   (b) It may simplify certain hardware implementations by emphasizing measurements of overactive operations.

3. **Limitations**

   (a) Requires a highly entangled initial state, which can be resource-intensive to generate.
   (b) Measurement errors can propagate, making error correction challenging.

4. **Comparison with Gate-Based Model**

   (a) Gate-Based: Operations are applied sequentially during computation.
   (b) MBQC: Computation is driven by measurements on a pre-prepared entangled state.

5.9 Developing and Running Quantum Algorithms

Designing and executing quantum algorithms involves translating classical computational problems into quantum terms. Engineers must leverage the unique properties of quantum mechanics, such as superposition and entanglement while addressing hardware-specific challenges like noise and limited coherence times.

Developing and running quantum algorithms requires careful problem representation, circuit design, and noise mitigation. Engineers must translate classical problems into quantum terms, design efficient circuits, and implement strategies to manage hardware imperfections. Tools like Qiskit provide a powerful framework for creating, testing, and optimizing quantum algorithms, enabling engineers to unlock the potential of quantum computing. This section outlines the core stages in developing quantum algorithms, providing practical examples and implementations.

5.9.1 Problem Representation

Quantum algorithms typically reformulate classical problems into unitary operations compatible with quantum systems. For example, optimization problems are often expressed using a cost function mapped onto a quantum Hamiltonian operator H:

$$H = \sum_{i,j} c_{ij}\sigma_i^z \sigma_j^z + \sum_i b_i \sigma_i^z$$

where σ_i^z represents the Pauli-Z operator acting on the i-th qubit, and c_{ij}, b_i are problem-specific coefficients.

Python Code Example: Hamiltonian Problem Representation

Consider a simple optimization problem where the objective is to minimize the function:

$$f(x_1, x_2) = x_1 x_2 - 2x_1 - 3x_2 + 5$$

Using quantum principles, this can be reformulated as a Hamiltonian acting on two qubits:

$$H = -2\sigma_1^z - 3\sigma_2^z + \left(\sigma_1^z \sigma_2^z\right) + 5I$$

where I is the identity operator.

This Hamiltonian can then be implemented in a quantum circuit for solving the optimization problem, as explained in the code below. Note that in Python the "^" symbol is the tensor operator.

```python
#---------------------------------------------------------------------
# Simple Hamiltonian Problem Representation
# Chapter 5 in the QUANTUM COMPUTING AND QUANTUM MACHINE LEARNING BOOK
#---------------------------------------------------------------------
# Version 1.0
# Qiskit changes frequently.
# We recommend using the latest version from the book code repository at:
# https://aqtinitiative.org/quantum-computing-for-engineers

# (c) 2025 Jesse Van Griensven, Roydon Fraser, and Jose Rosas
# License: MIT - Citation required
#---------------------------------------------------------------------
from qiskit.opflow import Z, I
# Hamiltonian: H = -2 * Z ^ I - 3 * I ^ Z + (Z ^ Z) + 5 * (I ^ I)
# Z ^ I: Pauli Z acts on qubit 0, and identity I acts on qubit 1.
# I ^ Z: Identity II acts on qubit 0, and Pauli Z acts on qubit 1.
# Z ^ Z: Pauli Z acts on both qubits.
# I ^ I: Identity acts on both qubits.
# "^" is the operator for tenscr product

# Define the Hamiltonian with consistent qubit alignment
H = ( -2 * (Z ^ I) ) + ( -3 * (I ^ Z) ) + (Z ^ Z) + ( 5 * (I ^ I) )

print("Hamiltonian Representation:")
print(H)
```

Table 5.3 Quantum circuit functions

Quantum circuit component	Function
Unitary gates	Define transformations on quantum states
Entanglement operations	Establish correlations between qubits
Measurement operations	Extract results from quantum states

5.9.2   Quantum Circuit Design

Quantum circuits graphically represent the sequence of operations in a quantum algorithm. Engineers must carefully design these circuit components. Some of these are summarized in Table 5.3.

5.9.3   Example: Grover's Algorithm

Grover's algorithm demonstrates how quantum circuits can solve unstructured search problems efficiently.

5.9.3.1   Steps of Grover's Algorithm

1. **Initialization**: Prepare the uniform superposition state:

$$|\psi\rangle = H^{\otimes n} |0\rangle^n$$

where $H^{\otimes n}$ applies the Hadamard gate to n qubits all initially in the $|0\rangle$ state:
2. **Oracle Application:** Encodes the problem into a unitary operator U_f:

$$U_f |x\rangle = (-1)^{f(x)} |x\rangle$$

3. **Diffusion Operator:** Amplifies the amplitude of the correct solution:

$$D = 2 |\psi\rangle\langle\psi| - I$$

4. **Iteration:** Repeat the oracle and diffusion steps approximately $\sqrt{N}$ times, where N is the size of the dataset.

Python Code Example: Grover's Algorithm
This example demonstrates Grover's algorithm on a 3-qubit system, marking $|110\rangle$ as the solution.

```python
#------------------------------------------------------------------
# Grover's Algorithm - More Complete
# Chapter 5 in the QUANTUM COMPUTING AND QUANTUM MACHINE LEARNING BOOK
#------------------------------------------------------------------
# Version 1.0
# Qiskit changes frequently.
# We recommend using the latest version from the book code repository at:
# https://aqtinitiative.org/quantum-computing-for-engineers

# (c) 2025 Jesse Van Griensven, Roydon Fraser, and Jose Rosas
# License: MIT - Citation required
#------------------------------------------------------------------
import matplotlib.pyplot as plt
from qiskit import QuantumCircuit, Aer, execute
from qiskit.visualization import circuit_drawer
from qiskit.circuit.library import GroverOperator
#------------------------------------------------------------------

# Define the oracle
n = 3  # Number of qubits
oracle = QuantumCircuit(n)

 # Mark |110⟩ as the solution
oracle.z(2)
oracle = GroverOperator(oracle)

# Initialize the quantum circuit
qc = QuantumCircuit(n)

# Apply Hadamard gates to place all qc gates in superposition
qc.h(range(n))

# Apply the oracle and diffusion operator
qc.append(oracle, range(n))

# Measure the result
qc.measure_all()

# Print the Quantum Circuit
qc.draw()

# Simulate the circuit
simulator = Aer.get_backend('qasm_simulator')
job  = execute(qc, simulator, shots=1000)
result = job.result()
```

```python
counts = result.get_counts()
#print("Grover's Algorithm Results:", result.get_counts())
print("Grover's Algorithm Results:", counts)


#-----------------------------------------------------------------
# Visualize Grover's Result
# Extract keys and values
labels = list(counts.keys())
values = list(counts.values())

# Create the histogram
# Note: using matplotlib directly - qiskit plot_histogram(counts) is
buggy
plt.bar(labels, values, color=['green', 'orange'])
plt.title("Grover's Algorithm Results")
plt.xlabel("Measurement Outcomes")
plt.grid(alpha=0.2)
plt.ylim(0, 180)
plt.ylabel("Counts")

# Add counts above the bars
for i, value in enumerate(values):
  plt.text(i, value + 5, str(value), ha='center', fontsize=10,
color='black')

plt.show()
```

Figure 5.2 displays the results of Grover's algorithm presented in the code above.

5.10   Hands-On Examples for Engineers

Practical examples provide a concrete understanding of quantum computing concepts and their implementation. This section focuses on step-by-step implementations of foundational quantum algorithms and their applications in engineering, optimization, and machine learning. These hands-on examples illustrate practical implementations of quantum algorithms for engineering applications. Deutsch–Jozsa highlights quantum speedup for decision problems, while variational algorithms like VQE and QAOA solve optimization challenges. Quantum machine learning techniques such as QSVMs and QNNs integrate quantum power into classical models, paving the way for future advances in engineering and technology.

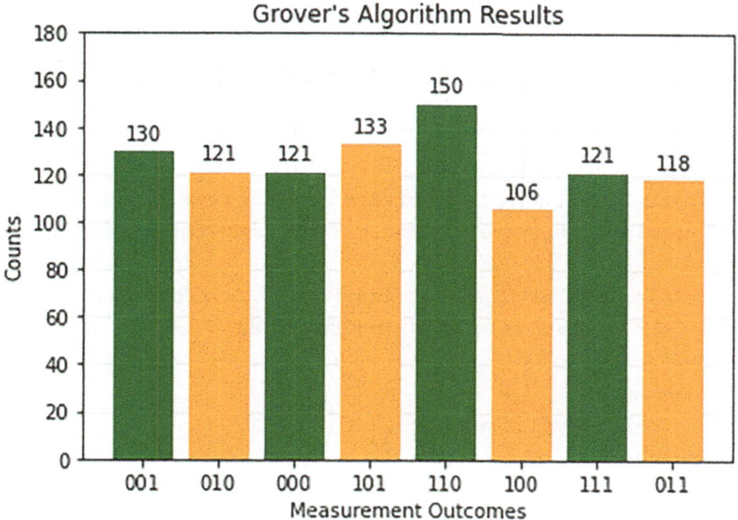

Fig. 5.2 Grover's algorithm results $|110\rangle$ solution. (By the authors)

5.10.1 Deutsch–Jozsa Algorithm Quantum Circuit

The Deutsch–Jozsa algorithm determines whether a function $f(x)$ is constant or balanced using quantum parallelism. A function $f(x)$ is constant if it outputs the same value for all inputs and balanced if it outputs an equal number of $|0\rangle$ and $|1\rangle$ solutions. For a function with n bits of input, classical evaluation requires $2^{n-1} + 1$ queries in the worst case, whereas the quantum algorithm solves it with one query.

Here is the "Step-by-Step" Deutsch–Jozsa algorithm implementation:

1. **Circuit Initialization**

   The circuit starts with two qubits, one in the superposition state for input and one as the auxiliary qubit for the oracle.

$$|\psi\rangle = H^{\otimes n} |0\rangle^{\otimes n} \otimes H|1\rangle$$

2. **Oracle Application**

   The oracle encodes the function $f(x)$.

3. **Result Measurement**

   Apply the Hadamard gate again and measure the output qubits.

Python Code Example: Deutsch–Jozsa Algorithm

This example demonstrates how a simple oracle identifies whether $f(x)$ is constant or balanced in one quantum query.

```python
#----------------------------------------------------------------
# Deutsch-Jozsa Algorithm
# Chapter 5 in the QUANTUM COMPUTING AND QUANTUM MACHINE LEARNING BOOK
#----------------------------------------------------------------
# Version 1.0
# Qiskit changes frequently.
# We recommend using the latest version from the book code repository at:
# https://aqtinitiative.org/quantum-computing-for-engineers

# (c) 2025 Jesse Van Griensven, Roydon Fraser, and Jose Rosas
# License: MIT - Citation of this work required
#----------------------------------------------------------------
import numpy as np
import matplotlib.pyplot as plt
import warnings
warnings.filterwarnings('ignore')


#----------------------------------------------------------------
from qiskit import QuantumCircuit, Aer, execute
from qiskit.visualization import circuit_drawer
#----------------------------------------------------------------


# Create the Deutsch-Jozsa circuit
qc = QuantumCircuit(2, 1)

# Step 1: Initialization
qc.h(0)     # Apply Hadamard gate to the input qubit
qc.x(1)     # Prepare the auxiliary qubit in |1)
qc.h(1)

# Step 2: Oracle application
qc.cx(0, 1) # Example oracle for f(x) = x

# Step 3: Measurement
qc.h(0)     # Apply Hadamard gate to the input qubit
qc.measure(0, 0)

# Simulate the circuit
simulator = Aer.get_backend('qasm_simulator')
result  = execute(qc, simulator, shots=1024).result()
counts  = result.get_counts()
print("Deutsch-Jozsa Results:", counts)

# Draw the circuit
display(circuit_drawer(qc, output='mpl', style="iqp"))
```

5.10.2 Variational Quantum Algorithms (VQAs)

Variational quantum algorithms (VQAs) use hybrid quantum-classical approaches to solve problems, such as optimization and eigenspectrum determination. The variational quantum eigensolver (VQE) is a type of VQA that minimizes the expected energy of a molecule's Hamiltonian H using parameterized quantum circuits.

$$E(\theta) = \langle \psi(\theta) \, | \, H \, | \, \psi(\theta) \rangle$$

The classical optimizer adjusts parameters θ to minimize $E(\theta)$.

Python Code Example: VQE

This example optimizes a parameterized circuit to minimize the energy of a two-qubit Hamiltonian, employing the Variational Quantum Eigensolver (VQE).

```python
#----------------------------------------------------------------------
# Variational Quantum Eigensolver - VQE
# Chapter 5 in the QUANTUM COMPUTING AND QUANTUM MACHINE LEARNING BOOK
#----------------------------------------------------------------------
# Version 1.0
# Qiskit changes frequently.
# We recommend using the latest version from the book code repository at:
# https://aqtinitiative.org/quantum-computing-for-engineers


# (c) 2025 Jesse Van Griensven, Roydon Fraser, and Jose Rosas
# License: MIT - Citation of this work required
#----------------------------------------------------------------------


from qiskit import Aer
from qiskit.opflow import Z, I
from qiskit.circuit.library import RealAmplitudes

from qiskit.algorithms import VQE
from qiskit.algorithms.optimizers import COBYLA
#----------------------------------------------------------------------


# Define the Hamiltonian
H = (Z ^ Z) + (Z ^ I) + (I ^ Z)

# Define the variational form
ansatz = RealAmplitudes(num_qubits=2, reps=1)

# Use Qiskit's VQE algorithm
simulator = Aer.get_backend('statevector_simulator')
```

```
optimizer = COBYLA(maxiter=200)  # Create the COBYLA optimizer object
vqe = VQE(ansatz, optimizer=optimizer, quantum_instance=simulator)

# Solve for the ground state energy
result = vqe.compute_minimum_eigenvalue(operator=H)
print("VQE Energy:", result.eigenvalue)
```

5.10.3   Quantum Approximate Optimization Algorithm (QAOA)

QAOA solves combinatorial optimization problems by iteratively applying the cost Hamiltonian H_C:

$$H_C = \sum_{i<j} J_{ij} Z_i Z_j + \sum_i h_i Z_i$$

The quantum circuit alternates between:

1. **Applying the Cost Hamiltonian**

$$U_C(\gamma) = e^{-i\gamma H_C}$$

2. **Mixing Operator**

$$U_M(\beta) = e^{-i\beta \sum_i X_i}$$

Figure 5.3 summarizes the QAOA's main components.

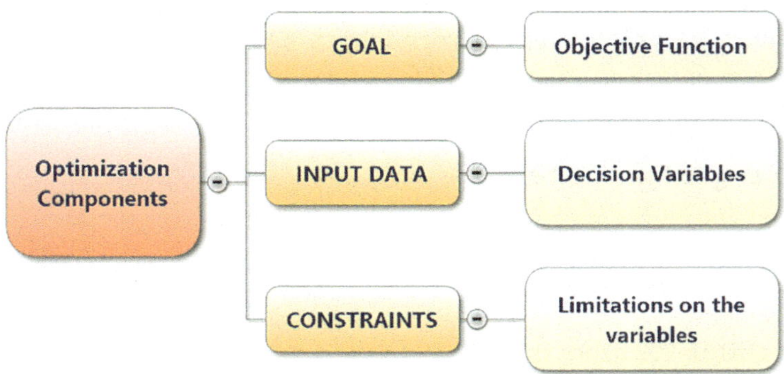

Fig. 5.3 QAOA: Optimization solution chart. (By the authors)

Python Code Example: Quantum Approximate Optimization Algorithm (QAOA)

This example demonstrates a single-layer QAOA circuit applied to a simple optimization problem.

```python
#------------------------------------------------------------------
# Quantum Approximate Optimization Algorithm - QAOA
# Chapter 5 in the QUANTUM COMPUTING AND QUANTUM MACHINE LEARNING BOOK
#------------------------------------------------------------------
# Version 1.0
# Qiskit changes frequently.
# We recommend using the latest version from the book code repository at:
# https://aqtinitiative.org/quantum-computing-for-engineers

# (c) 2025 Jesse Van Griensven, Roydon Fraser, and Jose Rosas
# License: MIT - Citation of this work required
#------------------------------------------------------------------
from qiskit import QuantumCircuit, Aer, execute
from qiskit.circuit import Parameter

import warnings
warnings.filterwarnings('ignore')
from qiskit.visualization import circuit_drawer
#------------------------------------------------------------------

# Define parameters
# gamma: A parameter associated with the cost Hamiltonian.
# gamma determines how strongly the cost Hamiltonian influences the
evolution of the quantum state during the optimization.
gamma = Parameter('γ')

# beta: A parameter associated with the mixing Hamiltonian.
# beta governs the amplitude mixing between different states in the
quantum superposition.
beta = Parameter('β')

# Create a QAOA circuit
qc = QuantumCircuit(2)
qc.h([0, 1])          # Initialize qubits in superposition
qc.cz(0, 1)           # Cost Hamiltonian
qc.rz(2 * gamma, [0, 1])
qc.rx(2 * beta, [0, 1]) # Mixing Hamiltonian
qc.measure_all()
```

```
# Simulate the circuit
simulator = Aer.get_backend('qasm_simulator')
result  = execute(qc.bind_parameters({gamma: 1.0, beta: 0.5}),
simulator, shots=1024).result()
counts  = result.get_counts()

# Print Results
print("QAOA Results:", counts)

# Draw the circuit
display(circuit_drawer(qc, output='mpl', style="iqp"))
```

5.11 Quantum Machine Learning

Quantum machine learning (QML) integrates quantum algorithms into classical machine learning frameworks, offering enhanced feature mapping, kernel methods, and optimization.

5.11.1 Quantum Support Vector Machines (QSVMs)

QSVMs utilize quantum kernels to perform data classification and uncover patterns in complex datasets. The kernel function is computed as:

$$K(x, y) = | \langle \psi(x) | \psi(y) \rangle |^2$$

Python Code Example: Quantum Support Vector Machines (QSVMs)
This example shows how QSVMs compute kernel matrices using quantum circuits.

```
#------------------------------------------------------------------
# Quantum Support Vector Machines - QSVM
# Chapter 5 in the QUANTUM COMPUTING AND QUANTUM MACHINE LEARNING BOOK
#------------------------------------------------------------------
# Version 1.0
# Qiskit changes frequently.
# We recommend using the latest version from the book code repository at:
# https://aqtinitiative.org/quantum-computing-for-engineers

# (c) 2025 Jesse Van Griensven, Roydon Fraser, and Jose Rosas
# License: MIT - Citation of this work required
#------------------------------------------------------------------
```

```python
from qiskit_machine_learning.kernels import QuantumKernel
from qiskit import Aer
from qiskit.circuit.library import ZZFeatureMap


#-------------------------------------------------------------
def sprint(Matrix, decimals=4):
    """ Prints a numpy Matrix in a nice format with sympy and specified
precision """
    # Define the SYMPY routines we need
    import sympy as sp
    # The input Matrix can be a Numpy or a Sympy Matrix
    SMatrix = sp.Matrix(Matrix)
    # Apply the rounding to each element
    SMatrix = SMatrix.applyfunc(lambda x: round(float(x), decimals) if
isinstance(x, (float, int, sp.Float)) else x)
    display(SMatrix)
    return
#-------------------------------------------------------------


# Define a feature map
feature_map = ZZFeatureMap(feature_dimension=2, reps=2,
entanglement='linear')


# Quantum kernel
quantum_kernel = QuantumKernel(feature_map=feature_map,
quantum_instance=Aer.get_backend('statevector_simulator'))


# Example data
data = [[0, 0], [1, 1]]
kernel_matrix = quantum_kernel.evaluate(x_vec=data)
print("Quantum Kernel Matrix:")
sprint(kernel_matrix)
```

5.11.2   Quantum Neural Networks (QNNs)

QNNs employ variational circuits for feature mapping and training. They are trained
by minimizing a cost function, typically the expectation value of a Hamiltonian.

Python Code Example: Quantum Neural Networks (QNNs)
This code demonstrates a simple quantum neural network architecture for evaluating
input data.

```python
#-------------------------------------------------------------------
# Quantum Neural Network - QNN
# Chapter 5 in the QUANTUM COMPUTING AND QUANTUM MACHINE LEARNING BOOK
#-------------------------------------------------------------------
# Version 1.0
# Qiskit changes frequently.
# We recommend using the latest version from the book code repository at:
# https://aqtinitiative.org/quantum-computing-for-engineers

# (c) 2025 Jesse Van Griensven, Roydon Fraser, and Jose Rosas
# License: MIT - Citation of this work required
#-------------------------------------------------------------------
import numpy as np

#-------------------------------------------------------------------
from qiskit_machine_learning.neural_networks import TwoLayerQNN
from qiskit import Aer
from qiskit.utils import QuantumInstance
#-------------------------------------------------------------------

quantum_instance=QuantumInstance(Aer.get_backend
('statevector_simulator'))

# Define a quantum neural network
qnn = TwoLayerQNN(num_qubits=2, quantum_instance=quantum_instance)

# Example input data and weights for the QNN
input_data = [0, 1]  # Input features

# Randomly initialize the weights
weights = np.random.rand(qnn.num_weights)

# Compute the output of the QNN
result = qnn.forward(input_data, weights)

# Print Result
print("QNN Output:", result)
```

5.12   Quantum Algorithms

Quantum algorithms exploit the unique properties of quantum mechanics, such as
superposition, entanglement, and interference, to solve computational problems
more efficiently than classical algorithms in specific scenarios. Unlike classical

algorithms, which operate sequentially on deterministic states, quantum algorithms process information in parallel by manipulating quantum states in a superposition. This capability provides exponential speedups for certain problems and has significant implications for fields, such as cryptography, optimization, and material science.

This chapter examines the fundamental principles underlying quantum algorithms, their mathematical foundations, and their practical applications in engineering.

5.13 Fundamental Principles of Quantum Algorithms

Quantum algorithms are constructed to exploit the principles of quantum mechanics, translating physical phenomena into computational advantages.

5.13.1 Superposition

Superposition allows quantum bits (qubits) to exist in a combination of $|0\rangle$ and $|1\rangle$ states simultaneously. For an n-qubit system, superposition creates a quantum state that encodes 2^n classical states simultaneously:

$$|\psi\rangle = \sum_{i=0}^{2^n-1} c_i \, |i\rangle$$

where c_i are complex amplitudes, and $\sum |c_i|^2 = 1$.

This property enables quantum parallelism, allowing quantum algorithms to explore multiple solutions concurrently.

5.13.2 Entanglement

Entanglement establishes correlations between qubits such that the state of one qubit depends on the state of another, regardless of their spatial separation. For example, the Bell state:

$$|\phi^+\rangle = \frac{1}{\sqrt{2}}(|00\rangle + |11\rangle)$$

creates an inseparable relationship between two qubits.

Entanglement is a critical resource in quantum algorithms, enabling global operations and information sharing across qubits.

5.13.3 Interference

Quantum algorithms manipulate the phases of quantum states to amplify the probability of correct solutions and suppress incorrect ones. For instance, Grover's search algorithm leverages constructive and destructive interference to find a marked element in an unstructured database with quadratic speedup.

5.14 Key Quantum Algorithms

Quantum algorithms address diverse computational problems, providing speedups over classical methods in areas, such as search, factorization, and simulation.

5.14.1 Shor's Algorithm

Shor's algorithm revolutionized computational mathematics by demonstrating that quantum computers can factorize integers exponentially faster than the best-known classical algorithms.

The algorithm uses the quantum Fourier transform (QFT) to find the periodicity of a function:

$$f(x) = a^x \mod N$$

where N is the number to be factorized, and a is a randomly chosen integer coprime to N.

The QFT maps a quantum state from the time domain to the frequency domain, identifying the period r:

$$|x\rangle \rightarrow \frac{1}{\sqrt{r}} \sum_{k=0}^{r-1} e^{2\pi ikx/r} |k\rangle$$

Once r is determined, N can be factorized by computing $\gcd(a^{r/2} \pm 1, N)$. This capability has profound implications for cryptography, as it threatens the security of widely used schemes like RSA.

5.14.2 Grover's Algorithm

Grover's algorithm provides a quadratic speedup for unstructured search problems. Given a database of N items, the algorithm finds a marked element in $O(\sqrt{N})$ queries, compared to $O(N)$ for classical search methods.

The algorithm initializes the system in an equal superposition of all states:

$$|\psi_0\rangle = \frac{1}{\sqrt{N}} \sum_{x=0}^{N-1} |x\rangle$$

It then iteratively applies two operations:

1. **Oracle Query**
   Flips the phase of the marked element.
2. **Amplitude Amplification**
   Enhances the probability amplitude of the marked state through a reflection operation.

After $O(\sqrt{N})$ iterations, the marked state dominates the probability distribution, allowing efficient measurement.

5.14.3 Quantum Approximate Optimization Algorithm (QAOA)

QAOA solves combinatorial optimization problems by encoding the cost function H_C into a Hamiltonian and using a mixing Hamiltonian H_M to explore the solution space:

$$|\psi\rangle = e^{-i\beta H_M} e^{-i\gamma H_C} |\psi_0\rangle$$

where $|\psi_0\rangle$ is the initial state, and β and γ are tunable parameters.

Applications of QAOA include scheduling, route optimization, and resource allocation in engineering.

5.14.4 Quantum Simulations

Quantum computers excel at simulating quantum systems and solving Schrödinger's equation for large molecules and materials. These simulations provide insights into chemical reactions, material properties, and quantum dynamics:

$$H \,|\,\psi\rangle = E\,|\,\psi\rangle$$

where H is the Hamiltonian, $|\psi\rangle$ is the quantum state, and E is the energy eigenvalue.

Quantum simulations have applications in drug discovery, catalyst design, and renewable energy development.

5.15 Applications in Engineering

Quantum algorithms enable transformative advances across engineering disciplines by addressing computational bottlenecks in optimization, simulation, and data analysis.

1. **Optimization in Logistics and Supply Chains**

   Quantum optimization algorithms like QAOA streamline complex logistical tasks, such as optimizing delivery routes, scheduling tasks, and managing supply chains. For example, Grover's algorithm can enhance search processes in large datasets for efficient decision-making.

2. **Material Science and Chemistry**

   Quantum simulations aid in discovering new materials with desired properties, such as superconductors and lightweight alloys. Engineers leverage these capabilities to design more efficient batteries, solar cells, and structural materials.

3. **Artificial Intelligence and Machine Learning**

   Quantum algorithms accelerate the training of machine learning models by optimizing parameters more efficiently than classical methods. Quantum-enhanced support vector machines and neural networks improve pattern recognition and predictive modeling in engineering applications.

4. **Cryptography and Security**

   Quantum-resistant cryptographic schemes, such as lattice-based cryptography, are being developed to counteract the threat posed by quantum algorithms like Shor's. Engineers play a critical role in implementing and testing these new protocols.

5.16 Grover's Algorithm: Solving Search Problems

Grover's algorithm is one of the most celebrated quantum algorithms due to its ability to solve unstructured search problems significantly faster than classical approaches. By leveraging quantum properties such as superposition and interference, Grover's algorithm reduces the time complexity of searching an unsorted database from $O(N)$ to $O\left(\sqrt{N}\right)$. This quadratic speedup has profound implications for fields ranging from cryptography to engineering optimization.

Grover's algorithm is a powerful tool for solving unstructured search problems, achieving a quadratic speedup over classical methods. Its steps, including initialization, oracle query, diffusion operation, and repetition, leverage the principles of quantum mechanics to amplify the probability of the target state. With applications in fields such as material selection and circuit design, Grover's algorithm exemplifies the transformative potential of quantum computing in engineering.

5.16.1 Overview of Grover's Algorithm

Grover's algorithm is designed to locate a specific item, x_s, in an unsorted database of size N. Classical algorithms require $O(N)$ time to search the database because each entry must be checked sequentially. In contrast, Grover's algorithm uses quantum amplitude amplification to identify the target item in $O(\sqrt{N})$ steps.

The algorithm achieves this speedup by initializing a superposition state over all possible database entries and iteratively applying two key operations:

1. An oracle query that marks the target state by flipping its phase.
2. The Grover diffusion operator, which amplifies the amplitude of the marked state while suppressing the amplitudes of all others.

After approximately $\sqrt{N}$ iterations, the probability of measuring the target state becomes near unity.

Grover's algorithm is optimal for unstructured search problems, meaning no quantum algorithm can solve the problem with fewer queries. This optimality underscores its importance in quantum computing research and applications.

5.16.2 Steps of Grover's Algorithm

Grover's algorithm consists of four main steps: initialization, oracle query, diffusion operation, and repetition.

5.16.2.1 Initialization

The algorithm begins by creating an equal superposition of all N database entries. This state is generated using Hadamard gates applied to n qubits, where $N = 2^n$. The initial state is:

$$|\psi\rangle = \frac{1}{\sqrt{N}} \sum_{x=0}^{N-1} |x\rangle$$

This superposition ensures that all database entries are equally likely to be observed at the start.

5.16.2.2 Oracle Query

The oracle, O, is a quantum operation that identifies the target state $|x_s\rangle$ by flipping its phase. Mathematically, the oracle is defined as:

$$O\,|x\rangle = \begin{cases} -\,|x\rangle & \text{if } x = x_s, \\ |x\rangle & \text{otherwise} \end{cases}$$

This phase inversion is a crucial step that distinguishes the marked state from all other states, setting the stage for amplitude amplification. The oracle is problem-specific and must be designed to encode the condition that identifies $|x_s\rangle$.

5.16.2.3 Grover Diffusion Operator

The Grover diffusion operator, D, amplifies the amplitude of the marked state while reducing the amplitudes of the unmarked states. It is defined as:

$$D = 2\,|\psi\rangle\langle\psi|\, - I$$

where $|\psi\rangle$ is the initial superposition state and I is the identity operator.

The diffusion operator reflects the quantum state about the average amplitude of all states, thereby increasing the probability of measuring the target state. Mathematically, this process is analogous to inverting about the mean in classical statistics.

5.16.2.4 Repetition

The algorithm alternates between the oracle and diffusion operations, amplifying the amplitude of the marked state with each iteration. The number of iterations required is approximately $\sqrt{N}$. After this process, the marked state $|x_s\rangle$ dominates the quantum state, ensuring a high probability of observing it upon measurement.

5.16.3 Example of Grover's Algorithm in Engineering

Grover's algorithm has practical applications in engineering, particularly in optimization and decision-making tasks.

5.16.3.1 Use Case: Optimal Material Selection

Consider a database containing N materials, each characterized by properties such as tensile strength, thermal conductivity, and cost. The goal is to identify a material that meets specific criteria, such as maximizing strength while minimizing cost. In a classical search, each material must be evaluated individually, requiring $O(N)$ comparisons. Using Grover's algorithm, the engineer initializes a superposition of all materials and defines an oracle that marks materials meeting the criteria.

The Grover diffusion operator amplifies the amplitude of these marked materials, allowing the optimal material to be identified in $O(\sqrt{N})$ steps. This quantum-enhanced filtering significantly reduces the computational resources required for material selection in complex engineering applications.

5.16.4 Grover's Quantum Circuit Representation

Grover's algorithm can be implemented using a quantum circuit composed of three main components: Hadamard gates, the oracle, and the diffusion operator.

5.16.4.1 Hadamard Gates

The circuit begins with n Hadamard gates, one for each qubit, to create the initial superposition state. For $n = 3$ (representing a database of $N = 8$ entries), the Hadamard gates transform the state $|000\rangle$ into:

$$|\psi\rangle = \frac{1}{\sqrt{8}}(|000\rangle + |001\rangle + |010\rangle + \ldots + |111\rangle)$$

5.16.4.2 Oracle Implementation

The oracle is implemented using quantum gates that encode the condition for identifying the target state. For example, if $|x_s\rangle = |101\rangle$, the oracle applies a phase flip to this state while leaving others unchanged.

5.16.4.3 Diffusion Operator

The diffusion operator is realized using a combination of Hadamard gates, phase shifts, and controlled gates. The reflection about the mean is achieved by applying a phase shift to the initial superposition state followed by an inversion operation.

5.16.4.4   Quantum Circuit Example

The circuit for Grover's algorithm includes:

(a) **Initialization Layer**
    Hadamard gates applied to all qubits.
(b) **Oracle Layer**
    A custom gate that flips the phase of the marked state.
(c) **Diffusion Layer**
    A sequence of gates implementing the reflection operation.
(d) **Repetition Block**
    The oracle and diffusion layers are repeated approximately $\sqrt{N}$ times.

The final measurement yields the marked state with high probability, completing the search process.

5.17   Classical Fourier Transform

The classical Fourier transform (FT) provides a powerful method for analyzing signals by transforming them into the frequency domain. Whether working with continuous or discrete data, the FT reveals the underlying frequency components, amplitudes, and phases that define the signal. Its wide range of properties, such as linearity, time shifting, convolution, and energy preservation, makes it indispensable in applications spanning signal processing, image analysis, quantum mechanics, and modern machine learning.

The Fourier transform (FT) provides an effective method for transforming signals or functions from their native time domain (or spatial domain) into the frequency domain. By representing signals as combinations of sinusoidal components of varying frequencies, the Fourier transform enables deeper insights into periodic behaviors, structures, and hidden features within signals.

A signal, whether continuous or discrete, can be thought of as a superposition of simpler sinusoidal waves. The Fourier transform decomposes any such signal into its constituent frequency components, represented by a combination of sine and cosine functions or equivalently as complex exponentials using Euler's formula:

$$e^{i\omega t} = \cos(\omega t) + i\sin(\omega t)$$

Here, ω is the angular frequency of the signal, expressed in radians per second. The FT identifies the amplitude and phase of these frequency components, providing a frequency representation of the signal.

5.17.1 Classical Fourier Transform Equations

The Fourier transform can be categorized into continuous Fourier transform (CFT) for continuous signals and discrete Fourier transform (DFT) for sampled signals. Each version serves a distinct purpose but shares a common goal of representing signals in the frequency domain.

5.17.1.1 Continuous Fourier Transform Equation

For a continuous signal $x(t)$, where t represents time, the Fourier Transform and its inverse are defined as follows:

1. **Forward Fourier Transform**

$$X(\omega) = \int_{-\infty}^{\infty} x(t)e^{-i\omega t}dt$$

where

- $x(t)$ is the signal in the time domain.
- $X(\omega)$ is the signal in the frequency domain.
- ω is the angular frequency, related to the standard frequency f by $\omega = 2\pi f$.
- $e^{-i\omega t}$ is the complex exponential that decomposes $x(t)$ into sinusoidal components.

2. Inverse Fourier Transform

$$x(t) = \frac{1}{2\pi}\int_{-\infty}^{\infty} X(\omega)e^{i\omega t}d\omega$$

This equation reconstructs the original signal $x(t)$ from its frequency components $X(\omega)$.

5.17.1.2 Discrete Fourier Transform (DFT) Equation

For digital signals sampled at discrete intervals, the discrete Fourier transform (DFT) replaces the continuous FT. Let $x[n]$ represent a signal sampled at N points, where $n = 0, 1, ..., N - 1$. The DFT is defined as:

1. **Forward DFT**

$$X[k] = \sum_{n=0}^{N-1} x[n]e^{-i\frac{2\pi}{N}kn}, \quad k = 0, 1, ..., N - 1$$

where

- $x[n]$ is the discrete input signal (time domain).
- $X[k]$ is the frequency domain representation.
- N is the total number of samples.
- $e^{-i\frac{2\pi}{N}kn}$ corresponds to the basis functions in the frequency domain.

2. **Inverse DFT**

$$x[n] = \frac{1}{N} \sum_{k=0}^{N-1} X[k] e^{i\frac{2\pi}{N}kn}, \quad n = 0, 1, \ldots, N-1$$

This equation reconstructs the original time-domain signal $x[n]$ from its frequency-domain representation $X[k]$.

The DFT is computationally implemented through the fast Fourier transform (FFT) algorithm, which reduces the computational complexity from $O(N^2)$ to $O(N \log N)$, enabling efficient analysis of large datasets.

5.17.2 Understanding the Classical FT Equations

The Fourier transform's exponential terms $e^{\pm i\omega t}$ are derived from Euler's formula:

$$e^{i\omega t} = \cos(\omega t) + i\sin(\omega t)$$

This formula represents the complex exponential as a combination of sinusoidal components. The real part ($\cos(\omega t)$) corresponds to the cosine wave, while the imaginary part ($\sin(\omega t)$) corresponds to the sine wave. In the frequency domain:

1. Each frequency component $X(\omega)$ represents the contribution (amplitude and phase) of the corresponding sinusoidal wave at angular frequency ω.
2. The Fourier transform essentially projects the signal $x(t)$ onto a set of orthogonal basis functions $e^{-i\omega t}$, where each basis function corresponds to a specific frequency.

5.17.3 Properties of the Classical Fourier Transform

The Fourier transform has several fundamental properties that make it useful for analysis and signal manipulation.

5.17.3.1 Linearity

$$\mathcal{F}\{ax_1(t) + bx_2(t)\} = aX_1(\omega) + bX_2(\omega)$$

where a and b are constants. This property allows the decomposition of complex signals into simpler components.

5.17.3.2 Time Shifting

If $x(t) \rightarrow X(\omega)$, then shifting the signal in time by t_0 results in:

$$x(t - t_0) \rightarrow e^{-i\omega t_0} X(\omega)$$

This introduces a phase shift in the frequency domain.

5.17.3.3 Frequency Shifting

If $x(t) \rightarrow X(\omega)$, then multiplying $x(t)$ by $e^{i\omega_0 t}$ shifts the frequency:

$$x(t)e^{i\omega_0 t} \rightarrow X(\omega - \omega_0)$$

5.17.3.4 Scaling

Scaling the time domain signal $x(at)$ results in the following:

$$\mathcal{F}\{x(at)\} = \frac{1}{|a|} X\left(\frac{\omega}{a}\right)$$

where a is a scaling factor.

5.17.3.5 Convolution Theorem

The Fourier transform of the convolution of two signals $x_1(t)$ and $x_2(t)$ is given by:

$$\mathcal{F}\{x_1(t) * x_2(t)\} = X_1(\omega)X_2(\omega)$$

where $*$ denotes convolution.

5.17.3.6   Parseval's Theorem

Parseval's theorem states that the total energy of a signal is preserved in both the time and frequency domains:

$$\int_{-\infty}^{\infty} |x(t)|^2 dt = \frac{1}{2\pi} \int_{-\infty}^{\infty} |X(\omega)|^2 d\omega$$

For discrete signals:

$$\sum_{n=0}^{N-1} |x[n]|^2 = \frac{1}{N} \sum_{k=0}^{N-1} |X[k]|^2$$

This property is particularly useful in analyzing signal energy and power.

5.18   Classical Fourier Transform in Python

The Fourier transform is a foundational tool in signal and data analysis, and Python provides efficient computational implementations through libraries like NumPy. Using the fast Fourier transform (FFT), which is a computationally efficient algorithm for the discrete Fourier transform (DFT), we can analyze signals in both the time and frequency domains.

This subsection explores the implementation of the Fourier transform in Python, explains each step, and highlights its real-world applications across various fields.

Python Code Example: Fourier Transform of a Signal
In this example, we generate a composite time-domain signal composed of two sine waves with frequencies of 50 Hz and 80 Hz. Using NumPy's FFT module, we compute the Fourier transform to analyze the signal in the frequency domain.

```
#-----------------------------------------------------------------------
# Fourier Transform of a Signal
# Chapter 5 in the QUANTUM COMPUTING AND QUANTUM MACHINE LEARNING BOOK
#-----------------------------------------------------------------------
# Version 1.0
# Qiskit changes frequently.
# We recommend using the latest version from the book code repository at:
# https://aqtinitiative.org/quantum-computing-for-engineers

# (c) 2025 Jesse Van Griensven, Roydon Fraser, and Jose Rosas
# License: MIT - Citation of this work required
#-----------------------------------------------------------------------
```

```python
import numpy as np
import matplotlib.pyplot as plt
#--------------------------------------------------------------------
# Step 1: Time Domain Signal
# Number of samples
N = 512

# Sampling interval (1 / sampling frequency)
T = 1.0 / 800.0

# Time vector
x = np.linspace(0.0, N * T, N, endpoint=False)
# Create a composite signal with two sine waves (50 Hz and 80 Hz)
y = np.sin(50.0 * 2.0 * np.pi * x) + 0.5 * np.sin(80.0 * 2.0 * np.pi * x)
#--------------------------------------------------------------------
# Step 2: Fourier Transform
# Compute the FFT of the signal
yf = np.fft.fft(y)

# Frequency vector for positive frequencies
xf = np.fft.fftfreq(N, T)[:N // 2]
#--------------------------------------------------------------------
# Step 3: Visualization
plt.figure(figsize=(10, 6))

# Plot the signal in the time domain
plt.subplot(2, 1, 1)
plt.plot(x, y)
plt.title("Time Domain Signal")
plt.xlabel("Time [s]")
plt.ylabel("Amplitude")

# Plot the frequency domain (Magnitude Spectrum)
plt.subplot(2, 1, 2)
plt.plot(xf, 2.0 / N * np.abs(yf[:N // 2]))
plt.title("Frequency Domain (Magnitude Spectrum)")
plt.xlabel("Frequency [Hz]")
plt.ylabel("Amplitude")

plt.tight_layout()
plt.show()
```

Figure 5.4 shows an original signal on the top and the result after applying the
Fourier transform. One can see that two main frequencies dominate the original
signal. One obvious application in this case is filtering for noise reduction. The

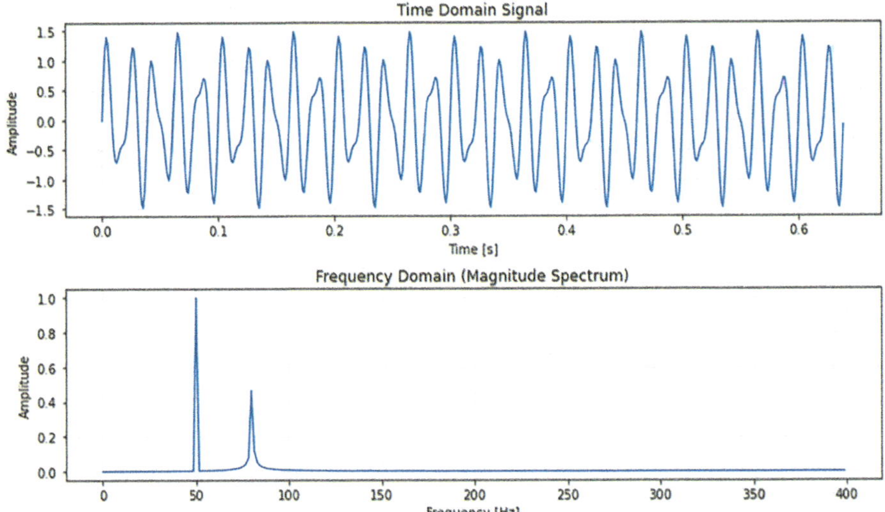

Fig. 5.4 Original signal and its transformation in the frequency domain. (By the authors)

frequencies above 150 Hz can be deleted, and the inversion of the Fourier transform can be applied. The signal would look the same but can be represented with way fewer variables. This is used both for noise reduction and signal compression.

5.18.1   Fourier Transform Code Explanation

This code is a bit complex for beginners. The following is an explanation of the steps in the code.

1. **Time-Domain Signal Generation**

   (a) Number of Samples (N): The total number of data points in the time domain.
   (b) Sampling Interval (T): The time gap between successive samples. It is the reciprocal of the sampling frequency.
   (c) Signal Definition: The composite signal y consists of:

      - A sine wave with a frequency of 50 Hz and an amplitude of 1.
      - A sine wave with a frequency of 80 Hz and an amplitude of 0.5.

   (d) The signal is generated using the sine function:

   $$y(t) = \sin(2\pi \cdot 50 \cdot t) + 0.5 \cdot \sin(2\pi \cdot 80 \cdot t).$$

2. **Fast Fourier Transform (FFT)**

   (a) The np.fft.fft() function computes the FFT of the signal, converting it from the time domain to the frequency domain.

   (b) The output yf is a complex array representing both the amplitude and phase of the frequency components.

3. **Frequency Representation**

   The np.fft.fftfreq() function generates the corresponding frequency bins. Since the FFT output is symmetric, we plot only the positive frequencies up to $N/2$.

4. **Visualization**

   (a) The first plot displays the signal in the time domain.

   (b) The second plot shows the magnitude spectrum, representing the amplitude of the frequency components.

5. **Time Domain**

   The time-domain signal clearly shows oscillations but does not directly reveal the underlying frequencies.

6. **Frequency Domain**

   The magnitude spectrum displays two sharp peaks at 50 Hz and 80 Hz, corresponding to the two sine waves in the composite signal. This confirms that the FFT successfully decomposed the signal into its frequency components.

The frequency domain representation is more insightful for understanding the signal's structure, identifying dominant frequencies, and isolating noise.

5.18.2   Applications of the Classical Fourier Transform

The classical Fourier transform and its computational implementation via FFT plays a critical role across various domains. Below are some key applications:

5.18.2.1   Signal Processing

Fourier analysis is a cornerstone of digital signal processing (DSP). By analyzing signals in the frequency domain, we can filter noise, compress data, and identify key frequency components.

(a) **Noise Filtering**

   Unwanted noise in signals often appears as specific high-frequency components. By applying frequency-domain filtering techniques, we can remove these components and reconstruct a clean signal.

(b) **Audio Analysis**

   The Fourier transform is widely used in analyzing sound signals to extract dominant frequencies in music, speech, or environmental noise. It is used in identifying the fundamental frequency of a musical note or removing background noise from an audio recording.

5.18.2.2   Image Processing

The Fourier transform is not limited to 1D signals; it extends to 2D images for frequency analysis. By analyzing and manipulating the frequency content of images, we can perform operations such as:

(a) **Frequency Filtering**
    Removing high-frequency noise to smooth images.
(b) **Edge Detection**
    Highlighting sharp edges by amplifying specific frequency ranges.
(c) **Pattern Recognition**
    Identifying repeating patterns or textures in images.
    Example: In medical imaging (e.g., MRI and CT scans), Fourier methods are used to reconstruct images from frequency-domain data, remove noise, and enhance visual features.

5.18.2.3   Communications

Fourier analysis is central to modern communication systems. In wireless communication, Fourier analysis ensures proper bandwidth allocation and noise suppression:

(a) **Modulation and Demodulation**
    Signals are modulated into higher frequencies for efficient transmission and demodulated back at the receiver using Fourier techniques.
(b) **Frequency-Domain Analysis**
    Wireless systems rely on analyzing frequency components for signal integrity and noise handling.

5.18.2.4   Fourier Transform in Quantum Computing

In quantum computing, the quantum Fourier transform (QFT) is the quantum analog of the classical Fourier transform. The QFT provides an exponential speedup over classical Fourier transform algorithms, making it essential for quantum applications. The QFT is a key component of several quantum algorithms, including:

(a) **Shor's Algorithm**
    Used for factoring large integers efficiently, with applications in cryptography.
(b) **Quantum Phase Estimation**
    QFT is used to estimate eigenvalues in quantum systems.

5.18.2.5 Machine Learning

In machine learning and data science, the Fourier transform is used to extract frequency-based features for various tasks:

(a) **Time-Series Analysis**

Analyzing periodic patterns and trends in time-series data (e.g., financial data, sensor readings).

(b) **Anomaly Detection**

Identifying anomalies in datasets by analyzing frequency components and detecting unexpected variations. For example, detecting seasonal patterns in energy consumption data or spotting rare deviations in network traffic.

5.19 Quantum Fourier Transform

The quantum Fourier transform (QFT) is a key algorithm in quantum computing, serving as the quantum analog of the classical discrete Fourier transform (DFT). It maps quantum states into the Fourier domain, enabling the efficient extraction of periodicity and frequency information. The QFT is a foundational component in several quantum algorithms, including Shor's algorithm, and has wide-ranging applications in engineering, signal processing, and simulations.

The quantum Fourier transform (QFT) is a fundamental operation in quantum computing, enabling efficient frequency analysis and periodicity detection. Its properties as a unitary transformation and its implementation using quantum gates make it a core component of algorithms like Shor's and phase estimation. With applications in signal processing, data compression, and wave simulations, the QFT exemplifies the transformative potential of quantum algorithms in engineering and computational sciences.

5.19.1 Overview of QFT

The QFT transforms a quantum state $|x\rangle$ into a superposition of states weighted by complex exponential coefficients. Mathematically, it is defined as:

$$\text{QFT}(|x\rangle) = \frac{1}{\sqrt{N}} \sum_{k=0}^{N-1} e^{2\pi i x k / N} |k\rangle$$

where N is the dimensionality of the quantum system, $|x\rangle$ is a basis state, and k indexes the Fourier components.

Unlike the classical DFT, which requires $O(N^2)$ operations, the QFT can be implemented on a quantum computer with $O(\log^2 N)$ gates, providing an exponential speedup for large-scale problems. This efficiency makes the QFT indispensable for quantum algorithms that rely on periodicity or frequency analysis.

5.19.2 QFT Properties

The QFT exhibits several mathematical and operational properties that underpin its utility in quantum computing.

5.19.2.1 Linear Transformation on Quantum States

The QFT is a unitary operation, meaning it preserves the total probability of the quantum state. For a multi-qubit system, the QFT acts as a linear transformation in the Hilbert space, satisfying:

$$U_{\mathrm{QFT}} U_{\mathrm{QFT}}^{\dagger} = I$$

where U_{QFT} is the QFT operator and I is the identity matrix.

This linearity ensures that the QFT can be decomposed into smaller operations, enabling efficient implementation in quantum circuits.

5.19.2.2 Core Component in Algorithms

The QFT is a crucial element in quantum algorithms that exploit periodicity. In Shor's algorithm, for instance, the QFT identifies the period r of the modular exponentiation function, enabling efficient integer factorization. Similarly, in the phase estimation algorithm, the QFT determines the eigenvalues of unitary operators, with applications in quantum chemistry and simulations.

5.19.2.3 Inverse QFT

The inverse QFT, denoted QFT^{-1}, reverses the Fourier transform:

$$\mathrm{QFT}^{-1}(\,|k\rangle) = \frac{1}{\sqrt{N}} \sum_{x=0}^{N-1} e^{-2\pi i x k / N} \,|x\rangle$$

The inverse QFT is used to transform frequency-domain states back into the time domain, completing algorithms that require bidirectional transformations.

5.19.3 QFT Quantum Circuit

The QFT can be implemented on a quantum computer using a sequence of controlled phase shift gates and Hadamard gates. This section describes the construction of a QFT circuit for a three-qubit system, illustrating the steps involved.

This is the "Step-by-Step" circuit representation:

1. **Initial State**

   The input state $|x\rangle = |x_2 x_1 x_0\rangle$, where x_2, x_1, and x_0 are the binary digits of x, is prepared.

2. **Hadamard Gate**

   A Hadamard gate is applied to the most significant qubit x_2, creating a superposition:

$$|x\rangle \rightarrow \frac{1}{\sqrt{2}}\left(|0\rangle + e^{2\pi i 0.x_2}|1\rangle\right)$$

3. **Controlled Phase Shifts**

   Controlled phase shift gates are applied between qubits to introduce phase factors based on their binary positions. For example:

$$\text{Controlled Phase Shift}: |x_2 x_1\rangle \rightarrow \frac{1}{\sqrt{2}}\left(|0\rangle + e^{2\pi i(0.x_2 + 0.x_1)}|1\rangle\right)$$

4. **Iterative Steps**

   The process is repeated for the next qubits, with phase shifts controlled by the higher-order qubits. Each qubit is transformed into a superposition weighted by the Fourier coefficients.

5. **Final State**

   The final state after the QFT is:

$$|k\rangle = \frac{1}{\sqrt{8}}\sum_{x=0}^{7} e^{\frac{2\pi i x k}{8}}|x\rangle$$

   where k encodes the frequency components of the original state.

Example QFT Circuit for 3 Qubits:
The circuit includes:

1. Hadamard gates on each qubit.

2. Controlled R_k gates, where R_k applies a phase shift of $e^{2\pi i/2^k}$.

3. A final reordering of qubits will be done to match the Fourier basis.

The QFT circuit demonstrates the modularity and efficiency of quantum algorithms, with each layer representing a distinct mathematical operation.

5.19.4   QFT Applications in Engineering

The QFT has significant applications in engineering, particularly in areas that require frequency analysis, signal processing, and simulations.

5.19.4.1   Signal Processing and Data Compression

Quantum signal processing leverages the QFT to analyze waveforms and extract frequency components with exponential speedup. Engineers use this capability for tasks such as:

1. **Speech and Audio Analysis**
   Identifying dominant frequencies in audio signals for compression and enhancement.
2. **Image Processing**
   Transforming image data into the frequency domain to perform edge detection and filtering.

Quantum algorithms based on the QFT can handle high-dimensional data efficiently, enabling real-time analysis of complex signals.

5.19.4.2   Quantum Simulations for Wave Analysis

In engineering applications involving wave mechanics, such as fluid dynamics and structural vibrations, the QFT is used to model and analyze wave phenomena. For example:

1. **Vibration Analysis**
   Engineers use the QFT to study the frequency response of materials and structures under oscillatory forces.
2. **Electromagnetic Waves**
   The QFT helps simulate the propagation of electromagnetic waves in antennas and communication systems.

By encoding waveforms into quantum states and applying the QFT, quantum simulations provide insights into dynamic systems that are computationally prohibitive for classical methods.

5.20   Shor's Algorithm: Factoring and Cryptographic Implications

Shor's algorithm represents one of the most groundbreaking advancements in quantum computing. By solving the problem of integer factorization exponentially faster than classical algorithms, it directly challenges the security of widely used cryptographic systems, such as RSA. Its ability to efficiently factorize large integers makes it a cornerstone of quantum computing with profound implications for cryptography, engineering, and secure communications.

Shor's algorithm represents a paradigm shift in computational mathematics and cryptography, providing an efficient solution to the integer factorization problem. Its reliance on the quantum Fourier transform and quantum parallelism offers exponential speedup over classical algorithms, posing a direct challenge to RSA-based systems. With applications in cryptographic security and engineering projects, Shor's algorithm underscores the importance of transitioning to post-quantum cryptographic protocols to protect sensitive information and maintain secure communication in a quantum-enabled world.

5.20.1   Overview of Shor's Algorithm

Shor's algorithm addresses the integer factorization problem, which involves decomposing a composite number N into its prime factors. Classical factorization algorithms, such as the general number field sieve (GNFS), have exponential time complexity for large N. Specifically, the classical runtime is:

$$O\left(e^{c \cdot (\log N)^{1/3} (\log\log N)^{2/3}}\right)$$

where c is a constant.

Shor's algorithm reduces the runtime to:

$$O\left((\log N)^2 (\log\log N)(\log\log\log N)\right)$$

offering an exponential speedup over classical methods. This improvement is achieved by leveraging quantum parallelism and the quantum Fourier transform (QFT) to determine the periodicity r of a function related to N.

The security of RSA cryptography, which underpins much of modern secure communication, relies on the difficulty of integer factorization. Shor's algorithm poses a direct threat to these systems, making the development of post-quantum cryptography an urgent priority.

5.20.2   Mathematical Foundation

The core idea behind Shor's algorithm is the reduction of integer factorization to the problem of finding the period r of the function:

$$f(x) = a^x \quad mod\ N$$

where a is a randomly chosen integer coprime to N.

The function $f(x)$ is periodic with period r, meaning:

$$f(x + r) = f(x)$$

Once the period r is determined, the factors of N can be extracted using the relationships:

$$p = \gcd\left(a^{r/2} - 1, N\right), \quad q = \gcd\left(a^{r/2} + 1, N\right)$$

where gcd denotes the greatest common divisor. These steps are efficient and can be performed classically. Note if r is odd, then Shor's algorithm is repeated with a different random number a until an even r is obtained.

To find r, Shor's algorithm employs the quantum Fourier transform (QFT), which maps the input state into the frequency domain, enabling the extraction of periodicity.

5.20.3   Steps of Shor's Algorithm

Shor's algorithm consists of four primary steps: quantum state preparation, modular exponentiation, the quantum Fourier transform, and classical post-processing.

5.20.3.1   Quantum State Preparation

The algorithm begins by initializing a quantum register in a superposition of all possible states x from 0 to $q - 1$, where q is a power of 2 such that $q > N^2$. The state is:

$$|\psi\rangle = \frac{1}{\sqrt{q}} \sum_{x=0}^{q-1} |x\rangle$$

This step ensures that the quantum system encodes all possible values of x, enabling parallel computation of $f(x)$.

5.20.3.2   Apply Modular Exponentiation

The next step computes $f(x) = a^x \mod N$ for each x in superposition. This is achieved using quantum gates designed to perform modular arithmetic. The resulting state is:

$$|\psi\rangle = \frac{1}{\sqrt{q}} \sum_{x=0}^{q-1} |x\rangle \, |f(x)\rangle$$

The function $f(x)$ imprints its periodicity r onto the quantum state.

5.20.3.3   Quantum Fourier Transform (QFT)

The QFT is applied to the first register to extract the periodicity r. The QFT maps the quantum state into the frequency domain, creating interference patterns that highlight the periodic structure of $f(x)$:

$$|\psi\rangle \rightarrow \frac{1}{\sqrt{q}} \sum_{k=0}^{q-1} \sum_{x=0}^{q-1} e^{2\pi i k x / q} \, |k\rangle$$

Measurement of the first register yields a value k proportional to q/r, from which r can be determined using continued fraction expansion.

5.20.3.4   Classical Post-Processing

After finding r, the factors of N are computed using:

$$p = \gcd\left(a^{r/2} - 1, N\right), \quad q = \gcd\left(a^{r/2} + 1, N\right)$$

If r is odd or if $a^{r/2} \equiv \pm 1 \mod N$, the process is repeated with a new value of a.

5.20.4   Cryptographic Implications

The ability of Shor's algorithm to factorize integers efficiently poses a significant threat to RSA-based cryptographic systems. RSA encryption relies on the difficulty of factorizing the product of two large primes, $N = pq$, to ensure the security of private keys. With Shor's algorithm, a sufficiently powerful quantum computer could break RSA encryption in polynomial time, compromising secure communication protocols, digital signatures, and engineering data pipelines.

5.20.4.1 Post-Quantum Cryptography

To address this threat, researchers are developing cryptographic schemes that are resistant to quantum attacks. These include lattice-based cryptography, code-based cryptography, and hash-based signatures, which rely on computational problems that are not susceptible to Shor's algorithm.

5.20.4.2 Applications in Engineering

Engineering fields reliant on secure data transmission, such as aerospace and telecommunications, must adopt post-quantum cryptographic protocols to safeguard sensitive information. For example, secure control of unmanned aerial vehicles (UAVs) and remote monitoring systems requires robust encryption that is resistant to quantum attacks.

5.20.5 *Shor's Algorithm in Securing Drone Communication Channels*

Consider an engineering project involving a fleet of drones communicating with a central command system. The communication channels use RSA-based encryption to protect against eavesdropping and unauthorized access.

If an adversary possesses a quantum computer capable of running Shor's algorithm, the RSA encryption can be broken, exposing critical commands and drone locations. To mitigate this risk, post-quantum cryptographic algorithms, such as lattice-based encryption, can be implemented to ensure secure communication. Quantum-resistant encryption ensures that even in the presence of quantum computers, the communication channels remain secure, preventing unauthorized access to the drones.

5.21 Variational Quantum Eigensolver for Optimization

The variational quantum eigensolver (VQE) is a hybrid quantum-classical algorithm designed to solve eigenvalue problems by leveraging quantum hardware for state preparation and measurement while using classical computers for optimization. VQE is particularly suitable for optimization problems in engineering, where it provides approximate solutions to complex problems that are computationally expensive for classical methods.

The variational quantum eigensolver (VQE) is a versatile and powerful algorithm for solving optimization problems in engineering. By combining quantum state preparation with classical optimization, VQE provides a practical approach to finding approximate solutions for complex problems. Its applications in material science, manufacturing, energy systems, and aerospace highlight its potential to revolutionize engineering practices, making it a cornerstone of quantum computing research and application.

5.21.1   Overview of VQE

VQE's flexibility makes it well-suited for solving optimization problems in engineering, where the Hamiltonian can represent an objective function or system dynamics. VQE is based on the variational principle in quantum mechanics, which states that for a given Hamiltonian H, the expectation value of H with respect to any trial wave function $|\psi(\theta)\rangle$ is an upper bound for the ground-state energy:

$$E(\theta) = \langle \psi(\theta) | H | \psi(\theta)\rangle \geq E_0$$

where E_0 is the ground-state energy.

The algorithm aims to approximate the ground state of H by iteratively updating the parameters θ of a trial wave function, also called the Ansatz, to minimize $E(\theta)$. This hybrid approach combines the quantum advantage of representing high-dimensional states with the classical efficiency of optimization algorithms. Figure 5.5 summarize the most commonly used types of Ansatz. An ansatz is a guess used in the process of finding a better solution, implemented by a parameterized quantum circuit (PQCircuit).

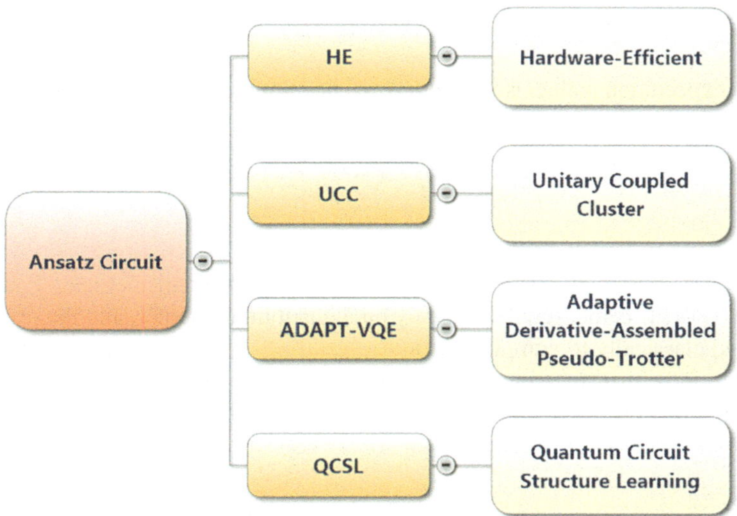

Fig. 5.5 ANSATZ circuit types. (By the authors)

5.21.2 VQE Steps

The VQE algorithm consists of three main steps: preparing the Ansatz, evaluating the cost function, and optimizing the parameters θ.

5.21.2.1 Prepare Ansatz

The first step is to construct a trial wave function $|\psi(\theta)\rangle$, parameterized by a set of classical variables $\theta = (\theta_1, \theta_2, \ldots)$. The ansatz is implemented on a quantum circuit designed to explore the relevant state space efficiently.

A common Ansatz used in VQE is the hardware-efficient Ansatz, which consists of alternating layers of parameterized single-qubit rotations and entangling gates. For example, the trial state for a two-qubit system can be written as:

$$|\psi(\theta)\rangle = U(\theta)\,|0\rangle = R_y(\theta_1) \otimes R_y(\theta_2) \cdot \text{CNOT} \cdot R_y(\theta_3) \otimes R_y(\theta_4)\,|00\rangle$$

where $R_y(\theta)$ represents a rotation around the y-axis of the Bloch sphere.

The choice of ansatz depends on the problem and the hardware constraints, balancing expressiveness with circuit depth to minimize decoherence effects.

5.21.2.2 Evaluate Cost Function

The cost function is the expectation value of the Hamiltonian H with respect to the trial state $|\psi(\theta)\rangle$:

$$E(\theta) = \langle \psi(\theta)\,|\,H\,|\,\psi(\theta)\rangle$$

This expectation value is computed by decomposing H into a sum of Pauli operators:

$$H = \sum_i c_i P_i$$

where c_i are coefficients, and P_i are tensor products of Pauli matrices. Each term $\langle \psi(\theta)\,|\,P_i\,|\,\psi(\theta)\rangle$ is measured separately on a quantum computer, and the results are combined classically to compute $E(\theta)$.

The iterative measurement process ensures that the quantum resources are used efficiently, making VQE feasible on noisy intermediate-scale quantum (NISQ) devices.

5.21.2.3 Optimization

The parameters θ are updated using classical optimization algorithms, such as gradient descent, Nelder–Mead, or COBYLA, to minimize $E(\theta)$. The optimization process iterates until convergence is achieved, producing the approximate ground state and energy:

$$\theta^* = \arg \min_{\theta} E(\theta)$$

This hybrid structure enables VQE to leverage the strengths of both quantum and classical computing, mitigating the limitations of current quantum hardware.

5.21.3 VQE Engineering Applications

VQE has diverse applications in engineering, where optimization is a fundamental task. Its ability to solve complex problems makes it a valuable tool for improving efficiency and performance across various domains.

5.21.3.1 Optimizing Material Properties

In material science, engineers aim to design materials with specific properties, such as high strength, thermal conductivity, or electrical resistance. The Hamiltonian in VQE represents the quantum interactions within the material, allowing the algorithm to compute properties like energy levels, reaction pathways, and structural stability.

For example, VQE can simulate molecular systems to identify optimal configurations for catalysis or battery materials. By approximating the ground state of a molecular Hamiltonian, VQE helps engineers develop more efficient and sustainable materials.

(a) **Optimal Alloys**
   Identifying compounds with strength-to-weight ratios suitable for aerospace applications.
(b) **Semiconductor Materials**
   Finding materials with desired electrical properties for electronic components.

5.21.3.2 Resource Allocation in Manufacturing Systems

Manufacturing systems often involve resource allocation problems, where limited resources must be distributed among competing tasks to maximize efficiency or minimize costs. These problems are typically NP-hard, making them challenging for classical solvers.

VQE can encode resource allocation as a Hamiltonian, where the ground-state energy corresponds to the optimal allocation. For example, in a production line, VQE can optimize the scheduling of tasks, reducing downtime and improving throughput.

5.21.3.3   Energy Grid Optimization

In power grid management, VQE can optimize the placement of renewable energy sources, balance load distribution, and minimize energy losses. By encoding grid dynamics into a Hamiltonian, the algorithm identifies optimal configurations that enhance grid stability and efficiency.

5.21.3.4   Applications in Aerospace

Aerospace engineering involves optimizing aerodynamic designs, structural materials, and fuel efficiency. VQE enables high-dimensional simulations of aerodynamic flows and material properties, helping engineers design lighter and more efficient aircraft.

5.21.3.5   Case Study: VQE in Battery Design

A research team used VQE to model lithium-air batteries, a promising technology for energy storage. By simulating the quantum interactions within the battery materials, the algorithm identified configurations that enhanced energy density by 15%. This improvement translates to lighter batteries with longer lifespans, addressing critical challenges in renewable energy storage and electric vehicles.

5.22   VQE Practical Engineering Use Cases

Quantum computing has immense potential to transform engineering by addressing computational bottlenecks and enabling new capabilities across multiple domains. Practical engineering applications of quantum computing include material discovery, supply chain optimization, secure communication, and advanced simulations. Quantum algorithms such as Grover's, VQE, and QFT provide engineers with tools to tackle complex problems more efficiently, paving the way for innovative designs and solutions.

This section explores practical use cases for quantum algorithms in engineering applications, including material design, supply chain optimization, secure communication, and quantum simulations.

5.22.1   Applications: Material Design

Material design involves searching for and optimizing compounds with specific properties for use in manufacturing, construction, and other industries. Quantum algorithms enhance these processes by accelerating computations and uncovering solutions that classical methods cannot efficiently find.

5.22.2   Grover's Algorithm for Compound Search

Grover's algorithm is effective for searching unsorted databases for optimal compounds. For example, in identifying a material with specific thermal and electrical properties, the algorithm searches through all possible compounds:

$$|\psi\rangle = H^{\otimes n} |0\rangle^{\otimes n}$$

where H is the Hadamard gate applied to n qubits to create a superposition state. The Grover iteration involves applying an oracle U_f and a diffusion operator D:

$$U_f |\psi\rangle = (-1)^{f(x)} |\psi\rangle, \quad D = 2|\psi\rangle\langle\psi| - I$$

Python Code Example: Grover for Material Search

```
#----------------------------------------------------------------------
# Quantum Grover for Material Search
# Chapter 5 in the QUANTUM COMPUTING AND QUANTUM MACHINE LEARNING BOOK
#----------------------------------------------------------------------
# Version 1.0
# Qiskit changes frequently.
# We recommend using the latest version from the book code repository at:
# https://aqtinitiative.org/quantum-computing-for-engineers

# (c) 2025 Jesse Van Griensven, Roydon Fraser, and Jose Rosas
# License: MIT - Citation required
#----------------------------------------------------------------------
import warnings
warnings.filterwarnings('ignore')
```

```python
from qiskit import QuantumCircuit, Aer, execute
from qiskit.visualization import circuit_drawer
from qiskit.circuit.library import GroverOperator
#---------------------------------------------------------------------

# Define the oracle
n = 3  # Number of qubits
oracle = QuantumCircuit(n)

 # Mark |110⟩ as the solution
oracle.z(2)
oracle = GroverOperator(oracle)

# Initialize the quantum circuit
qc = QuantumCircuit(n)

# Apply Hadamard gates to place all qc gates in superposition
qc.h(range(n))

# Apply the oracle and diffusion operator
qc.append(oracle, range(n))

# Measure the result
qc.measure_all()

# Visualize the circuit
print("Quantum Circuit:")
display(circuit_drawer(qc, output='mpl', style="iqp"))

# Simulate the circuit
simulator = Aer.get_backend('qasm_simulator')
result  = execute(qc, simulator, shots=1000).result()
print("Grover's Algorithm Results:", result.get_counts())
```

5.23   Optimization in Supply Chain

Efficient supply chain management is critical in engineering sectors like manufacturing and logistics. Quantum algorithms such as the variational quantum eigensolver (VQE) are particularly useful in optimizing transport logistics.

5.23.1   VQE for Transport Logistics

VQE minimizes a cost Hamiltonian representing the transportation network. For instance:

$$H_C = \sum_{i,j} c_{ij} Z_i Z_j$$

where c_{ij} is the cost between nodes i and j, and Z_i represents qubits encoding the route selection.

Python Code Example: VQE for Logistics

```python
#-------------------------------------------------------------------
# VQE for Logistics
# Chapter 5 in the QUANTUM COMPUTING AND QUANTUM MACHINE LEARNING BOOK
#-------------------------------------------------------------------
# Version 1.1 (Fixed Backend Issue)
# Qiskit changes frequently.
# We recommend using the latest version from the book code repository at:
# https://aqtinitiative.org/quantum-computing-for-engineers

# (c) 2025 Jesse Van Griensven, Roydon Fraser, and Jose Rosas
# License: MIT - Citation of this work required
#-------------------------------------------------------------------

import numpy as np
import matplotlib.pyplot as plt
from qiskit import Aer
from qiskit.opflow import Z, I
from qiskit.algorithms import VQE
from qiskit.circuit.library import RealAmplitudes
from qiskit.utils import QuantumInstance
from qiskit.algorithms.optimizers import COBYLA


#-------------------------------------------------------------------
# Define the cost Hamiltonian for logistics optimization
# This Hamiltonian represents the cost structure of a transportation
problem
H = (Z ^ Z) + (Z ^ I) + (I ^ Z)

# Define the variational form (ansatz)
# This circuit is used to parameterize the quantum state
ansatz = RealAmplitudes(num_qubits=2, reps=2)
```

```python
# Define a classical optimizer
optimizer = COBYLA(maxiter=100)

# Define a backend (QASM simulator)
backend = Aer.get_backend("aer_simulator")

# Create a quantum instance
quantum_instance = QuantumInstance(backend)

# Solve using Variational Quantum Eigensolver (VQE)
vqe = VQE(ansatz, optimizer=optimizer,
quantum_instance=quantum_instance)
result = vqe.compute_minimum_eigenvalue(operator=H)

print("Optimized Transport Cost:", result.eigenvalue.real)


#-------------------------------------------------------------------
# Visualization of the cost landscape
#-------------------------------------------------------------------
# Generate a range of phase values to visualize the cost function
phases = np.linspace(0, 2*np.pi, 100)
costs = np.cos(phases)  # Simplified representation of cost changes

# Plot the cost function landscape
plt.figure(figsize=(8, 5))
plt.plot(phases, costs, label="Cost Function", color='blue')
plt.axvline(x=result.eigenvalue.real, color='red', linestyle='--',
label='Optimal Cost')
plt.xlabel("Phase (radians)")
plt.ylabel("Cost")
plt.title("Quantum Optimization: Cost Function Landscape")
plt.legend()
plt.grid()
plt.show()


#-------------------------------------------------------------------
# Plot of ansatz parameters
#-------------------------------------------------------------------
# Extract ansatz parameters
parameters = np.linspace(0, 2*np.pi, len(ansatz.parameters))
values = np.sin(parameters)  # Example of how parameters might evolve

plt.figure(figsize=(8, 5))
plt.bar(range(len(ansatz.parameters)), values, color='green',
label='Parameter Values')
```

```
plt.xlabel("Parameter Index")
plt.ylabel("Value")
plt.title("Optimization Parameters in Ansatz")
plt.legend()
plt.grid()
plt.show()
```

5.23.2 Quantum-Inspired Optimization

Quantum-inspired techniques such as tensor networks bring quantum principles to classical systems, enabling supply chain managers to optimize delivery schedules and reduce costs even without quantum hardware.

5.24 Noise Mitigation in Quantum Systems

Quantum systems are inherently noisy, with errors arising from decoherence, gate imperfections, and environmental interactions. Engineers must incorporate noise mitigation strategies to enhance algorithm reliability.

5.24.1 Quantum Error Correction Codes

1. **Shor's Code**
   Encodes a single logical qubit into nine physical qubits to correct arbitrary single-qubit errors:

$$|0_L\rangle = \frac{1}{\sqrt{2}}(|000\rangle + |111\rangle) \otimes (|000\rangle + |111\rangle) \otimes (|000\rangle + |111\rangle)$$

2. **Steane Code**
   Encodes one logical qubit into seven physical qubits, offering a balance between redundancy and efficiency.

Python Code Example: Simple Error Correction
This example shows how Shor's code protects quantum information against single-qubit errors.

```python
#-------------------------------------------------------------------
# Simple Error Correction Code
# Chapter 5 in the QUANTUM COMPUTING AND QUANTUM MACHINE LEARNING BOOK
#-------------------------------------------------------------------
# This is a simplified version of Shor's error correction code.
# The objective is to protect a logical qubit against single-qubit
errors.
# Each logical qubit is encoded into 3 physical qubits.
# Ensures that a single error is detected and corrected by majority
voting.
#-------------------------------------------------------------------
# Version 1.0
# Qiskit changes frequently.
# We recommend using the latest version from the book code repository at:
# https://aqtinitiative.org/quantum-computing-for-engineers

# (c) 2025 Jesse Van Griensven, Roydon Fraser, and Jose Rosas
# License:  MIT - Citation required
#-------------------------------------------------------------------
import warnings
warnings.filterwarnings('ignore')

from qiskit import QuantumCircuit
from qiskit.visualization import circuit_drawer
#-------------------------------------------------------------------

# Define a circuit for Shor's code - Requires 9 qubits
qc = QuantumCircuit(9)

# Each logical qubit is encoded into 3 physical qubits
# Each qubit group
# Encode a logical |0⟩ state
for i in range(3):
    qc.h(i)
    qc.cx(i, i + 3) # qc.cx is a CNOT Gate
    qc.cx(i, i + 6)

# A deliberate X-error (bit flip) is introduced on qubit 2
qc.x(2)

# Decode and measure
# Decoding reverses the encoding process.
# If an error occurred, the redundancy allows the error to be detected
during this step.
for i in range(3):
```

```
  qc.cx(i, i + 3)
  qc.cx(i, i + 6)
qc.measure_all()

# Display circuit
#print(qc)

# Draw the circuit
display(circuit_drawer(qc, output='mpl', style="iqp"))
```

5.24.2   Error Mitigation Techniques

In the current NISQ era, quantum computers are susceptible to decoherence, which causes computational errors. Error mitigation techniques attempt to reduce the impact of this problem. Here are two of the promising techniques:

1. **Zero-Noise Extrapolation**
   Executes circuits at varying noise levels and extrapolates results to estimate noise-free outcomes.
2. **Purification**
   Increases fidelity by selectively discarding erroneous measurements.

Python Code Example: Noise Mitigation
This code demonstrates how to simulate and mitigate noise in quantum circuits.

```
#------------------------------------------------------------------
# Noise Mitigation Code
# Chapter 5 in the QUANTUM COMPUTING AND QUANTUM MACHINE LEARNING BOOK
#------------------------------------------------------------------
# Version 1.0
# Qiskit changes frequently.
# We recommend using the latest version from the book code repository at:
# https://aqtinitiative.org/quantum-computing-for-engineers


# (c) 2025 Jesse Van Griensven, Roydon Fraser, and Jose Rosas
# License: MIT - Citation required
#------------------------------------------------------------------
import warnings
warnings.filterwarnings('ignore')

from qiskit import Aer, QuantumCircuit, execute
from qiskit.visualization import circuit_drawer
```

```python
from qiskit.providers.aer.noise import NoiseModel,
depolarizing_error
#-------------------------------------------------------------------

# Define a quantum circuit
qc = QuantumCircuit(2)
qc.h(0)
qc.cx(0, 1)
qc.measure_all()

# Define a noise model
noise_model = NoiseModel()

# Add a 1-qubit depolarizing error to 'h' gate
noise_model.add_all_qubit_quantum_error(depolarizing_error(0.01,
1), ['h'])

# Add a 2-qubit depolarizing error to 'cx' gate
noise_model.add_all_qubit_quantum_error(depolarizing_error(0.02,
2), ['cx'])

# Simulate a noisy circuit
simulator = Aer.get_backend('qasm_simulator')
result  = execute(qc, simulator, noise_model=noise_model,
shots=1000).result()

# Print the results
print("Results with Noise:", result.get_counts())

# Draw the circuit
display(circuit_drawer(qc, output='mpl', style="iqp"))
```

5.25 Secure Communication

Secure communication is paramount in engineering, particularly for sensitive systems in defense, finance, and healthcare. Quantum algorithms ensure robust encryption and detection of cryptographic vulnerabilities.

5.25.1   *Quantum Fourier Transform (QFT) for Cryptographic Analysis*

The QFT identifies patterns in cryptographic keys, aiding in the detection of weaknesses:

$$\text{QFT}(|x\rangle) = \frac{1}{\sqrt{N}} \sum_{k=0}^{N-1} e^{2\pi i x k / N} |k\rangle$$

Python Code Example: QFT for Key Analysis

```python
#------------------------------------------------------------------
# QFT for Key Analysis
# Chapter 5 in the QUANTUM COMPUTING AND QUANTUM MACHINE LEARNING BOOK
#------------------------------------------------------------------
# Version 1.1 (Extended with More Comments & Visualizations)
# Qiskit changes frequently.
# We recommend using the latest version from the book code repository at:
# https://aqtinitiative.org/quantum-computing-for-engineers

# (c) 2025 Jesse Van Griensven, Roydon Fraser, and Jose Rosas
# License: MIT - Citation required
#------------------------------------------------------------------

import warnings
warnings.filterwarnings('ignore')

import numpy as np
import matplotlib.pyplot as plt
from qiskit import Aer, QuantumCircuit, execute
from qiskit.visualization import circuit_drawer, plot_histogram
from qiskit.circuit.library import QFT
#------------------------------------------------------------------
# Explanation:
#------------------------------------------------------------------
# 1. The Quantum Fourier Transform (QFT) is applied to n qubits.
# 2. The circuit is simulated using a quantum simulator.
# 3. The measurement results are collected and displayed as a histogram.
# 4. The output is analyzed in phase-space representation.
# 5. The quantum circuit visualization helps us understand the
structure.

# The QFT helps in analyzing cryptographic keys by transforming the state
# into a Fourier basis, useful for key analysis and phase estimation.
```

```python
#------------------------------------------------------------------
# Define QFT circuit for cryptographic key analysis
#------------------------------------------------------------------
# Number of qubits for the Quantum Fourier Transform (QFT)
n = 3  # You can increase n for more complex key analysis

# Create a quantum circuit with n qubits
qc = QuantumCircuit(n)

# Apply the Quantum Fourier Transform (QFT)
qc.append(QFT(num_qubits=n).to_instruction(), range(n))

# Add measurement operations to observe results
qc.measure_all()


#------------------------------------------------------------------
# Visualizing the Quantum Circuit
#------------------------------------------------------------------
# Display the circuit diagram
display(circuit_drawer(qc, output='mpl', style="iqp"))


#------------------------------------------------------------------
# Simulating the Quantum Circuit
#------------------------------------------------------------------
# Define the quantum simulator
simulator = Aer.get_backend('qasm_simulator')

# Execute the circuit using the simulator with 1024 shots
result = execute(qc, simulator, shots=1024).result()

# Retrieve the measured output probabilities (bitstring distribution)
counts = result.get_counts()

# Print the cryptographic key analysis results
print("Cryptographic Key Analysis Results:", counts)


#------------------------------------------------------------------
# Plot the output distribution
#------------------------------------------------------------------
# Visualize the probability distribution of measured keys
plt.figure(figsize=(8, 5))
plot_histogram(counts, title="QFT Output Distribution for
Cryptographic Key Analysis")
plt.show()
```

```
#-------------------------------------------------------------------
# Analyzing Phase Relationships
#-------------------------------------------------------------------
# Convert binary measurement outcomes to decimal values for better
analysis
decimal_keys = [int(k, 2) for k in counts.keys()]
probabilities = np.array(list(counts.values())) / sum(counts.values
())

# Plot the phase-space representation
plt.figure(figsize=(8, 5))
plt.scatter(decimal_keys, probabilities, color='red',
label="Measured Key Probabilities")
plt.xticks(decimal_keys)
plt.xlabel("Key (Decimal Representation)")
plt.ylabel("Probability")
plt.title("Phase-Space Representation of QFT Measured Keys")
plt.legend()
plt.grid(alpha = 0.2)
plt.show()
```

5.25.2 *Quantum Cryptography*

Quantum cryptographic methods such as Quantum Key Distribution (QKD) leverage quantum principles to ensure secure communication. These systems detect eavesdropping by measuring quantum states, as any interference collapses the state.

5.26 **Quantum Simulation**

Quantum simulation is critical for modeling complex physical systems, which is essential for advancing engineering designs. Quantum computers simulate molecular dynamics and physical interactions more efficiently than classical counterparts by solving the Schrödinger equation:

$$H\,|\psi\rangle = E\,|\psi\rangle$$

where H is the system Hamiltonian.

Python Code Example: Quantum Simulation

```python
#-----------------------------------------------------------------
# Quantum NATURE Simulation - H2 Molecule (Qiskit Nature 0.7.2)
#-----------------------------------------------------------------
# Qiskit changes frequently.
# We recommend using the latest version from the book code repository at:
# https://aqtinitiative.org/quantum-computing-for-engineers

# (c) 2025 Jesse Van Griensven, Roydon Fraser, and Jose Rosas
# License: MIT - Citation required
#-----------------------------------------------------------------

import matplotlib.pyplot as plt
import numpy as np

# For Qiskit Nature 0.7.2
from qiskit_nature.second_q.drivers import PySCFDriver
from qiskit_nature.second_q.problems import
ElectronicStructureProblem

def main():
    """

    Demonstrates how to run a basic H2 simulation in Qiskit Nature 0.7.2,
    extracting the second-quantized Hamiltonian, and plotting the
molecule
    without hitting import or attribute errors.
    """

    #-----------------------------------------------------------------
    # 1. Define the quantum chemistry driver for the H2 molecule
    #    - Coordinates in Angstroms, minimal STO-3G basis
    #-----------------------------------------------------------------
    driver = PySCFDriver(atom='H 0 0 0; H 0 0 0.74', basis='sto3g')

    #-----------------------------------------------------------------
    # 2. Run the driver to get an ElectronicStructureProblem
    #-----------------------------------------------------------------
    problem = driver.run()

    #-----------------------------------------------------------------
    # 3. Extract the second-quantized operators (the Hamiltonian) for H2
    #-----------------------------------------------------------------
    second_q_ops = problem.second_q_ops()
    hamiltonian = second_q_ops[0]  # Fix: Use index instead of string key
```

```python
print("=========================================")
print(" Second-Quantized Hamiltonian for H2 (STO-3G):")
print("=========================================\n")
print(hamiltonian)


#------------------------------------------------------------
# 4. Visualize the H2 molecule's geometry
#------------------------------------------------------------
molecule_geometry = driver.molecule.geometry  # Fix: Correct way to
access geometry
positions = np.array([[atom.x, atom.y, atom.z] for atom in
molecule_geometry])
labels = [atom.symbol for atom in molecule_geometry]

fig = plt.figure(figsize=(6, 6))
ax = fig.add_subplot(111, projection='3d')

for (x, y, z), lbl in zip(positions, labels):
    ax.scatter(x, y, z, color='red', s=100)
    ax.text(x, y, z, f" {lbl}", color='black', fontsize=12)

# If the molecule is diatomic (H2), connect the two atoms
if len(positions) == 2:
    ax.plot(*zip(positions[0], positions[1]), color='gray',
linestyle='--')

ax.set_title("H2 Molecule Geometry (STO-3G)")
ax.set_xlabel('X (Å)')
ax.set_ylabel('Y (Å)')
ax.set_zlabel('Z (Å)')

# Adjust aspect ratio for a better 3D view
max_range = np.array([positions[:, 0].max() - positions[:, 0].min(),
            positions[:, 1].max() - positions[:, 1].min(),
            positions[:, 2].max() - positions[:, 2].min()]).max()
mid_x = (positions[:, 0].max() + positions[:, 0].min()) * 0.5
mid_y = (positions[:, 1].max() + positions[:, 1].min()) * 0.5
mid_z = (positions[:, 2].max() + positions[:, 2].min()) * 0.5
ax.set_xlim(mid_x - max_range/2, mid_x + max_range/2)
ax.set_ylim(mid_y - max_range/2, mid_y + max_range/2)
ax.set_zlim(mid_z - max_range/2, mid_z + max_range/2)

plt.show()
```

```
#------------------------------------------------------------------
# 5. Next Steps
#   - 'hamiltonian' can be used within a VQE or any other solver to
#     approximate ground/excited states of H2.
#------------------------------------------------------------------


if __name__ == "__main__":
    main()
```

5.26.1 *Enhancing Design Processes*

Quantum simulations improve the design of aerospace and automotive systems by providing accurate models for fluid dynamics, material stress, and vibration analysis.

Exercise Questions: Quantum Algorithms

1. **Fundamental Principles of Quantum Algorithms**

   (a) Explain how the principles of superposition, entanglement, and interference contribute to the efficiency of quantum algorithms.

   (b) Write the mathematical expression for a superposition state of n qubits and explain its significance in quantum parallelism.

2. **Shor's Algorithm**

   (a) Describe the mathematical foundation of Shor's algorithm and explain its impact on classical cryptographic systems like RSA.

   (b) Using the function $f(x) = a^x \; mod \, N$, explain how the periodicity r helps in factorizing N.

3. **Grover's Algorithm**

   (a) Describe the steps of Grover's algorithm and explain how it achieves quadratic speedup for unstructured search problems.

   (b) Derive the mathematical expression for the Grover diffusion operator $D = 2 \, | \psi \rangle \langle \psi | \, - I$ and explain its role in amplitude amplification.

4. **Quantum Fourier Transform (QFT)**

   (a) Define the quantum Fourier transform and write its mathematical expression. Compare it with the classical discrete Fourier transform.

   (b) Explain how the QFT is used in Shor's algorithm to extract periodicity.

5. **Variational Quantum Eigensolver (VQE)**

   (a) Describe the steps involved in the VQE algorithm and explain the role of the Ansatz in this process.

   (b) Write the variational principle equation used in VQE and explain how it helps in solving optimization problems.

6. **Applications of Quantum Algorithms**

   (a) Discuss how quantum simulations can aid in drug discovery and renewable energy research.
   (b) Explain how QAOA can be used for route optimization in logistics.

7. **Quantum Circuit Representation**

   (a) Draw and describe the quantum circuit representation of Grover's algorithm for a 3-qubit system.
   (b) Explain the role of Hadamard gates in the initialization of quantum states for algorithms like Grover's and Shor's.

8. **Quantum Simulations**

   (a) Using the Schrödinger equation $H \mid \psi \rangle = E \mid \psi \rangle$, explain how quantum computers simulate physical systems.
   (b) Discuss how quantum simulations can improve material design for aerospace engineering.

9. **Engineering Applications of Quantum Algorithms**

   (a) Provide a detailed example of how QAOA can optimize engineering manufacturing processes.
   (b) Explain the importance of quantum cryptography in securing engineering communications.

10. **Comparing Algorithms**

   (a) Compare the computational complexities of Shor's and Grover's algorithms, explaining the types of problems each solves.
   (b) Discuss the trade-offs between using Grover's algorithm and classical search methods in database applications.

11. **Mathematical Tools in Quantum Algorithms**

   (a) Derive the expression for the phase applied by the QFT and explain its significance in periodicity detection.
   (b) Compute the tensor product of two 1-qubit states $|\psi_1\rangle = \mid 0 \rangle + \mid 1 \rangle$ and $|\psi_2\rangle = \mid 0 \rangle - \mid 1 \rangle$.

12. **Future Directions in Quantum Algorithms**

   (a) Describe the potential breakthroughs in quantum machine learning and their applications in engineering.
   (b) Discuss how quantum error correction is integrated into quantum algorithms to enhance their reliability.

Additional Bibliography

1. S. Aaronson, *Quantum Computing Since Democritus* (Cambridge University Press, 2013)
2. F. Arute et al., Quantum supremacy using a programmable superconducting processor. Nature **574**(7779), 505–510 (2019). https://doi.org/10.1038/s41586-019-1666-5
3. A. Aspuru-Guzik, A. Aspuru-Guzik, *Quantum Chemistry and Machine Learning* (Wiley, 2021)
4. A. Barenco et al., Elementary gates for quantum computation. Phys. Rev. A **52**(5), 3457–3467 (1995). https://doi.org/10.1103/PhysRevA.52.3457
5. P. Benioff, *The Computer as a Physical System* (Springer, 2016)
6. J. Biamonte, P. Wittek, *Quantum Machine Learning: An Introduction* (Cambridge University Press, 2017)
7. R. Cleve, *Quantum Algorithms Revisited* (Elsevier, 2001)
8. D. Deutsch, Quantum theory, the church–Turing principle, and the universal quantum computer. Proc. R. Soc. Lond. A Math. Phys. Sci. **400**(1818), 97–117 (1985). https://doi.org/10.1098/rspa.1985.0070
9. D. Deutsch, *The Fabric of Reality: The Science of Parallel Universes—And Its Implications* (Penguin Books, 1997)
10. E. Farhi, J. Goldstone, *A Quantum Approximate Optimization Algorithm* (Springer, 2014)
11. L.K. Grover, A fast quantum mechanical algorithm for database search, in *Proceedings of the 28th Annual ACM Symposium on Theory of Computing (STOC)*, (1996), pp. 212–219
12. P. Kaye, R. Laflamme, M. Mosca, *An Introduction to Quantum Computing* (Oxford University Press, 2007)
13. S. Lloyd, Universal quantum simulators. Science **273**(5278), 1073–1078 (1996). https://doi.org/10.1126/science.273.5278.1073
14. S. Lloyd, *Programming the Universe: A Quantum Computer Scientist Takes on the Cosmos* (Vintage, 2006)
15. S. Majidy, C. Wilson, R. Laflamme, *Building Quantum Computers—A Practical Introduction* (Cambridge Press, 2025)
16. A. Montanaro, Quantum algorithms: an overview. NPJ Quantum Inf. **2**, 15023 (2016). https://doi.org/10.1038/npjqi.2015.23
17. M.A. Nielsen, I.L. Chuang, *Quantum Computation and Quantum Information*, 10th edn. (Cambridge University Press, 2010)
18. J. Preskill, Quantum computing in the NISQ era and beyond. Quantum **2**, 79 (2018). https://doi.org/10.22331/q-2018-08-06-79
19. C. Rigetti, *Practical Quantum Computing for Developers* (Addison-Wesley Professional, 2020)
20. P.W. Shor, Algorithms for quantum computation: discrete logarithms and factoring, in *Proceedings of the 35th Annual Symposium on Foundations of Computer Science (FOCS)*, (1994), pp. 124–134. https://doi.org/10.1109/SFCS.1994.365700
21. P.W. Shor, Polynomial-time algorithms for prime factorization and discrete logarithms on a quantum computer. SIAM J. Comput. **26**(5), 1484–1509 (1997)
22. A. Vaswani et al., Attention is all you need. Adv. Neural Inf. Process. Syst. **30**, 5998–6008 (2017)
23. W.K. Wootters, W.H. Zurek, *Quantum Theory and Measurement* (Princeton University Press, 1983)
24. C. Zalka, Simulating quantum systems on a quantum computer. Proc. R. Soc. A Math. Phys. Eng. Sci. **454**(1969), 313–322 (1998). https://doi.org/10.1098/rspa.1998.0162
25. W.H. Zurek, *Decoherence and the Transition from Quantum to Classical* (Oxford University Press, 2003)
26. W.H. Zurek, Decoherence, einselection, and the quantum origins of the classical. Rev. Mod. Phys. **75**(3), 715–775 (2003). https://doi.org/10.1103/RevModPhys.75.715

Chapter 6
Quantum Optimization Techniques

Optimization is fundamental in engineering and other scientific disciplines, addressing problems, such as resource allocation, process efficiency, and optimal design. Classical optimization methods rely on iterative computations and often struggle with complex, high-dimensional problems due to exponential scaling. Quantum optimization leverages principles of quantum mechanics, such as superposition and entanglement, to explore solutions more efficiently. This section introduces the quantum approximate optimization algorithm (QAOA), compares classical and quantum optimization methods, and explores quantum-inspired techniques for classical systems.

Quantum optimization introduces transformative techniques for solving complex problems, leveraging principles of quantum mechanics to enhance efficiency and scalability. The quantum approximate optimization algorithm (QAOA) exemplifies the power of quantum approaches, while quantum-inspired techniques extend these advantages to classical systems. As quantum hardware evolves, quantum optimization will play an increasingly pivotal role in engineering and computational sciences.

6.1 Classical Optimization Methods

Classical optimization involves finding the minimum or maximum of an objective function $f(x)$, often subject to constraints.

6.1.1 Classical Gradient-Based Methods

These methods rely on the gradient of the objective function to iteratively update variables:

© The Author(s), under exclusive license to Springer Nature Switzerland AG 2025

J. Van Griensven Thé et al., *Quantum Computing and Quantum Machine Learning for Engineers and Developers*, https://doi.org/10.1007/978-3-031-98245-3_6

$$x_{k+1} = x_k - \eta \nabla f(x_k)$$

where η is the learning rate.

Challenges:

1. Difficulty handling non-convex functions.
2. Computationally expensive for large-scale problems.

6.1.2 Combinatorial Optimization

Combinatorial optimization solves discrete problems, such as the traveling salesman problem (TSP) or maximum cut (Max-Cut) problem. These problems are often NP-hard, requiring exponential time for exact solutions. For example, in the TSP, the objective is to minimize the total distance:

$$f(\pi) = \sum_{i=1}^{n} d_{\pi(i),\pi(i+1)}$$

where π is a permutation of cities and d_{ij} is the distance between cities i and j.

Python Code Example: Classical Brute-Force TSP Solver
This example demonstrates the computational complexity of solving combinatorial optimization problems classically.

```python
#-------------------------------------------------------------------
# Classical Brute Force TSP Solver
# Chapter 6 in the QUANTUM COMPUTING AND QUANTUM MACHINE LEARNING BOOK
#-------------------------------------------------------------------
# Version 1.0
# Qiskit changes frequently.
# We recommend using the latest version from the book code repository at:
# https://aqtinitiative.org/quantum-computing-for-engineers

# (c) 2025 Jesse Van Griensven, Roydon Fraser, and Jose Rosas
# License: MIT - Citation required
#-------------------------------------------------------------------
from itertools import permutations


#-------------------------------------------------------------------
# Brute-force solution for TSP
def tsp_brute_force(distance_matrix):
  n = len(distance_matrix)
  cities = range(n)
```

```python
    min_cost = float('inf')
    best_route = None
    for perm in permutations(cities):
      cost = sum(distance_matrix[perm[i]][perm[i + 1]] for i in range(n - 1))
      cost += distance_matrix[perm[-1]][perm[0]]  # Return to start
      if cost < min_cost:
        min_cost = cost
        best_route = perm
    return min_cost, best_route
#-------------------------------------------------------------------

# Example distance matrix
distance = [[0, 2, 9], [1, 0, 6], [7, 4, 0]]

cost, route = tsp_brute_force(distance)
print("Optimal Cost:", cost)
print("Optimal Route:", route)
```

6.2 Quantum Approximate Optimization Algorithm (QAOA)

The quantum approximate optimization algorithm (QAOA) is a variational quantum algorithm designed to solve combinatorial optimization problems. QAOA provides a promising framework for solving combinatorial optimization problems using quantum resources. It uses parameterized quantum circuits to approximate the solution by minimizing an objective function.

By encoding the cost function into a Hamiltonian, QAOA iteratively optimizes variational parameters to minimize the objective. Its classical-quantum hybrid nature makes QAOA well-suited for practical applications in machine learning, such as hyperparameter tuning and loss function optimization. The integration of QAOA into classical machine learning workflows highlights its potential to address complex optimization challenges in the NISQ era.

6.2.1 Mathematical Foundation of QAOA

The QAOA operates on a quantum state defined as:

$$|\psi\rangle = e^{-i\beta H_M} e^{-i\gamma H_C} |\psi_0\rangle$$

where

H_C is the cost Hamiltonian, encoding the objective function to be minimized.
H_M is the mixing Hamiltonian, which allows exploration of the solution space.
β and γ are variational parameters optimized during execution.
$|\psi_0\rangle$ is the initial quantum state, often chosen as the uniform superposition of all possible states.

The cost function is minimized iteratively by updating the parameters β and γ using classical optimization algorithms.

6.2.2 QAOA Mathematical Algorithm Steps

The quantum approximate optimization algorithm (QAOA) steps are delineated below:

6.2.2.1 Problem Encoding

The complete optimization problem is represented as a cost Hamiltonian H_C, where the eigenvalues correspond to the cost of specific solutions:

$$H_C = \sum_i C_i Z_i + \sum_{i,j} C_{ij} Z_i Z_j$$

where C_i, C_{ij} are the problem-sepcific weights, and Z_i is the Pauli-Z operator acting on qubit i.

For example, the maximum cut problem can be expressed as:

$$H_C = \sum_{(i,j)\in E} Z_i Z_j$$

where Z_i and Z_j are Pauli-Z operators acting on qubits representing vertices.

For example, in the Max-Cut problem, the objective is to partition the vertices of a graph $G(V, E)$ such that the number of edges crossing the partition is maximized. The cost Hamiltonian for Max-Cut is:

$$H_C = \frac{1}{2} \sum_{(i,j)\in E} \left(1 - Z_i Z_j\right)$$

6.2.2.2 Ansatz Construction

Initialize a quantum state $|\psi\rangle$ and apply alternating layers of the cost and mixing Hamiltonians:

$$|\psi(\gamma,\beta)\rangle = \prod_{k=1}^{p} e^{-i\beta_k H_M} e^{-i\gamma_k H_C} |\psi_0\rangle$$

where γ and β are variational parameters, H_M is the mixing Hamiltonian, and p is the number of layers.

6.2.2.3 Parameter Optimization

Use a classical optimizer to minimize the expectation value:

$$\langle H_C \rangle = \langle \psi(\gamma,\beta) | H_C | \psi(\gamma,\beta) \rangle$$

6.2.2.4 Solution Extraction

Measure the quantum state to obtain the solution.

Python Code Example: QAOA Simple Implementation
This example defines a QAOA circuit with parameterized gates for optimization.

```python
#-----------------------------------------------------------------
# QAOA Simple Implementation
# Chapter 6 in the QUANTUM COMPUTING AND QUANTUM MACHINE LEARNING BOOK
#-----------------------------------------------------------------
# Version 1.0
# Qiskit changes frequently.
# We recommend using the latest version from the book code repository at:
# https://aqtinitiative.org/quantum-computing-for-engineers

# (c) 2025 Jesse Van Griensven, Roydon Fraser, and Jose Rosas
# License: MIT - Citation required
#-----------------------------------------------------------------
import numpy as np
from scipy.optimize import minimize

from qiskit import Aer, execute
from qiskit import QuantumCircuit
```

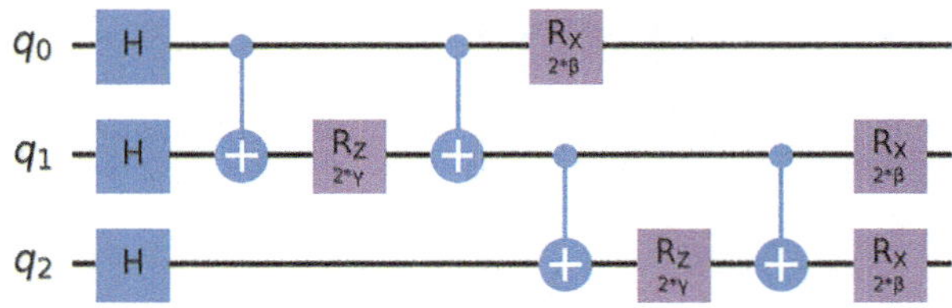

Fig. 6.1 A simple QAOA circuit. (By the authors)

```python
from qiskit.circuit import Parameter
from qiskit.visualization import circuit_drawer
#-----------------------------------------------------------------

# Define QAOA parameters
num_qubits = 3
gamma = Parameter('γ')
beta  = Parameter('β')

# Create QAOA circuit
qc = QuantumCircuit(num_qubits)

# Initialize in superposition
qc.h(range(num_qubits))
for i in range(num_qubits - 1):  # Apply cost Hamiltonian
    qc.cx(i, i + 1)
    qc.rz(2 * gamma, i + 1)
    qc.cx(i, i + 1)

qc.rx(2 * beta, range(num_qubits))  # Apply mixing Hamiltonian

print("Simple QAOA Circuit:")
print(qc)
```

The circuit generated in the code above is shown graphically in Fig. 6.1.

6.2.3 Implementation Steps in QAOA

The following are the standard steps to perform a QAOA optimization:

6.2.3.1 Initialization

Prepare the initial quantum state $|\psi_0\rangle$, which is typically the uniform superposition state:

$$|\psi_0\rangle = \frac{1}{\sqrt{2^n}} \sum_{z \in \{0,1\}^n} |z\rangle$$

6.2.3.2 Quantum Evolution

Apply alternating layers of the cost Hamiltonian H_C and mixing Hamiltonian H_M:

$$|\psi(\beta,\gamma)\rangle = \prod_{l=1}^{p} e^{-i\beta_l H_M} e^{-i\gamma_l H_C} |\psi_0\rangle$$

where p is the number of layers.

6.2.3.3 Measurement and Cost Evaluation

Measure the quantum state in the computational basis to estimate the expectation value of the cost Hamiltonian:

$$\langle H_C \rangle = \langle \psi(\beta,\gamma)|H_C|\psi(\beta,\gamma)\rangle$$

6.2.3.4 Optimization

Use classical optimization algorithms, such as COBYLA, Nelder–Mead, or SPSA, to update the variational parameters β and γ and minimize $\langle H_C \rangle$.

6.2.4 QAOA Implementation in Qiskit

The following is a more complete Python implementation of QAOA for a simple 2-qubit combinatorial optimization problem using Qiskit.

Python Code Example: More Complete Implementation

```
#-----------------------------------------------------------------
# QAOA More Complete Implementation
# Chapter 6 in the QUANTUM COMPUTING AND QUANTUM MACHINE LEARNING BOOK
#-----------------------------------------------------------------
# Version 1.0
# Qiskit changes frequently.
```

```python
# We recommend using the latest version from the book code repository at:
# https://aqtinitiative.org/quantum-computing-for-engineers

# (c) 2025 Jesse Van Griensven, Roydon Fraser, and Jose Rosas
# License: MIT - Citation required
#------------------------------------------------------------------
import numpy as np
import matplotlib.pyplot as plt

from qiskit import Aer, QuantumCircuit, transpile
from qiskit.algorithms.optimizers import COBYLA
from qiskit.utils import QuantumInstance
from qiskit.opflow import PauliSumOp, StateFn, AerPauliExpectation,
CircuitSampler
from qiskit.visualization import plot_histogram
#------------------------------------------------------------------


#------------------------------------------------------------------
def qaoa_circuit(gamma, beta):
# QAOA ansatz circuit (no classical register, no measurements)
  qc = QuantumCircuit(2)
  # Apply cost Hamiltonian (ZZ) as an RZZ rotation
  qc.rzz(2 * gamma, 0, 1)
  # Apply mixing Hamiltonian (global X rotations)
  qc.rx(2 * beta, 0)
  qc.rx(2 * beta, 1)
  return qc


#------------------------------------------------------------------
def cost_function(params):
# Cost function:
# 1. Build an operator ~StateFn(H_C) @ StateFn(qaoa_circuit).
# 2. Convert it to a measurable circuit using CircuitSampler.
# 3. Evaluate the expectation value (return a float).

  gamma, beta = params
  # Build the QAOA ansatz circuit
  qc = qaoa_circuit(gamma, beta)

  # StateFn of the circuit
  ansatz_op = StateFn(qc)

  # We want <H_C> = <psi|H_C|psi>, so we create the operator:
  # ~StateFn(H_C) @ StateFn(psi)
  operator = ~StateFn(H_C) @ ansatz_op
```

```python
    # Convert the operator to a circuit and evaluate using the sampler
    sampler = CircuitSampler(quantum_instance).convert(operator)
    value = sampler.eval().real  # Expectation value as a real number

    return value


#--------------------------------------------------------------------
def qaoa_circuit_with_measurements(gamma, beta):
# Create a circuit with measurements
# to visualize results via histogram (counts).

    qc = qaoa_circuit(gamma, beta)
    qc.measure_all()
    return qc
#--------------------------------------------------------------------

# Define the cost Hamiltonian for a 2-qubit ZZ problem.
# For instance: H_C = ZZ
H_C = PauliSumOp.from_list([('ZZ', 1.0)])

# Create a QuantumInstance using the statevector simulator
# (for cost function evaluation).
quantum_instance = QuantumInstance(backend=Aer.get_backend
('statevector_simulator'))

# Optimize gamma and beta using COBYLA
optimizer = COBYLA()
initial_params = [0.5, 0.5]
result = optimizer.minimize(fun=cost_function, x0=initial_params)

# Print Results
print(f"Optimal parameters: gamma={result.x[0]:.4f}, beta={result.x
[1]:.4f}")

# Create the measurement circuit using the optimal parameters
qc_meas = qaoa_circuit_with_measurements(result.x[0], result.x[1])

# Use the QASM simulator to generate counts (i.e., measure many shots)
backend = Aer.get_backend('qasm_simulator')
compiled_circuit = transpile(qc_meas, backend)
job = backend.run(compiled_circuit, shots=1024)
counts = job.result().get_counts()
```

```
# Plot histogram of measured states
plot_histogram(counts)
plt.show()
```

The QAOA iteratively improves the parameters β and γ to minimize the cost function. The state evolution can be visualized using the Bloch sphere for single-qubit problems or computational basis probabilities for multi-qubit cases.

6.2.5   Applications of QAOA in Machine Learning

The following are examples of QAOA applications:

1. **Hyperparameter Optimization**
   QAOA can be applied to optimize hyperparameters in classical machine learning models. The optimization process translates the hyperparameter search into a combinatorial problem solvable via QAOA.
2. **Loss Function Optimization**
   In neural network training, minimizing the loss function can be framed as an optimization problem. QAOA can be integrated as a subroutine to find optimal parameters more efficiently.
3. **Feature Selection**
   Selecting the most relevant features in a dataset can also be modeled as a combinatorial optimization problem. QAOA can identify the subset of features that minimizes a given cost function.

6.2.6   QAOA Phase Separation

Apply the cost Hamiltonian H_C to encode the problem into the quantum state:

$$U(C, \gamma) = e^{-i\gamma H_C}$$

For a simple example with $H_C = Z_1 Z_2$, the phase separation operator is:

$$U(C, \gamma) = \exp(-i\gamma Z_1 Z_2)$$

Python Code Example: QAOA Phase Separation
This example demonstrates applying the phase separation step for two qubits.

```
#---------------------------------------------------------------------
# QAOA Phase Separation
# Chapter 6 in the QUANTUM COMPUTING AND QUANTUM MACHINE LEARNING BOOK
```

```python
# Code demonstrates applying the phase separation step for two qubits
#-----------------------------------------------------------------------
# Version 1.0
# Qiskit changes frequently.
# We recommend using the latest version from the book code repository at:
# https://aqtinitiative.org/quantum-computing-for-engineers

# (c) 2025 Jesse Van Griensven, Roydon Fraser, and Jose Rosas
# License: MIT - Citation of this work is required
#-----------------------------------------------------------------------
import warnings
warnings.filterwarnings('ignore')

import numpy as np
from qiskit import QuantumCircuit
from qiskit.visualization import circuit_drawer
#-----------------------------------------------------------------------

# Define phase separation
def phase_separation(qc, gamma):
    qc.cx(0, 1)
    qc.rz(2 * gamma, 1)
    qc.cx(0, 1)
#-----------------------------------------------------------------------

# Apply phase separation
qc = QuantumCircuit(2)
qc.h([0, 1]) # Initialize
phase_separation(qc, np.pi / 4)

# Output Results
print("Phase Separation Circuit:")
print(qc)

# Draw the circuit
display(circuit_drawer(qc, output='mpl', style="iqp"))
```

6.2.7   QAOA Mixing

Apply the mixing Hamiltonian $H_M = \sum_i |X_i$, which ensures exploration of the solution space:

$$U(M,\beta) = e^{-i\beta H_M}$$

In practice, this corresponds to applying $R_X(2\beta)$ rotations:

$$R_X(2\beta) = \begin{bmatrix} \cos(\beta) & -i\sin(\beta) \\ -i\sin(\beta) & \cos(\beta) \end{bmatrix}$$

Python Code Example: QAOA Mixing

```python
#-------------------------------------------------------------------
# QAOA Mixing
# Chapter 6 in the QUANTUM COMPUTING AND QUANTUM MACHINE LEARNING BOOK
# Code demonstrates applying the phase separation step for two qubits
#-------------------------------------------------------------------
# Version 1.0
# Qiskit changes frequently.
# We recommend using the latest version from the book code repository at:
# https://aqtinitiative.org/quantum-computing-for-engineers

# (c) 2025 Jesse Van Griensven, Roydon Fraser, and Jose Rosas
# License: MIT - Citation of this work is required
#-------------------------------------------------------------------
import warnings
warnings.filterwarnings('ignore')

import numpy as np
from qiskit import QuantumCircuit
from qiskit.visualization import circuit_drawer
#-------------------------------------------------------------------

# Define mixing operation
def mixing_layer(qc, beta):
  qc.rx(2. * beta, [0, 1])
  return qc
#-------------------------------------------------------------------

# Apply mixing layer
qc = QuantumCircuit(2)
qc.h([0, 1]) # Initialize
beta = np.pi / 6.
mixing_layer(qc, beta)

# Output Results
print("Mixing Circuit:")
print(qc)
```

```python
# Draw the circuit
display(circuit_drawer(qc, output='mpl', style="iqp"))
```

Repeat the above steps for p layers, alternating phase separation and mixing. Measure the final quantum state to obtain the optimized solution.

Python Code Example: QAOA Full Circuit
This full circuit demonstrates the implementation of one layer of QAOA and simulates the results.

```python
#-------------------------------------------------------------------
# QAOA Full Circuit
# Chapter 6 in the QUANTUM COMPUTING AND QUANTUM MACHINE LEARNING BOOK
# Code demonstrates applying the phase separation step for two qubits
#-------------------------------------------------------------------
# Version 1.0
# Qiskit changes frequently.
# We recommend using the latest version from the book code repository at:
# https://aqtinitiative.org/quantum-computing-for-engineers

# (c) 2025 Jesse Van Griensven, Roydon Fraser, and Jose Rosas
# License: MIT - Citation of this work is required
#-------------------------------------------------------------------
import warnings
warnings.filterwarnings('ignore')

import numpy as np
from qiskit import QuantumCircuit
from qiskit import Aer, execute
from qiskit.visualization import circuit_drawer

#-------------------------------------------------------------------
# Define mixing operation
def mixing_layer(qc, beta):
    qc.rx(2 * beta, [0, 1])
    return qc
#-------------------------------------------------------------------
# Define phase separation
def phase_separation(qc, gamma):
    qc.cx(0, 1)
    qc.rz(2 * gamma, 1)
    qc.cx(0, 1)
#-------------------------------------------------------------------
```

```python
def qaoa_circuit(gamma, beta):
    """ Full QAOA circuit """
    # Create the Circuit
    qc = QuantumCircuit(2)

    # Initialize Circuit qubits in superposition
    qc.h([0, 1])

    # Prepare Phase Separation Circuit
    phase_separation(qc, gamma)

    # Apply mixing layer
    mixing_layer(qc, beta)

    qc.measure_all()
    return qc
#-----------------------------------------------------------------------

# Simulate QAOA circuit
gamma = np.pi / 4.
beta  = np.pi / 6

qc = qaoa_circuit(gamma, beta)

# Display the complete circuit
display(circuit_drawer(qc, output='mpl', style="iqp"))

# Execute the circuit
simulator = Aer.get_backend('aer_simulator')
result = execute(qc, simulator).result()

# Output Results
print("Measurement Results:", result.get_counts())
```

6.2.8   Advantages of QAOA

The quantum approximate optimization algorithms have the following advantages.

1. **Scalability with Problem Size**

   QAOA can scale to large problem sizes as the number of qubits n increases. The algorithm explores multiple solutions simultaneously due to quantum superposition. For this reason:

(a) QAOA is NISQ-friendly and can be executed on near-term quantum hardware.

(b) It offers a hybrid quantum-classical approach, leveraging quantum power for optimization.

2. **Efficient Exploration**

By leveraging quantum coherence and entanglement, QAOA efficiently navigates the solution space, avoiding local minima common in classical optimization.

3. **Example: Quantum Speedup**

For a problem of size N, classical algorithms may require $O(N)$ operations, while QAOA achieves approximate solutions in $O(p)$, where p is the number of layers.

6.3   Quantum Kernel Method in Support Vector Machines (SVMs)

The quantum kernel method (QKM) is a quantum-enhanced approach to machine learning that integrates quantum computing principles into classical support vector machines (SVMs). SVMs are powerful tools for supervised classification, where the objective is to find an optimal hyperplane to separate data points in high-dimensional feature spaces.

In classical SVMs, a kernel function computes the inner product between data points in a transformed feature space, often requiring significant computational resources. Quantum kernels exploit the ability of quantum systems to efficiently compute these inner products in exponentially large Hilbert spaces. By leveraging quantum feature maps and kernel matrices, quantum kernels enhance the ability to classify complex datasets.

The Qiskit implementations in this section demonstrate how to compute quantum kernels and train SVMs using quantum circuits. Despite challenges such as noise and scalability, quantum kernels hold significant promise for improving machine learning performance, particularly for high-dimensional and nonlinear data.

6.3.1   Support Vector Machines: A Brief Review

In SVMs, the classification problem is formulated as an optimization problem. Given a set of labeled training data $\{(x_i, y_i)\}_{i=1}^{N}$, where x_i represents the feature vectors and $y_i \in \{-1, 1\}$ represents the class labels, the objective is to find a hyperplane defined as:

$$f(x) = \text{sgn}(\mathbf{w} \cdot \phi(x) + b)$$

where

- $\mathbf{w}$ is the weight vector.
- b is the bias term.
- $\phi(x)$ is a mapping of x into a higher-dimensional feature space.

The SVM minimizes the following objective function:

$$\min_{\mathbf{w}, b} \frac{1}{2} \parallel \mathbf{w} \parallel^2 \quad \text{subject to} \quad y_i(\mathbf{w} \cdot \phi(x_i) + b) \geq 1 \quad \forall i.$$

The kernel trick avoids explicitly computing $\phi(x)$ by defining a kernel function $K(x_i, x_j) = \phi(x_i) \cdot \phi(x_j)$, which computes the inner product in the high-dimensional feature space.

6.3.2   Quantum Kernels: Enhancing SVMs

The quantum kernel method replaces the classical kernel $K(x_i, x_j)$ with a quantum kernel, where the quantum circuit encodes data into a quantum state. The quantum kernel is defined as:

$$K(x_i, x_j) = |\langle \psi(x_i) | \psi(x_j) \rangle|^2$$

where $|\psi(x)\rangle$ is the quantum state corresponding to the input x.

The key steps in a quantum-enhanced SVM are:

1. **Feature Encoding**

    Map classical data x into quantum states $|\psi(x)\rangle$ using a feature map $\mathcal{U}(x)$:

$$|\psi(x)\rangle = \mathcal{U}(x) |0\rangle^{\otimes n}$$

    where $\mathcal{U}(x)$ is a parameterized quantum circuit acting on n-qubits.

2. **Kernel Calculation**

    Use a quantum computer to compute the kernel $K(x_i, x_j)$, which is the overlap between two quantum states:

$$K(x_i, x_j) = |\langle \psi(x_i) | \psi(x_j) \rangle|^2$$

3. **Classical SVM**

    Input the quantum kernel into a classical SVM for training and prediction.

6.3.3   *Mathematical Representation of Quantum Kernels*

The kernel function $K(x_i, x_j)$ can be expressed using quantum states as follows:

$$K(x_i, x_j) = |\langle \psi(x_i) | \psi(x_j) \rangle|^2 = |\langle 0 | \mathcal{U}^\dagger(x_i) \mathcal{U}(x_j) | 0 \rangle|^2$$

The quantum kernel leverages quantum feature maps $\mathcal{U}(x)$ to project the input data into an exponentially large Hilbert space.

6.3.4   *Qiskit Implementation: Quantum Kernel in SVM*

Here is an implementation of the quantum kernel method for SVM using Qiskit:

```python
#-------------------------------------------------------------------
# Quantum Kernel in SVM
# Chapter 6 in the QUANTUM COMPUTING AND QUANTUM MACHINE LEARNING BOOK
# Implementation of the Quantum Kernel Method for SVM
#-------------------------------------------------------------------
# Version 1.0
# (c) 2025 Jesse Van Griensven, Roydon Fraser, and Jose Rosas
# License: MIT - Citation of this work is required
#-------------------------------------------------------------------
import numpy as np

# Import the Scikit Learn Machine Learning libraries
from sklearn.svm import SVC
from sklearn.datasets import make_classification
from sklearn.model_selection import train_test_split
from sklearn.metrics import accuracy_score

# Import the Qiskit Libraries
from qiskit import Aer
from qiskit.circuit.library import ZZFeatureMap
from qiskit_machine_learning.kernels import QuantumKernel
#-------------------------------------------------------------------

# Generate synthetic data
X, y = make_classification(n_samples=100, n_features=2, n_classes=2,
n_redundant=0)
y = 2 * y - 1 # Convert to {-1, 1}
X_train, X_test, y_train, y_test = train_test_split(X, y,
test_size=0.2)
```

```python
# Define a quantum feature map
feature_map = ZZFeatureMap(feature_dimension=2, reps=2)

# Quantum kernel
quantum_kernel = QuantumKernel(feature_map=feature_map,
quantum_instance=Aer.get_backend('statevector_simulator'))

# Compute kernel matrix
kernel_matrix_train = quantum_kernel.evaluate(X_train)
kernel_matrix_test  = quantum_kernel.evaluate(X_test, X_train)

# Train classical SVM with quantum kernel
svm = SVC(kernel='precomputed')
svm.fit(kernel_matrix_train, y_train)

# Predict and evaluate
y_pred = svm.predict(kernel_matrix_test)
print("Accuracy:", accuracy_score(y_test, y_pred))

# Visualization of Quantum Kernels
plt.imshow(kernel_matrix_train, cmap='hot',
interpolation='nearest')
plt.title("Quantum Kernel Matrix (Training)")
plt.colorbar()
plt.show()
```

6.3.5 Visualization of Quantum Kernels

Quantum kernels can be visualized to understand the separability of data in the feature space. Figure 6.2 visualizes the kernel matrix to observe the pairwise similarities between data points. The kernel matrix illustrates how well the quantum feature map separates the data points. High values in the matrix indicate strong overlap between states, while low values suggest orthogonality.

6.3.6 Advantages of Quantum Kernels

The following list presents some advantages of quantum kernels:

1. **Exponential Feature Space**
   Quantum feature maps project classical data into exponentially large Hilbert spaces, enabling better separability for complex data.

Fig. 6.2 Quantum kernel matrix for the Sect. 6.3.4 Python Code Example. (By the authors)

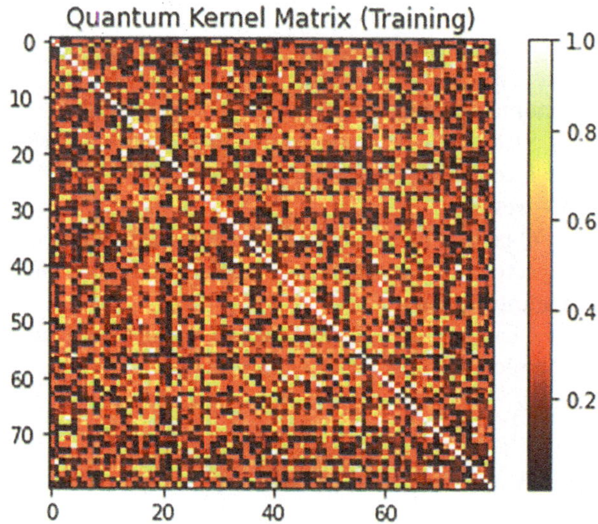

2. **Efficiency**

   Quantum kernels leverage quantum parallelism to compute overlaps, potentially offering a speedup over classical methods for large datasets.
3. **No Explicit Feature Mapping**

   Similar to the classical kernel trick, quantum kernels compute inner products without explicitly constructing the quantum feature space.

6.3.7   Challenges of Quantum Kernels

The following list presents some of quantum kernels' challenges:

1. **NISQ Hardware Limitations**

   Quantum hardware introduces noise, which can affect the accuracy of kernel evaluations.
2. **Scalability**

   Computing kernel matrices requires repeated evaluations of quantum circuits, which may become inefficient for very large datasets.
3. **Choice of Feature Map**

   The performance of quantum kernels depends heavily on the choice of quantum feature maps $\mathcal{U}(x)$.

6.3.8   Applications of Quantum Kernel Methods

These are some examples of quantum kernel method applications:

1. **Classification Tasks**

   Quantum kernels can enhance the performance of SVMs for binary and multi-class classification problems.

2. **Anomaly Detection**

   SVMs with quantum kernels can identify anomalies in high-dimensional data by leveraging the quantum feature space.

3. **Regression**

   Quantum kernels can also be extended to support regression tasks, where the objective is to fit a model to predict continuous outputs.

4. **Quantum Chemistry**

   Quantum kernels are used to predict molecular properties by embedding chemical features into quantum states.

6.4 Variational Quantum Classifier (VQC)

Variational quantum classifiers (VQCs) are hybrid quantum-classical algorithms that use parameterized quantum circuits to optimize decision boundaries for classification tasks. These classifiers leverage the unique properties of quantum mechanics, such as superposition and entanglement, to enhance the representational power of the model.

In a VQC, the loss function $L(\theta)$ is minimized using classical gradient-based optimization techniques. The general form of the loss function is expressed as:

$$L(\theta) = \sum_{i=1}^{n} (y_i - \mathrm{Pred}_\theta(x_i))^2$$

where θ represents the parameters of the quantum circuit, y_i are the true labels, and $\mathrm{Pred}_\theta(x_i)$ is the prediction from the quantum model for input x_i.

6.4.1 Components of a Variational Quantum Classifier (VQC)

The following list is a presentation of the components of a variational quantum classifier (VQC):

1. **Parameterized Quantum Circuit (PQC)**

   The PQC serves as the core of a VQC. It is a quantum circuit with adjustable parameters, typically implemented as rotation angles of quantum gates, such as $R_X(\theta)$, $R_Y(\theta)$, or $R_Z(\theta)$. These parameters are optimized during training to minimize the loss function.

2. **Feature Encoding**

   Classical input data x_i must be encoded into a quantum state. A common approach is to use amplitude encoding or angle encoding, where features are mapped to the amplitudes or angles of quantum gates.

3. **Measurement**

   After applying the PQC, measurements are performed on the quantum system to extract information. The measurement results are used as the model's predictions.

4. **Classical Optimization**

   The parameters θ of the PQC are optimized using classical methods, such as gradient descent or Adam. Gradients can be computed using the parameter-shift rule.

Python Code Example in Qiskit: Variational Quantum Classifier (VQC)

Below is a Python implementation of a simple variational quantum classifier using Qiskit. This example demonstrates the training of a VQC on a synthetic dataset.

```python
#-----------------------------------------------------------------
# Variational Quantum Classifier (VQC)
# Chapter 6 in the QUANTUM COMPUTING AND QUANTUM MACHINE LEARNING BOOK
# Code demonstrates applying the phase separation step for two qubits
#-----------------------------------------------------------------
# Version 1.0
# Qiskit changes frequently.
# We recommend using the latest version from the book code repository at:
# https://aqtinitiative.org/quantum-computing-for-engineers

# (c) 2025 Jesse Van Griensven, Roydon Fraser, and Jose Rosas
# License: MIT - Citation of this work is required
#-----------------------------------------------------------------
import numpy as np
from qiskit import QuantumCircuit
from qiskit.circuit import Parameter
from qiskit.opflow import Z, StateFn, CircuitStateFn
from qiskit.algorithms.optimizers import COBYLA
#-----------------------------------------------------------------

def create_pqc(theta_val):
    """ Create a simple parameterized quantum circuit (PQC) """
    qc = QuantumCircuit(1)
    qc.rx(theta_val, 0)
    return qc
```

```python
#----------------------------------------------------------------

def loss_function(theta_values, data, labels):
    """ Define the loss function """
    # Extract the single parameter value from the array
    theta_val = theta_values[0]
    qc = create_pqc(theta_val)
    results = []

    # Loop over each data point and perform feature encoding via an extra
rotation.
    for x in data:
        qc_data = qc.copy()
        qc_data.rx(x, 0)  # Feature encoding
        # Build a StateFn for measurement using the Z operator.
        result = StateFn(Z, is_measurement=True) @ CircuitStateFn
(qc_data)
        results.append(result.eval())

    # Convert the results to real predictions and compute the mean squared
error.
    predictions = np.array([np.real(r) for r in results])
    return np.mean((predictions - labels) ** 2)
#----------------------------------------------------------------

# Generate synthetic data
data  = np.linspace(0, 2. * np.pi, 10)
labels = np.sin(data)

# Optimize the circuit parameters using COBYLA
optimizer = COBYLA(maxiter=100)
initial_theta = np.array([0.1])
optimal_result = optimizer.minimize(
    fun=lambda theta: loss_function(theta, data, labels),
    x0=initial_theta
)

# The optimizer returns an object containing the solution in the 'x'
attribute.
print(f"Optimal θ: {optimal_result.x}")
```

6.4.2   Variational Quantum Circuit Design

Designing an effective PQC involves choosing a suitable structure and gate layout. A layered structure is commonly used, where each layer consists of entangling gates followed by parameterized single-qubit rotations.

For example, a layered PQC with two qubits can be designed as follows:

$$U(\theta) = \text{Layer}_1(\theta_1) \cdot \text{EntanglingGates} \cdot \text{Layer}_2(\theta_2)$$

Python Code Example in Qiskit: Variational Quantum Circuit

```
#------------------------------------------------------------------
# Variational Quantum Circuit
# Chapter 6 in the QUANTUM COMPUTING AND QUANTUM MACHINE LEARNING BOOK
# Code demonstrates applying the phase separation step for two qubits
#------------------------------------------------------------------
# Version 1.0
# Qiskit changes frequently.
# We recommend using the latest version from the book code repository at:
# https://aqtinitiative.org/quantum-computing-for-engineers

# (c) 2025 Jesse Van Griensven, Roydon Fraser, and Jose Rosas
# License: MIT - Citation of this work is required
#------------------------------------------------------------------
import warnings
warnings.filterwarnings('ignore')

import numpy as np
from qiskit import QuantumCircuit
from qiskit.circuit import Parameter
from qiskit.visualization import circuit_drawer
#------------------------------------------------------------------

def layered_pqc(num_qubits, num_layers):
    qc = QuantumCircuit(num_qubits)
    for layer in range(num_layers):
        for qubit in range(num_qubits):
            # Create a unique parameter for each layer and qubit
            theta = Parameter(f'θ_{layer}_{qubit}')
            qc.rx(theta, qubit)
        for i in range(num_qubits - 1):
            qc.cx(i, i + 1)  # Entangling gate
    return qc
#------------------------------------------------------------------
```

```
# Example of a layered PQC with 2 qubits and 3 layers
qc_example = layered_pqc(2, 3)
print(qc_example)

# Display the complete circuit
display(circuit_drawer(qc_example, output='mpl', style="iqp"))
```

6.4.3 Parameter Optimization in VQCs

Parameter optimization in VQCs can be performed using the parameter-shift rule, which is a quantum gradient computation method. The gradient of the loss function with respect to a parameter θ can be obtained by using central differencing given by:

$$\frac{\partial L(\theta)}{\partial \theta} = \frac{L(\theta + \pi/2) - L(\theta - \pi/2)}{2}$$

This method allows efficient gradient estimation on quantum hardware.

Python Code Example: Binary Classification
A practical example is the classification of linearly inseparable data, such as XOR data. Using a VQC, the decision boundary can be optimized to correctly classify such data.

```
#--------------------------------------------------------------
# Variational Quantum Classifier (VQC) Example
# Example: classification of nonlinearly separable data (XOR-like data)
#
# This script:
#  1. The "moons dataset" is a "toy dataset", standard in use in ML tests.
#  2. Defines a VQC with a simple feature map and variational layer.
#  3. Trains the classifier using a mean-squared error loss.
#  4. Evaluates and prints the test accuracy.
#  5. Plots the decision boundary along with training/test points.
##-------------------------------------------------------------

# Version 1.0
# Qiskit changes frequently.
# We recommend using the latest version from the book code repository at:
# https://aqtinitiative.org/quantum-computing-for-engineers

# (c) 2025 Jesse Van Griensven, Roydon Fraser, and Jose Rosas
# License: MIT - Citation of this work is required.
#--------------------------------------------------------------
```

```python
import numpy as np
import matplotlib.pyplot as plt

# Scikit-learn for data generation and preprocessing
from sklearn.datasets import make_moons
from sklearn.preprocessing import StandardScaler
from sklearn.model_selection import train_test_split

# Qiskit imports for building and simulating the quantum circuit
from qiskit import QuantumCircuit, Aer
from qiskit.circuit import Parameter
from qiskit.quantum_info import Statevector, Operator
from qiskit.algorithms.optimizers import COBYLA
#-------------------------------------------------------------------

def create_vqc():
    """Creates a parameterized quantum circuit for binary
classification"""
    qc = QuantumCircuit(2)
    # Feature encoding: use RY rotations to encode each input feature.
    qc.ry(x0, 0)
    qc.ry(x1, 1)
    # Variational (trainable) layer: apply RX rotations with trainable
angles.
    qc.rx(theta0, 0)
    qc.rx(theta1, 1)
    # Entanglement: apply a CNOT gate.
    qc.cx(0, 1)
    return qc


#-------------------------------------------------------------------
def quantum_predict(x, theta, circuit):
    """
    Given a 2D input x and trainable parameters theta (an array with
2 elements),
    bind the parameters into the circuit, simulate the statevector,
    and return a prediction probability computed from the expectation
value of Z on qubit 0.
    """
    # Bind the input features and variational parameters.
    bound_circuit = circuit.bind_parameters({x0: x[0], x1: x[1],
                        theta0: theta[0], theta1: theta[1]})
    # Simulate the circuit to get the statevector.
    state = Statevector.from_instruction(bound_circuit)
    # Compute the expectation value of the observable Z on qubit 0 (and
```

```python
identity on qubit 1).
  op = Operator.from_label('ZI')
  expectation = np.real(state.expectation_value(op))
  # Map the expectation (which lies in [-1, 1]) to a probability in [0, 1].
  prob = (expectation + 1) / 2.
  return prob
#-----------------------------------------------------------------------

def loss_function(theta, X_data, y_data):
    """
    Computes the mean squared error between the predicted probabilities
(via the quantum circuit)
    and the true binary labels.
    """
    predictions = np.array([quantum_predict(x, theta, vqc) for x in
X_data])
    # Mean squared error loss.
    MSEL = np.mean((predictions - y_data) ** 2)
    return MSEL
#-----------------------------------------------------------------------

# --------------------------------
# 1. Generate and Preprocess Data
# --------------------------------
# Here we generate 100 samples from a moons dataset (nonlinear, XOR-like
structure).
X, y = make_moons(n_samples=100, noise=0.1, random_state=42)
# Standardize features for better performance.
X = StandardScaler().fit_transform(X)
# Split data into training and testing sets.
X_train, X_test, y_train, y_test = train_test_split(X, y,
test_size=0.2, random_state=42)

# --------------------------------
# 2. Define the Variational Quantum Circuit
# --------------------------------
# We use 2 qubits. The circuit encodes the 2 input features via rotations
(feature map)
# and then applies a variational layer with trainable parameters.
# Finally, the circuit will be simulated to obtain the expectation value
of Z on qubit 0.
#
# Define the parameters for feature encoding and variational (trainable)
gates.
x0 = Parameter('x0')
```

```python
x1 = Parameter('x1')
theta0 = Parameter('theta0')
theta1 = Parameter('theta1')

# Create the parameterized circuit.
vqc = create_vqc()

# Use the statevector simulator for obtaining expectation values.
backend = Aer.get_backend('statevector_simulator')


# -------------------------------
# 3. Train the VQC
# -------------------------------
# We use the COBYLA optimizer (a gradient-free optimizer) to minimize the
loss.
optimizer = COBYLA(maxiter=100)

# Initialize the variational parameters.
initial_theta = np.array([0.1, 0.1])

# Optimize: note that the optimizer minimizes the loss_function over the
training data.
optimal_result = optimizer.minimize(fun=lambda theta: loss_function
(theta, X_train, y_train),
                    x0=initial_theta)
optimal_theta = optimal_result.x

print("Optimal variational parameters found:", optimal_theta)


# -------------------------------
# 4. Evaluate the VQC on Test Data
# -------------------------------
# For each test sample, compute the probability prediction and assign a
label.
test_probs = np.array([quantum_predict(x, optimal_theta, vqc) for x in
X_test])

# Apply a threshold of 0.5 to obtain binary predictions.
test_preds = (test_probs > 0.5).astype(int)

# Compute the test accuracy.
accuracy = np.mean(test_preds == y_test)
print("Test accuracy:", accuracy)
```

```python
# --------------------------------
# 5. Plot the Decision Boundary
# --------------------------------
# Create a grid of points covering the input feature space.
x_min, x_max = X[:, 0].min() - 1, X[:, 0].max() + 1
y_min, y_max = X[:, 1].min() - 1, X[:, 1].max() + 1
xx, yy = np.meshgrid(np.linspace(x_min, x_max, 100),
            np.linspace(y_min, y_max, 100))
grid_points = np.c_[xx.ravel(), yy.ravel()]

# Compute the predicted probability for each point on the grid.
grid_preds = np.array([quantum_predict(point, optimal_theta, vqc)
for point in grid_points])
grid_preds = grid_preds.reshape(xx.shape)

# Plot the decision boundary as a contour map.
plt.figure(figsize=(8, 6))

# Contour: areas with prediction below 0.5 will be one color, above 0.5
another.
plt.contourf(xx, yy, grid_preds, levels=[0, 0.5, 1], alpha=0.3,
cmap='coolwarm')

# Scatter plot the training points.
plt.scatter(X_train[:, 0], X_train[:, 1], c=y_train, edgecolors='k',
marker='o', label='Train')

# Scatter plot the test points.
plt.scatter(X_test[:, 0], X_test[:, 1], c=y_test, edgecolors='k',
marker='x', s=100, label='Test')
plt.title(" Decision Boundary of the VQC on make_moons Data")
plt.xlabel("Feature 1")
plt.ylabel("Feature 2")
plt.legend()
plt.show()
```

6.4.4 Evaluation of VQC Algorithms

VQE has many advantages. These are the main ones:

1. Enhanced representational power due to quantum mechanics.
2. Ability to leverage high-dimensional Hilbert spaces for complex decision
   boundaries.

VQE also has challenges, especially with NISQ quantum computers. Here are the main ones:

1. Noise and errors in quantum hardware.
2. Scalability with increasing qubits and circuit depth.

6.5 Comparison of Classical and Quantum Optimization

Quantum optimization algorithms, like QAOA, leverage quantum superposition to explore multiple solutions simultaneously, offering a potential speedup over classical algorithms. However, quantum optimization is limited by the number of qubits available, while classical methods can scale with increased computational resources. One more limitation of quantum optimizations is that classical methods provide exact solutions for small problems, whereas quantum methods are inherently probabilistic and provide approximate solutions.

6.5.1 *Quantum-Inspired Techniques for Classical Systems*

Quantum-inspired optimization algorithms simulate the principles of quantum mechanics on classical systems. Techniques such as tensor networks and simulated annealing have shown promise in solving complex problems without requiring quantum hardware.

6.5.2 *Simulated Annealing*

Simulated annealing mimics quantum tunneling by probabilistically escaping local minima during optimization. The energy state E transitions based on a temperature parameter T:

$$P(\Delta E) = e^{-\Delta E/k_B T}$$

6.5.3 *Tensor Networks*

Tensor networks approximate high-dimensional data structures, reducing the computational cost of solving optimization problems.

Python Code Example: Simulated Annealing

This code demonstrates simulated annealing for a simple cost function.

```python
#-------------------------------------------------------------------
# Simulated Annealing
# Chapter 6 in the QUANTUM COMPUTING AND QUANTUM MACHINE LEARNING BOOK
#-------------------------------------------------------------------
# Version 1.0
# Qiskit changes frequently.
# We recommend using the latest version from the book code repository at:
# https://aqtinitiative.org/quantum-computing-for-engineers

# (c) 2025 Jesse Van Griensven, Roydon Fraser, and Jose Rosas
# License: MIT - Citation of this work is required
#-------------------------------------------------------------------
import numpy as np

#-------------------------------------------------------------------
def simulated_annealing(cost_function, bounds, max_iter, temp):
  """ Simulated annealing example"""
  current_state = np.random.uniform(bounds[0], bounds[1])
  current_cost = cost_function(current_state)
  for _ in range(max_iter):
    next_state = current_state + np.random.uniform(-1, 1) * temp
    next_cost = cost_function(next_state)
    if next_cost < current_cost or np.random.rand() < np.exp(-
(next_cost - current_cost) / temp):
        current_state, current_cost = next_state, next_cost
    temp *= 0.99  # Decrease temperature

  return current_state, current_cost
#-------------------------------------------------------------------

# Example cost function
cost_function = lambda x: x**2 + 4. * np.sin(2. * x)
result = simulated_annealing(cost_function, bounds=(-10, 10),
max_iter=1000, temp=10)

# Print the output variables with 4 decimal places using an f-string.
print(f"Optimal Solution: x = {result[0]:.4f}, cost = {result
[1]:.4f}")
```

6.5.4   The Quantum Approximate Optimization Algorithm (QAOA)

The quantum approximate optimization algorithm (QAOA) is a powerful hybrid quantum-classical approach designed to solve combinatorial optimization problems. By leveraging quantum superposition and entanglement, QAOA provides an efficient framework to approximate solutions to problems that are computationally intensive for classical methods.

The quantum approximate optimization algorithm (QAOA) provides a versatile framework for solving combinatorial optimization problems. Its hybrid quantum-classical approach combines the strengths of quantum superposition with classical parameter optimization. QAOA is poised to revolutionize optimization in engineering, data science, and beyond as quantum hardware continues to improve.

6.6   Quantum Phase Estimation

Quantum phase estimation is a foundational quantum algorithm that enables efficient phase determination. Its ability to solve classically intractable problems makes it essential in quantum computing applications like cryptography, chemistry, and machine learning.

6.6.1   Introduction to Quantum Phase Estimation

Quantum phase estimation (QPE) is a fundamental algorithm in quantum computing used to determine the eigenvalue (phase) associated with an eigenvector of a unitary operator U. The algorithm forms the backbone for applications like Shor's algorithm, quantum simulation, and quantum algorithms for solving linear systems.

Given a unitary operator U and an eigenvector $|u\rangle$, the corresponding eigenvalue can be written as:

$$U\,|u\rangle = e^{2\pi i\phi}\,|u\rangle$$

where ϕ is the phase to be estimated. The QPE algorithm estimates ϕ to high precision using quantum Fourier transform (QFT) and controlled operations.

6.6.2   Algorithm Overview

The QPE algorithm follows these steps:

1. **Initialize the qubits**

   (a) n-qubits for phase estimation, initialized to $|0\rangle^{\otimes n}$.
   (b) One auxiliary qubit prepared in the eigenvector $|u\rangle$.

2. **Apply Hadamard Gates**
   Place the n-qubit register into superposition.

3. **Controlled-U gates**
   Apply U^{2^j} to the auxiliary qubit, controlled by the j-th qubit in the n-qubit register.

4. **Quantum Fourier Transform (QFT)**
   Apply the inverse QFT to the n-qubit register to extract the phase.

5. **Measure**
   Measure the n-qubit register to obtain an estimate of $\boldsymbol{\phi}$.

6.6.3 QPE Mathematical Foundation

The unitary operator U acts on an eigenvector $|u\rangle$ as:

$$U\,|u\rangle = e^{2\pi i\phi}\,|u\rangle$$

We expand the phase ϕ in binary:

$$\phi = 0.\phi_1\phi_2...\phi_n \quad \text{where} \quad \phi_k \in \{0, 1\}$$

yielding a fraction for ϕ, for example, for $n{=}4$, ϕ could be 0.1001.
The controlled-U gates apply U^{2^j}, leading to:

$$U^{2^j}\,|u\rangle = e^{2\pi i 2^j \phi}\,|u\rangle$$

Thus, the state of the system after the controlled operations is as follows:

$$\frac{1}{2^{n/2}} \sum_{k=0}^{2^n-1} e^{2\pi i k\phi}\,|k\rangle \otimes |u\rangle$$

The QFT maps this state to a computational basis state proportional to ϕ.

6.6.4 Quantum Phase Estimation Circuit

The QPE circuit comprises three main parts:

1. **Initialization**

   Prepare the n-qubit register and auxiliary qubit:

   $$|0\rangle^{\otimes n} \otimes |u\rangle$$

2. **Apply Hadamard Gates**

   Place the n-qubit register into a uniform superposition:

   $$\frac{1}{2^{n/2}} \sum_{k=0}^{2^n-1} |k\rangle \otimes |u\rangle$$

3. **Controlled Unitary Operations**

   For each qubit j in the n-qubit register, apply U^{2^j} controlled on that qubit:

   $$\text{State}: \frac{1}{2^{n/2}} \sum_{k=0}^{2^n-1} |k\rangle \otimes U^{2^k} |u\rangle$$

   Since $U|u\rangle = e^{2\pi i \phi}|u\rangle$, we get:

   $$U^{2^k}|u\rangle = e^{2\pi i 2^k \phi}|u\rangle$$

   Thus, the state becomes:

   $$\frac{1}{2^{n/2}} \sum_{k=0}^{2^n-1} e^{2\pi i 2^k \phi} |k\rangle \otimes |u\rangle$$

4. **Inverse Quantum Fourier Transform**

   Apply the inverse QFT on the n-qubit register to transform the phase information into a measurable form.

5. **Measurement**

   Measure the n-qubit register to obtain an approximation of ϕ in binary.

6.6.5 Analytical Example: Single-Qubit Case

1. Let $U = R_z(\theta)$, the Z-axis rotation gate:
2. $R_z(\theta) = \begin{bmatrix} 1 & 0 \\ 0 & e^{i\theta} \end{bmatrix}$
3. Suppose $|u\rangle = |1\rangle$. Then:
4. $U|1\rangle = e^{i\theta}|1\rangle \quad \Rightarrow \quad \phi = \frac{\theta}{2\pi}.$

5. We estimate ϕ using 3 qubits ($n = 3$). The algorithm computes θ from the binary approximation of ϕ.

Python Code Example: QPE on a General Unitary

The following code is an extended implementation of QPE to clarify implementation steps.

```python
#------------------------------------------------------------------------
# QPE on a General Unitary
# Chapter 6 in the QUANTUM COMPUTING AND QUANTUM MACHINE LEARNING BOOK
#------------------------------------------------------------------------
# Version 1.0
# Qiskit changes frequently.
# We recommend using the latest version from the book code repository at:
# https://aqtinitiative.org/quantum-computing-for-engineers

# (c) 2025 Jesse Van Griensven, Roydon Fraser, and Jose Rosas
# License:  MIT - Citation of this work is required
#------------------------------------------------------------------------
import numpy as np
from qiskit import QuantumCircuit, Aer, transpile
from qiskit.visualization import plot_histogram
from qiskit.circuit.library import QFT, ZGate
#------------------------------------------------------------------------


def qpe_general(U, n_qubits):
  # Create a circuit with n_qubits counting qubits and 1 target qubit.
  qc = QuantumCircuit(n_qubits + 1, n_qubits)

  # Step 1: Apply Hadamard gates to the counting qubits.
  for i in range(n_qubits):
    qc.h(i)

  # Step 2: Apply controlled-U gates.
  # Note: In a typical QPE, U should be raised to powers of 2^i.
  for i in range(n_qubits):
    qc.append(U.control(), [i, n_qubits])

  # Step 3: Apply the inverse Quantum Fourier Transform (QFT).
  # Use QFT from qiskit.circuit.library, with inverse=True.
  inv_qft = QFT(n_qubits, inverse=True, do_swaps=True)
  qc.append(inv_qft.to_gate(), range(n_qubits))
```

```
# Step 4: Measure the counting qubits.
qc.measure(range(n_qubits), range(n_qubits))


    return qc
#------------------------------------------------------------------


# Define a unitary gate. Here we use ZGate as an example (its eigenvalue is
e^(i*pi)).
U = ZGate()
qc = qpe_general(U, 3)


# Draw the circuit using matplotlib
qc.draw('mpl')
```

6.6.6   Applications of Quantum Phase Estimation

Here are some of the main applications of QPE.

1. **Shor's Algorithm**
   QPE is a key subroutine in factoring integers and finding discrete logarithms.
2. **Quantum Chemistry**
   Estimating the ground-state energy of molecules using the phase associated with a unitary evolution.
3. **Solving Linear Systems: The HHL Algorithm**
   QPE enables solving $Ax = b$ using quantum methods.
4. **Quantum Simulation**
   Simulating the time evolution of quantum systems by estimating eigenvalues of Hamiltonians.

6.6.7   QPE Accuracy and Complexity

The precision of the phase estimation is determined by the number of qubits n. The error in the phase is approximately:

$$\epsilon \sim \frac{1}{2^n}$$

Thus, increasing n improves the accuracy exponentially.
The complexity of QPE is dominated by:

1. Controlled unitary operations, which scale with $O(2^n)$.
2. QFT, which scales with $O(n^2)$.

6.6.8 QPE Variants

There are a few variations on the QPE algorithm. Table 6.1 summarizes three of these alternate implementations.

6.7 Solving Linear Systems with the HHL Algorithm

The Harrow–Hassidim–Lloyd (HHL) algorithm demonstrates the power of quantum computing for solving linear systems exponentially faster than classical methods under specific conditions. It has profound implications for optimization, machine learning, and scientific simulations, representing a foundation of many quantum algorithms.

6.7.1 HHL Introduction

The Harrow–Hassidim–Lloyd (HHL) algorithm is a quantum algorithm for solving linear systems of equations of the form:

$$Ax = b$$

where

- A is an $N \times N$ Hermitian matrix.
- b is a known vector.
- x is the solution vector.

While classical algorithms for solving such systems scale as $O(N^3)$ (Gaussian elimination) or $O(N^2)$ for sparse systems, the HHL algorithm achieves an exponential speedup under certain conditions, with runtime $O(\log N)$. This exponential improvement is valuable for applications in optimization, quantum machine learning, and computational science.

Table 6.1 Alternate implementations of quantum phase estimations

QPE variant	Description
Iterative QPE	Uses fewer qubits by iteratively estimating ϕ bit-by-bit
Quantum chemistry applications	QPE is used to find eigenvalues of Hamiltonians to determine ground-state energies
Hybrid algorithms	Combine QPE with classical optimization to improve precision

6.7.2 HHL Problem Setup and Assumptions

The HHL algorithm requires the following setup:

1. Hermitian Matrix A

    The HHL algorithm assumes A is Hermitian. If A is not Hermitian, it can be transformed into a Hermitian matrix A' using the following:

$$A' = \begin{bmatrix} 0 & A \\ A^\dagger & 0 \end{bmatrix}$$

2. State Preparation

    The input vector b is encoded as a quantum state $|b\rangle$.
3. Eigenvalue Decomposition

    The matrix A has an eigenvalue decomposition:

$$A = \sum_j \lambda_j \, |\lambda_j\rangle\langle\lambda_j|$$

    where λ_j are eigenvalues, and $|\lambda_j\rangle$ are corresponding eigenvectors.

The goal of the algorithm is to prepare a quantum state $|x\rangle$ proportional to the solution x, such that:

$$|x\rangle = \frac{A^{-1}\,|b\,\vec{b}\,\rangle}{\|\,A^{-1}\,|b\,\vec{b}\,\rangle\,\|}$$

6.7.3 HHL Algorithm Steps

The HHL algorithm consists of **three main steps**:

1. **Quantum Phase Estimation (QPE)**

    Estimate the eigenvalues λ_j of A using QPE.
2. **Controlled Rotation**

    Apply rotations conditioned on the eigenvalues λ_j to encode the reciprocal $1/\lambda_j$.
3. **Inverse Quantum Phase Estimation**

    Undo the phase estimation and extract the solution $|x\rangle$.

6.7.4   Detailed HHL Quantum Circuit

The HHL circuit includes three registers, as explained in Table 6.2.

The quantum circuit can be broken into the following steps:

Step 1: Quantum Phase Estimation (QPE)

The QPE process decomposes A into its eigenvalues and eigenvectors. Starting with the system in $|b\rangle$, we apply QPE:

$$|b\,\vec{b}\rangle \rightarrow \sum_j \beta_j \,|\lambda_j\rangle \otimes |\tilde{\lambda}_j\rangle$$

where $|\tilde{\lambda}_j\rangle$ is the quantum register encoding the eigenvalues λ_j, and β_j are the coefficients of $|b\rangle$ in the eigenbasis.

Step 2: Controlled Rotation

To solve $xA = b$, we require the reciprocal $1/\lambda_j$. This is implemented using a controlled rotation gate acting on an ancilla qubit:

$$|\tilde{\lambda}_j\rangle \otimes |0\rangle \rightarrow |\tilde{\lambda}_j\rangle \otimes \left(\frac{\lambda_{\max}}{\lambda_j} \, |1\rangle + \sqrt{1 - \left(\frac{\lambda_{\max}}{\lambda_j}\right)^2} \, |0\rangle \right)$$

where $\lambda_{\max}$ is the largest eigenvalue of A.

Step 3: Inverse Quantum Phase Estimation

We undo the QPE to disentangle the eigenvalue register, leaving the system register in a state proportional to the solution:

$$|x\rangle = \sum_j \frac{\beta_j}{\lambda_j} \, |\lambda_j\rangle$$

6.7.5   HHL Mathematical Derivation

The system evolves as follows:

Table 6.2   Registers in the HHL circuit

HHL circuit register	Objective
Eigenvalue register	To store eigenvalues via QPE
System register	To represent the input vector $|b\rangle$
Ancilla qubit	To control rotations

1. After QPE:

$$|b\,\vec{b}\rangle \rightarrow \sum_j \beta_j |\lambda_j\rangle \otimes |\tilde{\lambda}_j\rangle$$

2. Controlled rotation:

$$\sum_j \beta_j |\lambda_j\rangle \otimes |\tilde{\lambda}_j\rangle \otimes |0\rangle \rightarrow \sum_j \beta_j |\lambda_j\rangle \otimes |\tilde{\lambda}_j\rangle \otimes \left(\frac{C}{\lambda_j} |1\rangle + \sqrt{1 - \left(\frac{C}{\lambda_j}\right)^2} \, |0\rangle \right)$$

3. Measurement of the ancilla qubit in state $|1\rangle$ collapses the system register to:

$$|x\rangle = \sum_j \frac{\beta_j}{\lambda_j} |\lambda_j\rangle$$

Normalization ensures:

$$\| x \| = \frac{\| b\,\vec{b} \|}{C}$$

6.7.6 HHL Complexity Analysis

The HHL algorithm achieves exponential speedup under the following conditions:

1. The matrix A is sparse.
2. The condition number $\kappa = \frac{\lambda_{\max}}{\lambda_{\min}}$ is bounded.
3. Efficient state preparation of $|b\rangle$.

The runtime scales as follows:

$$O\left(\kappa^2 \log N\right)$$

compared to $O(N^3)$ for classical methods.

Python Code Example: HHL Implementation

The following code is an implementation of the HHL algorithm using Qiskit.

```
#-------------------------------------------------------------------
# HHL Implementation
```

```python
# Chapter 6 in the QUANTUM COMPUTING AND QUANTUM MACHINE LEARNING BOOK
#-------------------------------------------------------------------
# Version 1.0
# Qiskit changes frequently.
# We recommend using the latest version from the book code repository at:
# https://aqtinitiative.org/quantum-computing-for-engineers

# (c) 2025 Jesse Van Griensven, Roydon Fraser, and Jose Rosas
# License: MIT - Citation of this work is required
#-------------------------------------------------------------------
from qiskit import Aer, transpile
from qiskit import QuantumCircuit
from qiskit.algorithms import HHL
from qiskit.providers.aer import AerSimulator
from qiskit.circuit.library import HGate, RYGate
from qiskit.visualization import plot_histogram
import numpy as np


#-------------------------------------------------------------------

def sprint(Matrix, decimals=4):
  """ Prints a Matrix with real and imaginary parts rounded to 'decimals'
"""
    import sympy as sp
    SMatrix = sp.Matrix(Matrix)  # Convert to Sympy Matrix if it's not
already

    def round_complex(x):
      """Round real and imaginary parts of x to the given number of
decimals."""
        c = complex(x)  # handle any real or complex Sympy expression
        r = round(c.real, decimals)
        i = round(c.imag, decimals)
        # If imaginary part is negligible, treat as purely real
        if abs(i) < 10**(-decimals): return sp.Float(r)
        else: return sp.Float(r) + sp.Float(i)*sp.I

    # Display the rounded Sympy Matrix
    display(SMatrix.applyfunc(round_complex))
    return


#-------------------------------------------------------------------
# Solving AX = b, where Solution X is [0.9487, 0.3162]
# Define matrix A and vector b
```

```
A_matrix = [[1., -1./3.], [-1./3., 1.]]
b_vector = [1., 0.]

# Convert A_matrix and b_vector to numpy arrays
A_matrix = np.array(A_matrix)
b_vector = np.array(b_vector)

# Initialize HHL algorithm
hhl = HHL()

# Solve the linear system
solution = hhl.solve(A_matrix, b_vector)

# Extract the solution vector from the quantum state
solution_vector = solution.state

# Print the result
print("Solution vector:", solution_vector)
```

6.7.7 HHL Applications

The HHL algorithm is valuable to many quantum computing methods, which is summarized in Table 6.3.

6.8 Applications of Quantum Optimization in Engineering

Quantum optimization has transformative potential across various engineering domains. By leveraging quantum mechanics principles, such as superposition and entanglement, quantum algorithms like QAOA provide innovative solutions to complex optimization problems. This section delves into three primary applications: energy grid optimization, supply chain and logistics, and engineering design. As quantum hardware matures, quantum optimization will play a pivotal role in engineering advancements.

Table 6.3 Sample of applications for HHL

HHL application	Solver
Machine learning	Solves systems in quantum support vector machines
Quantum chemistry	Solves systems of linear equations arising in chemical simulations
Optimization problems	Solves linear programming and convex optimization

6.8.1   Energy Grid Optimization

Energy grid optimization involves minimizing energy loss while ensuring stable and efficient power distribution. Classical methods face challenges in balancing power loads across complex grids, especially under dynamic conditions.

6.8.1.1   Quantum Optimization in Power Flow

Quantum optimization algorithms optimize power flow in electrical grids by minimizing energy losses. The cost Hamiltonian H_C for power flow optimization is:

$$H_C = \sum_{i,j} L_{ij} P_i P_j$$

where

- L_{ij}: Loss coefficients represent energy dissipation between nodes i and j.
- P_i, P_j: Power flow variables associated with nodes i and j.

Example: Load Balancing

For a simple grid with three nodes, the loss coefficients L_{ij} form a matrix:

$$L = \begin{bmatrix} 0 & 0.1 & 0.2 \\ 0.1 & 0 & 0.15 \\ 0.2 & 0.15 & 0 \end{bmatrix}$$

The optimization goal is to minimize:

$$H_C = 0.1 P_1 P_2 + 0.2 P_1 P_3 + 0.15 P_2 P_3.$$

Python Code Example: Energy Grid Optimization

This code demonstrates optimizing power flow in a simplified grid using classical optimization as a surrogate for QAOA.

```
#------------------------------------------------------------------
# Energy Grid Optimization
# Chapter 6 in the QUANTUM COMPUTING AND QUANTUM MACHINE LEARNING BOOK
#------------------------------------------------------------------
# Version 1.0
# Qiskit changes frequently.
# We recommend using the latest version from the book code repository at:
# https://aqtinitiative.org/quantum-computing-for-engineers
```

```
# (c) 2025 Jesse Van Griensven, Roydon Fraser, and Jose Rosas
# License: MIT - Citation of this work is required
#-------------------------------------------------------------------

from qiskit import QuantumCircuit
from scipy.optimize import minimize
import numpy as np


#-------------------------------------------------------------------
def cost_function(params):
    """ Define QAOA parameters """
    P = params # Power flow variables
    cost = 0
    for i in range(len(P)):
        for j in range(len(P)):
            cost += loss_coefficients[i, j] * P[i] * P[j]
    return cost
#-------------------------------------------------------------------

# Define cost Hamiltonian coefficients
loss_coefficients = np.array([[0, 0.1, 0.2], [0.1, 0, 0.15], [0.2, 0.15,
0]])

# Optimize power flow
initial_guess = [1, 1, 1]
result = minimize(cost_function, initial_guess, method="COBYLA")
print("Optimized Power Flow:", result.x)
```

6.8.2   *Supply Chain and Logistics*

Quantum optimization significantly improves routing and scheduling in logistics by solving combinatorial problems like the traveling salesman problem (TSP) and vehicle routing problem (VRP).

6.8.2.1   Traveling Salesman Problem (TSP)

In TSP, the goal is to find the shortest route to visit all nodes exactly once and return to the start. The cost Hamiltonian is:

$$H_C = \sum_{i=1}^{n} \sum_{j=1}^{n} d_{ij} x_{ij}$$

where

- d_{ij}: Distance between nodes i and j.
- x_{ij}: Binary variable indicating whether the route includes $i \to j$.

Quantum optimization can explore multiple TSP routes simultaneously, reducing computation time compared to classical brute-force methods.

Python Code Example: TSP with QAOA

This code outlines how to represent TSP as a QAOA problem, encoding distances into the quantum circuit.

```python
#------------------------------------------------------------------
# TSP with QAOA - Traveling Salesman Problem
# Chapter 6 in the QUANTUM COMPUTING AND QUANTUM MACHINE LEARNING BOOK
#------------------------------------------------------------------
# Version 1.0
# Qiskit changes frequently.
# We recommend using the latest version from the book code repository at:
# https://aqtinitiative.org/quantum-computing-for-engineers

# (c) 2025 Jesse Van Griensven, Roydon Fraser, and Jose Rosas
# License: MIT - Citation of this work is required
#------------------------------------------------------------------
from qiskit import Aer, execute
from qiskit import QuantumCircuit
from qiskit.circuit.library import QAOAAnsatz
from qiskit.visualization import circuit_drawer
#------------------------------------------------------------------


def tsp_cost_circuit(qc, distances, gamma):
    """ Cost Hamiltonian as phase separation """
    for i in range(len(distances)):
        for j in range(len(distances)):
            if distances[i][j] != 0:
                qc.cx(i, j)
                qc.rz(2. * gamma * distances[i][j], j)
                qc.cx(i, j)
#------------------------------------------------------------------


# Define distances
distances = [[0, 10, 15], [10, 0, 20], [15, 20, 0]]
```

```
# Create QAOA circuit for TSP
num_qubits = len(distances)
qc = QuantumCircuit(num_qubits)
qc.h(range(num_qubits))  # Initialize in superposition

# Cost Hamiltonian
tsp_cost_circuit(qc, distances, gamma=0.5)

# Output the results
print("QAOA Circuit for TSP:')
print(qc)
```

6.8.2.2 Vehicle Routing Problem (VRP)

In VRP, vehicles must serve multiple destinations while minimizing the total cost. Quantum optimization provides an efficient method for solving the VRP by simultaneously evaluating multiple routing configurations.

6.8.3 Engineering Design

Structural optimization in engineering involves exploring multiple configurations to identify the best design while meeting specific constraints. Classical methods often rely on finite element analysis (FEA), which becomes computationally intensive for large systems.

6.8.3.1 Quantum Structural Optimization

Quantum algorithms like QAOA allow simultaneous evaluation of multiple configurations, accelerating the search for optimal designs.

6.8.3.2 Example: Beam Design

Consider optimizing the design of a beam for maximum strength-to-weight ratio. The cost Hamiltonian H_C can encode material distribution:

$$H_C = \sum_{i=1}^{n} (w_i \cdot \sigma_i)$$

where w_i is the weight contribution of section i, and σ_i is the stress in section i.

Python Code Example: Beam Design Optimization

This example optimizes the material distribution in a beam using classical techniques analogous to QAOA.

```python
#------------------------------------------------------------------
# Beam Design Optimization (Classical vs Quantum)
# Chapter 6 in the QUANTUM COMPUTING AND QUANTUM MACHINE LEARNING BOOK
#------------------------------------------------------------------
# Version 1.0
# Qiskit changes frequently.
# We recommend using the latest version from the book code repository at:
# https://aqtinitiative.org/quantum-computing-for-engineers

# (c) 2025 Jesse Van Griensven, Roydon Fraser, and Jose Rosas
# License: MIT - Citation of this work is required
#------------------------------------------------------------------
"""
Beam Design Optimization (Classical vs Quantum)

This script:
  - Uses classical optimization (COBYLA) to minimize cost.
  - Uses Qiskit's QAOA (Quantum Approximate Optimization Algorithm) for
quantum optimization.
  - Converts the problem into a binary representation for QAOA
compatibility.
"""


import numpy as np
import qiskit
import scipy
from scipy.optimize import minimize


# Check for Qiskit Optimization installation
try:
    from qiskit_optimization import QuadraticProgram
    from qiskit_optimization.algorithms import MinimumEigenOptimizer
    from qiskit.algorithms import QAOA
    from qiskit import Aer
    from qiskit.utils import QuantumInstance
    qiskit_optimization_available = True
except ImportError:
    qiskit_optimization_available = False
```

```python
#------------------------------------------------------------------
# Print Library Versions
#------------------------------------------------------------------
print("\nLibrary Versions:")
print("Qiskit version:", qiskit.__version__)
print("NumPy version:", np.__version__)
print("SciPy version:", scipy.__version__)
print("Qiskit Optimization installed:",
qiskit_optimization_available)


#------------------------------------------------------------------
# Define Problem Parameters
#------------------------------------------------------------------
weights = np.array([2, 3, 1, 4])   # Material weights
stresses = np.array([0.8, 1.2, 0.6, 1.5])  # Corresponding stresses


#------------------------------------------------------------------
# Classical Optimization (COBYLA)
#------------------------------------------------------------------
def beam_cost(params):
    """ Cost function for classical optimization """
    return np.sum(params * weights * stresses)
#------------------------------------------------------------------

# Classical optimization using COBYLA
initial_guess = [1, 1, 1, 1]
result_classical = minimize(beam_cost, initial_guess,
method="COBYLA")
print("\nClassical Optimization Result:")
print("Optimized Material Distribution:", result_classical.x)
print("Minimum Cost:", result_classical.fun)


#------------------------------------------------------------------
# Quantum Optimization (QAOA via Qiskit), only if available
#------------------------------------------------------------------
if qiskit_optimization_available:
    # Define a Quadratic Program
    qp = QuadraticProgram()
    num_vars = len(weights)


    # **Binary Encoding: Convert Continuous Variables into Binary
Variables**
    # - Each material variable x_i (originally continuous) is now
represented by 3 binary bits (b_i1, b_i2, b_i3).
    # - The sum of these binary variables approximates the continuous
```

```python
value.
  bin_bits = 3 # Number of binary bits per variable
  scale_factor = 2**bin_bits - 1 # Scale continuous values to [0, 1]
range

  for i in range(num_vars):
    for j in range(bin_bits):
      qp.binary_var(name=f"x{i}_b{j}")

  # **Define the objective function in a binary-compatible way**
  linear_objective = {}
  for i in range(num_vars):
    for j in range(bin_bits):
      coeff = (2**j) / scale_factor # Scale binary representation
     linear_objective[f"x{i}_b{j}"] = coeff * weights[i] * stresses[i]

  qp.minimize(linear=linear_objective)  # Correct way to define the
objective

  # Run QAOA Optimization
  backend = Aer.get_backend("aer_simulator")
  quantum_instance = QuantumInstance(backend)

  qaoa = QAOA(quantum_instance=quantum_instance, reps=3)  # Use 3 QAOA
repetitions
  qaoa_optimizer = MinimumEigenOptimizer(qaoa)
  result_qaoa = qaoa_optimizer.solve(qp)


#----------------------------------------------------------------------
  # Compare Classical and Quantum Results

#----------------------------------------------------------------------
  print("\nQuantum Optimization Result (QAOA):")

  # Decode binary solution back to approximate continuous values
  qaoa_solution = []
  for i in range(num_vars):
   value = sum(result_qaoa.x[f"x{i}_b{j}"] * (2**j) / scale_factor for
j in range(bin_bits))
    qaoa_solution.append(value)

  print("Optimized Material Distribution (QAOA Approximation):",
qaoa_solution)
  print("Minimum Cost (QAOA):", result_qaoa.fval)
```

```
else:
  print("\nQiskit Optimization is not installed. Skipping QAOA
optimization.")
```

6.9 Comparing Classical and Quantum Optimization Methods

Classical optimization methods such as gradient descent, simulated annealing, and genetic algorithms remain effective for many practical problems. However, quantum optimization offers unique advantages like parallel exploration, quantum tunneling, and reduced complexity for large-scale problems. As quantum hardware continues to improve, quantum methods are expected to surpass classical techniques in scalability and efficiency, revolutionizing optimization in engineering and beyond.

Optimization is a central problem in engineering and science, where finding the best solution within constraints often involves computationally intensive processes. Classical optimization techniques have been the foundation of this domain for decades, while quantum optimization offers novel approaches leveraging quantum mechanics. This section provides an in-depth comparison of classical and quantum optimization methods.

6.9.1 Classical Optimization Techniques

Classical optimization relies on mathematical and heuristic methods to find minima or maxima of cost functions. One powerful technique is the that of gradient descent (GS). It is a widely used method that iteratively minimizes a cost function $f(x)$ by updating variables along the negative gradient direction:

$$x_{k+1} = x_k - \alpha \nabla f(x_k)$$

where α is the learning rate, and $\nabla f(x_k)$ is the gradient at iteration k.

Classical optimization techniques still have many challenges. For example:

1. Sensitive to the choice of α.
2. Prone to getting stuck in local minima for non-convex functions.

Python Code Example: Gradient Descent
This code demonstrates minimizing a quadratic cost function using gradient descent.

```
#------------------------------------------------------------------
# Gradient Descent
```

```
# Chapter 6 in the QUANTUM COMPUTING AND QUANTUM MACHINE LEARNING BOOK
#-----------------------------------------------------------------
# Version 1.0
# Qiskit changes frequently.
# We recommend using the latest version from the book code repository at:
# https://aqtinitiative.org/quantum-computing-for-engineers

# (c) 2025 Jesse Van Griensven, Roydon Fraser, and Jose Rosas
# License: MIT - Citation of this work is required
#-----------------------------------------------------------------

import numpy as np

# Define a quadratic cost function
def cost_function(x):
    return x**2 + 4 * x + 4.

# Gradient of the cost function
def gradient(x):
    return 2 * x + 4.

# Perform gradient descent
x = 10     # Initial guess
alpha = 0.1 # Learning rate
for _ in range(100):
    x = x - alpha * gradient(x)

print("Optimal x:  {:.4f}".format(x))
print("Minimum value: {:.4f}".format(cost_function(x)))
```

6.9.1.1 Simulated Annealing

Simulated annealing mimics the physical annealing process to find a global minimum by allowing probabilistic escapes from local minima:

$$P(E) = e^{-E/kT}$$

where E is the energy of the state, T is the temperature, and k is Boltzmann's constant. The temperature decreases over iterations, reducing the probability of accepting worse solutions.

Python Code Example: Another Simulated Annealing
This code simulates annealing to minimize a non-convex function.

```python
#------------------------------------------------------------------
# Another Simulated Annealing
# Chapter 6 in the QUANTUM COMPUTING AND QUANTUM MACHINE LEARNING BOOK
#------------------------------------------------------------------
# Version 1.0
# Qiskit changes frequently.
# We recommend using the latest version from the book code repository at:
# https://aqtinitiative.org/quantum-computing-for-engineers

# (c) 2025 Jesse Van Griensven, Roydon Fraser, and Jose Rosas
# License: MIT - Citation of this work is required
#------------------------------------------------------------------
import numpy as np


#------------------------------------------------------------------
def simulated_annealing(cost_function, bounds, max_iter, temp):
    """ Classical Simulated Annealing Example """
    current_state = np.random.uniform(bounds[0], bounds[1])
    current_cost = cost_function(current_state)

    for _ in range(max_iter):
        next_state = current_state + np.random.uniform(-1, 1) * temp
        next_cost = cost_function(next_state)
        if next_cost < current_cost or np.random.rand() < np.exp(-
(next_cost - current_cost) / temp):
            current_state, current_cost = next_state, next_cost
        temp *= 0.99 # Decrease temperature

    return current_state, current_cost
#------------------------------------------------------------------

# Example cost function
cost_function = lambda x: x**2 + 4. * np.sin(2. * x)
result = simulated_annealing(cost_function, bounds=(-10, 10),
max_iter=1000, temp=10)

# Print the output variables with 4 decimal places using an f-string.
print(f"Optimal Solution: x = {result[0]:.4f}, cost = {result
[1]:.4f}")
```

6.9.1.2 Genetic Algorithms

Genetic algorithms (GAs) evolve solutions by simulating biological processes like selection, crossover, and mutation. The steps in genetic algorithms are summarized below:

1. Selection: Choose the fittest individuals.
2. Crossover: Combine solutions to produce offspring.
3. Mutation: Introduce random changes for diversity.

Python Code Example: Genetic Algorithm
The following example demonstrates finding the maximum of a quadratic function using a genetic algorithm.

```python
#-------------------------------------------------------------------
# Genetic Algorithm
# Chapter 6 in the QUANTUM COMPUTING AND QUANTUM MACHINE LEARNING BOOK
# Finding the maximum of a quadratic function using a genetic algorithm
#-------------------------------------------------------------------
# Version 1.0
# Qiskit changes frequently.
# We recommend using the latest version from the book code repository at:
# https://aqtinitiative.org/quantum-computing-for-engineers

# (c) 2025 Jesse Van Griensven, Roydon Fraser, and Jose Rosas
# License: MIT - Citation of this work is required
#-------------------------------------------------------------------
import numpy as np

# Define fitness function
def fitness(x):
  return -1 * (x**2 - 5 * x + 6)  # Maximize this function


#-------------------------------------------------------------------
def genetic_algorithm(pop_size, generations):
  # Genetic Algorithm
  population = np.random.uniform(0, 5, pop_size)
  for _ in range(generations):
    fitness_values = fitness(population)

    # Select top half
    selected = population[np.argsort(fitness_values)[-pop_size//2:]]
```

```
  # Mutate
  offspring = np.mean(selected) + np.random.randn(pop_size) * 0.1
  population = np.concatenate((selected, offspring))
return population[np.argmax(fitness(population))]


#--------------------------------------------------------------------
optimal_solution = genetic_algorithm(pop_size=20, generations=50)
print("Optimal solution:", optimal_solution)
```

6.10 Advantages of Quantum Optimization

Quantum optimization introduces novel mechanisms that can outperform classical methods in certain cases.

6.10.1 Parallel Exploration of Solutions

Quantum superposition enables the simultaneous evaluation of multiple states. In QAOA, for example, all possible solutions are encoded in the initial quantum state:

$$|\psi_0\rangle = \frac{1}{\sqrt{2^n}} \sum_{x=0}^{2^n-1} |x\rangle$$

6.10.2 Quantum Tunneling

Quantum systems can "tunnel" through barriers in the energy landscape, allowing escape from local minima that would trap classical methods.

6.10.3 Reduced Complexity

For large-scale problems, quantum algorithms such as QAOA can provide approximate solutions with fewer resources than classical methods.

Python Code Example: Quantum Tunneling Simulation
This code simulates tunneling through quantum interference.

```python
#------------------------------------------------------------------------
# Quantum Tunneling Simulation
# Chapter 6 in the QUANTUM COMPUTING AND QUANTUM MACHINE LEARNING BOOK
#------------------------------------------------------------------------
# Version 1.0
# Qiskit changes frequently.
# We recommend using the latest version from the book code repository at:
# https://aqtinitiative.org/quantum-computing-for-engineers

# (c) 2025 Jesse Van Griensven, Roydon Fraser, and Jose Rosas
# License: MIT - Citation of this work is required
#------------------------------------------------------------------------
import warnings
warnings.filterwarnings('ignore')

from qiskit import QuantumCircuit, Aer, execute
from qiskit.visualization import circuit_drawer
#------------------------------------------------------------------------

# Quantum tunneling example
qc = QuantumCircuit(2)

# Create superposition
qc.h([0, 1])

# Introduce phase flip
qc.cz(0, 1)

# Measure
qc.measure_all()

# Draw the circuit
display(circuit_drawer(qc, output='mpl', style="iqp"))

# Execute the Circuit
simulator = Aer.get_backend('aer_simulator')
result   = execute(qc, simulator).result()
print("Quantum Tunneling Results:", result.get_counts())
```

6.10.4   Practical Comparison: Classical Vs. Quantum

Classical algorithms generally outperform quantum methods for small problems due to the noise and limitations of current quantum hardware. However, as problem sizes grow, quantum optimization excels in scalability and efficiency.

1. **Scalability**
   Classical algorithms face exponential scaling in problems, such as the traveling salesman problem (TSP), while quantum methods like QAOA can explore large solution spaces efficiently.
2. **Resource Requirements**
   Quantum methods require fewer resources for problems that naturally map onto quantum hardware, such as Hamiltonian optimization. For a problem size N, the complexity of classical optimization is $O(N^2)$, while QAOA complexity is $O(p)$, where p is the number of layers.

6.11   Quantum-Inspired Optimization for Classical Systems

Quantum-inspired optimization takes principles of quantum computing, such as superposition, entanglement, and adiabatic evolution, and applies them to classical systems. These approaches provide scalable and efficient solutions for problems that are difficult to solve using conventional methods. Quantum annealing provides a framework for energy minimization, tensor networks reduce dimensionality in high-dimensional datasets, and quantum-inspired neural networks enhance machine learning architectures.

These methods pave the way for scalable solutions in engineering and data science, bridging the gap between classical and quantum paradigms. This section explores quantum annealing, tensor networks, and quantum-inspired neural networks.

6.11.1   Quantum Annealing

Quantum annealing is inspired by adiabatic quantum computation. It evolves a quantum system from an easily solvable initial Hamiltonian H_0 to a problem-specific Hamiltonian H_C:

$$H(t) = (1 - t)H_0 + tH_C$$

where $t \in [0, 1]$ represents time progression.

The key steps in quantum annealing are summarized in Table 6.4.

Table 6.4 Quantum annealing summary

Key steps in quantum annealing	Description
Initialization	Start with H_0, typically representing a superposition of all possible states
Evolution	Slowly evolve $H(t)$ while maintaining the system in its ground state
Measurement	At $t = 1$, measure the system to find the optimal solution encoded in the ground state of H_C

Python Code Example: Simulating Quantum Annealing

This code demonstrates the creation of a time-dependent Hamiltonian for quantum annealing.

```python
#------------------------------------------------------------------
# Simulating Quantum Annealing
# Chapter 6 in the QUANTUM COMPUTING AND QUANTUM MACHINE LEARNING BOOK
# Simulating a time-dependent Hamiltonian Quantum Annealing
#------------------------------------------------------------------
# Version 1.0
# Qiskit changes frequently.
# We recommend using the latest version from the book code repository at:
# https://aqtinitiative.org/quantum-computing-for-engineers

# (c) 2025 Jesse Van Griensven, Roydon Fraser, and Jose Rosas
# License: MIT - Citation of this work is required
#------------------------------------------------------------------
import numpy as np

#------------------------------------------------------------------
def sprint(Matrix):
    """ Prints a numpy Matrix in a nice format with sympy """
    import sympy as sp
    # Jan 2024 by Jesse Thé
    Smatrix = sp.Matrix(Matrix)
    display( Smatrix )
    return

#------------------------------------------------------------------
def H0(n):
# Define initial and cost Hamiltonians
    return np.eye(2**n)  # Superposition state

#------------------------------------------------------------------
def HC(weights):
```

```
  n = len(weights)
  H = np.zeros((2**n, 2**n))
  for i in range(n):
    for j in range(n):
      if weights[i][j] != 0:
        H += weights[i][j] * np.kron(np.eye(2**i), np.kron([[1, 0], [0,
-1]], np.eye(2**(n-i-1))))
  return H


#----------------------------------------------------------------------
def H(t, H0, HC):
# Linear interpolation of H(t)
  return (1 - t) * H0 + t * HC


#----------------------------------------------------------------------
# Example usage
n = 3
weights = [[0, 1, 1], [1, 0, 1], [1, 1, 0]]
H0_matrix = H0(n)
HC_matrix = HC(weights)
H_t = H(0.5, H0_matrix, HC_matrix)


# Print Results
print("Interpolated Hamiltonian at t=0.5:")
sprint(H_t)
```

6.11.2   Tensor Networks

Tensor networks efficiently represent large datasets or high-dimensional objects by factorizing their structure into interconnected tensors. Matrix product states (MPS) is a specific type of tensor network that represents quantum states compactly:

$$|\psi\rangle = \sum_{i_1, i_2, \ldots, i_n} A^{[1]}_{i_1} A^{[2]}_{i_2} \ldots A^{[n]}_{i_n} |i_1 i_2 \ldots i_n\rangle$$

where $A^{[k]}$ are tensors.

Applications of Tensor Networks in Classical Systems:

1. Data Compression: Representing large datasets efficiently.
2. Optimization Problems: Solving high-dimensional linear algebra problems.

Python Code Example: Tensor Networks

This code demonstrates an MPS decomposition to reduce the dimensionality of classical data.

```python
#-----------------------------------------------------------------
# MPS Tensor Networks
# Chapter 6 in the QUANTUM COMPUTING AND QUANTUM MACHINE LEARNING BOOK
# MPS decomposition to reduce the dimensionality of classical data
# MPS: Matrix Product States
#-----------------------------------------------------------------
# Version 1.0
# Qiskit changes frequently.
# We recommend using the latest version from the book code repository at:
# https://aqtinitiative.org/quantum-computing-for-engineers

# (c) 2025 Jesse Van Griensven, Roydon Fraser, and Jose Rosas
# License: MIT - Citation of this work is required
#-----------------------------------------------------------------
import numpy as np


#-----------------------------------------------------------------
def mps_representation(data, max_rank):
  # Example Matrix Product State representation
  n = len(data)
  tensors = []
  for i in range(n - 1):
    u, s, vh = np.linalg.svd(data[i], full_matrices=False)
    u = u[:, :max_rank]
    s = np.diag(s[:max_rank])
    vh = vh[:max_rank, :]
    tensors.append(u)
    data[i + 1] = np.dot(s, vh)
  tensors.append(data[-1])
  return tensors


#-----------------------------------------------------------------
# Example data and usage
data  = [np.random.rand(4, 4) for _ in range(3)]
tensors = mps_representation(data, max_rank=2)
print("Tensor Network Representation:")
print(tensors)
```

6.11.3   Quantum-Inspired Neural Networks

Quantum-inspired neural networks integrate quantum concepts like amplitude encoding and entanglement into classical architectures to improve efficiency and performance.

6.11.3.1   Amplitude Encoding

Classical data is encoded into quantum amplitudes, enabling high-dimensional representation:

$$|x\rangle = \frac{1}{\|x\|} \sum_i x_i \, |i\rangle$$

6.11.3.2   Quantum-Inspired Layers

Quantum-inspired layers simulate unitary transformations or entanglement using classical operations. For example, orthogonal matrices can be used in place of quantum gates.

Python Code Example: Quantum-Inspired Neural Network
This code implements a quantum-inspired neural network using TensorFlow.

```python
#-----------------------------------------------------------------
# Quantum-Inspired Neural Network
# Chapter 6 in the QUANTUM COMPUTING AND QUANTUM MACHINE LEARNING BOOK
#-----------------------------------------------------------------
# Version 1.0
# Qiskit changes frequently.
# We recommend using the latest version from the book code repository at:
# https://aqtinitiative.org/quantum-computing-for-engineers

# (c) 2025 Jesse Van Griensven, Roydon Fraser, and Jose Rosas
# License: MIT - Citation required
#-----------------------------------------------------------------
import tensorflow as tf
#-----------------------------------------------------------------

# Define a quantum-inspired layer
class QuantumInspiredLayer(tf.keras.layers.Layer):
  def __init__(self, units):
```

```python
    super().__init__()
    self.units = units

  def build(self, input_shape):
    self.weight = self.add_weight(shape=(input_shape[-1], self.
units),
                    initializer="orthogonal",
                    trainable=True)

  def call(self, inputs):
    return tf.matmul(inputs, self.weight)

# Example model
model = tf.keras.Sequential([
  QuantumInspiredLayer(16),
  tf.keras.layers.Activation('relu'),
  tf.keras.layers.Dense(1, activation='sigmoid')
])

# Compile and train the model
model.compile(optimizer='adam', loss='binary_crossentropy')
print("Quantum-Inspired Neural Network Model:")
model.summary()
```

6.11.4 Implementation Challenges in Quantum Optimization

The implementation challenges in quantum optimization stem from both hardware limitations and algorithmic complexities. Limited coherence times and high noise levels constrain the capabilities of current quantum processors, while parameter tuning and scalability remain significant hurdles for algorithms like QAOA. Addressing these challenges requires advancements in quantum hardware, error correction techniques, and hybrid frameworks, paving the way for practical applications of quantum optimization in engineering and science.

This subsection discusses the key hurdles in deploying quantum optimization systems, focusing on hardware constraints and algorithmic challenges.

6.11.5 Hardware Constraints

The effectiveness of quantum optimization algorithms, such as the quantum approximate optimization algorithm (QAOA), heavily depends on the capabilities of quantum hardware. Current quantum processors face several limitations.

6.11.5.1   Limited Qubit Coherence Times

Qubits must maintain coherence for the duration of the computation to produce reliable results. However, in contemporary quantum hardware, coherence times are limited, often ranging from microseconds to milliseconds.

The coherence time T_2 of a qubit represents how long it can maintain its quantum state:

$$T_2 \propto \frac{1}{\text{noise power density}}$$

Short coherence times restrict the depth of quantum circuits and the number of QAOA layers p that can be applied. Increasing circuit depth leads to decoherence and incorrect results. For example, a simple QAOA circuit with $p = 2$ may work on current quantum hardware, but $p = 10$ exceeds coherence limits.

6.11.5.2   Noise and Error Rates

Quantum gates are inherently noisy, and errors accumulate as more gates are applied. Below are the quantum error sources (Table 6.5).

Fidelity F, the measure of how accurately a gate performs, must satisfy:

$$F = |\langle \psi_{\text{ideal}} | \psi_{\text{real}} \rangle|^2$$

Python Code Example: Another Noise Simulation
This code demonstrates how noise affects quantum circuit execution, leading to reduced fidelity.

```
#------------------------------------------------------------------
# Another Noise Simulation
# Chapter 6 in the QUANTUM COMPUTING AND QUANTUM MACHINE LEARNING BOOK
#------------------------------------------------------------------
# Version 1.0
# Qiskit changes frequently.
# We recommend using the latest version from the book code repository at:
# https://aqtinitiative.org/quantum-computing-for-engineers
```

Table 6.5 Type of quantum errors

Quantum error source	Description
Gate errors	Imperfect implementation of quantum gates
Measurement errors	Inaccuracies in reading qubit states
Environmental interference	External factors disrupting qubit states

```python
# (c) 2025 Jesse Van Griensven, Roydon Fraser, and Jose Rosas
# License: MIT - Citation required
#--------------------------------------------------------------------
import warnings
warnings.filterwarnings('ignore')

from qiskit import QuantumCircuit
from qiskit.visualization import circuit_drawer

from qiskit.providers.aer import AerSimulator
from qiskit.providers.aer.noise import NoiseModel,
depolarizing_error
#--------------------------------------------------------------------

# Define a noise model
noise_model = NoiseModel()

# Place "Noise" in Hadamard gate
noise_model.add_all_qubit_quantum_error(depolarizing_error(0.01,
1), ['h'])

# Create a simple QAOA circuit
qc = QuantumCircuit(2)

# Apply the Hadamard gate to place qubit zero in superposition
qc.h(0)

# Apply the CNOT gate
qc.cx(0, 1)

# Now measure
qc.measure_all()

# Simulate the circuit with noise
simulator = AerSimulator(noise_model=noise_model)
result   = simulator.run(qc).result()
print("Noisy Results:", result.get_counts())
```

6.11.6 Algorithmic Challenges

In addition to hardware constraints, quantum optimization algorithms face significant challenges in parameter tuning and scalability.

6.11.6.1   Parameter Tuning for QAOA Layers

QAOA involves optimizing parameters γ and β for each layer p:

$$|\psi(\gamma,\beta)\rangle = \prod_{k=1}^{p} e^{-i\beta_k H_M} e^{-i\gamma_k H_C} |\psi_0\rangle$$

Finding the optimal values for γ and β requires classical optimization over a high-dimensional parameter space, which becomes increasingly complex as p grows.

Python Code Example: Parameter Optimization
This example demonstrates optimizing γ and β for a simple cost function.

```python
#----------------------------------------------------------------------
# Customized Cost QAOA
# Chapter 6 in the QUANTUM COMPUTING AND QUANTUM MACHINE LEARNING BOOK
#----------------------------------------------------------------------
# Version 1.0
# Qiskit changes frequently.
# We recommend using the latest version from the book code repository at:
# https://aqtinitiative.org/quantum-computing-for-engineers

# (c) 2025 Jesse Van Griensven, Roydon Fraser, and Jose Rosas
# License: MIT - Citation required
#----------------------------------------------------------------------
from scipy.optimize import minimize
import numpy as np

# Example cost function for QACA
def qaoa_cost(params):
    gamma, beta = params
    # Placeholder for QAOA simulation; replace with actual quantum
computation
    return np.sin(gamma) ** 2 + np.cos(beta) ** 2

# Optimize parameters
initial_guess = [np.pi / 4., np.pi / 6.]
result = minimize(qaoa_cost, initial_guess, method="COBYLA")
print("Optimal Parameters:", result.x)
```

6.11.6.2   Scalability of Hybrid Quantum-Classical Frameworks

Quantum optimization often relies on hybrid frameworks, where quantum circuits compute intermediate results, and classical algorithms optimize parameters. As problem size increases, the classical component becomes a bottleneck.

Scalability Issues:

1. **Quantum Resource Limits**
   Current hardware supports only tens of qubits, limiting the size of problems that can be solved.
2. **Classical Feedback Loops**
   Repeated execution of quantum circuits during optimization slows the process.

Proposed Solutions:

1. **Layer-Wise Optimization**
   Optimize parameters for each QAOA layer incrementally to reduce complexity.
2. **Surrogate Models**
   Machine learning can be used to predict cost function values, reducing the need for quantum circuit evaluations.

Python Code Example: Layer-Wise Optimization
This code optimizes parameters incrementally for each layer, improving scalability.

```
#-----------------------------------------------------------------
# Layer-Wise Optimization
# Chapter 6 in the QUANTUM COMPUTING AND QUANTUM MACHINE LEARNING BOOK
#-----------------------------------------------------------------
# Version 1.0
# Qiskit changes frequently.
# We recommend using the latest version from the book code repository at:
# https://aqtinitiative.org/quantum-computing-for-engineers


# (c) 2025 Jesse Van Griensven, Roydon Fraser, and Jose Rosas
# License: MIT - Citation of this work is required
#-----------------------------------------------------------------
import numpy as np
import qiskit
import scipy
from scipy.optimize import minimize
#-----------------------------------------------------------------
# Define a placeholder QAOA circuit simulation
```

```python
def qaoa_layer_cost(layer, gamma, beta):
    # Placeholder: actual quantum simulation should go here
    Layer = np.sin(gamma + layer) ** 2 + np.cos(beta + layer) ** 2
    return Layer
#-----------------------------------------------------------------

# Layer-wise optimization
layers = 3
params = []
pi_4 = np.pi / 4.
pi_6 = np.pi / 6.

for layer in range(layers):
    result = minimize(lambda p: qaoa_layer_cost(layer, p[0], p[1]),
[pi_4, pi_6], method="COBYLA")
    params.append(result.x)

print("Layer-Wise Optimized Parameters:", params)
```

6.12 Future Directions in Quantum Optimization

The future of quantum optimization lies in overcoming current hardware limitations, developing domain-specific algorithms, and integrating quantum solutions into industrial workflows. Industries such as aerospace, logistics, and materials science will be among the first to experience the transformative potential of quantum optimization, paving the way for broader adoption as quantum hardware matures.

Quantum optimization holds significant promise for solving complex problems that are intractable to classical approaches. As quantum technologies continue to advance, future directions in research and industrial applications will shape how quantum optimization is integrated into various fields. This section explores key research trends and anticipated industrial adoption.

6.12.1 Research Trends in Quantum Optimization

Noise resilience is critical for advancing the capabilities of quantum optimization algorithms. Current quantum devices are limited by decoherence and gate errors, which degrade the fidelity of computations. Future research focuses on improving noise tolerance through better hardware design and advanced error correction techniques.

Quantum error correction (QEC) encodes logical qubits into multiple physical qubits to detect and correct errors without destroying quantum information. Surface codes are among the most promising QEC methods.

6.12.1.1 Surface Code Example

Logical qubits are encoded in a 2D lattice of physical qubits, where stabilizers detect errors:

$$\text{Stabilizers} : \langle Z_i Z_j, X_k X_l \rangle$$

The error threshold for fault-tolerant quantum computation is given by:

$$p_{\text{error}} < p_{\text{threshold}}$$

where $p_{\text{threshold}}$ is a hardware-dependent value.

Python Code Example: Another Error Correction Simulation
This example demonstrates how noise affects quantum circuits and underscores the importance of error correction techniques.

```python
#-------------------------------------------------------------------
# Another Error Correction Simulation
# Chapter 6 in the QUANTUM COMPUTING AND QUANTUM MACHINE LEARNING BOOK
#-------------------------------------------------------------------
# Version 1.0
# Qiskit changes frequently.
# We recommend using the latest version from the book code repository at:
# https://aqtinitiative.org/quantum-computing-for-engineers

# (c) 2025 Jesse Van Griensven, Roydon Fraser, and Jose Rosas
# License: MIT - Citation of this work is required
#-------------------------------------------------------------------
import warnings
warnings.filterwarnings('ignore')

from qiskit.providers.aer.noise import NoiseModel, pauli_error
from qiskit import QuantumCircuit, Aer, execute
from qiskit.visualization import circuit_drawer
#-------------------------------------------------------------------

# Define a simple noise model
noise_model = NoiseModel()
```

```
error = pauli_error([('X', 0.1), ('I', 0.9)])
noise_model.add_all_qubit_quantum_error(error, 'measure')

# Define a quantum circuit
qc = QuantumCircuit(3)
qc.h(0)
qc.cx(0, 1)
qc.cx(0, 2)
qc.measure_all()

# Simulate with noise
simulator = Aer.get_backend('aer_simulator')
result = execute(qc, simulator, noise_model=noise_model).result()
print("Noisy Results:", result.get_counts())

# Display the complete circuit
display(circuit_drawer(qc, output='mpl', style="iqp"))
```

6.12.1.2   Hardware Improvements

Future quantum hardware aims to extend coherence times and reduce gate errors through better materials, cryogenic systems, and qubit connectivity. Research in topological qubits and photonic quantum systems also shows promise for inherently noise-resistant architectures.

6.12.1.3   Development of Domain-Specific Quantum Algorithms

Another significant trend is the development of quantum algorithms tailored to specific domains. By customizing algorithms to exploit the structure of domain-specific problems, researchers aim to achieve exponential or quadratic speedups. For applications such as logistics or energy optimization, variations of the quantum approximate optimization algorithm (QAOA) can incorporate problem-specific Hamiltonians. For instance, in the vehicle routing problem (VRP), cost Hamiltonians may include constraints such as vehicle capacities:

$$H_C = \sum_{i,j} d_{ij} x_{ij} + \lambda \sum_i \left(\sum_j x_{ij} - C_i \right)^2$$

where C_i represents vehicle capacity and λ is a penalty factor.

Python Code Example: QAOA for Logistics Optimization

This code demonstrates a problem-specific variation of QAOA for logistics optimization.

```python
#-----------------------------------------------------------------
# QAOA for logistics optimization
# Chapter 6 in the QUANTUM COMPUTING AND QUANTUM MACHINE LEARNING BOOK
#-----------------------------------------------------------------
# Version 1.0
# Qiskit changes frequently.
# We recommend using the latest version from the book code repository at:
# https://aqtinitiative.org/quantum-computing-for-engineers

# (c) 2025 Jesse Van Griensven, Roydon Fraser, and Jose Rosas
# License: MIT - Citation required
#-----------------------------------------------------------------
import warnings
warnings.filterwarnings('ignore')

from qiskit import QuantumCircuit
from qiskit.visualization import circuit_drawer
#-----------------------------------------------------------------

# Customized Cost Hamiltonian for Vehicle Routing Problem - VRP
def custom_vrp_cost(qc, distances, penalties, gamma):
  for i in range(len(distances)):
    for j in range(len(distances)):
      if distances[i][j] != 0:
        qc.cx(i, j)
        qc.rz(2 * gamma * distances[i][j], j)
        qc.cx(i, j)

# Example circuit
qc = QuantumCircuit(3)
distances = [[0, 1, 2], [1, 0, 3], [2, 3, 0]]
penalties = [1, 1, 1]
custom_vrp_cost(qc, distances, penalties, gamma=0.5)
print("Customized QAOA Circuit:")
print(qc)
```

6.12.2   Industrial Adoption

Industries with computationally intensive optimization needs are poised to benefit significantly from advancements in quantum optimization. For example, in aerospace, quantum optimization can improve flight scheduling, air traffic management, and spacecraft trajectory planning. These applications involve solving high-dimensional problems with constraints that classical methods struggle to handle efficiently.

Quantum optimization can calculate the optimal trajectory for a spacecraft by minimizing fuel consumption:

$$H_C = \sum_i (F_i - m_i a_i)^2$$

where F_i is the force, m_i is the mass, and a_i is the acceleration at time step i.

Quantum optimization is expected to revolutionize logistics by enhancing route planning, supply chain management, and warehouse optimization. Algorithms like QAOA and quantum annealing can solve the traveling salesman problem (TSP) and vehicle routing problem (VRP) more efficiently.

Python Code Example: Quantum-Enhanced Logistics
This example highlights how quantum superposition and entanglement can explore multiple routing configurations simultaneously.

```
#--------------------------------------------------------------------
# Quantum-Enhanced Logistics
# Chapter 6 in the QUANTUM COMPUTING AND QUANTUM MACHINE LEARNING BOOK
#--------------------------------------------------------------------
# Version 1.0
# Qiskit changes frequently.
# We recommend using the latest version from the book code repository at:
# https://aqtinitiative.org/quantum-computing-for-engineers

# (c) 2025 Jesse Van Griensven, Roydon Fraser, and Jose Rosas
# License: MIT - Citation of this work is required
#--------------------------------------------------------------------
import warnings
warnings.filterwarnings('igncre')

from qiskit import QuantumCircuit
from qiskit.visualization import circuit_drawer
from qiskit import Aer, execute
#--------------------------------------------------------------------
```

```
# Simple quantum circuit for logistics optimization
qc = QuantumCircuit(3)
qc.h([0, 1, 2])  # Initialize superposition
qc.cz(0, 1)  # Phase inversion for constraint
qc.measure_all()

# Simulate the circuit
simulator = Aer.get_backend('aer_simulator')
result = execute(qc, simulator).result()
print("Logistics Optimization Results:", result.get_counts())
```

6.12.3   Materials Science

In materials science, quantum optimization can predict material properties and design novel compounds. By simulating molecular systems with quantum Hamiltonians, researchers can accelerate the discovery of materials for batteries, semiconductors, and pharmaceuticals.

New molecules can be obtained by simulating the ground-state energy of a molecule:

$$H = \sum_i h_i P_i + \sum_{i<j} h_{ij} P_i P_j$$

where P_i are Pauli operators and h_i, h_{ij} are coefficients from electronic structure calculations.

Python Code Example: Molecular Simulation
This example uses Qiskit Nature to simulate the molecular structure of hydrogen.

```
#------------------------------------------------------------------------
# Molecular Simulation
# Chapter 6 in the QUANTUM COMPUTING AND QUANTUM MACHINE LEARNING BOOK
#------------------------------------------------------------------------
# Version 1.0
# Qiskit changes frequently.
# We recommend using the latest version from the book code repository at:
# https://aqtinitiative.org/quantum-computing-for-engineers

# (c) 2025 Jesse Van Griensven, Roydon Fraser, and Jose Rosas
# License: MIT - Citation of this work is required
#------------------------------------------------------------------------
#from qiskit_nature.drivers import PySCFDriver
from qiskit_nature.second_q.drivers import PySCFDriver
```

```
# The Beauty from IBM qiskit:
# Replaced the location to make the code even messier:
# Qiskit Nature 0.7.2 no longer includes qiskit_nature.drivers.
# In newer versions, the module has moved to qiskit_nature.second_q.
drivers.

# Define a molecular structure
driver  = PySCFDriver(atom="H 0 0 0; H 0 0 0.74", basis="sto3g")
molecule = driver.run()

# Print molecular properties
print("Number of Orbitals:", molecule.num_orbitals)
print("Nuclear Repulsion Energy:", molecule.
nuclear_repulsion_energy)
```

6.13   Final Notes on Quantum Optimizations

Quantum optimization represents a set of significant methods to solve complex engineering problems, and they bridge the gap between classical limitations and quantum capabilities. As quantum computer hardware matures, its impact on real-world applications will only grow, making it an essential area for engineers to understand and harness.

Exercise Questions: Quantum Optimization Techniques
1. **Introduction to Quantum Optimization**

    (a) Compare classical and quantum optimization approaches in terms of scalability and efficiency.
    (b) Explain how quantum properties like superposition and entanglement are leveraged in quantum optimization algorithms.

2. **Classical Optimization Methods**

    (a) Write the gradient descent update rule and describe its limitations for non-convex functions.
    (b) Explain how combinatorial optimization problems, such as the traveling salesman problem (TSP), are solved classically.

3. **Quantum Approximate Optimization Algorithm (QAOA)**

    (a) Define the cost Hamiltonian in QAOA and explain its role in representing optimization problems.
    (b) Describe the alternating application of cost and mixing Hamiltonians in QAOA and its impact on solution exploration.

4. **Python Implementation of QAOA**

(a) Write Python code to initialize a QAOA circuit for a 3-qubit system. Explain the purpose of each step in the circuit.

(b) Using the phase separation operator, encode a simple cost Hamiltonian for a graph with three vertices.

5. **Simulated Annealing**

(a) Compare simulated annealing with quantum annealing in terms of their approach to escaping local minima.

(b) Write Python code to minimize a quadratic cost function using simulated annealing.

6. **Tensor Networks**

(a) Explain the role of tensor networks in optimizing high-dimensional datasets. Provide an example application.

(b) Write Python code to demonstrate matrix product state (MPS) decomposition for data compression.

7. **Quantum Optimization Applications**

(a) Discuss how quantum optimization can improve load balancing in energy grids.

(b) Explain how the traveling salesman problem (TSP) is encoded into a quantum circuit for optimization.

8. **Challenges in Quantum Optimization**

(a) Identify two hardware constraints that affect the scalability of quantum optimization algorithms.

(b) Describe the parameter optimization challenges in QAOA and propose strategies to address them.

9. **Quantum-Inspired Optimization**

(a) Explain how quantum-inspired techniques like tensor networks mimic quantum mechanics to solve classical problems.

(b) Describe the key steps in quantum annealing and their significance in solving optimization problems.

10. **Industrial Applications**

(a) Provide an example of how quantum optimization enhances logistics, such as vehicle routing or warehouse management.

(b) Explain how quantum optimization accelerates materials science research by simulating molecular systems.

11. **Future Trends in Quantum Optimization**

(a) Discuss the role of error correction techniques, such as surface codes, in improving quantum optimization reliability.

(b) Propose a future research direction to enhance the noise resilience of quantum hardware.

12. **Hybrid Quantum-Classical Frameworks**

(a) Explain the benefits of hybrid quantum-classical approaches in quantum optimization.

(b) Write Python code to optimize parameters in a QAOA circuit using a classical feedback loop.

Additional Bibliography

1. F. Arute, K. Arya, R. Babbush, et al., Quantum supremacy using a programmable superconducting processor. Nature **574**(7779), 505–510 (2019)
2. G. Benenti, G. Casati, G. Strini, *Principles of Quantum Computation and Information: Basic Concepts* (World Scientific Publishing Company, 2007)
3. D. Bouwmeester, A. Ekert, A. Zeilinger, *The Physics of Quantum Information: Quantum Cryptography, Quantum Teleportation, Quantum Computation* (Springer, 2000)
4. S. Bravyi, D. Gosset, R. König, Quantum advantage with shallow circuits. Science **362**(6412), 308–311 (2018)
5. E. Farhi, J. Goldstone, S. Gutmann, A quantum approximate optimization algorithm, in *arXiv preprint arXiv:1411.4028*, (2000)
6. L.K. Grover, A fast quantum mechanical algorithm for database search, in *Proceedings of the 28th Annual ACM Symposium on Theory of Computing*, (1996)
7. J. Gruska, *Quantum Computing* (McGraw-Hill, 1999)
8. A.W. Harrow, A. Hassidim, S. Lloyd, Quantum algorithm for linear Systems of Equations. Phys. Rev. Lett. **103**(15), 150502 (2009)
9. M. Hirvensalo, *Quantum Computing* (Springer, 2004)
10. A.Y. Kitaev, A.H. Shen, M.N. Vyalyi, *Classical and Quantum Computation* (American Mathematical Society, 2002)
11. S. Lloyd, Universal Quantum Simulators. Science **273**(5278), 1073–1078 (1996)
12. S. Majidy, C. Wilson, R. Laflamme, *Building Quantum Computers—A Practical Introduction* (Cambridge Press, 2025)
13. N.D. Mermin, *Quantum Computer Science: An Introduction* (Cambridge University Press, 2007)
14. A. Montanaro, Quantum algorithms: an overview. NPJ Quantum Inf. **2**, 15023 (2016)
15. M.A. Nielsen, I.L. Chuang, *Quantum Computation and Quantum Information* (Cambridge University Press, 2010)
16. A. Peruzzo, J. McClean, P. Shadbolt, et al., A variational eigenvalue solver on a photonic quantum processor. Nat. Commun. **5**, 4213 (2014)
17. J. Preskill, Quantum computing in the NISQ era and beyond, in *arXiv preprint arXiv:1801.00862*, (2018)
18. E.G. Rieffel, W.H. Polak, *Quantum Computing: A Gentle Introduction* (MIT Press, 2011)
19. P.W. Shor, Algorithms for quantum computation: discrete logarithms and factoring, in *IEEE Symposium on Foundations of Computer Science*, (1997)
20. V. Vedral, *Introduction to Quantum Information Science* (Oxford University Press, 2006)
21. J. Watrous, *The Theory of Quantum Information* (Cambridge University Press, 2018)
22. N.S. Yanofsky, M.A. Mannucci, *Quantum Computing for Computer Scientists* (Cambridge University Press, 2008)

Chapter 7
Classical Artificial Intelligence

This section introduced the concept of artificial intelligence (AI), its historical development, and its differentiation from machine learning and deep learning. AI's transformative potential, particularly in engineering, is evident in its ability to automate tasks, improve accuracy, and foster innovation. As AI technologies continue to evolve, their integration into engineering applications will further redefine the boundaries of what is possible.

7.1 Definition of Artificial Intelligence

Artificial intelligence (AI) refers to the simulation of human intelligence in machines that are designed to think, learn, and make decisions autonomously. At its core, AI involves creating systems that can perform tasks typically requiring human intelligence, such as recognizing patterns, understanding natural language, solving problems, and making predictions.

AI can be broadly categorized into three types, as shown in Table 7.1.

Key characteristics of AI systems include:

1. **Adaptability**

    The ability to improve performance through learning from data or experiences.
2. **Autonomy**

    Performing tasks without human intervention.
3. **Perception**

    Interpreting data from the environment, such as images, text, or speech.

© The Author(s), under exclusive license to Springer Nature Switzerland AG 2025 451
J. Van Griensven Thé et al., *Quantum Computing and Quantum Machine Learning for Engineers and Developers*, https://doi.org/10.1007/978-3-031-98245-3_7

Table 7.1 Types of AI

Type of artificial intelligence	Description
Narrow AI (weak AI)	Specialized systems designed to perform a single task or a narrow set of tasks (e.g., virtual assistants like Siri or recommendation algorithms on Netflix)
General AI (strong AI)	Hypothetical systems capable of performing any intellectual task that a human can do, demonstrating generalized cognitive abilities (not yet achieved)
Superintelligent AI	A theoretical stage where AI surpasses human intelligence in all fields

7.2 Brief History and Milestones in AI Development

Below is an expanded chronological overview of notable milestones in AI development, including major breakthroughs in deep neural networks for vision and natural language processing. These chronological milestones highlight how AI, driven by breakthroughs in deep learning and computational power, transformed fields ranging from healthcare and finance to creative industries.

1. **1943: McCulloch and Pitts'Neural Model**
   Warren McCulloch and Walter Pitts proposed the first mathematical model of artificial neurons, forming the basis for modern neural network research.
2. **1950: Turing Test**
   Alan Turing introduced the concept of the Turing Test in his paper "Computing Machinery and Intelligence," suggesting that a machine could be considered intelligent if its responses could not be distinguished from those of a human.
3. **1956: The Birth of AI**
   The term "artificial intelligence" was coined at the Dartmouth Conference, attended by pioneers, such as John McCarthy and Marvin Minsky, to explore the potential of creating machines capable of reasoning.
4. **1960s: Early AI Systems**
   Researchers created simple AI programs like ELIZA (an early chatbot that emulated a psychotherapist) and SHRDLU (a system capable of understanding and executing commands in a virtual "blocks world").
5. **1980s: Expert Systems**
   Knowledge-based expert systems gained commercial popularity by applying rule-based logic to solve specialized industrial problems, such as medical diagnosis and financial forecasting.
6. **1997: Deep Blue Defeats Garry Kasparov**
   IBM's Deep Blue beat the reigning world chess champion Garry Kasparov, showcasing how brute-force search algorithms combined with strategic heuristics could outperform human experts.

7. **2011: Watson Wins Jeopardy!**

   IBM's Watson triumphed against human champions in the quiz show Jeopardy! through advanced natural language processing, massive data repositories, and sophisticated search algorithms.

8. **2012: AlexNet and the Deep Learning Revolution**

   A significant turning point for modern AI research came when AlexNet, a deep convolutional neural network, won the ImageNet Large Scale Visual Recognition Challenge by a wide margin. Its success demonstrated the power of GPUs and deep learning for computer vision tasks.

9. **2014: Generative Adversarial Networks (GANs)**

   Ian Goodfellow introduced GANs, a framework in which two neural networks (a generator and a discriminator) compete against each other. GANs have since been used to create realistic images, art, and even synthetic training data.

10. **2016: AlphaGo Triumphs Over Go Champion**

    Google DeepMind's AlphaGo defeated Lee Sedol, a world champion in the game of Go, by integrating deep neural networks with reinforcement learning, marking a new era in AI capabilities.

11. **2017: The Transformer Architecture**

    The paper "Attention Is All You Need" by Google introduced the Transformer architecture, a breakthrough that relies on attention mechanisms rather than recurrent or convolutional structures. Transformers have become the foundation for many state-of-the-art language models.

12. **2018: BERT (Bidirectional Encoder Representations from Transformers)**

    Google's BERT demonstrated the power of large-scale pretraining using Transformers, significantly improving performance on a wide range of natural language understanding tasks like question answering and text classification.

13. **2020: GPT-3 (Generative Pretrained Transformer 3)**

    OpenAI's GPT-3, with 175 billion parameters, showcased remarkable capabilities in text generation, question answering, and language translation, heralding a new wave of AI applications and prompting global attention to large language models.

14. **2022: ChatGPT**

    The publication of GPT-3.5 revolutionized interactive AI by providing fluid, context-aware conversations with users. It rapidly gained widespread use in writing assistance, coding support, customer service, and more, demonstrating the growing impact of large language models on everyday tasks.

15. **2022–2023: Expansion of Multimodal and Next-Generation Models.**

    Advances in image-generating models (such as DALL-E and Stable Diffusion) alongside subsequent GPT evolutions (e.g., GPT-4) illustrate AI's accelerating progress across text, image, and other modalities, promising even more robust and integrated applications in the near future.

Figure 7.1 summarizes this chronological evolution.

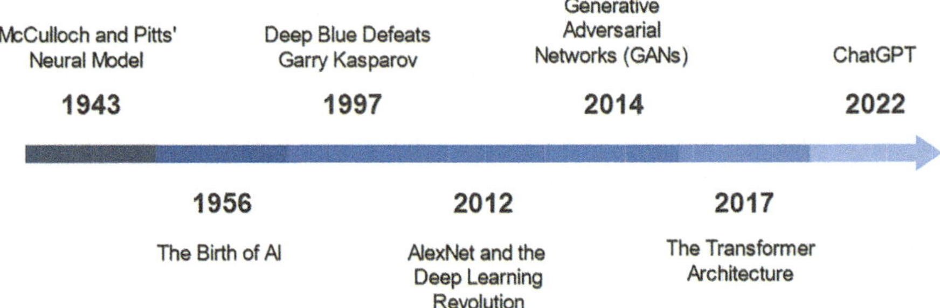

Fig. 7.1 Key milestones in the AI development. (By the authors)

Table 7.2 Comparison table of AI, ML, and DL

Feature	Artificial intelligence	Machine learning	Deep learning
Focus	General intelligence tasks	Learning from data	Complex pattern recognition
Techniques used	Rule-based, statistical models	Statistical learning algorithms	Neural networks
Data requirements	Low to medium	Medium to high	High (big data)
Computational power	Moderate	High	Very high (GPUs/TPUs required)

7.3 Differences Between AI, Machine Learning, and Deep Learning

AI, machine learning (ML), and deep learning (DL) are often used interchangeably, but they represent distinct concepts within the broader field of artificial intelligence (Table 7.2):

1. **Artificial Intelligence (AI)**

    (a) Scope: The overarching discipline focused on creating intelligent machines.
    (b) Methods: Encompasses a wide range of techniques, including rule-based systems, optimization, and machine learning.
    (c) Example: A chess-playing AI designed using decision trees and heuristics.

2. **Machine Learning (ML)**

    (a) Scope: A subset of AI that focuses on systems that learn from data to improve performance without being explicitly programmed.
    (b) Methods: Regression, classification, clustering, and reinforcement learning.
    (c) Example: Predicting house prices based on historical data.

3. **Deep Learning (DL)**

   (a) Scope: A specialized subset of ML that uses deep neural networks to model complex patterns in large datasets.
   (b) Applications: Image recognition, speech synthesis, and natural language understanding.
   (c) Example: Facial recognition systems in smartphones.

7.4   Importance of AI for Engineers

AI is revolutionizing engineering by automating processes, optimizing designs, and enabling smarter decision-making. Table 7.3 summarizes some applications of AI in different fields.

Key reasons why AI is essential for engineers include:

1. **Efficiency and Automation**

   AI-powered systems can automate repetitive tasks, reducing manual effort and increasing productivity. For example, AI in computer-aided design (CAD) software speeds up design iterations.

2. **Data-Driven Decision-Making**

   Engineers often deal with vast amounts of data. AI tools, such as predictive analytics, help extract actionable insights, improving the reliability of engineering solutions.

3. **Improved Accuracy**

   AI minimizes errors in critical applications, such as structural health monitoring or predictive maintenance in manufacturing.

4. **Innovation in Design**

   AI-driven generative design enables engineers to explore innovative solutions by suggesting designs based on performance criteria.

7.5   Foundational Concepts in AI

The foundational concepts of AI, such as intelligent systems, search algorithms, optimization, and knowledge representation, form the building blocks for solving complex problems. These tools are vital for creating systems that can reason, adapt,

Table 7.3 Applications across disciplines

Discipline	Applications
Civil engineering	AI in traffic management systems
Mechanical engineering	Optimizing robotic control systems
Electrical engineering	AI in smart grid optimization

and operate autonomously across various engineering applications. As we advance further into the quantum age, these classical techniques will serve as a stepping stone to understanding quantum-enhanced AI.

7.5.1 Overview of Intelligent Systems

An intelligent system is any computational system designed to perform tasks typically requiring human intelligence, such as reasoning, learning, decision-making, and adapting to new information. These systems can operate autonomously in complex and dynamic environments.

Key Features of Intelligent Systems

Intelligent systems have various features that make them attractive to all aspects of science and business. A few examples of these features are presented in Table 7.4.

Examples of AI Systems

1. **Expert Systems**

   Use domain-specific knowledge to solve problems (e.g., medical diagnosis systems like MYCIN).

2. **Robotics**

   Physical systems capable of performing tasks in real-world environments (e.g., autonomous vehicles).

3. **Decision Support Systems**

   Tools that assist in complex decision-making by analyzing data and presenting actionable insights.

Intelligent systems are foundational to AI and provide the architecture for building solutions across disciplines, such as healthcare, manufacturing, and transportation.

Table 7.4 Features of intelligent systems

Intelligent system feature	Description
Adaptability	The ability to learn and adjust behavior based on past experiences or new data
Autonomy	Operating without continuous human intervention
Interactivity	Communicating and responding to users or other systems in real time
Goal-oriented behavior	Acting to achieve specific objectives efficiently and effectively

7.5.2 AI as a Problem-Solving Framework

At its core, AI is a framework for solving problems that are computationally challenging or impractical to address with traditional algorithms. AI approaches emulate human problem-solving by employing search strategies, optimization techniques, and knowledge representation.

7.5.3 Search Algorithms and Optimization

Search algorithms are fundamental to AI because they enable systems to explore possible solutions to a problem systematically. These algorithms operate within a state space, which represents all possible configurations of a problem. Common types of search are presented in the list below.

1. **Uninformed Search**

   (a) Explores the state space without any domain-specific knowledge.
   (b) Examples:

   - Breadth-First Search (BFS): Explores all nodes at the current depth before moving to the next level.
   - Depth-First Search (DFS): Explores as far down a branch as possible before backtracking.

2. **Informed Search**

   (a) Uses heuristic functions to guide the search, improving efficiency.
   (b) Examples:

   - Greedy Best-First Search: Select the node that appears closest to the goal based on a heuristic.
   - A* Search: Combines cost-so-far (g) and estimated cost-to-go (h) to find the optimal path.

3. **Applications in AI**

   (a) Route planning (e.g., Google Maps).
   (b) Game playing (e.g., chess, Go).
   (c) Problem-solving in robotics (e.g., pathfinding for autonomous vehicles).

7.5.4 Optimization in AI

Optimization focuses on finding the best solution from a set of possible solutions by minimizing or maximizing an objective function. Optimization has applications in many commercial and scientific fields, such as scheduling tasks in manufacturing, designing efficient networks, and optimizing engineering processes.

The following is a list of key optimization techniques:

1. **Gradient Descent**

   (a) Iteratively adjusts parameters to minimize a loss function.
   (b) Widely used in machine learning for training models.

2. **Genetic Algorithms**
   Inspired by natural selection, these algorithms evolve solutions over generations using crossover and mutation.

3. **Simulated Annealing**
   Mimics the cooling process of metals to find a global optimum by allowing occasional worse solutions to escape local minima.

4. **Swarm Intelligence**
   Techniques like particle swarm optimization (PSO) and ant colony optimization (ACO) use collective behavior for optimization.

7.5.5 *Knowledge Representation and Reasoning*

To solve problems effectively, AI systems must represent and reason about knowledge in a structured manner. Knowledge representation provides a way to model the world so that AI systems can interpret and manipulate it. This includes facts, concepts, and relationships between entities. The following is a list of common knowledge representation methods:

1. **Logic-Based Representation**

   (a) Uses formal logic (propositional or predicate logic) to define rules and relationships.
   (b) Example: "If it rains, the ground will be wet."
   (c) Expressed as: Rain $\rightarrow$ WetGround.

2. **Semantic Networks**

   (a) Graph structures represent entities as nodes and relationships as edges.
   (b) Example: Representing "A dog is an animal" with a "is-a" relationship.

3. **Frames**

   (a) Structured templates for representing stereotypical situations.
   (b) Example: Representing the attributes of a "car" as a frame with slots for make, model, color, etc.

4. **Ontologies**

   (a) Hierarchical structures that define the concepts and relationships in a domain.
   (b) Widely used in natural language processing and semantic web applications.

7.5.6 Reasoning

Reasoning is the process of deriving conclusions from known facts or rules. It enables AI systems to make informed decisions and solve problems. The following is a list of types of reasoning:

1. **Deductive Reasoning.**

   (a) Derives specific conclusions from general principles.
   (b) Example: "All humans are mortal. Socrates is human. Therefore, Socrates is mortal."

2. **Inductive Reasoning.**

   (a) Generalizes rules based on observations or patterns.
   (b) Example: Observing that the sun rises every day and concluding it will rise tomorrow.

3. **Probabilistic Reasoning.**

   (a) Deals with uncertainty using probability theory.
   (b) Example: Bayesian networks model dependencies between variables to make predictions.

7.5.7 Applications of Search, Optimization, and Knowledge Representation

Search, optimization, and knowledge representation are foundational pillars of AI, enabling systems to tackle complex problem spaces, discover optimal or near-optimal solutions, and encode domain-specific information for advanced reasoning. By systematically combining computational heuristics with structured data models, these techniques drive a wide range of applications, from designing intricate engineering systems to guiding intelligent control and troubleshooting processes. Their synergy underlies the development of robust, adaptive, and efficient AI solutions that can navigate real-world challenges across multiple industries.

1. **Engineering Design**

   AI systems can explore vast design spaces to optimize structures or circuits, assessing numerous configurations more rapidly and accurately than human designers. Generative designs are applied in mechanical engineering, where algorithms iteratively refine models to meet criteria, such as weight reduction, load-bearing capacity, or material constraints.

2. **Control Systems**

   Search and reasoning methods are pivotal in robotics for navigation and manipulation, determining feasible paths and actions under dynamic conditions.

For example, control systems are employed for autonomous drone path planning, which rely on algorithms to avoid obstacles and efficiently reach their target destinations.

3. **Diagnostics and Troubleshooting**

   AI systems leverage knowledge representation and reasoning to identify faults in industrial processes or equipment, drawing on extensive rule sets or probabilistic models. Expert systems are the tools in aviation maintenance, where structured diagnostic knowledge helps technicians pinpoint and resolve issues in aircraft systems swiftly and safely.

7.6   Core Components of AI Systems

The core components of artificial intelligence (AI) systems revolve around how they represent knowledge, draw conclusions, and learn from data. In other words, the core components of AI, which are knowledge representation, reasoning and inference, and learning paradigms, form the basis of how intelligent systems process, understand, and adapt to data. Logical frameworks enable AI to represent complex relationships, inference techniques empower decision-making, and learning paradigms allow systems to improve continuously. Together, these elements equip AI with the tools necessary to address challenges across diverse domains, from healthcare to robotics and beyond. This section explores knowledge representation, reasoning and inference, and learning paradigms, which together form the foundation of classical AI.

7.6.1   *Logical Knowledge Representation*

Knowledge representation is how an AI system organizes and stores information about the world to make it accessible for reasoning and decision-making. Effective representation is essential for enabling AI systems to solve problems, understand relationships, and adapt to new situations.

Logical representation uses formal logic to encode knowledge. This approach enables reasoning by applying rules to derive new information.

1. **Propositional Logic**

   Represents statements as propositions, which can either be true or false. For example:

   - P : It is raining.
   - Q : The ground is wet.
   - Rule: $P \rightarrow Q$ (If it is raining, then the ground is wet).
   - Propositional logic is limited in expressing complex relationships because it cannot capture variables or quantifiers.

2. **Predicate Logic (First-Order Logic).**

   Extends propositional logic by introducing quantifiers ($\forall$, $\exists$) and predicates, allowing for richer representations. Example:

   - $Rain(x) \rightarrow WetGround(x)$
   - "If it rains in location x, the ground in x will be wet."

7.6.1.1   Semantic Networks and Frames

Semantic networks and frames are two classic approaches to knowledge representation that highlight the importance of structuring information in meaningful, efficient ways. By capturing the relationships between concepts and organizing attributes into logical templates, these methods enable AI systems to store, retrieve, and manipulate knowledge more effectively. They serve as the backbone for many advanced reasoning systems, serving as instruments for modeling complex domains and handling real-world data with clarity.

1. **Semantic Networks**

   Semantic networks represent knowledge as a graph where nodes stand for concepts and edges define the relationships between them. This visual representation makes it easier to understand how different ideas interconnect and also facilitates inference through graph traversal algorithms.

   Nodes and links naturally capture hierarchical and associative relationships, making it straightforward to represent ideas, such as inheritance ("mammal" to "dog") or properties ("dog" to "barks"). For example, a semantic network for animals could link "dog" to "mammal" (type of) and "barks" to "dog" (action performed by), illustrating how multiple concepts can be woven into a cohesive framework.

2. **Frames**

   Frames encapsulate knowledge in a structured template, similar to objects in programming. Each frame holds "slots" for default or variable attributes, which can be updated or inherited by more specialized frames, enabling efficient reuse and extensibility. Frames allow default values and hierarchical organization, ensuring that common attributes can be inherited across multiple instances without unnecessary repetition. For example, a "Car" frame may have slots for "Color," "Model," and "Owner," which can be filled with specific information. Derived frames, like "Electric Car," could inherit these slots and introduce additional ones, such as "Battery Capacity." This is akin to "Objects" in object-oriented computer languages.

7.6.2   Ontologies in AI

Ontologies formalize knowledge by defining concepts, relationships, and constraints in a specific domain. Ontologies are widely used in natural language processing (NLP) and semantic web technologies to enable interoperability and reasoning. For example:

In healthcare, an ontology might define relationships between "Disease," "Symptoms," and "Treatment."

7.6.3   Reasoning and Inference

Reasoning enables AI systems to derive conclusions based on known information. It allows machines to mimic human cognitive processes, such as solving puzzles, making decisions, or diagnosing issues.

1. **Deductive Reasoning**

   Deductive reasoning applies general rules to specific cases to reach logical conclusions. Deductive reasoning is precise but requires complete and accurate rules, limiting its applicability in uncertain or complex domains. For example:

   - Rule: All humans are mortal.
   - Fact: Socrates is a human.
   - Conclusion: Socrates is mortal.

2. **Inductive Reasoning**

   Inductive reasoning infers general rules from specific observations, making it essential for learning from data. While inductive reasoning is useful for pattern recognition, it is inherently uncertain and prone to errors when extrapolated beyond observed data. For example:

   (a) Observation: The sun rises in the east every day.
   (b) Inference: The sun always rises in the east.

7.6.4   Probabilistic Reasoning (Bayesian Networks)

Probabilistic reasoning deals with uncertainty by representing knowledge as probabilities. Graphical models represent variables and their conditional dependencies using a directed acyclic graph. For example, a Bayesian network for disease diagnosis might link "Fever," "Cough," and "Flu" with probabilities reflecting how likely one leads to another:

- $P(\text{Flu} \mid \text{Fever}, \text{Cough}) = 0.8$ (If a patient has a fever and a cough, there is an 80% chance they have the flu).

Probabilistic reasoning excels in real-world applications where data is incomplete or noisy, such as medical diagnosis and predictive modeling.

7.7 Supervised Learning

Supervised learning trains a model using labeled data, where each input has a corresponding output. For example:

1. Spam detection in emails.
2. Face recognition systems.
3. Predicting house prices based on features like size, location, and age.

There are many techniques employed in supervised learning. Those that are not based on deep learning are:

1. Linear regression for continuous outputs.
2. Classification algorithms (e.g., decision trees, SVMs) for categorical outputs.

7.8 Unsupervised Learning

Unsupervised learning discovers patterns or structures in unlabeled data. This is particularly important since most available data are not preprocessed, requiring machine learning techniques to extract meaningful insights. Typical applications include grouping customers into clusters based on purchasing behavior, market segmentation, and anomaly detection.

Two primary categories of unsupervised learning techniques are clustering (e.g., K-means, affinity propagation, mean shift, and Gaussian mixture models) and dimensionality reduction (e.g., PCA). Additionally, neural-based methods such as autoencoders are also employed in unsupervised settings. Their association with other unsupervised algorithms is illustrated in Fig. 7.2.

7.9 Reinforcement Learning

Reinforcement learning trains an agent to take actions in an environment to maximize cumulative rewards. This is important, for example, when training a robot to navigate a maze by rewarding it for reaching the exit. Other applications include autonomous driving (e.g., vehicles and drones) and game playing (e.g., AlphaGo).

In reinforcement learning, the core concepts are given in Table 7.5.

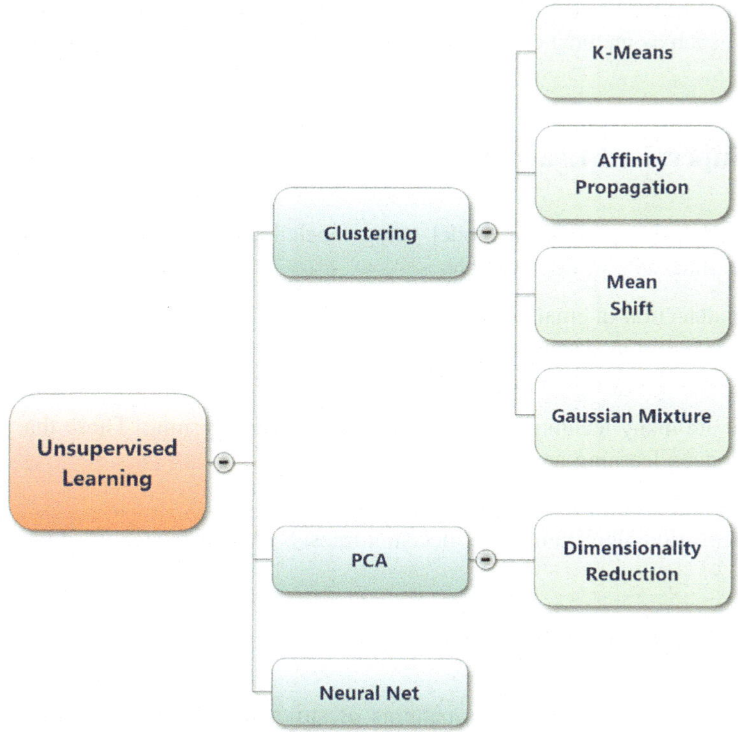

Fig. 7.2 Algorithms employed for unsupervised machine learning. (By the authors)

Table 7.5 Reinforcement learning key concepts

Core concept	Utility
State	The current situation (e.g., robot's position)
Action	Possible moves (e.g., left, right)
Reward	Feedback for an action (e.g., +10 for finding the exit)

7.10   Transfer Learning

Transfer learning involves leveraging knowledge gained from one task to improve performance on a related task. Applications include healthcare imaging and natural language processing (NLP). For example, a model trained on a large image dataset can be used to classify medical images with minimal additional training.

7.11   Search and Optimization in AI

Search strategies provide a framework for systematically exploring solution spaces, while optimization techniques refine solutions to achieve desired objectives. Together, these methods empower AI systems to solve problems across diverse fields, from robotics to logistics and strategic games. Their effectiveness depends on choosing the right technique for the problem, balancing computational efficiency and solution quality. While search strategies focus on exploring possible solutions systematically or heuristically, optimization techniques refine solutions to achieve the best outcomes. This section delves into key strategies and methods that underpin intelligent problem-solving in AI.

7.11.1   Search Strategies

Search strategies explore a solution space to find the best or most feasible solution to a problem. They can be categorized as uninformed (blind) or informed (heuristic).

7.11.1.1   Breadth-First Search (BFS)

BFS is an uninformed search strategy that explores all nodes at a given depth before moving to the next level. As a significant advantage, BFS guarantees the shortest path in an unweighted graph, and its systematic exploration ensures completeness. However, BFS is very memory-intensive, as it stores all nodes at a given level. Some of the applications of BFS include pathfinding in robotics and computer network analysis.

The BFS algorithm steps are:

1. Start with the root node.
2. Enqueue the root and explore all its neighbors.
3. Dequeue the explored node and repeat for the next level.

7.11.1.2   Depth-First Search (DFS)

Depth-first search (DFS) is another uninformed search strategy that explores as far as possible along a branch before backtracking. It has some advantages over BFS, such as better use of memory, as it uses a stack instead of storing all nodes. However, DFS may get stuck in infinite loops without proper termination checks, and it does not guarantee the shortest path. Some of the DFS applications include solving puzzles (e.g., mazes) and game state exploration.

The DFS algorithm steps are:

1. Start with the root node.
2. Recursively explore each branch until a goal is found or a dead end is reached.
3. Backtrack to explore unexplored branches.

7.11.1.3   Heuristic Search

Many problems in the real world are not viable to be solved by classical computers. This way, heuristic search strategies use domain-specific knowledge to guide the search process, improving efficiency. Most heuristic search algorithms consider that the shortest path of $h(x)$ is admissible (does not overestimate). Heuristic search is used in navigation systems, route selection, and scheduling.

Heuristic searches combine a path cost ($g(x)$) and a heuristic estimate ($h(x)$) to prioritize exploration:

$$f(x) = g(x) + h(x)$$

7.11.1.4   Minimax Algorithm with Alpha-Beta Pruning

The minimax algorithm is a game theory-based strategy for decision-making in two-player games.

1. **Minimax**

   (a) Assumes both players play optimally.
   (b) Alternates between maximizing the current player's score and minimizing that of the opponent's.

2. **Alpha-Beta Pruning**
   Optimizes minimax by eliminating branches that cannot affect the outcome, reducing computational overhead.

3. **Applications**

   (a) Chess, Go, and other strategic games.
   (b) AI agents in adversarial environments.

7.11.2   *Optimization Techniques*

Optimization is the process of iteratively improving a solution to achieve the best possible outcome under given constraints.

7.11.2.1   Genetic Algorithms (GAs)

Genetic algorithms are inspired by the principles of natural selection and evolution. These algorithms have many advantages, such as being effective for complex, nonlinear problems and not requiring gradient information. However, they are computationally expensive, and performance depends on parameter tuning. Genetic algorithms have many applications in engineering design optimization and feature selection in machine learning.

Genetic algorithms follow the key steps presented in Table 7.6.

7.11.2.2   Simulated Annealing

Simulated annealing (SA) is inspired by the physical process of annealing, where materials are heated and slowly cooled to reach a low-energy state. It has many advantages, including effective for escaping local optima and requires minimal domain knowledge. However, simulated annealing has slow convergence and the overall performance depends on the "cooling schedule" adopted.

The SA algorithm steps are:

(a) Start with an initial solution.
(b) Iteratively modify the solution.
(c) Accept worse solutions with a probability P that decreases over time:

$$P = e^{-\Delta E/T}$$

where ΔE is the change in energy (objective function), and T is the temperature.

7.11.2.3   Particle Swarm Optimization (PSO)

Particle swarm optimization (PSO) algorithms are inspired by the collective behavior of swarms, such as birds, ant, or fish. Their main advantage is that they are simple to implement and are effective for continuous optimization problems. However, PSO algorithms are prone to premature convergence and are sensitive to parameter

Table 7.6 Genetic algorithms steps

Genetic algorithms steps	Procedure
Initialization	Generate a population of candidate solutions
Selection	Choose the fittest individuals based on a fitness function
Crossover	Combine pairs of solutions to create offspring
Mutation	Introduce small random changes to maintain diversity
Iteration	Repeat until convergence or a stopping criterion is met

choices. PSO is frequently employed in neural network training and energy optimization.

1. **PSO Mechanism**

    (a) A population of particles (candidate solutions) moves through the solution space. Each particle adjusts its position based on:

       • Its own best-known position.
       • The best-known position of the entire swarm.

2. **PSO Update Rules**

    (a) Velocity update, the velocity of particle i is updated as:

$$v_i(t+1) = wv_i(t) + c_1r_1(p_i - x_i(t)) + c_2r_2(g - x_i(t))$$

    (b) Position update, the new position of particle i is given by:

$$x_i(t+1) = x_i(t) + v_i(t+1)\Delta t$$

where
$v_i(t)$: velocity of particle i at iteration t
$x_i(t)$: current position of particle i.
p_i: best-known position of particle i
g: best-known global position among all particles
w: inertia weight controlling momentum
c_1, c_2: acceleration coefficients for cognitive and social components
r_1, r_2: random numbers drawn from a uniform distribution in $[0, 1]$
Δt: time step (typically set to 1 in standard PSO)

7.11.3 Applications across Domains

Search and optimization methods are central to solving real-world problems in AI. Some examples of their applications are contained in Table 7.7.

Table 7.7 Search and optimization methods

Application field	Objective
Logistics	Routing delivery vehicles using heuristic search
Healthcare	Optimizing radiation therapy plans with genetic algorithms
Robotics	Path planning with A* and PSO
Gaming	Implementing minimax for AI opponents

7.12   Classical Machine Learning Basics

This section introduced the mathematical foundations of machine learning. Linear algebra provides the tools for manipulating data, probability, and statistics to enable reasoning under uncertainty, and gradient-based optimization techniques empower efficient model training. These concepts are the backbone of classical machine learning and essential for understanding quantum machine learning's enhancements. This section introduces these fundamental concepts with examples, equations, and Python implementations to help engineers understand how ML models work under the hood.

7.13   Review of Linear Algebra for AI

Linear algebra provides the language and tools to manipulate data in machine learning, particularly for operations on vectors, matrices, and tensors.

7.13.1   Vectors and Matrices

In ML, data points are often represented as vectors, and collections of data points form matrices.

- **Vector Representation**: A vector is a one-dimensional array of numbers:

$$\mathbf{v} = \begin{bmatrix} v_1 \\ v_2 \\ v_n \end{bmatrix}$$

- **Matrix Representation**: A matrix is a two-dimensional array of numbers:

$$\mathbf{A} = \begin{bmatrix} a_{11} & a_{12} & \cdots & a_{1n} \\ a_{21} & a_{22} & \cdots & a_{2n} \\ & & \ddots & \\ a_{m1} & a_{m2} & \cdots & a_{mn} \end{bmatrix}$$

Python Code Example: Matrix Operations in Classical AI

```
#------------------------------------------------------------------
# Matrix Operations in Classical AI
# Chapter 7 in the QUANTUM COMPUTING AND QUANTUM MACHINE LEARNING BOOK
#------------------------------------------------------------------
# Version 1.0
# (c) 2025 Jesse Van Griensven, Roydon Fraser, and Jose Rosas
```

```
# License: MIT - Citation of this work required
#----------------------------------------------------------------

import numpy as np

# Define vectors and matrices
vector = np.array([1, 2, 3])
matrix = np.array([[1, 2, 3], [4, 5, 6], [7, 8, 9]])

# Operations
dot_product = np.dot(vector, vector)  # Dot product of a vector
matrix_mult = np.dot(matrix, vector)  # Matrix-vector multiplication

# Print Results
print("Input Vector:", vector)
print("Input Matrix:", matrix)
print("dot product:", dot_product)
print("Matrix multiplications:", matrix_mult)
```

7.14　Review of Probability and Statistics

Probability and statistics are essential for understanding uncertainty in data and model predictions.

1. **Probability Distribution**

   A distribution assigns probabilities to outcomes. Common distributions in ML include the Gaussian distribution, shown below:

$$P(x) = \frac{1}{\sqrt{2\pi\sigma^2}} \exp\left(-\frac{(x-\mu)^2}{2\sigma^2}\right)$$

   where μ is the mean, and σ^2 is the variance.

2. **Bayes' Theorem.**

   A cornerstone of probabilistic reasoning:

$$P(A\,|\,B) = \frac{P(B\,|\,A)P(A)}{P(B)}$$

   An example of Bayes' theorem problem is the spam detection model:

- P(Spam | Contains "Offer"): Probability an email is spam given it contains "Offer."
- P(Contains "Offer" | Spam): Probability "Offer" appears in spam emails.

Python Code Example: Probability Distributions

```python
#---------------------------------------------------------------------
# Probability Distributions
# Chapter 7 in the QUANTUM COMPUTING AND QUANTUM MACHINE LEARNING BOOK
#---------------------------------------------------------------------
# Version 1.0
# (c) 2025 Jesse Van Griensven, Roydon Fraser, and Jose Rosas
# License: MIT - Citation of this work required
#---------------------------------------------------------------------
import numpy as np
import matplotlib.pyplot as plt
from scipy.stats import norm
#---------------------------------------------------------------------

# Gaussian Distribution Parameters
mean, std_dev = 0, 1

# Generate x values from -3 to 3
x = np.linspace(-3, 3, 100)

# Calculate the PDF of the Gaussian distribution
pdf = norm.pdf(x, mean, std_dev)

# Create the plot
plt.figure(figsize=(8, 6))
plt.plot(x, pdf, color='blue', lw=2, label=f'Normal Distribution\n$
\mu={mean}$, $\sigma={std_dev}$')

# Add title and labels
plt.title('Gaussian (Normal) Distribution', fontsize=16)
plt.xlabel('x', fontsize=14)
plt.ylabel('Probability Density', fontsize=14)

# Add grid
plt.grid(alpha=0.2)

# Add legend
plt.legend(loc='upper right', fontsize=12)

# Show the plot
plt.show()
```

7.14.1 Statistical Measures

Statistical measures provide insights into the central tendencies, spread, and relationships within datasets. By quantifying these attributes, engineers and software developers can make informed decisions, identify patterns, and uncover meaningful correlations across a wide range of applications, from simple descriptive assessments to sophisticated machine learning algorithms. Three fundamental statistical measures are the mean (μ), the variance (σ^2), and the covariance ($Cov(X, Y)$). These measures are important, as they capture the most essential aspects of data distribution and interdependence.

1. **Mean (μ)**

    The mean offers a simple yet powerful representation of the central point of a dataset:

$$\mu = \frac{1}{n} \sum_{i=1}^{n} x_i$$

    By averaging all observations, the mean (μ) gives an immediate sense of where most data points lie. It is extensively used in numerous statistical methods, such as regression and hypothesis testing, due to its straightforward interpretation.

2. **Variance (σ^2)**

    Variance measures the average squared deviation of each data point from the mean (μ):

$$\sigma^2 = \frac{1}{n} \sum_{i=1}^{n} (x_i - \mu)^2$$

    A high variance indicates that data points are spread out over a broad range of values, while a low variance suggests that observations cluster close to the mean (μ). The variance (σ^2) is essential when evaluating the reliability of predictions and assessing risks in fields like finance, engineering, and scientific research.

3. **Covariance ($Cov(X, Y)$).**

    Covariance captures the degree to which two variables X and Y change together:

$$Cov(X, Y) = \frac{1}{n} \sum_{i=1}^{n} \left(x_i - \bar{x}\right)\left(y_i - \bar{y}\right)$$

    A positive covariance suggests that when X increases, Y tends to increase as well, whereas a negative covariance indicates an inverse relationship. In data analysis, covariance helps identify potential dependencies or correlations between variables, setting the stage for more advanced techniques, such as correlation coefficients and linear regression.

7.15   Gradient-Based Optimization Techniques

Optimization is the process of finding model parameters that minimize a loss function. Gradient-based methods are fundamental to training ML models.

7.15.1   Gradient Descent

Gradient descent is extensively used in machine learning and statistical modeling to find the parameter set θ that minimizes a given loss function $L(\theta)$. The method proceeds iteratively, moving θ in the direction opposite to the gradient of the loss function. This ensures that each step leads, on average, to a reduction in the value of $L(\theta)$, driving the parameters closer to an optimal solution.

Formally, the update rule at each iteration is:

$$\theta \leftarrow \theta - \eta \nabla L(\theta)$$

where $\nabla L(\theta)$ is the gradient (a vector of partial derivatives indicating how $L(\theta)$ changes with respect to each parameter), and η is the learning rate, controlling the size of each update step.

Choosing η too large, the algorithm may overshoot the minimum; choosing it too small, the algorithm may take excessively long to converge.

Python Code Example: Gradient Descent

```python
#---------------------------------------------------------------------
# Gradient Descent
# Chapter 7 in the QUANTUM COMPUTING AND QUANTUM MACHINE LEARNING BOOK
#---------------------------------------------------------------------
# Version 1.1
# (c) 2025 Jesse Van Griensven, Roydon Fraser, and Jose Rosas
# License: MIT - Citation of this work required
#---------------------------------------------------------------------
import numpy as np
import matplotlib.pyplot as plt
#---------------------------------------------------------------------

# Loss function: L(theta) = theta^2
def loss(theta):
    return theta ** 2

# Gradient of the loss function
def gradient(theta):
    return 2 * theta
```

```python
# Gradient Descent Parameters
initial_theta = 10   # Initial parameter
learning_rate = 0.1   # Learning rate
num_iterations = 100   # Number of iterations

# Lists to store the history of theta and loss
theta_history = [initial_theta]
loss_history = [loss(initial_theta)]

# Gradient Descent Loop
theta = initial_theta
for i in range(1, num_iterations + 1):
  grad = gradient(theta)
  theta = theta - learning_rate * grad
  theta_history.append(theta)
  loss_history.append(loss(theta))

# Iteration numbers
iterations = np.arange(0, num_iterations + 1)

# Plot 1: Theta vs. Iteration
plt.figure(figsize=(12, 4))

plt.subplot(1, 3, 1)
plt.plot(iterations, theta_history, marker='o', color='blue')
plt.title('Theta vs. Iteration')
plt.xlabel('Iteration')
plt.ylabel(r'$\theta$')
plt.grid(True)

# Plot 2: Loss vs. Iteration
plt.subplot(1, 3, 2)
plt.plot(iterations, loss_history, marker='o', color='red')
plt.title('Loss vs. Iteration')
plt.xlabel('Iteration')
plt.ylabel(r'$L(\theta)$')
plt.grid(True)

# Plot 3: Loss Function with Gradient Descent Path
theta_values = np.linspace(-12, 12, 400)
loss_values = loss(theta_values)

plt.subplot(1, 3, 3)
plt.plot(theta_values, loss_values, color='green', label=r'$L
(\theta) = \theta^2$')
```

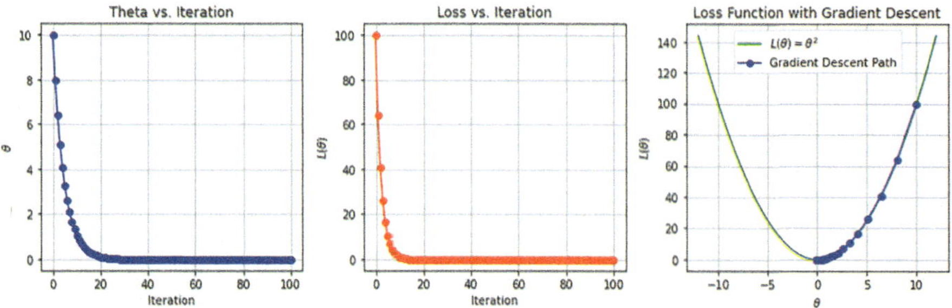

Fig. 7.3 Demonstration of the gradient descent code. (By the authors)

```python
plt.plot(theta_history, loss_history, marker='o', color='blue',
linestyle='-', label='Gradient Descent Path')
plt.title('Loss Function with Gradient Descent')
plt.xlabel(r'$\theta$')
plt.ylabel(r'$L(\theta)$')
plt.legend()
plt.grid(True)

plt.tight_layout()
plt.show()
```

Figure 7.3 presents the results of the code above.

7.15.2 Stochastic Gradient Descent (SGD)

Stochastic gradient descent (SGD) is a more efficient variant of the gradient descent algorithm and, for this reason, is extensively used in deep neural network training. SGD updates model parameters based on a single data point (or a small batch) at a time. This approach significantly reduces computational overhead compared to standard (batch) gradient descent, which requires calculating gradients across the entire dataset for every update. By processing fewer examples per iteration, SGD enables quicker updates and is particularly well-suited for large-scale problems where loading or computing on the full dataset is impractical. However, the use of fewer data in each gradient calculation introduces higher variance in parameter updates, which can lead to a noisier convergence process and the need for tuning additional hyperparameters, such as "momentum" and "adaptive learning rates."

Python Code Example: SGD

```python
#-----------------------------------------------------------------
# SGD for linear regression
# Chapter 7 in the QUANTUM COMPUTING AND QUANTUM MACHINE LEARNING BOOK
#-----------------------------------------------------------------
```

```python
# Version 1.1
# (c) 2025 Jesse Van Griensven, Roydon Fraser, and Jose Rosas
# License: MIT - Citation of this work required
#----------------------------------------------------------------
# Importing necessary libraries
import matplotlib.pyplot as plt
import numpy as np
from sklearn.linear_model import SGDRegressor
#----------------------------------------------------------------

# Data for regression
X = np.array([[1], [2], [3]])
y = np.array([2, 4, 6])

# Train the model
model = SGDRegressor(max_iter=1000, tol=1e-3)
model.fit(X, y)

# Predictions
X_range = np.linspace(0, 4, 100).reshape(-1, 1)
y_pred = model.predict(X_range)

# Plotting the results
plt.figure(figsize=(8, 6))
plt.scatter(X, y, color="blue", label="Training Data")
plt.plot(X_range, y_pred, color="red", label="SGD Regression Line")
plt.title("Stochastic Gradient Descent for Linear Regression")
plt.xlabel("Feature (X)")
plt.ylabel("Target (y)")
plt.grid(alpha=0.5)
plt.legend()
plt.show()
```

7.15.3 Advanced Optimization Methods

Advanced optimization techniques address issues such as slow convergence and
sensitivity to hyperparameters, which are common in training deep neural networks.
By incorporating historical gradient information and adapting learning rates on the
fly, these methods can significantly improve training speed and stability, especially
for complex, high-dimensional models. This approach can reduce oscillations and
lead to faster convergence compared to vanilla gradient descent.

1. **Momentum**

  Momentum accelerates gradient descent by accumulating an exponentially decaying average of past gradients, effectively smoothing updates:

$$v_t = \beta v_{t-1} + \eta \nabla L(\theta_t), \quad \theta_{t+1} = \theta_t - v_t$$

  where,

- v_t is the "velocity" that helps maintain a consistent update direction.
- β is a damping factor (often between 0.9 and 0.99), and η is the learning rate.

2. **Adam Optimizer**

  Adam ("adaptive moment estimation") combines the advantages of momentum and RMSProp by computing adaptive learning rates for each parameter, using first- and second-moment estimates of the gradients. This adaptability makes Adam particularly effective for sparse datasets or problems with noisy and nonstationary objectives. The following equation describes the parameter update rule for θ:

$$\theta_t = \theta_{t-1} - \eta \frac{\widehat{m}_t}{\sqrt{\widehat{v}_t} + \epsilon}$$

  where,

  θ_t is the parameter vector (or scalar) being optimized at iteration t.

  $\widehat{m}_t = \frac{m_t}{1 - \beta_1^t}$ is the bias-corrected first moment estimate (mean of gradients).

  $\widehat{v}_t = \frac{v_t}{1 - \beta_2^t}$ is the bias-corrected second-moment estimate (uncentered variance of gradients).

  β_1, β_2 are the exponential decay rates for the moment estimates (typically set to 0.9 and 0.999, respectively).

  ϵ is a small constant added to improve numerical stability.

  η is the learning rate.

  t is the iteration index or time step.

Python Code Example: Adam Optimizer

```python
# Importing necessary libraries
import matplotlib.pyplot as plt
import numpy as np
from tensorflow.keras.optimizers import Adam


# Example: Visualizing the impact of Adam optimizer on a quadratic loss
function
# Define a quadratic loss function
def loss_function(x):
  return (x - 3) ** 2
```

```python
# Simulate optimization steps
x_values = np.linspace(-1, 7, 100)  # Input values for loss function
learning_rate = 0.1
steps = [5, 10, 20, 50, 100]  # Simulated steps

# Generate plots for different optimization steps
plt.figure(figsize=(10, 6))
for step in steps:
    updated_values = [3 + (x - 3) * (1 - 0.1 / step) for x in x_values]
    plt.plot(x_values, loss_function(x_values), label=f"Step {step}")

plt.title("Impact of Adam Optimizer on Loss Function")
plt.xlabel("X values")
plt.ylabel("Loss function")
plt.legend()
plt.show()
```

7.16　Key Algorithms in Machine Learning

Machine learning (ML) relies on diverse algorithms to solve problems, such as prediction, classification, clustering, and optimization. By mastering the mathematical foundations of these algorithms and utilizing code, such as Python-based implementations, engineers and software developers can develop resilient and efficient machine-learning solutions. These methods lie at the core of numerous ML applications, equipping engineers with powerful tools for thorough data analysis and interpretation. This section explores key algorithms widely used in classical ML, including the ones listed below (as shown in Fig. 7.4):

1. K-nearest neighbors (KNN)
2. Decision trees and random forests *(TREE, Random Forest)*
3. Gradient boosting *(GBR, XGBoost)*
4. Support vector machines (SVMs) *(SVC, SVR)*
5. Dimensionality reduction techniques
6. Additional regressors and classifiers (e.g., *GBR*, *SVR*) associated with each algorithmic family as shown in the figure

7.16.1　Regression Models

Regression models are fundamental in supervised learning and are used to predict continuous or discrete outcomes based on input variables.

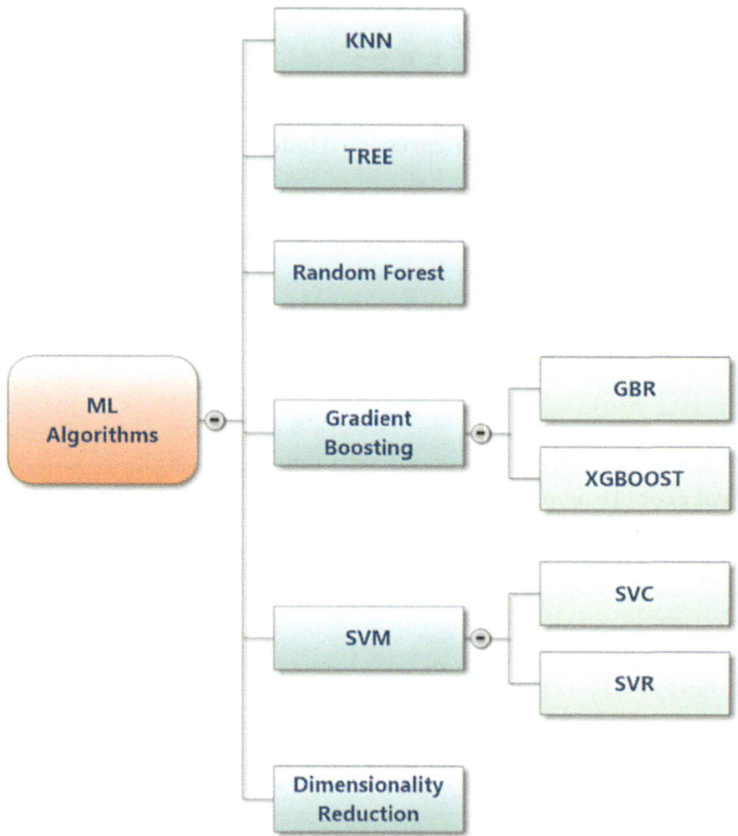

Fig. 7.4 Classical machine learning algorithms, non-deep learning. (By the authors)

7.16.1.1 Linear Regression

Linear regression models the relationship between a dependent variable (y) and one or more independent variables (x_1, x_2, ..., x_n) using a linear equation:

$$y = \beta_0 + \beta_1 x_1 + \beta_2 x_2 + \cdots + \beta_n x_n + \epsilon$$

where β_0 is the y-intercept, β_i are the coefficients of independent variables, and ϵ is the error term.

The goal in linear regression is: to minimize the mean squared error (MSE):

$$MSE = \frac{1}{n} \sum_{i=1}^{n} (y_i - \widehat{y}_i)^2$$

Python Code Example: Linear Regression Model

```python
#----------------------------------------------------------------
# Linear Regression Model
# Chapter 7 in the QUANTUM COMPUTING AND QUANTUM MACHINE LEARNING BOOK
#----------------------------------------------------------------
# Version 1.0
# (c) 2025 Jesse Van Griensven, Roydon Fraser, and Jose Rosas
# License: MIT - Citation of this work required
#----------------------------------------------------------------

# Importing necessary libraries
import matplotlib.pyplot as plt
import numpy as np
from sklearn.linear_model import LinearRegression
#----------------------------------------------------------------

# Example data
X = np.array([[1], [2], [3], [4], [5]])
y = np.array([2, 4, 5, 4, 5])

# Linear regression model
model = LinearRegression()
model.fit(X, y)
predictions = model.predict(X)

# Plotting the data and regression line
plt.figure(figsize=(8, 6))
plt.scatter(X, y, color="blue", label="Training Data")
plt.plot(X, predictions, color="red", label="Regression Line")
plt.title("Linear Regression Model")
plt.xlabel("Feature (X)")
plt.ylabel("Target (y)")
plt.grid(alpha=0.5)
plt.legend()
plt.show()
```

7.16.1.2   Logistic Regression

Logistic regression is: used for binary classification problems. It predicts the probability that an input belongs to a specific class using the sigmoid function:

$$P(y = 1 \mid x) = \frac{1}{1 + e^{-(\beta_0 + \beta_1 x_1 + \cdots + \beta_n x_n)}}$$

The decision boundary is determined by the threshold (commonly $P = 0.5$).

Python Code Example: Logistic Regression

```python
#-----------------------------------------------------------------
# Logistic Regression Model
# Chapter 7 in the QUANTUM COMPUTING AND QUANTUM MACHINE LEARNING BOOK
#-----------------------------------------------------------------
# Version 1.0
# (c) 2025 Jesse Van Griensven, Roydon Fraser, and Jose Rosas
# License: MIT - Citation of this work required
#-----------------------------------------------------------------
# Importing necessary libraries
import matplotlib.pyplot as plt
import numpy as np
from sklearn.linear_model import LogisticRegression
#-----------------------------------------------------------------

# Example data
X = np.array([[1], [2], [3], [4], [5]])
y = np.array([0, 0, 1, 1, 1])

# Logistic regression model
model = LogisticRegression()
model.fit(X, y)

# Generate probabilities for the logistic function
X_range = np.linspace(0, 6, 100).reshape(-1, 1)
probabilities = model.predict_proba(X_range)[:, 1]  # Probability of
class 1

# Plotting the data and logistic function
plt.figure(figsize=(8, 6))
plt.scatter(X, y, color="blue", label="Training Data")
plt.plot(X_range, probabilities, color="red", label="Logistic
Regression Curve")
plt.title("Logistic Regression Model")
plt.xlabel("Feature (X)")
```

```
plt.ylabel("Probability of Class 1")
plt.grid(alpha=0.2)
plt.legend()
plt.show()
```

7.16.2　Decision Trees and Random Forests

Decision trees and random forests stand as robust, classic machine learning techniques known for their resilience in managing unbalanced datasets and their capacity to offer interpretable decision processes. By clearly illustrating which features drive predictions, these models allow engineers and software developers to trace and understand the reasoning behind the output results.

7.16.2.1　Decision Trees

Decision trees work by partitioning the dataset into subsets based on feature values and then making decisions via a tree-like structure of nodes and branches. At each split, the criterion aims to minimize impurity. Smaller impurity values indicate more homogeneous subsets, improving classification accuracy. This hierarchical approach enables a straightforward visualization of the decision path from root to leaf, enhancing model interpretability. The definition for "Gini Impurity" and Entropy is presented below:

- **Gini Impurity**:

$$G = 1 - \sum_{i=1}^{k} P_i^2$$

- **Entropy**:

$$H = -\sum_{i=1}^{k} P_i \log_2(P_i)$$

where,
P_i is the probability of class i.

Python Code Example: Decision Tree Model

```
#---------------------------------------------------------------------------------
# Decision Tree Model
# Chapter 7 in the QUANTUM COMPUTING AND QUANTUM MACHINE LEARNING BOOK
#---------------------------------------------------------------------------------
```

```python
# Version 1.0
# (c) 2025 Jesse Van Griensven, Roydon Fraser, and Jose Rosas
# License:  MIT - Citation of this work required
#-------------------------------------------------------------------


# Importing necessary libraries
import matplotlib.pyplot as plt
import numpy as np
from sklearn.tree import DecisionTreeClassifier
from sklearn.tree import plot_tree


# Example data
X = np.array([[1], [2], [3], [4], [5]])
y = np.array([0, 0, 1, 1, 1])


# Decision tree model
model = DecisionTreeClassifier(criterion='gini')
model.fit(X, y)


# Predictions
predictions = model.predict(X)


# Plotting decision boundaries
x_range = np.linspace(0, 6, 100).reshape(-1, 1)
predicted_classes = model.predict(x_range)

plt.figure(figsize=(8, 6))
plt.scatter(X, y, color="blue", label="Training Data")
plt.plot(x_range, predicted_classes, color="red", label="Decision
Boundaries")
plt.title("Decision Tree Classifier")
plt.xlabel("Feature (X)")
plt.ylabel("Class (y)")
plt.grid(alpha=0.5)
plt.legend()
plt.show()


# Plotting the decision tree
plt.figure(figsize=(10, 8))
plot_tree(model, feature_names=["Feature X"], class_names=["Class
0", "Class 1"], filled=True)
plt.title("Decision Tree Visualization")
plt.show()
```

7.16.2.2　Random Forests

A random forest is an ensemble of decision trees that improves accuracy and reduces overfitting by averaging predictions or voting for classifications. Each tree is trained on a random subset of data and features. The diversity among individual trees allows the model to capture intricate patterns and minimize variance, resulting in robust, accurate predictions across a broad range of applications.

Python Code Example: Random Forest Model

```python
#------------------------------------------------------------------
# Random Forest Model
# Chapter 7 in the QUANTUM COMPUTING AND QUANTUM MACHINE LEARNING BOOK
#------------------------------------------------------------------
# Version 1.0
# (c) 2025 Jesse Van Griensven, Roydon Fraser, and Jose Rosas
# License: MIT - Citation of this work required
#------------------------------------------------------------------
# Importing necessary libraries
import matplotlib.pyplot as plt
import numpy as np
from sklearn.ensemble import RandomForestClassifier
from sklearn.tree import plot_tree
#------------------------------------------------------------------

# Example data
X = np.array([[1], [2], [3], [4], [5]])
y = np.array([0, 0, 1, 1, 1])

# Random forest model
model = RandomForestClassifier(n_estimators=100, random_state=42)
model.fit(X, y)

# Predictions
x_range = np.linspace(0, 6, 100).reshape(-1, 1)
predicted_classes = model.predict(x_range)

# Plotting decision boundaries
plt.figure(figsize=(8, 6))
plt.scatter(X, y, color="blue", label="Training Data")
plt.plot(x_range, predicted_classes, color="red", label="Decision
Boundaries")
plt.title("Random Forest Classifier")
plt.xlabel("Feature (X)")
plt.ylabel("Class (y)")
plt.grid(alpha=0.5)
```

```
plt.legend()
plt.show()

# Plotting a single decision tree from the Random Forest
plt.figure(figsize=(10, 8))
plot_tree(model.estimators_[0], feature_names=["Feature X"],
class_names=["Class 0", "Class 1"], filled=True)
plt.title("Random Forest - Single Tree Visualization")
plt.show()
```

7.16.3   Clustering Techniques

Clustering algorithms group similar data points into clusters without predefined labels, making them fundamental to unsupervised learning.

7.16.3.1   K-Means Clustering

K-means minimizes the within-cluster variance by iteratively assigning points to clusters and updating cluster centroids:

1. Initialize k cluster centroids.
2. Assign each point to the nearest centroid.
3. Update centroids based on the mean of points in each cluster.

The objective function:

$$J = \sum_{i=1}^{k} \sum_{x \in C_i} ||x - \mu_i||^2$$

Python Code Example: K-Means Clustering

```
#---------------------------------------------------------------------
# K-Means Clustering
# Chapter 7 in the QUANTUM COMPUTING AND QUANTUM MACHINE LEARNING BOOK
#---------------------------------------------------------------------
# Version 1.0
# (c) 2025 Jesse Van Griensven, Roydon Fraser, and Jose Rosas
# License: MIT - Citation of this work required
#---------------------------------------------------------------------
# Importing necessary libraries
import matplotlib.pyplot as plt
import numpy as np
```

```python
from sklearn.cluster import KMeans
#-----------------------------------------------------------------

# Example data
X = np.array([[1], [2], [3], [10], [11], [12]])

# K-means clustering
model = KMeans(n_clusters=2, random_state=42)
model.fit(X)
labels = model.labels_
centroids = model.cluster_centers_

# Plotting the clusters
plt.figure(figsize=(8, 6))
plt.scatter(X, np.zeros_like(X), c=labels, cmap="viridis",
edgecolors="k", s=100, label="Data Points")
plt.scatter(centroids, np.zeros_like(centroids), c="red",
marker="X", s=200, label="Centroids")
plt.title("K-Means Clustering")
plt.xlabel("Feature (X)")
plt.ylabel("Cluster Assignment")
plt.grid(alpha=0.2)
plt.legend()
plt.show()
```

7.16.3.2   Hierarchical Clustering

Hierarchical clustering builds a tree of clusters using either:

1. Agglomerative (Bottom-Up): Start with each point as its own cluster and merge the closest clusters iteratively.
2. Divisive (Top-Down): Start with one cluster and recursively split it.

   The similarity metric can be:

1. Euclidean distance.
2. Cosine similarity.

Python Code Example: Hierarchical Clustering

```python
#-----------------------------------------------------------------
# Hierarchical Clustering
# Chapter 7 in the QUANTUM COMPUTING AND QUANTUM MACHINE LEARNING BOOK
#-----------------------------------------------------------------
# Version 1.0
```

```
# (c) 2025 Jesse Van Griensven, Roydon Fraser, and Jose Rosas
# License: MIT - Citation of this work required
#-----------------------------------------------------------------
import numpy as np
import matplotlib.pyplot as plt
from scipy.cluster.hierarchy import dendrogram, linkage
#-----------------------------------------------------------------

# Example data
X = np.array([[1], [2], [3], [10], [11], [12]])

# Hierarchical clustering
Z = linkage(X, method='ward')

# Plot the dendrogram with improved labels and title
plt.figure(figsize=(8, 6))
dendrogram(Z)
plt.title("Hierarchical Clustering Dendrogram")
plt.xlabel("Sample Index")
plt.ylabel("Distance")
plt.grid(alpha=0.2)
plt.show()
```

7.17 Support Vector Machines (SVMs)

Support vector machines (SVMs) are powerful classifiers that find the hyperplane
separating data points of different classes with the maximum margin.

7.17.1 Linear SVM

For linearly separable data, the hyperplane satisfies:

$$w \cdot x + b = 0$$

The margin is maximized by minimizing:

$$L(w, b) = \frac{1}{2} ||w||^2$$

Although the loss function depends only on w, the variable b is part of the
optimization problem through the constraint:

$$y_i(w \cdot x_i + b) \geq 1$$

which jointly influences the optimal solution for both w and b,

where
w is the weight vector (defines the orientation of the hyperplane).
b is the bias term (defines the offset of the hyperplane).
x_i is the input vector for sample i.
$y_i \in \{-1, +1\}$ is the class label for sample i.

7.17.2   *Kernel SVM*

For nonlinear data, SVMs use kernels to map data to higher-dimensional spaces:

- Polynomial kernel:

$$K(x, x') = (x \cdot x' + 1)^d$$

- Radial basis function (RBF) kernel:

$$K(x, x') = \exp\left(-\gamma \, || x - x' ||^2\right)$$

Python Code Example: SVC

```
#--------------------------------------------------------------------
# Support Vector Classifier - SVC
# Chapter 7 in the QUANTUM COMPUTING AND QUANTUM MACHINE LEARNING BOOK
#--------------------------------------------------------------------
# Version 1.0
# (c) 2025 Jesse Van Griensven, Roydon Fraser, and Jose Rosas
# License: MIT - Citation of this work required
#--------------------------------------------------------------------
# Importing necessary libraries
import numpy as np
import matplotlib.pyplot as plt
from sklearn.svm import SVC
#--------------------------------------------------------------------

# Example data
X = np.array([[1, 1], [2, 2], [3, 3], [10, 10], [11, 11], [12, 12]])
y = np.array([0, 0, 0, 1, 1, 1])
```

```python
# SVM with RBF kernel
model = SVC(kernel='rbf')
model.fit(X, y)

# Create a mesh grid for plotting decision boundaries
x_min, x_max = X[:, 0].min() - 1, X[:, 0].max() + 1
y_min, y_max = X[:, 1].min() - 1, X[:, 1].max() + 1
xx, yy = np.meshgrid(np.linspace(x_min, x_max, 100), np.linspace
(y_min, y_max, 100))

# Predict the classification for each grid point
Z = model.predict(np.c_[xx.ravel(), yy.ravel()])
Z = Z.reshape(xx.shape)

# Plot decision boundary and data points
plt.figure(figsize=(8, 6))
plt.contourf(xx, yy, Z, alpha=0.3, cmap="coolwarm")
plt.scatter(X[:, 0], X[:, 1], c=y, cmap="coolwarm", edgecolors="k",
s=100, label="Training Data")
plt.title("Support Vector Classifier (SVC) with RBF Kernel")
plt.xlabel("Feature 1")
plt.ylabel("Feature 2")
plt.grid(alpha=0.5)
plt.legend()
plt.show()
```

7.18 Deep Learning and Neural Networks

Deep learning, a subfield of machine learning, is revolutionizing artificial intelligence by enabling machines to learn from vast amounts of data through neural networks. Neural networks mimic the structure and function of the human brain to process information, recognize patterns, and make decisions.

Deep neural network architectures such as MLPs, CNNs, RNNs, and autoencoders form the building blocks of advanced AI systems. Each architecture addresses specific challenges, enabling AI to process complex data types like images, sequences, and high-dimensional datasets. By mastering these techniques, engineers can design solutions for diverse applications, from medical diagnostics to natural language processing.

This section explores the fundamentals of neural networks, their key architectures, and applications, providing an in-depth understanding of their functionality and implementation.

7.18.1 *Overview of Classical Neural Networks*

Classical neural networks (NNs) have evolved significantly since their inception. The journey began with the development of the perceptron in the late 1950s by Frank Rosenblatt, which marked the first attempt to mimic biological neurons mathematically. Over the decades, advances in computing power and algorithm design have enabled the growth of modern deep learning models, including the ones presented in Table 7.8.

Classical neural networks have found widespread applications, as summarized in Table 7.9.

7.18.1.1 Perceptron

The perceptron is the simplest type of artificial neural network, introduced by Frank Rosenblatt in 1958. It consists of a single neuron with adjustable weights and a bias.

1. **The Perceptron Computes**

$$y = \text{step}(\mathbf{w} \cdot \mathbf{x} + b)$$

where,

- $\mathbf{w}$: weight vector
- $\mathbf{x}$: input vector
- b: bias term
- step(z): activation function that outputs 1 if $z > 0$, otherwise 0.

Table 7.8 Summary of classical neural networks

Neural network architecture	Description
Convolutional neural networks (CNNs)	For image recognition, object detection, and computer vision tasks
Recurrent neural networks (RNNs) and transformers	For natural language processing (NLP), speech recognition, and machine translation
Generative models	Variational autoencoders (VAEs) and generative adversarial networks (GANs) for generating synthetic data

Table 7.9 Summary of applications of neural networks

Neural net applications	Action
Image processing	Face detection, medical image analysis
Natural language understanding	Chatbots, language translation
Autonomous systems	Robotics, self-driving cars

2. **Limitations**:

- Can only solve linearly separable problems.
- Example: It cannot handle the XOR problem.

Python Code Example: Simple Perceptron

```python
#-------------------------------------------------------------------
# Simple Perceptron
# Chapter 7 in the QUANTUM COMPUTING AND QUANTUM MACHINE LEARNING BOOK
#-------------------------------------------------------------------
# Version 1.0
# (c) 2025 Jesse Van Griensven, Roydon Fraser, and Jose Rosas
# License: MIT - Citation of this work required
#-------------------------------------------------------------------


# Importing necessary libraries
import numpy as np
import matplotlib.pyplot as plt


# Step function
def step_function(z):
    return 1 if z > 0 else 0


# Perceptron function
def perceptron(x, w, b):
    z = np.dot(w, x) + b
    return step_function(z)


# Example data points
X = np.array([[1, -1], [-1, 1], [2, 2], [-2, -2], [3, 1], [-3, -1]])
y = np.array([1, 1, 1, 0, 1, 0])  # Labels


# Perceptron weights and bias
w = np.array([0.5, 0.5])
b = -0.2


# Generate a mesh grid for decision boundary
x_min, x_max = X[:, 0].min() - 1, X[:, 0].max() + 1
y_min, y_max = X[:, 1].min() - 1, X[:, 1].max() + 1
xx, yy = np.meshgrid(np.linspace(x_min, x_max, 100), np.linspace
(y_min, y_max, 100))


# Compute perceptron output over the grid
Z = np.array([perceptron(np.array([a, b]), w, b) for a, b in zip(xx.
```

```
ravel(), yy.ravel())])
Z = Z.reshape(xx.shape)

# Plot decision boundary and data points
plt.figure(figsize=(8, 6))
plt.contourf(xx, yy, Z, alpha=0.3, cmap="coolwarm")
plt.scatter(X[:, 0], X[:, 1], c=y, cmap="coolwarm", edgecolors="k",
s=100, label="Training Data")
plt.title("Perceptron Decision Boundary")
plt.xlabel("Feature 1")
plt.ylabel("Feature 2")
plt.grid(alpha=0.5)
plt.legend()
plt.show()
```

7.18.1.2   Multi-Layer Perceptrons (MLPs)

Multi-layer perceptrons (MLPs) extend the perceptron by introducing multiple layers of neurons, enabling them to learn complex patterns.

1. **Structure**

    (a) Input Layer: Receives input data.
    (b) Hidden Layers: Perform intermediate computations.
    (c) Output Layer: Produces predictions.

2. **Activation Functions**

    (a) Sigmoid: $\sigma(z) = \frac{1}{1+e^{-z}}$
    (b) ReLU: $\text{ReLU}(z) = \max(0, z)$
    (c) Softmax: Used in classification tasks.

3. **Learning Process: Forward Propagation**

    • Computes predictions.

4. **Loss Calculation:**

    • Measure error using loss functions like mean squared error (MSE).

5. **Backpropagation**:
    Adjust weights using gradients computed by the chain rule.

Python Code Example: MLP Classifier

```python
#----------------------------------------------------------------
# MLP Classifier
# Chapter 7 in the QUANTUM COMPUTING AND QUANTUM MACHINE LEARNING BOOK
#----------------------------------------------------------------
# Version 1.0
# (c) 2025 Jesse Van Griensven, Roydon Fraser, and Jose Rosas
# License: MIT - Citation of this work required
#----------------------------------------------------------------
# Importing necessary libraries
import numpy as np
import matplotlib.pyplot as plt
from sklearn.neural_network import MLPClassifier
#----------------------------------------------------------------

# Define dataset (XOR problem)
X = np.array([[0, 0], [0, 1], [1, 0], [1, 1]])
y = np.array([0, 1, 1, 0])  # XOR labels

# Train an MLP
mlp = MLPClassifier(hidden_layer_sizes=(5,), activation='relu',
max_iter=1000, random_state=42)
mlp.fit(X, y)

# Generate a mesh grid for decision boundary
x_min, x_max = X[:, 0].min() - 0.5, X[:, 0].max() + 0.5
y_min, y_max = X[:, 1].min() - 0.5, X[:, 1].max() + 0.5
xx, yy = np.meshgrid(np.linspace(x_min, x_max, 100), np.linspace
(y_min, y_max, 100))

# Compute MLP predictions over the grid
Z = mlp.predict(np.c_[xx.ravel(), yy.ravel()])
Z = Z.reshape(xx.shape)

# Plot decision boundary and data points
plt.figure(figsize=(8, 6))
plt.contourf(xx, yy, Z, alpha=0.3, cmap="coolwarm")
plt.scatter(X[:, 0], X[:, 1], c=y, cmap="coolwarm", edgecolors="k",
s=100, label="Training Data')
plt.title("MLP Classifier Decision Boundary (XOR Problem)")
plt.xlabel("Feature 1")
plt.ylabel("Feature 2")
plt.grid(alpha=0.5)
plt.legend()
plt.show()
```

7.19   Convolutional Neural Networks (CNNs)

Convolutional neural networks (CNNs) are specialized deep learning architectures designed for image data. They mimic the human visual system by learning hierarchical patterns from raw pixels, making them the backbone of modern computer vision. This section explores the fundamental components of CNNs, their operation, mathematical formulation, and applications in image processing.

7.19.1   Fundamentals of CNNs

A CNN is a type of neural network explicitly designed for grid-like data, such as images. Unlike traditional neural networks, CNNs use convolutional layers to extract features from local regions of the input.

CNNs are the cornerstone of modern image processing, capable of automatically learning complex patterns from raw data. With components like convolutional layers, pooling, and fully connected layers, CNNs power applications in object detection, medical imaging, and more. Despite challenges such as data dependence and computational cost, advances in transfer learning and data augmentation continue to enhance their capabilities, making CNNs an indispensable tool in the AI landscape.

7.19.2   CNN Benefits for Image Processing

CNNs are extensively employed for image processing. The following is a list of reasons for its application in this field.

1. **Parameter Efficiency**
   CNNs use shared weights, significantly reducing the number of parameters compared to fully connected layers.
2. **Spatial Hierarchies**
   They learn hierarchical representations, capturing low-level features like edges in early layers and complex structures like shapes in deeper layers.

7.19.3   Components of CNNs

This subsection explains the components of a convolutional neural network.

1. **Convolutional Layer**
   The convolutional layer is the core of a CNN. It applies filters (kernels) to extract features from the input. The output of the convolution operation at position (i, j) is:

$$Y[i,j] = \sum_{m=0}^{k-1} \sum_{n=0}^{k-1} K[m,n] \cdot X[i+m, j+n]$$

where $Y[i,j]$ is the output feature map, $[m,n]$ is the kernel (filter) weights of size $k \times k$, and $X[i+m, j+n]$ are the input pixels within the receptive field.

Python Code Example: Convolution Layer

```python
#----------------------------------------------------------------
# Convolution Layer
# Chapter 7 in the QUANTUM COMPUTING AND QUANTUM MACHINE LEARNING BOOK
#----------------------------------------------------------------
# Version 1.0
# (c) 2025 Jesse Van Griensven, Roydon Fraser, and Jose Rosas
# License: MIT - Citation of this work required
#----------------------------------------------------------------
import numpy as np
import matplotlib.pyplot as plt
from scipy.signal import convolve2d
#----------------------------------------------------------------

# Define a sample image (3x3) and filter (2x2)
image = np.array([[1, 2, 3], [4, 5, 6], [7, 8, 9]])
filter = np.array([[1, 0], [0, -1]])

# Perform convolution
output = convolve2d(image, filter, mode='valid')

# Plot original image, filter, and convolution result
fig, ax = plt.subplots(1, 3, figsize=(12, 4))

# Original image
ax[0].imshow(image, cmap='gray', aspect='auto')
ax[0].set_title("Original Image")
ax[0].set_xticks([])
ax[0].set_yticks([])

# Filter
ax[1].imshow(filter, cmap='gray', aspect='auto')
ax[1].set_title("Filter (Kernel)")
ax[1].set_xticks([])
ax[1].set_yticks([])
```

```
# Convolution output
ax[2].imshow(output, cmap='gray', aspect='auto')
ax[2].set_title("Convolution Output")
ax[2].set_xticks([])
ax[2].set_yticks([])

plt.show()
```

2. **Activation Function**

   After convolution, nonlinear activation functions are applied to introduce nonlinearity, enabling CNNs to learn complex patterns. Common activation functions are:

   (a) ReLU: Rectified linear unit:

$$\mathrm{ReLU}(z) = \max(0, z)$$

   (b) Sigmoid: Used in specific tasks like binary classification.

3. **Pooling Layer**

   The pooling layer reduces the spatial dimensions of the feature maps, retaining the most important information while reducing computational cost. Types of pooling include:

   (a) **Max Pooling**: Select the maximum value from the pooling window:

$$Y[i,j] = \max(X_{m,n} \quad \text{for } m, n \in \text{window})$$

   (b) **Average Pooling**: Averages all values in the pooling window.

4. **Fully Connected Layer**

   The fully connected (also known as dense) layer connects all neurons in the previous layer to the next, enabling classification. The role of the fully connected layer in CNN is to convert high-level feature maps into predictions, such as class probabilities.

Python Code Example: Pooling

```
#-----------------------------------------------------------------
# Pooling
# Chapter 7 in the QUANTUM COMPUTING AND QUANTUM MACHINE LEARNING BOOK
#-----------------------------------------------------------------
# Version 1.0
# (c) 2025 Jesse Van Griensven, Roydon Fraser, and Jose Rosas
# License: MIT - Citation of this work required
#-----------------------------------------------------------------
import numpy as np
```

```python
import matplotlib.pyplot as plt
from tensorflow.keras.layers import MaxPooling2D
import tensorflow as tf
#-------------------------------------------------------------------

# Define a simple input (4x4 image) and pooling layer
input_data  = np.random.rand(1, 4, 4, 1)  # Batch size: 1, Input size:
4x4, Channels: 1
pooling_layer = MaxPooling2D(pool_size=(2, 2))
pooled_output = pooling_layer(input_data)

# == Convert to 2D arrays for visualization ==
# Extracting 4x4 matrix
input_image  = input_data[0, :, :, 0]

# Extracting 2x2 pooled result
pooled_image = pooled_output.numpy()[0, :, :, 0]

# Plot original input and pooled output
fig, ax = plt.subplots(1, 2, figsize=(10, 5))

# Original image
ax[0].imshow(input_image, cmap='Blues', aspect='auto')
ax[0].set_title("Original 4x4 Image")
ax[0].set_xticks([])
ax[0].set_yticks([])

# Pooled output
ax[1].imshow(pooled_image, cmap='Blues', aspect='auto')
ax[1].set_title("Pooled 2x2 Output")
ax[1].set_xticks([])
ax[1].set_yticks([])

plt.show()
```

7.20 CNN Architectures

As previously identified, convolutional neural networks (CNNs) have been a cornerstone of artificial intelligence, particularly in the domain of image processing and computer vision. These architectures are designed to mimic the way humans perceive visual information, breaking down images into smaller, manageable parts before reconstructing them to identify patterns and features. Over the years, several key CNN architectures have emerged, each introducing innovative ideas and pushing the boundaries of what machines can achieve in visual recognition tasks.

7.20.1 LeNet

LeNet, proposed by Yann LeCun in 1998, is considered one of the pioneering architectures in deep learning, specifically in the realm of computer vision. LeNet was initially designed to recognize handwritten digits for processing bank checks and postal codes. This model demonstrated the potential of convolutional layers to automatically extract meaningful features from raw input, eliminating the need for manual feature engineering. Its success laid the groundwork for the widespread adoption of CNNs in various industries.

LeNet's architecture consists of multiple convolutional layers interspersed with pooling layers, followed by fully connected layers for classification. This sequential design allowed the model to learn hierarchical features, from simple edges in earlier layers to complex shapes in later ones. Despite its simplicity by modern standards, LeNet was revolutionary for its time and proved that neural networks could outperform traditional methods in pattern recognition tasks.

7.20.2 AlexNet

AlexNet marked a turning point in deep learning by demonstrating the scalability of CNNs for large-scale image classification. Developed by Alex Krizhevsky, Geoffrey Hinton, and Ilya Sutskever, AlexNet achieved widespread recognition after winning the 2012 ImageNet Large Scale Visual Recognition Challenge (ILSVRC) with a significant margin over traditional methods. This achievement not only showcased the power of CNNs but also brought deep learning into mainstream AI research and application.

AlexNet introduced several innovations that are now standard in deep learning. First, it utilized the rectified linear unit (ReLU) activation function, which alleviates the vanishing gradient problem and accelerates training. Second, AlexNet implemented dropout regularization to combat overfitting by randomly disabling neurons during training. Lastly, it leveraged GPU acceleration to handle the computational demands of processing millions of high-resolution images. With its eight layers, including convolutional, pooling, and fully connected layers, AlexNet set the stage for modern deep learning architectures.

7.20.3   VGGNet

VGGNet, developed by researchers at the University of Oxford, built upon the success of AlexNet by exploring the impact of network depth on model performance. Karen Simonyan and Andrew Zisserman designed VGGNet, in 2014, to achieve high accuracy in image classification tasks while maintaining a simple and modular design. Their work demonstrated that deeper networks, when carefully constructed, could significantly improve performance in computer vision tasks.

The hallmark of VGGNet is its use of small convolutional filters (3×3), which are stacked in increasing depth to enhance the model's capacity for feature extraction. This approach allows the network to capture fine-grained details while maintaining computational efficiency. However, the increased depth comes at the cost of higher memory and computational requirements, making VGGNet more demanding than its predecessors. Despite this, VGGNet's clean and straightforward design has made it a favorite for transfer learning and other applications.

7.20.4   ResNet

ResNet, or residual network, addressed one of the critical challenges in deep learning: the vanishing gradient problem, which makes training very deep networks difficult. ResNet introduced the concept of residual connections, revolutionizing the design of deep neural networks. ResNet won the ILSVRC 2015 competition, solidifying its position as a milestone in CNN development.

The primary innovation in ResNet is the use of skip connections or residual connections, which allow information to bypass one or more layers. These connections mitigate the problem of vanishing gradients, enabling the effective training of networks with hundreds or even thousands of layers. This breakthrough made it possible to construct extremely deep networks that maintain high accuracy without degradation. ResNet's modularity and robustness have made it a "building bock" in various domains, from image recognition to natural language processing.

7.21   CNN Applications in Image Processing

By leveraging hierarchical feature extraction, CNNs can understand images in depth, identifying patterns, structures, and relationships within visual data. This section explores various applications of CNNs in image processing, highlighting their significance and the architectures that have driven these advancements.

7.21.1 Object Detection

Object detection is a critical application of CNNs, aiming to identify objects within an image and localize them using bounding boxes. This task goes beyond classification by determining both the presence and location of multiple objects in an image. Object detection has broad applications, including autonomous driving, security surveillance, and augmented reality.

The following are two relevant object detection algorithms.

1. **YOLO (You Only Look Once)**

   YOLO was introduced by Joseph Redmon et al. in 2016 as a groundbreaking real-time object detection system. Unlike traditional methods that apply a classifier to multiple regions of an image, YOLO processes the entire image in a single forward pass, enabling fast and accurate detection. Its unified architecture divides the image into a grid, predicting bounding boxes and class probabilities simultaneously, making it a popular choice for time-critical applications.

2. **Faster R-CNN**

   Developed in 2015 by Shaoqing Ren and colleagues, Faster R-CNN builds on the success of R-CNN and Fast R-CNN. It introduced the region proposal network (RPN), which generates candidate object proposals directly from feature maps, eliminating the need for computationally expensive selective search algorithms. Faster R-CNN delivers state-of-the-art accuracy and has been widely adopted for applications requiring high precision.

7.21.2 Image Segmentation

Image segmentation assigns a label to each pixel in an image, effectively dividing it into meaningful regions. This pixel-level understanding is crucial for tasks requiring fine-grained image analysis, such as autonomous driving (lane detection) or medical diagnostics (tumor segmentation). Unlike object detection, which outputs bounding boxes, segmentation provides detailed spatial information about objects and their boundaries.

The following are relevant image segmentation architectures:

1. **U-Net**

   U-Net was proposed by Olaf Ronneberger et al. in 2015, specifically for biomedical image segmentation. Its unique encoder–decoder architecture enables it to capture both spatial and contextual information. Skip connections between encoder and decoder layers preserve spatial resolution, making U-Net highly effective for delineating intricate structures in images.

2. **SegNet**

   Introduced by Vijay Badrinarayanan et al. in 2016, SegNet is an encoder–decoder architecture designed for semantic segmentation. It stands out for its use of max-pooling indices from the encoder during upsampling in the decoder, preserving spatial accuracy while being computationally efficient. SegNet is widely used to understand urban scene and monitor environmental conditions.

7.21.3   CNN for Medical Image Analysis

Medical image analysis is one of the most transformative applications of convolutional neural networks (CNNs), offering unprecedented accuracy and efficiency in healthcare diagnostics. By leveraging the ability of CNNs to analyze complex patterns in visual data, medical professionals can detect and classify abnormalities in medical images that may be difficult for human observers to identify. This application not only accelerates diagnosis but also improves outcomes through early detection and treatment. CNN-based models are becoming vital tools in radiology, pathology, and other medical imaging fields, contributing significantly to precision medicine.

The following are two relevant examples of medical image analysis with CNNs:

1. **Tumor Detection in MRI Scans**

   Tumor detection involves identifying abnormal growths in tissues, such as the brain, liver, or breast, using MRI scans. CNNs are particularly well-suited for this task due to their ability to process high-resolution images and detect intricate patterns indicative of tumors. Models like ResNet and U-Net have been successfully employed to identify and segment tumor regions in MRI scans, significantly reducing the risk of human error and enabling earlier interventions. By automating the analysis process, these systems provide radiologists with faster and more accurate diagnostic support.

2. **Diabetic Retinopathy Classification from Retinal Images**

   Diabetic retinopathy is a severe complication of diabetes that can lead to blindness if untreated. CNNs have been widely adopted to analyze retinal fundus images and classify the presence and severity of the disease. Pretrained architectures such as InceptionV3 and DenseNet have demonstrated high accuracy in identifying subtle features like microaneurysms, hemorrhages, and neovascularization in retinal images. This capability not only assists ophthalmologists in making informed decisions but also facilitates large-scale screening programs in areas with limited access to specialized care.

The integration of CNNs into medical image analysis is revolutionizing healthcare by enabling faster, more accurate, and scalable diagnostic capabilities, ultimately improving patient outcomes and reducing healthcare costs.

7.21.4 CNN Style Transfer

Style transfer is a fascinating application of CNNs that merges the artistic world with advanced computational techniques. It enables the transformation of images by applying the style of one image, such as a famous painting, to another image while preserving the original content. This capability relies on the deep learning model's understanding of artistic features and image composition, which are derived from its convolutional layers. Style transfer has found applications in digital art, graphic design, and even video production, where it transforms ordinary visuals into captivating pieces of art.

The primary objective of style transfer is to reconstruct an image by blending the content of one image with the stylistic elements of another. For instance, a photograph of a cityscape can be reimagined in the style of Van Gogh's "Starry Night," resulting in a unique artistic rendition. This process leverages a pretrained CNN, such as VGGNet, to extract the content features from one image and the style features from another and then combine them using optimization techniques.

Style transfer techniques often utilize models like the neural style transfer (NST) framework, which calculates the Gram matrix of feature maps to capture style patterns. The resulting images are used in artistic endeavors, advertising, and social media applications. By democratizing access to advanced artistic tools, CNN-based style transfer empowers individuals and industries to create visually stunning content with minimal effort, bridging the gap between creativity and technology.

7.21.5 Limitations of CNNs

While convolutional neural networks (CNNs) have achieved remarkable success, they face several challenges. These are presented in Table 7.10.

7.21.6 CNN Data Augmentation

Data augmentation artificially expands datasets by applying transformations like rotation, flipping, and scaling. This is employed extensively in the following applications:

Table 7.10 CNN limitations

Limitation factor	Description
Data dependence	Require large datasets for effective training
Computational cost	Training CNNs demands significant hardware resources (e.g., GPUs)
Lack of explainability	Difficult to interpret the features learned by CNNs

1. Object recognition.
2. Medical image analysis.
3. Autonomous vehicles (e.g., road sign detection).

Python Code Example: Data Augmentation

```python
#--------------------------------------------------------------------
# Data Augmentation
# Chapter 7 in the QUANTUM COMPUTING AND QUANTUM MACHINE LEARNING BOOK
#--------------------------------------------------------------------
# Version 1.0
# (c) 2025 Jesse Van Griensven, Roydon Fraser, and Jose Rosas
# License: MIT - Citation of this work required
#--------------------------------------------------------------------
import numpy as np
import matplotlib.pyplot as plt
from tensorflow.keras.preprocessing.image import ImageDataGenerator,
array_to_img, img_to_array, load_img
#--------------------------------------------------------------------

# Load a sample image (replace with a valid image path)
sample_image_path = "sample_image.jpg" # Replace with a valid image file
in your directory

try:
    # Load and preprocess the image
    image = load_img(sample_image_path) # Load image
    image_array = img_to_array(image) # Convert to array
    image_array = np.expand_dims(image_array, axis=0) # Expand
dimensions for batch processing

    # Define Data Augmentation
    datagen = ImageDataGenerator(rotation_range=40,
width_shift_range=0.2,
                height_shift_range=0.2, shear_range=0.2,
                zoom_range=0.2, horizontal_flip=True,
                fill_mode='nearest')

    # Generate augmented images
    augmented_images = datagen.flow(image_array, batch_size=1)

    # Plot original and augmented images
    fig, axes = plt.subplots(1, 5, figsize=(15, 5))
    axes[0].imshow(image)
```

```
axes[0].set_title("Original Image")
axes[0].axis("off")

# Generate and plot 4 augmented images
for i in range(1, 5):
  augmented_image = next(augmented_images)[0].astype("uint8")
  axes[i].imshow(augmented_image)
  axes[i].set_title(f"Augmented {i}")
  axes[i].axis("off")

plt.show()

except FileNotFoundError:
  print("Error: Please provide a valid image path for data
augmentation.")
```

7.22 Recurrent Neural Networks

Recurrent neural networks (RNNs) are a class of artificial neural networks designed to process sequence data by maintaining memory of previous inputs. Unlike feedforward neural networks, RNNs have loops in their architecture, allowing them to process sequences of varying lengths and remember temporal relationships between data points. This makes them particularly useful for tasks like time-series analysis, natural language processing, and speech recognition.

Recurrent neural networks are powerful tools for sequence modeling, leveraging hidden states to capture temporal dependencies. While basic RNNs face challenges like vanishing gradients, advanced variants such as LSTMs and GRUs overcome these limitations, enabling robust performance across a variety of applications. From text generation to stock price prediction, RNNs remain an indispensable architecture in the deep learning toolbox.

7.22.1 Hidden State

The defining feature of an RNN is its hidden state, which captures information from previous time steps and enables the network to maintain context over a sequence.

The mathematical formulation is described below:

At each time step t, the hidden state h_t is updated based on the current input x_t, the hidden state from the previous step h_{t-1}, and a bias term b:

$$h_t = f(W_x x_t + W_h h_{t-1} + b)$$

where W_x is the weight matrix for the input, W_h is the weight matrix for the hidden state, and f is the activation function (e.g., tanh or ReLU).

7.22.2   Backpropagation through Time (BPTT)

RNNs use a modified version of backpropagation called backpropagation through time (BPTT) to compute gradients for all time steps in the sequence. This allows the network to adjust weights based on the entire sequence but also introduces challenges like vanishing gradients.

7.22.3   RNN in Natural Language Processing (NLP)

Natural language processing (NLP) is one of the most prominent domains in which RNNs have had a significant impact. By processing text as sequences of words or characters, RNNs can capture the contextual meaning necessary for understanding language. Advanced models, such as long short-term memory (LSTM) and gated recurrent unit (GRU), address the vanishing gradient problem in traditional RNNs, enabling better performance on complex language tasks.

1. **Text Generation**

   Text generation involves creating coherent sequences of text based on an input prompt or dataset. RNNs are trained on large text corpora to learn language structures, grammar, and contextual dependencies. Applications include generating creative content, such as poetry and storylines, and practical uses, like chatbot responses and auto-completion in text editors. A notable example is OpenAI's GPT series, which, while more advanced, builds on the foundational principles of RNNs for text generation.

2. **Sentiment Analysis**

   Sentiment analysis uses RNNs to classify the emotional tone of a given text. By analyzing sequential relationships between words, RNNs can detect positive, negative, or neutral sentiments in product reviews, social media posts, or customer feedback. This application is widely used in marketing, customer service, and social media monitoring to gauge public opinion and improve user engagement.

3. **Machine Translation**

   Machine translation involves converting text from one language to another while preserving meaning and grammar. RNNs, often paired with attention mechanisms, have been instrumental in enabling systems like Google Translate to handle translations more accurately. These models learn to map sentence structures between languages, ensuring syntactic and semantic fidelity.

7.22.4   *RNN in Time-Series Analysis*

Time-series analysis is another area where RNNs excel, as they can model temporal dependencies in sequential data. This capability makes them invaluable for predicting future events based on historical trends. An example of applications of time series employing RNN is the prediction of stock prices, weather patterns, and sensor data.

RNNs analyze time-series data, such as financial market trends, climate records, or IoT sensor outputs, to predict future values. For example, in stock price prediction, an RNN can capture recurring patterns and fluctuations over time to forecast price movements. Similarly, in meteorology, RNNs help predict weather conditions by analyzing historical weather data. These predictions enhance decision-making in finance, logistics, and energy management.

7.22.5   *RNN in Speech Recognition*

Speech recognition is a transformative application of RNNs, enabling the conversion of spoken language into written text. The sequential nature of speech data, where each sound depends on the preceding context, makes RNNs particularly effective for this task. An example of the application of speech recognition is converting audio signals into text.

In voice assistants like Siri, Alexa, and Google Assistant, RNNs process audio waveforms to identify phonemes, words, and sentences. Advanced architectures such as LSTMs and GRUs enhance recognition accuracy by maintaining contextual memory over long speech sequences. This application is not only integral to personal assistant devices but also plays a critical role in accessibility tools for individuals with hearing or speech impairments.

7.22.6   *RNN in Video Analysis*

Video analysis extends the capabilities of RNNs to the domain of visual data by analyzing sequential frames to extract meaningful information about actions and events. RNNs process video data by examining the temporal relationships between consecutive frames. This approach is essential for tasks such as action recognition in surveillance systems, where identifying activities like running, fighting, or loitering is critical. Additionally, in autonomous vehicles, video analysis powered by RNNs helps understand dynamic environments, such as detecting obstacles or predicting the movements of other vehicles.

7.22.7   *Challenges in RNNs*

Despite their advantages, RNNs are not without challenges. Several limitations hinder their efficiency and effectiveness, particularly in large-scale or complex tasks. Below are some of the notable limitations of RNNs:

1. **Vanishing Gradients**
   Gradients of earlier time steps become exponentially smaller during backpropagation, causing the network to "forget" long-term dependencies.
2. **Exploding Gradients**
   Gradients grow uncontrollably, destabilizing the training process. This can often be mitigated by gradient clipping.
3. **Limited Context Retention.**
   While RNNs can theoretically model long sequences, their practical effectiveness diminishes as sequence length increases due to the difficulty of maintaining relevant context.
4. **Overfitting**
   One of the primary issues with RNNs is their susceptibility to overfitting, especially when working with small datasets. RNNs may memorize input data instead of learning generalizable patterns, resulting in poor performance on unseen data. This limitation can be mitigated through techniques like dropout regularization, data augmentation, and careful model tuning.
5. **Interpretability**
   RNNs, like many deep learning models, are often criticized for being "black boxes." While they can learn latent representations that capture complex temporal patterns, these representations are not easily interpretable. This lack of transparency poses challenges in domains where explainability is critical, such as healthcare or legal systems. For example, in medical diagnosis, understanding why an RNN flagged a sequence of symptoms as indicative of a specific condition is: essential for gaining trust in the model's predictions.
6. **Computational Cost**
   Training RNNs is computationally intensive, particularly for long sequences or large datasets. The feedback loop in RNNs necessitates sequential processing of data, making parallelization more challenging compared to feedforward networks. This results in longer training times and higher hardware requirements.

7.22.8 RNN Variants

Advanced variants such as long short-term memory (LSTM) networks and gated recurrent units (GRUs) were introduced to address the limitations of basic RNNs.

7.22.8.1 Long Short-Term Memory (LSTM)

LSTMs are designed to handle long-term dependencies by introducing gates that regulate the flow of information. The key components are:

1. Input Gate: Decides how much new information to store in the cell state.
2. Forget Gate: Decides which information to discard from the cell state.
3. Output Gate: Controls the output of the hidden state.
4. Mathematical Formulation:

$$f_t = \sigma\left(W_f x_t + U_f h_{t-1} + b_f\right) \quad \text{(Forget Gate)}$$

$$i_t = \sigma(W_i x_t + U_i h_{t-1} + b_i) \quad \text{(Input Gate)}$$

$$\tilde{C}_t = \tanh(W_c x_t + U_c h_{t-1} + b_c) \quad \text{(Candidate State)}$$

$$C_t = f_t \odot C_{t-1} + i_t \odot \tilde{C}_t \quad \text{(Cell State)}$$

$$o_t = \sigma(W_o x_t + U_o h_{t-1} + b_o) \quad \text{(Output Gate)}$$

$$h_t = o_t \odot \tanh(C_t) \quad \text{(Hidden State)}$$

where
x_t is the input vector at time step t.
h_{t-1} is the hidden state from previous time step.
f_t is the forget gate vector.
i_t is the input gate vector.
o_t is the output gate vector.
C_t is the cell state at time t.
C_{t-1} is the cell state from previous time step.
$\tilde{C}_t$ is the candidate cell state.
h_t is the current hidden state output.
W_*, U_*, b_* are the weight matrices and bias vectors for gates ($* = $ forget f, input i, output o, and candidate c).
$\sigma(\cdot)$ is the sigmoid activation function.
$tanh(\cdot)$ is the hyperbolic tangent function.
$\odot$ is the element-wise (Hadamard) product.

5. **Advantages**

   (a) Effectively captures long-term dependencies.
   (b) Robust against vanishing gradient issues.

Python Code Example: LSTM

```python
#-----------------------------------------------------------------
# LSTM
# Chapter 7 in the QUANTUM COMPUTING AND QUANTUM MACHINE LEARNING BOOK
#-----------------------------------------------------------------
# Version 1.0
# (c) 2025 Jesse Van Griensven, Roydon Fraser, and Jose Rosas
# License: MIT - Citation of this work required
#-----------------------------------------------------------------
import numpy as np
import matplotlib.pyplot as plt
from tensorflow.keras.models import Sequential
from tensorflow.keras.layers import LSTM, Dense
from tensorflow.keras.utils import plot_model
#-----------------------------------------------------------------

# Define the LSTM model
model = Sequential([
  LSTM(50, activation='tanh', input_shape=(10, 1),
return_sequences=False),
  Dense(1)
])

# Display model summary
model.summary()

# Plot the LSTM model architecture
plt.figure(figsize=(10, 6))
try:
  plot_model(model, to_file="lstm_model.png", show_shapes=True,
show_layer_names=True)
  img = plt.imread("lstm_model.png")
  plt.imshow(img)
  plt.axis("off")
  plt.title("LSTM Model Architecture")
  plt.show()
except Exception as e:
  print("Error: Unable to display model architecture. Ensure Graphviz
is installed.")
```

7.22.8.2   Gated Recurrent Units (GRUs)

Gated recurrent units (GRUs) simplify the LSTM architecture by combining the
input and forget gates into a single update gate and eliminating the cell state.

Advantages include fewer parameters compared to LSTMs and faster training while retaining similar performance.

The key components of GRUs are:

(a) Update Gate: Decides the amount of past information to retain.
(b) Reset Gate: Decides how much past information to ignore.
(c) Mathematical Formulation:

$$z_t = \sigma(W_z x_t + U_z h_{t-1} + b_z) \qquad \text{(Update Gate)}$$

$$r_t = \sigma(W_r x_t + U_r h_{t-1} + b_r) \qquad \text{(Reset Gate)}$$

$$\tilde{h}_t = \tanh\left(W_h x_t + U_h\left(r_t \odot h_{t-1}\right) + b_h\right) \quad \text{(Candidate State)}$$

$$h_t = z_t \odot h_{t-1} + (1 - z_t) \odot \tilde{h}_t \qquad \text{(Hidden State)}$$

Python Code Example: GRU

```python
#------------------------------------------------------------------
# GRU
# Chapter 7 in the QUANTUM COMPUTING AND QUANTUM MACHINE LEARNING BOOK
#------------------------------------------------------------------
# Version 1.0
# (c) 2025 Jesse Van Griensven, Roydon Fraser, and Jose Rosas
# License:  MIT - Citation of this work required
#------------------------------------------------------------------
import numpy as np
import matplotlib.pyplot as plt
from tensorflow.keras.models import Sequential
from tensorflow.keras.layers import GRU, Dense
from tensorflow.keras.utils import plot_model
#------------------------------------------------------------------

# Define the GRU model
model = Sequential([
  GRU(50, activation='tanh', input_shape=(10, 1),
return_sequences=False),
  Dense(1)
])

# Display model summary
model.summary()

# Plot the GRU model architecture
plt.figure(figsize=(10, 6))
```

```
try:
  plot_model(model, to_file='gru_model.png', show_shapes=True,
show_layer_names=True)
  img = plt.imread("gru_model.png")
  plt.imshow(img)
  plt.axis("off")
  plt.title("GRU Model Architecture")
  plt.show()
except Exception as e:
  print("Error: Unable to display model architecture. Ensure Graphviz
is installed.")
```

7.22.9 *Autoencoders for Feature Extraction*

Autoencoders are a class of neural networks designed for unsupervised learning. They are used to encode input data into a compressed representation (latent space) and reconstruct the input from this representation. The primary goal is to capture the most important features of the data while minimizing reconstruction errors. Autoencoders are widely used in dimensionality reduction, anomaly detection, data compression, and feature extraction.

Autoencoders are versatile models for unsupervised learning and feature extraction. By encoding data into a compressed latent space and decoding it back, they enable applications, such as dimensionality reduction, anomaly detection, and data denoising. Advanced variants like VAEs and denoising autoencoders expand their utility in real-world tasks, making them an essential tool in deep learning.

An autoencoder consists of three main components: the encoder, the latent space, and the decoder.

7.22.10 *The Encoder*

The encoder maps the input data into a compressed latent representation. This process involves reducing the input's dimensionality while preserving as much relevant information as possible. The encoder mathematical formulation is:

For an input vector $\mathbf{x}$, the encoder produces a latent representation $\mathbf{z}$:

$$\mathbf{z} = f_{\text{encoder}}(\mathbf{x}) = \sigma(W_e \mathbf{x} + \mathbf{b}_e)$$

where W_e is the weight matrix of the encoder, $\mathbf{b}_e$ is the bias vector of the encoder, and σ is the activation function (e.g., ReLU, sigmoid).

7.22.11   The Latent Space

The latent space, often represented as $\mathbf{z}$, is a compact and meaningful representation of the input. It acts as a bottleneck that forces the model to learn a reduced yet informative encoding of the data.

7.22.12   The Decoder

The decoder reconstructs the input data from the latent space, attempting to approximate the original input as closely as possible. The decoder mathematical formulation is:

For a latent representation $\mathbf{z}$, the decoder produces the reconstructed input $\widehat{\mathbf{x}}$:

$$\widehat{\mathbf{x}} = f_{\text{decoder}}(\mathbf{z}) = \sigma(\mathbf{W}_d \mathbf{z} + \mathbf{b}_d)$$

where $\mathbf{W}_d$ is the weight matrix of the decoder, and $\mathbf{b}_d$ is the bias vector of the decoder.

Here, $\widehat{\mathbf{x}}$ denotes the reconstructed input, a common notation in autoencoders where the "hat" indicates an estimated or regenerated value.

7.22.13   Autoencoder Loss Function

The reconstruction loss measures how well the autoencoder can reproduce the input. A common loss function is the mean squared error (MSE):

$$L = \frac{1}{n} \sum_{i=1}^{n} (x_i - \widehat{x}_i)^2$$

where x_i is the original input value, and $\widehat{x}_i$ is the reconstructed value.

Python Code Example: Autoencoder

```
#-------------------------------------------------------------------
# GRU
# Chapter 7 in the QUANTUM COMPUTING AND QUANTUM MACHINE LEARNING BOOK
#-------------------------------------------------------------------
# Version 1.0
# (c) 2025 Jesse Van Griensven, Roydon Fraser, and Jose Rosas
# License: MIT - Citation of this work required
#-------------------------------------------------------------------
import numpy as np
```

```python
import matplotlib.pyplot as plt
from tensorflow.keras.models import Sequential
from tensorflow.keras.layers import GRU, Dense
from tensorflow.keras.utils import plot_model
#-----------------------------------------------------------------

# Define the GRU model
model = Sequential([
  GRU(50, activation='tanh', input_shape=(10, 1),
return_sequences=False),
  Dense(1)
])

# Display model summary
model.summary()

# Plot the GRU model architecture
plt.figure(figsize=(10, 6))
try:
  plot_model(model, to_file="gru_model.png", show_shapes=True,
show_layer_names=True)
  img = plt.imread("gru_model.png")
  plt.imshow(img)
  plt.axis("off")
  plt.title("GRU Model Architecture")
  plt.show()
except Exception as e:
  print("Error: Unable to display model architecture. Ensure Graphviz
is installed.")
```

7.22.14 Variants of Autoencoders

Autoencoders come in various forms, each tailored for specific tasks.

7.22.14.1 Sparse Autoencoders

Sparse autoencoders impose a sparsity constraint on the latent space, forcing the model to use only a few active neurons.

1. Objective: Encourage the encoder to learn distinct features by activating a limited subset of neurons.
2. Loss Function: Adds a sparsity penalty to the reconstruction loss:

$$L = \frac{1}{n} \sum_{i=1}^{n} (x_i - \widehat{x}_i)^2 + \beta \sum_{j=1}^{k} KL\big(\rho \,|\, |\widehat{\rho}_j\big)$$

where *KL* is the Kullback–Leibler divergence between the desired sparsity ρ and the average activation $\widehat{\rho}_j$.

7.22.14.2 Denoising Autoencoders

Denoising autoencoders are trained to reconstruct the original input from a corrupted (noisy) version, thereby improving robustness.

1. Objective: Learn to remove noise from data by reconstructing the original (clean) input.
2. Training Process: Introduce noise to the input data **x**, resulting in a corrupted version ~**x** (e.g., via additive Gaussian noise), and train the autoencoder to output a reconstruction $\widehat{x}$ that approximates the original input **x**.
   where
   x is the original (clean) input data.
   ~**x** is the corrupted (noisy) input data.
   $\widehat{x}$ is the reconstructed output produced by the autoencoder.

Python Code Example: Denoising Autoencoders

```python
#-------------------------------------------------------------------
# Denoising Autoencoders
# Chapter 7 in the QUANTUM COMPUTING AND QUANTUM MACHINE LEARNING BOOK
#-------------------------------------------------------------------
# Version 1.0
# (c) 2025 Jesse Van Griensven, Roydon Fraser, and Jose Rosas
# License: MIT - Citation of this work required
#-------------------------------------------------------------------
import numpy as np
import matplotlib.pyplot as plt
from tensorflow.keras.models import Model
from tensorflow.keras.datasets import mnist
from tensorflow.keras.utils import plot_model
from tensorflow.keras.layers import Input, Dense
#-------------------------------------------------------------------


# Add Gaussian noise function
def add_noise(x, noise_factor=0.5):
  Noise = x + noise_factor*np.random.normal(loc=0.0, scale=1.0,
size=x.shape)
```

```python
  return Noise
#----------------------------------------------------------------------

# Load MNIST dataset (as an example)
(x_train, _), (x_test, _) = mnist.load_data()
x_train = x_train.astype('float32') / 255.0
x_test = x_test.astype('float32') / 255.0
x_train = x_train.reshape(-1, 784)  # Flattening
x_test = x_test.reshape(-1, 784)

# Add noise to the dataset
x_train_noisy = add_noise(x_train, noise_factor=0.5)
x_test_noisy = add_noise(x_test, noise_factor=0.5)

# Ensure values are within [0,1]
x_train_noisy = np.clip(x_train_noisy, 0.0, 1.0)
x_test_noisy = np.clip(x_test_noisy, 0.0, 1.0)

#----------------------------------------------------------------------
# Define Autoencoder Structure
input_dim   = 784 # MNIST images (28x28 flattened)
encoding_dim = 32 # Latent space dimensionality

# Encoder
input_layer = Input(shape=(input_dim,))
encoded = Dense(encoding_dim, activation='relu')(input_layer)

# Decoder
decoded = Dense(input_dim, activation='sigmoid')(encoded)

# Autoencoder Model
autoencoder = Model(inputs=input_layer, outputs=decoded)
autoencoder.compile(optimizer='adam', loss='mse')

# Train the autoencoder
autoencoder.fit(x_train_noisy, x_train, epochs=10, batch_size=256,
shuffle=True, validation_data=(x_test_noisy, x_test))

#----------------------------------------------------------------------
# Visualizing the Results
n = 5 # Number of images to display
decoded_imgs = autoencoder.predict(x_test_noisy)

plt.figure(figsize=(10, 6))
for i in range(n):
```

```python
# Noisy images
ax = plt.subplot(3, n, i + 1)
plt.imshow(x_test_noisy[i].reshape(28, 28), cmap='gray')
plt.axis("off")
if i == 0:
  ax.set_title("Noisy Images")

# Original images
ax = plt.subplot(3, n, i + 1 + n)
plt.imshow(x_test[i].reshape(28, 28), cmap='gray')
plt.axis("off")
if i == 0:
  ax.set_title("Original Images")

# Reconstructed (Denoised) images
ax = plt.subplot(3, n, i + 1 + 2 * n)
plt.imshow(decoded_imgs[i].reshape(28, 28), cmap='gray')
plt.axis("off")
if i == 0:
  ax.set_title("Denoised Images")

plt.show()
```

7.22.15   Variational Autoencoders (VAEs)

VAEs are probabilistic models that learn a latent space with a continuous distribution, enabling generative tasks.

1. Objective: Encode input data as a probability distribution and reconstruct by sampling from this distribution.
2. Loss Function: Combines reconstruction loss and a Kullback–Leibler divergence term:

$$L = \text{Reconstruction Loss} + KL(q(z\,|\,x)\,|\,|\,p(z))$$

7.22.16   Contractive Autoencoders

Contractive autoencoders add a regularization term to penalize changes in the encoder's output with respect to the input. The objective of this autoencoder is to learn robust representations that are invariant to small changes in the input.

7.22.17   Applications of Autoencoders

Autoencoders are a class of neural networks that are trained to reconstruct their input. By encoding input data into a compressed latent space representation and then decoding it back, autoencoders learn to capture the most important features of the data. This makes them versatile tools for tasks like dimensionality reduction, anomaly detection, data denoising, feature extraction, and even image generation. Unlike traditional dimensionality reduction techniques like principal component analysis (PCA), autoencoders offer the added advantage of being able to learn nonlinear transformations, making them more powerful for complex datasets.

7.22.17.1   Dimensionality Reduction

Dimensionality reduction is one of the most common applications of autoencoders, enabling the visualization and analysis of high-dimensional data in lower dimensions. By encoding data into a compact latent space, autoencoders can reduce the complexity of the dataset while preserving its most critical information.

1. **How It Works**

   The encoder component of the autoencoder learns to map high-dimensional input data into a lower-dimensional latent representation. The decoder then reconstructs the original data from this compressed representation. Unlike PCA, which only captures linear relationships, autoencoders can model complex, nonlinear patterns in the data.

2. **Use Case**

   A key use case is the visualization of high-dimensional datasets. For example, in genomics, researchers often work with gene expression data that has thousands of dimensions. Autoencoders can reduce this complexity to two or three dimensions, making it easier to visualize and interpret patterns, such as clustering of similar samples. Similarly, in marketing analytics, dimensionality reduction helps visualize customer segmentation based on purchase behavior.

7.22.17.2   Anomaly Detection

Autoencoders are highly effective for anomaly detection, leveraging their reconstruction capabilities to identify outliers in data. Since autoencoders are trained to reconstruct normal data, they tend to exhibit higher reconstruction errors for anomalous or unseen patterns.

1. **How It Works**

   During training, the autoencoder learns to minimize reconstruction error for the typical patterns in the dataset. When presented with anomalous data, the reconstruction error increases significantly as the autoencoder struggles to represent features it has not learned.

2. Use Case

In manufacturing, autoencoders are employed to detect defective products. Sensors on production lines collect data about product features, which is then fed to an autoencoder trained on data from normal products. High reconstruction errors indicate defects, allowing manufacturers to identify issues early and maintain quality standards. Similarly, in cybersecurity, autoencoders are used to detect anomalies in network traffic that may signal a security breach.

7.22.17.3 Data Denoising

Data denoising is a critical application of autoencoders, particularly in scenarios where data quality is compromised by noise or corruption. Denoising autoencoders are specifically designed to remove noise from input data while preserving the underlying information.

1. How It Works

A denoising autoencoder is trained by adding noise to the input data and teaching the network to reconstruct the original, noise-free data. This encourages the autoencoder to focus on the most relevant features and ignore random noise.

2. Use Case

In medical imaging, denoising autoencoders enhance image quality by removing artifacts caused by imaging equipment or environmental factors. For example, MRI and CT scans often contain noise that can obscure important diagnostic details. Autoencoders can clean these images, improving their clarity and aiding in more accurate diagnoses. Similarly, in audio processing, denoising autoencoders are used to clean speech signals by removing background noise, improving the quality of communication in noisy environments.

7.22.17.4 Feature Extraction

Feature extraction is a powerful application of autoencoders, as the compressed latent space representations they learn can serve as meaningful features for various downstream tasks. By capturing the essential characteristics of the input data, autoencoders enable more efficient and accurate analysis.

1. How It Works

The encoder generates a compact latent representation of the input data. These latent features can then be used as input for other machine learning models, such as classifiers or clustering algorithms.

2. Use Case

In classification tasks, autoencoders can reduce the dimensionality of the dataset and provide high-quality features for training models. For instance, in image recognition, an autoencoder trained on a large dataset can generate latent features that capture essential visual patterns, such as edges, textures, and shapes. These features can then be used to train a classification model with improved accuracy. Similarly, in clustering, autoencoders help create more distinct clusters by providing compact and meaningful representations of the data.

7.22.17.5 Image Generation

Autoencoders, particularly their variational variant (VAEs), have opened new possibilities in generative modeling. By learning the underlying distribution of the data, VAEs can generate new, realistic samples that resemble the original data.

1. **How it Works**

   VAEs extend traditional autoencoders by introducing a probabilistic framework. The encoder learns to map input data to a distribution in the latent space, and new samples are generated by sampling from this distribution and decoding them back into the data space.
2. **Use Case**

   A prominent application is generating synthetic images for training deep learning models. In scenarios where labeled data is scarce, VAEs can create realistic synthetic data to augment the training set, improving model performance. For instance, in autonomous driving, VAEs can generate synthetic road scenarios to train vehicle perception systems. In the creative industry, VAEs are used to generate unique art, such as creating new variations of existing paintings or designs.

7.22.18 Advanced Topics in Autoencoders

Autoencoders have evolved beyond their foundational applications of dimensionality reduction, anomaly detection, and feature extraction. By combining autoencoders with other advanced techniques, researchers have unlocked new capabilities that expand their utility into more complex and specialized tasks. This section explores semi-supervised learning, the integration of autoencoders with generative adversarial networks (GANs), and the development of multimodal autoencoders, highlighting how these advancements push the boundaries of what autoencoders can achieve.

7.22.18.1 Semi-Supervised Learning

Semi-supervised learning bridges the gap between supervised and unsupervised learning by leveraging both labeled and unlabeled data to improve model performance. Autoencoders, with their ability to learn meaningful latent representations from unlabeled data, are well-suited for semi-supervised tasks. This approach is particularly valuable in domains where obtaining labeled data is expensive or time-consuming.

1. **How It Works**

In semi-supervised learning, an autoencoder is first trained on the available dataset, including both labeled and unlabeled data. The encoder learns a compact latent representation of the input data, capturing its essential features. These representations are then used to train a supervised model, such as a classifier, on the labeled subset of the data. Leveraging features learned from the entire dataset, the model achieves better generalization and higher accuracy compared to training solely on the labeled data.

2. **Applications**

In healthcare, semi-supervised learning with autoencoders has been applied to diagnose rare diseases where labeled medical data is scarce. For instance, autoencoders can pre-train models on large sets of unlabeled medical images and then use limited labeled examples to fine-tune a classifier for specific diagnoses. Similarly, in fraud detection, semi-supervised learning helps identify fraudulent transactions in datasets where labeled fraudulent cases are minimal.

This approach not only enhances performance in scenarios with limited labeled data but also reduces the dependency on extensive manual labeling efforts, making it a powerful tool for real-world applications.

7.22.18.2 Generative Adversarial Networks (GANs) with Autoencoders

Generative adversarial networks (GANs) are renowned for their ability to generate realistic data by pitting two networks, the generator and the discriminator, against each other in a zero-sum game. When combined with autoencoders, GANs benefit from the structured latent space representations provided by the autoencoder, leading to improved generation quality and control over the output.

1. **How It Works**

In this hybrid approach, the encoder of the autoencoder compresses input data into a latent representation, while the GAN's generator maps random noise or sampled latent vectors to output data. The decoder of the autoencoder ensures that the generated samples conform to the underlying data distribution. This collaboration enhances the quality and diversity of generated samples by combining the strengths of both architectures.

2. **Applications**

In the creative industries, autoencoder-GAN hybrids are used for high-quality image generation, including creating realistic human faces or artistic renditions of existing images. In medicine, these models generate synthetic medical images for training deep learning systems in scenarios where real patient data is limited due to privacy concerns. For instance, synthetic MRI scans generated through this technique provide diverse training data for brain tumor detection systems.

By integrating autoencoders and GANs, researchers achieve greater control over generative processes, making these hybrid systems valuable for applications requiring both precision and creativity.

7.22.18.3 Multimodal Autoencoders

Datasets often contain multiple types of data, such as images, text, or audio, requiring models capable of understanding and integrating these different modalities. Multimodal autoencoders address this challenge by learning joint representations across multiple data types, enabling coherent analysis and synthesis of complex datasets.

1. **How It Works**
   Multimodal autoencoders use separate encoder–decoder pairs for each data modality, while a shared latent space combines information from all modalities. For example, an image-text dataset may have one encoder for images and another for text, with their outputs mapped to a unified latent space. This shared representation allows the model to capture relationships between modalities, enabling tasks like cross-modal generation or analysis.
2. **Applications**
   Multimodal autoencoders have shown promise in applications like video captioning, where visual and textual data must be integrated. The model learns to associate visual frames with corresponding descriptive text, enabling automatic generation of captions for videos. Another application is multimodal medical diagnosis, where patient data includes medical images, text-based clinical notes, and laboratory results. By combining these modalities, the model provides a holistic view of the patient's condition, improving diagnostic accuracy.

Multimodal autoencoders also facilitate translation between modalities, such as converting speech to text or generating images from textual descriptions. This versatility makes them essential for applications in robotics, healthcare, and multimedia processing.

7.23 Applications of Deep Learning

Deep learning has become an essential technology, driving advancements across diverse industries. Its architectures, such as convolutional neural networks (CNNs), recurrent neural networks (RNNs), and autoencoders, provide unparalleled capabilities for solving complex problems. This section explores key applications of deep learning in healthcare, finance, and autonomous systems.

7.23.1 Healthcare: CNNs for Disease Diagnosis from X-Rays

Healthcare is one of the most transformative domains for deep learning applications. Among the various architectures, convolutional neural networks (CNNs) have demonstrated exceptional accuracy in analyzing medical images and revolutionizing disease diagnosis and treatment planning.

1. **How CNNs Work in Medical Imaging**

   CNNs excel at detecting patterns in visual data, making them ideal for medical imaging tasks. They use convolutional layers to extract features such as edges, textures, and shapes from X-rays, CT scans, and MRIs. These features are then processed through pooling and fully connected layers to classify diseases or identify anomalies.

2. **Applications in Disease Diagnosis**

   (a) In lung cancer detection, CNNs analyze chest X-rays to identify nodules indicative of lung cancer. Early detection significantly improves survival rates, and CNN-powered systems have achieved diagnostic accuracy comparable to radiologists.

   (b) In pneumonia diagnosis, CNNs help identify pneumonia with high precision, aiding in timely treatment decisions by detecting opacity in lung regions.

   (c) In bone fracture detection, CNNs process X-ray images to identify fractures, streamlining emergency care.

3. **Real-World Impact**

   (a) CNN-based systems like CheXNet have been deployed in hospitals, providing second opinions to radiologists and reducing diagnostic errors.

   (b) These models also enable telemedicine services, where patients in remote areas can receive accurate diagnoses without visiting specialized healthcare facilities.

7.23.1.1 Finance: RNNs for Fraud Detection in Transaction Sequences

In the finance sector, recurrent neural networks (RNNs) have proven to be invaluable for tasks requiring the analysis of sequential data, such as transaction sequences. Their ability to capture temporal dependencies makes them ideal for identifying fraudulent activities.

1. **How RNNs Work in Fraud Detection**

   RNNs process transaction sequences by analyzing patterns and detecting anomalies. Their feedback loops allow them to consider the context of past transactions when evaluating the current one, enabling precise detection of irregularities.

2. **Applications in Fraud Detection**

   (a) RNNs analyze credit card transaction patterns, such as location, amount, and frequency, to identify suspicious behavior. For example, a sudden large purchase in a different geographic location might trigger a fraud alert.
   (b) RNNs detect inconsistencies or unusual patterns that indicate fraudulent insurance claims fraud, by analyzing claim sequences and amounts.
   (c) RNNs process trading data to identify unusual trading patterns indicative of market manipulation.

3. **Real-World Impact**

   (a) Financial institutions like PayPal and Visa use RNN-based models to monitor millions of transactions daily, preventing billions of dollars in fraud.
   (b) These systems provide real-time alerts, enabling swift action to block suspicious transactions and protect customer accounts.

4. **Challenges**

   RNNs can suffer from issues, such as vanishing gradients, which hinder their ability to process long transaction histories. Advanced variants like LSTMs and GRUs address these limitations. Future innovations aim to integrate RNNs with graph-based methods and unsupervised learning for enhanced fraud detection.

7.23.1.2 Autonomous Systems: Autoencoders for Sensor Data Analysis in Robotics

Autonomous systems, such as robots and self-driving cars, rely on deep learning to process and interpret vast amounts of sensor data. Autoencoders play a critical role in analyzing and compressing this data, enabling efficient operation and decision-making.

1. **How Autoencoders Work in Sensor Data Analysis**

   Autoencoders reduce the dimensionality of sensor data by encoding it into a compact latent representation and decoding it back. This process highlights the most essential features while discarding noise and facilitating real-time analysis.

2. **Applications in Autonomous Systems**

   (a) In self-driving cars, autoencoders process data from LiDAR, cameras, and radar to detect obstacles, lane markings, and traffic signs. Compressing this high-dimensional data enables faster decision-making and navigation.
   (b) In industrial robotics, autoencoders analyze data from robotic arms to monitor performance and detect anomalies. This helps ensure consistent quality and prevent equipment failures.
   (c) In drones, autoencoders process visual and environmental data collected by drones, enabling tasks such as terrain mapping and object tracking.

3. **Real-World Impact**

   (a) Companies like Tesla and Boston Dynamics use autoencoders to optimize sensor data processing, enhancing the performance and reliability of their autonomous systems.

   (b) Autoencoders also enable predictive maintenance, reducing downtime and operational costs in industries reliant on robotics.

4. **Challenges**

   While autoencoders are effective, challenges include ensuring robustness in dynamic environments and integrating data from multiple sensor types. Future advancements may focus on combining autoencoders with reinforcement learning to enhance adaptability in complex, real-world scenarios.

7.24　Natural Language Processing

Natural language processing (NLP) is a subfield of artificial intelligence (AI) that focuses on the interaction between computers and human language. The primary goal of NLP is to enable machines to understand, interpret, and generate human language in a meaningful way. NLP bridges the gap between computational techniques and linguistics, making it possible to automate tasks involving text or speech.

By leveraging techniques like tokenization, stemming, lemmatization, and part-of-speech tagging (POS) tagging, NLP facilitates tasks, such as sentiment analysis, machine translation, and chatbot development. These tools and applications are integral to modern AI systems and provide engineers with the capability to harness the power of human language in diverse domains.

7.24.1　Tokenization

Tokenization is the process of splitting a text into smaller units, called tokens, which can be words, phrases, or even characters. Tokenization is essential for breaking down sentences into manageable parts for analysis.

Mathematical Representation

If T is the text, the tokenization function f_{tokenize} can be expressed as:

$$f_{\text{tokenize}}(T) = [t_1, t_2, \ldots, t_n]$$

where t_i represents the i-th token.

Python Code Example: Tokenization

```
#----------------------------------------------------------------
# Tokenization
```

```python
# Chapter 7 in the QUANTUM COMPUTING AND QUANTUM MACHINE LEARNING BOOK
#----------------------------------------------------------------------
# Version 1.0
# (c) 2025 Jesse Van Griensven, Roydon Fraser, and Jose Rosas
# License: MIT - Citation of this work required
#----------------------------------------------------------------------
import numpy as np
import matplotlib.pyplot as plt

import nltk
from nltk.tokenize import word_tokenize
#----------------------------------------------------------------------


# Ensure the necessary tokenizer is downloaded
nltk.download('punkt')

# Example text
text = "Natural Language Processing is fascinating!"

# Tokenize the text
tokens = word_tokenize(text)

# Print tokens
print("Tokens:", tokens)

# Visualizing token frequencies
token_counts = {token: tokens.count(token) for token in set(tokens)}

# Plot token frequencies
plt.figure(figsize=(8, 6))
plt.bar(token_counts.keys(), token_counts.values(),
color='skyblue')
plt.title("Token Frequency Distribution")
plt.xlabel("Tokens")
plt.ylabel("Frequency")
plt.grid(alpha=0.5)
plt.show()
```

7.24.2 Stemming

Stemming reduces words to their root forms by removing suffixes or prefixes, often resulting in non-standard words. It is useful for simplifying text without worrying about grammatical correctness. A stemming example is "running," "runner," and "runs" all reduce to "run."

Python Code Example: Stemming

```python
#-------------------------------------------------------------------
# Stemming
# Chapter 7 in the QUANTUM COMPUTING AND QUANTUM MACHINE LEARNING BOOK
#-------------------------------------------------------------------
# Version 1.0
# (c) 2025 Jesse Van Griensven, Roydon Fraser, and Jose Rosas
# License: MIT - Citation of this work required
#-------------------------------------------------------------------
import numpy as np
import matplotlib.pyplot as plt
from nltk.stem import PorterStemmer
#-------------------------------------------------------------------

# Initialize stemmer
stemmer = PorterStemmer()

# Example words
words = ["running", "runner", "runs", "played", "playing", "plays",
"better", "best"]

# Apply stemming
stemmed_words = [stemmer.stem(word) for word in words]

# Print results
print("Original Words:", words)
print("Stemmed Words:", stemmed_words)

# Visualizing the stemming process
plt.figure(figsize=(10, 6))
plt.bar(words, range(len(words)), label="Original Words",
color="blue", alpha=0.6)
plt.bar(stemmed_words, range(len(words)), label="Stemmed Words",
color="red", alpha=0.6)
plt.xlabel("Words")
plt.ylabel("Index")
plt.title("Stemming Visualization")
plt.legend()
plt.grid(alpha=0.5)
plt.show()
```

7.24.3   *Lemmatization*

Lemmatization is similar to stemming but ensures that the root word is a valid word in the language. It considers the context and uses a vocabulary to produce standard forms of words.

Example
"better" becomes "good," while "running" becomes "run."

Python Code Example: Lemmatization

```python
#------------------------------------------------------------------
# Lemmatization
# Chapter 7 in the QUANTUM COMPUTING AND QUANTUM MACHINE LEARNING BOOK
#------------------------------------------------------------------
# Version 1.0
# (c) 2025 Jesse Van Griensven, Roydon Fraser, and Jose Rosas
# License: MIT - Citation of this work required
#------------------------------------------------------------------
import numpy as np
import matplotlib.pyplot as plt

import nltk
from nltk.stem import WordNetLemmatizer
#------------------------------------------------------------------

# Ensure necessary resources are downloaded
nltk.download('wordnet')
nltk.download('omw-1.4')

# Initialize lemmatizer
lemmatizer = WordNetLemmatizer()

# Example words
words = ["better", "running", "flies", "played", "playing", "happier",
"stronger"]
lemmatized_words = [lemmatizer.lemmatize(word, pos="a") for word in
words]  # Adjective lemmatization

# Print results
print("Original Words:", words)
print("Lemmatized Words:", lemmatized_words)

# Visualizing the lemmatization process
plt.figure(figsize=(10, 6))
plt.bar(words, range(len(words)), label="Original Words",
```

```
color="blue", alpha=0.6)
plt.bar(lemmatized_words, range(len(words)), label="Lemmatized
Words", color="red", alpha=0.6)
plt.xlabel("Words")
plt.ylabel("Index")
plt.title("Lemmatization Visualization")
plt.legend()
plt.grid(alpha=0.2)
plt.show()
```

7.24.4   Part-of-Speech Tagging

Part-of-speech (POS) tagging assigns grammatical labels (e.g., noun, verb, and adjective) to each token in a sentence. This tagging provides structural context for understanding sentence composition.

Mathematical Representation

For a sequence of tokens $[t_1, t_2, ..., t_n]$, the POS tagging function f_{POS} outputs:

$$f_{\mathrm{POS}}([t_1, t_2, ..., t_n]) = [(t_1, \mathrm{POS}_1), (t_2, \mathrm{POS}_2), ..., (t_n, \mathrm{POS}_n)]$$

Python Code Example: Tagging

```
#------------------------------------------------------------------------
# Tagging
# Chapter 7 in the QUANTUM COMPUTING AND QUANTUM MACHINE LEARNING BOOK
#------------------------------------------------------------------------
# Version 1.0
# (c) 2025 Jesse Van Griensven, Roydon Fraser, and Jose Rosas
# License: MIT - Citation of this work required
#------------------------------------------------------------------------
import numpy as np
import matplotlib.pyplot as plt

import nltk
from nltk import pos_tag
from nltk.tokenize import word_tokenize
#------------------------------------------------------------------------

# Ensure necessary resources are downloaded
nltk.download('punkt')
nltk.download('averaged_perceptron_tagger')
```

```python
# Example text
text = "Natural Language Processing is fascinating!"

# Tokenize and apply POS tagging
tokens = word_tokenize(text)
pos_tags = pos_tag(tokens)

# Print POS tags
print("POS Tags:", pos_tags)

# Extract POS tags for visualization
words, tags = zip(*pos_tags)

# Visualizing POS tags as a bar chart
plt.figure(figsize=(10, 6))
plt.bar(words, range(len(words)), label="Words", color="blue",
alpha=0.6)
plt.bar(tags, range(len(tags)), label="POS Tags", color="red",
alpha=0.6)
plt.xlabel("Words")
plt.ylabel("Index")
plt.title("Part-of-Speech (POS) Tagging Visualization")
plt.legend()
plt.grid(alpha=0.2)
plt.show()
```

7.24.5   *Applications in AI*

NLP powers many real-world AI applications, from sentiment analysis to conversational agents.

7.24.5.1   Sentiment Analysis

Sentiment analysis determines the emotional tone of the text, categorizing it as positive, negative, or neutral.

Mathematical Representation

For a text T, sentiment analysis is a classification problem:

$$f_{\text{sentiment}}(T) \rightarrow \{\text{positive}, \text{negative}, \text{neutral}\}$$

Python Code Example: NPL

```python
#------------------------------------------------------------------
# NLP - Natural Language Processing
# Chapter 7 in the QUANTUM COMPUTING AND QUANTUM MACHINE LEARNING BOOK
#------------------------------------------------------------------
# Version 1.0
# (c) 2025 Jesse Van Griensven, Roydon Fraser, and Jose Rosas
# License: MIT - Citation of this work required
#------------------------------------------------------------------
import numpy as np
import matplotlib.pyplot as plt
from textblob import TextBlob
#------------------------------------------------------------------


# Example text for sentiment analysis
text = "I love learning about NLP, but sometimes it is challenging."

# Perform sentiment analysis
analysis = TextBlob(text)
sentiment_polarity = analysis.sentiment.polarity
sentiment_subjectivity = analysis.sentiment.subjectivity

# Print sentiment results
print("Sentiment Polarity:", sentiment_polarity)
print("Sentiment Subjectivity:", sentiment_subjectivity)

# Visualizing sentiment scores
labels = ["Polarity", "Subjectivity"]
values = [sentiment_polarity, sentiment_subjectivity]

plt.figure(figsize=(8, 6))
plt.bar(labels, values, color=["blue", "red"])
plt.title("Sentiment Analysis Results")
plt.ylabel("Score")
plt.ylim(-1, 1)  # Sentiment polarity ranges from -1 to 1
plt.grid(alpha=0.5)
plt.show()
```

7.24.5.2   Machine Translation

Machine translation automatically translates text from one language to another.
Models like Google Translate use neural networks to understand the context and
meaning of sentences.

Example

Translating "How are you?" into Spanish as "¿Cómo estás?"

Python Code Example: Using Hugging Face

```python
#------------------------------------------------------------------
# Hugging Face
# Chapter 7 in the QUANTUM COMPUTING AND QUANTUM MACHINE LEARNING BOOK
#------------------------------------------------------------------
# Version 1.0
# (c) 2025 Jesse Van Griensven, Roydon Fraser, and Jose Rosas
# License: MIT - Citation of this work required
#------------------------------------------------------------------
import numpy as np
import matplotlib.pyplot as plt
from transformers import pipeline
#------------------------------------------------------------------

# Load translation pipeline
translator = pipeline("translation", model="Helsinki-NLP/opus-mt-
en-es")

# Example English sentence
text = "How are you?"

# Perform translation
translation = translator(text)[0]['translation_text']

# Print translated text
print("Original:", text)
print("Translated:", translation)

#------------------------------------------------------------------
# Visualizing translation accuracy using word alignment (basic)
words_en = text.split()
words_es = translation.split()

# Pad words for alignment visualization
max_length = max(len(words_en), len(words_es))
words_en += [""] * (max_length - len(words_en))
words_es += [""] * (max_length - len(words_es))

# Plot word alignment
plt.figure(figsize=(6, 4))
plt.scatter(range(len(words_en)), [1] * len(words_en), color="blue",
label="English Words")
```

```python
plt.scatter(range(len(words_es)), [0] * len(words_es), color="red",
label="Spanish Words")

# Connect corresponding words
for i in range(len(words_en)):
  plt.plot([i, i], [1, 0], "k-", alpha=0.6)

plt.xticks(range(len(words_en)), words_en, rotation=45)
plt.yticks([0, 1], ["Spanish", "English"])
plt.title("Word Alignment (Translation)")
plt.legend()
plt.grid(alpha=0.3)
plt.show()
```

7.24.6   Conversational Agents and Chatbots

Chatbots use NLP to understand and respond to user queries. Advanced models integrate context and intent recognition to provide meaningful interactions.

Example
User: "What's the weather like?"
    Bot: "Today's weather is sunny with a high of 25°C."

Python Code Example: Transformers

```python
#-----------------------------------------------------------------
# Transformers
# Chapter 7 in the QUANTUM COMPUTING AND QUANTUM MACHINE LEARNING BOOK
#-----------------------------------------------------------------
# Version 1.0
# (c) 2025 Jesse Van Griensven, Roydon Fraser, and Jose Rosas
# License: MIT - Citation of this work required
#-----------------------------------------------------------------
import matplotlib.pyplot as plt
from transformers import pipeline
from transformers import ConversationalPipeline, Conversation
#-----------------------------------------------------------------

# Load chatbot pipeline
chatbot = pipeline("conversational", model="facebook/blenderbot-
400M-distill")

# Example conversation
user_input  = "What's the weather like?"
```

```python
conversation = Conversation(user_input)
response    = chatbot(conversation)

# Extract chatbot response
chatbot_response = response.generated_responses[-1]

# Print conversation
print("User:", user_input)
print("Chatbot:", chatbot_response)

#--------------------------------------------------------------------
# Visualizing chatbot interaction
history = [("User", user_input), ("Chatbot", chatbot_response)]

# Extract text for visualization
speakers, messages = zip(*history)

# Plot chatbot conversation
plt.figure(figsize=(6, 4))
plt.scatter(range(len(speakers)), [1] * len(speakers), color="blue",
label="User")
plt.scatter(range(len(messages)), [0] * len(messages), color="red",
label="Chatbot")

# Connect messages in conversation
for i in range(len(speakers)):
  plt.plot([i, i], [1, 0], "k-", alpha=0.6)

plt.xticks(range(len(speakers)), speakers, rotation=45)
plt.yticks([0, 1], ["Chatbot", "User"])
plt.title("Chatbot Conversation Flow")
plt.legend()
plt.grid(alpha=0.3)
plt.show()
```

7.24.7   *Case Study: Combining Techniques*

Consider a system that summarizes product reviews, identifies sentiment, and provides actionable insights.

Steps
1. Tokenize and clean reviews.
2. Apply sentiment analysis to classify emotions.
3. Use machine translation to make reviews accessible in multiple languages.

Python Code Example: Review Analysis

```python
#----------------------------------------------------------------
# Review Analysis
# Chapter 7 in the QUANTUM COMPUTING AND QUANTUM MACHINE LEARNING BOOK
#----------------------------------------------------------------
# Version 1.0
# (c) 2025 Jesse Van Griensven, Roydon Fraser, and Jose Rosas
# License: MIT - Citation of this work required
#----------------------------------------------------------------
import numpy as np
import matplotlib.pyplot as plt
from textblob import TextBlob
#----------------------------------------------------------------

# Sample reviews
reviews = [
    "The product is fantastic! I absolutely love it.",
    "It's okay, but could be better.",
    "I hated the product. It broke on the first day."
]

# Perform sentiment analysis
polarities   = [TextBlob(review).sentiment.polarity   for review in
reviews]
subjectivities = [TextBlob(review).sentiment.subjectivity for review
in reviews]

# Print sentiment results
for review, polarity, subjectivity in zip(reviews, polarities,
subjectivities):
    print(f"Review: {review}\nPolarity: {polarity}, Subjectivity:
{subjectivity}\n")

#----------------------------------------------------------------
# Visualizing Sentiment Analysis
fig, ax = plt.subplots(figsize=(8, 6))
bar_width = 0.4

# Plot polarity
ax.bar(range(len(reviews)), polarities, bar_width,
label="Polarity", color="blue", alpha=0.7)

# Plot subjectivity
ax.bar([p + bar_width for p in range(len(reviews))], subjectivities,
bar_width, label="Subjectivity", color="red", alpha=0.7)
```

```
# Formatting the graph
ax.set_xlabel("Reviews")
ax.set_ylabel("Sentiment Score")
ax.set_title("Sentiment Analysis of Product Reviews")
ax.set_xticks([p + bar_width / 2 for p in range(len(reviews))])
ax.set_xticklabels(["Review 1", "Review 2", "Review 3"], rotation=30)
ax.legend()
ax.grid(alpha=0.5)
plt.show()
```

7.25 AI in Computer Vision

Artificial intelligence (AI) has revolutionized the field of computer vision, enabling machines to interpret and analyze visual data with human-like precision. By leveraging advanced algorithms and deep learning models, computer vision systems can perform tasks ranging from basic image processing to complex scene understanding. This section explores foundational image processing techniques and their applications in engineering, showcasing the transformative impact of AI in various industries.

7.25.1 Image Processing Techniques

Image processing forms the backbone of computer vision, involving the manipulation and analysis of digital images to extract useful information. Traditional techniques such as edge detection and feature extraction have been enhanced with AI, allowing systems to process images with higher accuracy and adaptability.

1. **Edge Detection**

    Edge detection is a fundamental image processing technique that identifies boundaries within an image. It works by detecting abrupt changes in intensity, which correspond to the edges of objects. Algorithms like the Sobel, Prewitt, and Canny edge detectors have been widely used. With the advent of AI, convolutional neural networks (CNNs) have significantly improved edge detection by learning to identify edges in more complex and noisy environments. Applications include object localization, medical imaging, and industrial quality control.

2. **Feature Extraction**

    Feature extraction involves identifying distinct patterns or attributes in an image, such as corners, textures, or shapes. Traditional methods like scale-invariant feature transform (SIFT) and histogram of oriented gradients (HOG) have been replaced or complemented by AI-based approaches, where deep learning models automatically learn hierarchical features. For example, CNNs extract high-level features that capture the essence of an object, enabling tasks like face recognition and scene understanding.

3. **Object Recognition and Classification**

   Object recognition and classification are critical applications of image processing in computer vision. By analyzing an image, AI systems can identify objects and classify them into predefined categories. For instance, a model trained on the ImageNet dataset can recognize thousands of object classes. Real-world applications include retail inventory management, where objects on store shelves are detected and categorized, and wildlife monitoring, where cameras automatically identify and count animal species.

7.25.2 AI Vision Applications in Engineering

The integration of AI into engineering has opened up new possibilities for automating processes and improving efficiency. From ensuring product quality to enabling autonomous vehicle navigation, AI-driven computer vision is a cornerstone of modern engineering solutions.

1. **Defect Detection**

   In manufacturing, defect detection ensures that products meet quality standards by identifying imperfections in real-time. Traditional methods relied on rule-based image processing, which struggled with variability in product shapes, sizes, and textures. AI-based systems, particularly those using CNNs, excel at detecting subtle defects, such as scratches, dents, or misalignments. For example, in semiconductor manufacturing, computer vision systems inspect wafers for microscopic defects, reducing waste and improving production efficiency. Similarly, in textile production, AI identifies irregularities in patterns or fabric textures, ensuring consistent quality.

2. **Autonomous Vehicle Navigation**

   Autonomous vehicles rely heavily on AI-powered computer vision for navigation and decision-making. Cameras mounted on vehicles capture real-time images of the surroundings, which are processed by AI models to interpret the environment. Techniques like object detection and semantic segmentation are used to identify and classify objects such as pedestrians, traffic signs, and other vehicles. Deep learning models like YOLO (You Only Look Once) and Faster R-CNN play a pivotal role in enabling vehicles to detect objects in real-time and make split-second decisions. Additionally, lane detection algorithms guide the vehicle's path, while obstacle detection ensures safe navigation. Autonomous navigation systems are also applied in drones and robotic delivery systems, expanding their utility beyond passenger vehicles.

7.26   Reinforcement Learning and Control Systems

Reinforcement learning (RL) is a type of machine learning where an agent learns to make decisions by interacting with an environment. The learning process is driven by the principle of trial and error, where the agent takes action and receives feedback in the form of rewards or penalties. The goal of the agent is to maximize cumulative rewards over time, developing a strategy or policy that maps states of the environment to the best actions to take in those states.

The core elements of RL are summarized in Table 7.11.

The RL process involves exploration, where the agent tries new actions to discover their effects, and exploitation, where it uses known actions that yield high rewards. Striking the right balance between these two is crucial for efficient learning. RL differs from supervised learning in that it does not require labeled input–output pairs but instead relies on feedback from the environment. This makes RL particularly suited for dynamic and complex scenarios where direct supervision is not feasible.

7.26.1   RL Markov Decision Processes (MDPs)

Markov decision processes provide a mathematical framework for modeling decision-making scenarios in RL. Markov decision process steps are defined in Table 7.12.

An agent's objective in an MDP is to find a policy $\pi(s)$, which defines the best action to take in each state to maximize the cumulative discounted reward:

$$G_t = \sum_{k=0}^{\infty} \gamma^k R_{t+k}$$

Here, G_t is the total reward starting at time t, γ ensures the convergence of the infinite sum, and R_{t+k} is the reward at time $t + k$.

MDPs assume the Markov property, meaning the future state depends only on the current state and action, not on the sequence of past states. This assumption simplifies the modeling and computation of optimal policies.

Table 7.11   Reinforcement learning elements

Core element in RL	Description
Agent	The decision-maker
Environment	Everything the agent interacts with
State	A representation of the current situation in the environment
Action	A decision made by the agent
Reward	Feedback indicating the success of an action

Table 7.12 Markov decision process steps

Step	Action
States (S)	The set of all possible situations the agent may encounter
Actions (A)	The set of all possible decisions the agent can make
Transition probability (P)	The probability of moving from one state to another given a specific action
Reward function (R)	The immediate reward received after transitioning from one state to another due to an action
Discount factor (γ)	A value between 0 and 1 that determines the importance of future rewards compared to immediate rewards

Table 7.13 Robotics RL applications

RL applications in robotics	How it works
Motion planning	RL algorithms enable robots to learn how to navigate environments, avoid obstacles, and reach goals efficiently
Manipulation tasks	Robots can learn to pick and place objects, assemble components, and handle delicate materials
Autonomous vehicles	RL underpins decision-making in self-driving cars, allowing them to adapt to diverse traffic scenarios and optimize routes in real time
Robotic arms for industrial automation	These "intelligent" robotic arms can optimize movements for speed and precision, while minimizing energy consumption

7.26.2 RL Applications in Control Systems

Reinforcement learning has transformative applications in control systems, where the goal is to regulate the behavior of dynamic systems. Two major areas of application are robotics and energy optimization.

7.26.2.1 Robotics

In robotics, RL is employed to enable autonomous systems to perform complex tasks without explicit programming. RL algorithms, such as proximal policy optimization (PPO) or deep Q-networks (DQN), enable these advancements. Example application of RL is presented in Table 7.13.

7.26.2.2 RL in Energy Optimization

Reinforcement learning is revolutionizing energy management systems by providing intelligent control strategies that adapt to fluctuating demands and resource availability. For example, RL-based controllers in smart homes can learn to adjust lighting, heating, and appliance usage dynamically based on occupancy patterns

Table 7.14 RL applications in energy systems

RL applications in energy systems	How it works
Smart grids	RL algorithms optimize energy distribution in power grids, balancing supply and demand while minimizing losses
HVAC systems	RL is used to control heating, ventilation, and air conditioning systems in buildings, improving energy efficiency and reducing costs
Renewable energy management	RL helps in maximizing the utilization of renewable energy sources by efficiently scheduling storage and usage

Table 7.15 Reinforcement learning challenges

RL current challenges	Description
Sample efficiency	RL often requires a large number of interactions with the environment to learn effective policies, which may be impractical in real-world systems
Stability and safety	Ensuring that RL algorithms operate reliably and safely in critical applications is a significant concern
Scalability	Extending RL methods to handle high-dimensional state and action spaces remains a complex task

and energy prices, leading to significant energy savings and enhanced user comfort. Key applications are presented in Table 7.14.

7.26.2.3   RL Challenges

Research in reinforcement learning (RL) is ongoing to address these challenges, with promising advancements in model-based RL, transfer learning, and safe RL. These developments are expected to expand the applicability of RL in control systems, paving the way for more adaptive and intelligent solutions.

Table 7.15 summarizes the challenges that reinforcement learning is currently facing.

7.27   Bridging Classical AI to Quantum AI

Artificial intelligence (AI) has made remarkable strides in solving complex problems, but as we advance further into the era of data-intensive computations, the limitations of classical AI systems are becoming increasingly evident. Quantum AI, which leverages the principles of quantum computing, holds promise to overcome many of these challenges, potentially revolutionizing the field. This section explores the limitations of classical AI, provides an overview of quantum-enhanced machine learning, and highlights key areas where quantum computing could surpass classical AI.

7.27.1   Limitations of Classical AI in High-Complexity Problems

Classical AI systems rely on traditional computing hardware, which uses binary logic and deterministic algorithms to process data. While classical AI has excelled in numerous applications, it struggles with:

1. **Combinatorial Explosion**

   Classical AI algorithms face exponential growth in computational complexity when solving problems with large solution spaces, such as optimization problems in logistics or protein folding in biology. The resources required often become prohibitive, even for the most advanced supercomputers.

2. **Resource Constraints**

   Many AI models, especially deep learning networks, demand immense computational power and memory. This creates bottlenecks in processing massive datasets or training models efficiently.

3. **Noise Sensitivity and Approximation Errors**

   Classical systems are sensitive to noise in data and often rely on approximations for solving problems like nonlinear equations, leading to suboptimal solutions.

4. **Inefficiencies in Parallelism**

   While classical computers can simulate parallelism using multi-core processors, true parallelism is difficult to achieve, limiting the scalability of AI algorithms.

These limitations necessitate a paradigm shift in computational techniques, a shift that quantum computing is uniquely positioned to address.

7.27.2   Overview of Quantum-Enhanced Machine Learning

Quantum computing leverages principles of quantum mechanics, such as superposition, entanglement, and quantum interference, to perform computations that are infeasible for classical systems. Quantum-enhanced machine learning (QML) combines the strengths of quantum computing with traditional AI techniques to unlock new capabilities in problem-solving.

Key features of QML include:

1. **Superposition for Parallelism**

   Unlike classical systems that process one computation at a time, quantum computers can evaluate multiple states simultaneously, drastically increasing computational speed.

2. **Entanglement for Correlated Data Processing**

   Quantum entanglement enables the processing of interdependent data points with high efficiency, making it ideal for analyzing complex, high-dimensional datasets.

3. **Quantum Speedup**

   Algorithms such as quantum support vector machines (QSVM), quantum principal component analysis (QPCA), and Grover's search algorithm offer exponential or quadratic speedups for specific AI tasks.

4. **Enhanced Optimization**

   Quantum annealing, a specialized form of quantum computation, is particularly effective for solving optimization problems, which are central to AI.

5. **Improved Sampling**

   Quantum systems excel in generating random samples efficiently, which can be leveraged for probabilistic models and Bayesian inference in AI.

7.27.3   *Key Areas Where Quantum Computing Could Surpass Classical AI*

The integration of quantum computing into AI opens new frontiers in several domains where classical methods fall short. Below are key areas where quantum AI could achieve unprecedented breakthroughs:

1. **Optimization Problems**

   (a) Classical algorithms struggle with large-scale optimization problems due to their reliance on heuristic methods or brute-force searches.

   (b) Quantum algorithms, such as the quantum approximate optimization algorithm (QAOA), can find optimal solutions more efficiently, benefiting fields like supply chain logistics, financial portfolio optimization, and drug discovery.

2. **Big Data Analysis**

   (a) Classical systems face bottlenecks in processing and extracting insights from massive datasets, especially in real-time.

   (b) Quantum computing's ability to handle and analyze high-dimensional data efficiently makes it invaluable for fields like genomics, social network analysis, and market predictions.

3. **Natural Language Processing (NLP)**

   (a) Classical NLP models struggle with context comprehension and require enormous datasets for training.

   (b) Quantum-enhanced NLP could revolutionize language understanding by efficiently encoding and processing the probabilistic nature of human language.

4. **Pattern Recognition in Complex Systems**

   (a) Tasks such as recognizing patterns in chaotic systems or predicting outcomes in turbulent environments are computationally intensive for classical AI.

   (b) Quantum systems can process multiple pathways simultaneously, enabling superior pattern recognition in fields like weather forecasting and financial modeling.

5. **Cybersecurity and Cryptography**

   (a) Classical encryption systems, such as RSA and ECC, are increasingly vulnerable to advances in computing power.

   (b) Quantum AI can bolster cybersecurity through quantum-safe cryptographic techniques and advanced anomaly detection algorithms.

6. **Drug Discovery and Material Science**

   (a) Simulating molecular interactions at the quantum level is computationally prohibitive for classical systems.

   (b) Quantum AI can accurately model these interactions, leading to faster drug discovery and the development of advanced materials.

7. **Reinforcement Learning for Dynamic Environments**

   (a) Classical reinforcement learning (RL) algorithms require extensive trial-and-error cycles, which can be resource-intensive.

   (b) Quantum-enhanced RL can explore multiple pathways simultaneously, accelerating learning in dynamic and complex environments, such as robotics and autonomous systems.

8. **Quantum Neural Networks (QNNs)**

   (a) Classical neural networks face scalability and training challenges for extremely complex tasks.

   (b) QNNs combine the adaptability of neural networks with quantum speedups, offering powerful tools for tasks like image recognition and medical diagnostics.

Exercise Questions: Classical Artificial Intelligence

1. Foundations of Artificial Intelligence

   (a) Define artificial intelligence and differentiate between narrow AI, general AI, and superintelligent AI.

   (b) What are the key characteristics of an AI system? Provide examples for each characteristic.

2. Historical Development of AI

   (a) Discuss the significance of the Dartmouth Conference in the development of AI.

   (b) How did IBM's Deep Blue and Watson systems demonstrate advancements in AI capabilities?

3. AI, Machine Learning, and Deep Learning

   (a) Compare artificial intelligence, machine Learning, and deep learning, highlighting their differences in focus, techniques, and data requirements.

   (b) Provide an example of how deep learning outperforms machine learning and explain why.

4. Importance of AI for Engineers

   (a) Explain how AI improves accuracy and innovation in engineering applications.

   (b) Describe a real-world example where AI optimizes engineering processes, such as predictive maintenance or generative design.

5. Intelligent Systems

   (a) What are the four key features of intelligent systems, and how do they enable autonomy?

   (b) Provide an example of an intelligent system in healthcare or robotics, explaining its functionality.

6. Search Algorithms

   (a) Compare breadth-first search (BFS) and depth-first search (DFS) in terms of strategy, advantages, and limitations.

   (b) Write the pseudocode for the A* search algorithm and explain its efficiency using heuristic functions.

7. Optimization Techniques

   (a) Describe the steps of the genetic algorithm and explain its advantages over traditional optimization methods.

   (b) What is simulated annealing, and how does it differ from particle swarm optimization in solving optimization problems?

8. Knowledge Representation

   (a) Explain how semantic networks and frames are used in knowledge representation. Provide examples.
   (b) Discuss the role of ontologies in AI and their application in natural language processing.

9. Reasoning and Inference

   (a) Differentiate between deductive, inductive, and probabilistic reasoning. Provide examples for each.
   (b) Using a Bayesian network, explain how conditional probabilities are used to make predictions in a diagnostic system.

10. Learning Paradigms

   (a) Compare supervised, unsupervised, and reinforcement learning. Provide examples of each paradigm in real-world applications.
   (b) Explain the concept of transfer learning and its advantages in reducing training requirements for AI models.

11. Applications of AI in Engineering

   (a) How is AI applied in traffic management systems and smart grid optimization?
   (b) Discuss how search and optimization techniques assist in generative design for engineering applications.

12. The Future of AI in Quantum Computing

   (a) Discuss how classical AI techniques lay the foundation for quantum-enhanced AI.
   (b) Provide an example of a classical AI method that could benefit significantly from quantum computing advancements.

Additional Bibliography

1. V. Badrinarayanan et al., SegNet: A deep convolutional encoder-decoder architecture for scene segmentation. IEEE Trans. Pattern Anal. Mach. Intell. **39**(12), 2481–2495 (2017). https://doi.org/10.1109/TPAMI.2016.2644615
2. Y. Bengio, *Learning Deep Architectures for AI* (Now Publishers Inc., 2009)
3. I. Goodfellow et al., *Deep Learning* (MIT Press, 2016)
4. I. Goodfellow, Et al. "generative adversarial nets.". Adv. Neural Inf. Proces. Syst. **27**, 2672–2680 (2014)
5. Google Research. "Attention Is All You Need." A. Vaswani, et al. *Advances in Neural Information Processing Systems*, vol. 30, 2017, pp. 5998–6008. https://doi.org/10.48550/arXiv.1706.03762

6. G. Hinton et al., Deep learning. Nature **521**(7553), 436–444 (2015). https://doi.org/10.1038/nature14539

7. A. Krizhevsky et al., ImageNet classification with deep convolutional neural networks. Commun. ACM **60**(6), 84–90 (2017). https://doi.org/10.1145/3065386

8. Y. LeCun et al., Deep learning. Nature **521**(7553), 436–444 (2015). https://doi.org/10.1038/nature14539

9. W. McCulloch, W. Pitts, A logical calculus of the ideas immanent in nervous activity. Bull. Math. Biophys. **5**(4), 115–133 (1943). https://doi.org/10.1007/BF02478259

10. J. McCarthy et al., A proposal for the Dartmouth summer research project on artificial intelligence. AI Mag. **27**(4), 12–14 (2006). https://doi.org/10.1609/aimag.v27i4.1904

11. OpenAI. "Language Models Are Few-Shot Learners." A. Brown, et al. *Advances in Neural Information Processing Systems*, vol. 33, 2020, pp. 1877–1901. https://doi.org/10.48550/arXiv.2005.14165

12. S. Ren et al., Faster R-CNN: Towards real-time object detection with region proposal networks. IEEE Trans. Pattern Anal. Mach. Intell. **39**(6), 1137–1149 (2017). https://doi.org/10.1109/TPAMI.2016.2577031

13. Ronneberger, Olaf, et al. U-Net: Convolutional Networks for Biomedical Image Segmentation. *Medical Image Computing and Computer-Assisted Intervention*, vol. 9351, 2015, pp. 234–241. https://doi.org/10.1007/978-3-319-24574-4_28

14. S. Russell, P. Norvig, *Artificial Intelligence: A Modern Approach*, 4th edn. (Pearson, 2020)

15. R.S. Sutton, A.G. Barto, *Reinforcement Learning: An Introduction*, 2nd edn. (MIT Press, 2018)

16. A. Turing, Computing machinery and intelligence. Mind **59**(236), 433–460 (1950). https://doi.org/10.1093/mind/LIX.236.433

17. A. Vaswani et al., Attention is all you need. Adv. Neural Inf. Proces. Syst. **30**, 5998–6008 (2017)

18. IBM, *Deep Blue: Computer Chess Champion* (IBM Research, 1997)

Chapter 8
Introduction to Quantum AI

This chapter combines theoretical insights, practical applications, and mathematical foundations to empower engineers with a robust understanding of quantum AI.

8.1 Quantum Machine Learning Overview

Quantum machine learning (QML) combines the principles of quantum computing with machine learning (ML) techniques, aiming to address computational challenges that classical ML struggles to overcome. By leveraging quantum properties like superposition, entanglement, and interference, QML algorithms can process large datasets, perform complex computations in high-dimensional spaces, and potentially achieve exponential speedups for specific tasks.

Quantum machine learning leverages quantum mechanics to enhance classical ML techniques, offering new algorithms for classification, dimensionality reduction, and optimization. By combining quantum and classical systems, QML addresses challenges in scalability and computational complexity, opening avenues for advancements across industries. The use of quantum feature maps, variational circuits, and quantum kernels demonstrates the transformative potential of QML in solving real-world problems.

This section delves into the key aspects of QML, including quantum-enhanced algorithms, hybrid quantum-classical approaches, and quantum feature mapping, supported by equations and practical examples.

8.2 Quantum-Enhanced Algorithms

Quantum-enhanced algorithms exploit quantum properties to improve classical ML tasks, such as classification, clustering, and dimensionality reduction.

© The Author(s), under exclusive license to Springer Nature Switzerland AG 2025
J. Van Griensven Thé et al., *Quantum Computing and Quantum Machine Learning for Engineers and Developers*, https://doi.org/10.1007/978-3-031-98245-3_8

8.2.1 Quantum Support Vector Machines (QSVMs)

Quantum support vector machines (QSVMs) extend classical support vector machines (SVMs) by employing quantum kernels to enhance feature mapping. The quantum kernel function computes the similarity between two data points x and x' in a quantum feature space:

$$K(x, x') = |\langle \psi(x) | \psi(x') \rangle|^2$$

where $|\psi(x)\rangle$ and $|\psi(x')\rangle$ are the quantum states representing the data points.

The quantum kernel enhances the classification of nonlinearly separable data by leveraging high-dimensional quantum spaces.

Python Code Example: QSVM using Qiskit
This code example demonstrates the use of quantum kernels to classify data in high-dimensional quantum feature spaces.

```
#------------------------------------------------------------------
# Quantum Support Vector Machines - QSVM
# Chapter 8 in the QUANTUM COMPUTING AND QUANTUM MACHINE LEARNING BOOK
#------------------------------------------------------------------
# Version 1.0
# Qiskit changes frequently.
# We recommend using the latest version from the book code repository at:
# https://aqtinitiative.org/quantum-computing-for-engineers

# (c) 2025 Jesse Van Griensven, Roydon Fraser, and Jose Rosas
# License: MIT - Citation of this work required
#------------------------------------------------------------------
import numpy as np
import matplotlib.pyplot as plt

from sklearn.model_selection import train_test_split
from sklearn.datasets import make_classification
from sklearn.metrics import ConfusionMatrixDisplay

from qiskit import Aer, QuantumCircuit
from qiskit.circuit.library import ZZFeatureMap
from qiskit_machine_learning.kernels import QuantumKernel
from qiskit_machine_learning.algorithms import QSVC

#------------------------------------------------------------------
# Graphical Visualization Functions
#------------------------------------------------------------------
```

```python
def plot_decision_boundary(model, X, y):
    """ Plot decision boundary """
    x_min, x_max = X[:, 0].min() - 1, X[:, 0].max() + 1
    y_min, y_max = X[:, 1].min() - 1, X[:, 1].max() + 1
    xx, yy = np.meshgrid(np.linspace(x_min, x_max, 100),
                np.linspace(y_min, y_max, 100))

    Z = model.predict(np.c_[xx.ravel(), yy.ravel()])
    Z = Z.reshape(xx.shape)

    plt.figure(figsize=(8, 6))
    plt.contourf(xx, yy, Z, alpha=0.3, cmap="coolwarm")
    plt.scatter(X[:, 0], X[:, 1], c=y, cmap="coolwarm", edgecolors="k",
s=100)
    plt.xlabel("Feature 1")
    plt.ylabel("Feature 2")
    plt.title("QSVM Decision Boundary")
    plt.grid(alpha=0.5)
    plt.show()

# Plot Confusion Matrix
def plot_confusion(qsvc, X_test, y_test):
    """ Plot Confusion Matrix """
    y_pred = qsvc.predict(X_test)
    fig, ax = plt.subplots(figsize=(6,6))
    ConfusionMatrixDisplay.from_predictions(y_test, y_pred,
cmap="Blues", ax=ax)
    plt.title("Confusion Matrix - QSVM")
    plt.grid(False)
    plt.show()
#---------------------------------------------------------------------

def feature_map(x):
    """ Define a quantum feature map """
    qc = QuantumCircuit(2)
    qc.h(0)
    qc.rx(x[0], 0)
    qc.rx(x[1], 1)
    qc.cx(0, 1)
    return qc
#---------------------------------------------------------------------

# Generate synthetic data
X, y = make_classification(n_samples=100, n_features=2,
```

```
n_informative=2,
            n_redundant=0, n_clusters_per_class=1, random_state=42)

# Split Test / Train
X_train, X_test, y_train, y_test = train_test_split(X, y,
test_size=0.2, random_state=42)

# Create a quantum kernel
feature_map = ZZFeatureMap(feature_dimension=2, reps=2)
quantum_kernel = QuantumKernel(feature_map=feature_map,
quantum_instance=Aer.get_backend('aer_simulator'))
#quantum_kernel = QuantumKernel(feature_map=feature_map,
quantum_instance=Aer.get_backend('aer_simulator'))

# Train QSVM
qsvc = QSVC(quantum_kernel=quantum_kernel)
qsvc.fit(X_train, y_train)

# Evaluate QSVM
accuracy = qsvc.score(X_test, y_test)
print("QSVM Accuracy:", accuracy)

# Execute visualization functions
plot_decision_boundary(qsvc, X, y)
plot_confusion(qsvc, X_test, y_test)
```

8.2.2   Quantum Principal Component Analysis (QPCA)

QPCA utilizes quantum systems to perform dimensionality reduction by finding the principal components of a dataset. The quantum PCA algorithm operates on a quantum state representing the covariance matrix and estimates eigenvalues and eigenvectors using quantum phase estimation:

$$\text{Covariance Matrix}: \Sigma = \frac{1}{n} \sum_{i=1}^{n} x_i x_i^{\mathsf{T}}$$

where x_i are the data vectors in $\mathbb{R}^d$, and x_i^{T} is the transpose of vector x_i.

8.3 Hybrid Quantum-Classical Approaches

Hybrid quantum-classical approaches combine quantum processors for computationally expensive tasks with classical processors for optimization and data handling.

Variational Quantum Classifier (VQC)

VQCs are hybrid classifiers that use parameterized quantum circuits to optimize decision boundaries. The loss function $L(\theta)$ is minimized using classical gradient-based methods:

$$L(\theta) = \sum_{i=1}^{n} \left(y_i - \mathsf{Pred}_\theta(\boldsymbol{x_i}) \right)^2$$

where

x_i is the input data vector.
y_i is the corresponding label (scalar).
θ are the parameters of the quantum circuit.

Python Code Example: VQC

This code example demonstrates how a parameterized quantum circuit can act as a hybrid quantum-classical model for classification tasks.

```
#------------------------------------------------------------------
# Variational Quantum Classifier (VQC)
# Chapter 8 in the QUANTUM COMPUTING AND QUANTUM MACHINE LEARNING BOOK
#------------------------------------------------------------------
# Version 1.0
# Qiskit changes frequently.
# We recommend using the latest version from the book code repository at:
# https://aqtinitiative.org/quantum-computing-for-engineers

# (c) 2025 Jesse Van Griensven, Roydon Fraser, and Jose Rosas
# License: MIT - Citation of this work required
#------------------------------------------------------------------
import numpy as np
import matplotlib.pyplot as plt
from sklearn.datasets import make_classification
from sklearn.model_selection import train_test_split
from sklearn.preprocessing import OneHotEncoder
from sklearn.metrics import ConfusionMatrixDisplay

from qiskit import Aer, QuantumCircuit
from qiskit.circuit.library import ZZFeatureMap, RealAmplitudes
```

```python
from qiskit.algorithms.optimizers import COBYLA
from qiskit_machine_learning.algorithms import VQC
from qiskit_machine_learning.kernels import QuantumKernel


#--------------------------------------------------------------------
# Visualization Functions
#--------------------------------------------------------------------


def plot_decision_boundary(model, X, y):
    """ Function to plot decision boundary """
    x_min, x_max = X[:, 0].min() - 1, X[:, 0].max() + 1
    y_min, y_max = X[:, 1].min() - 1, X[:, 1].max() + 1
    xx, yy = np.meshgrid(np.linspace(x_min, x_max, 100),
                np.linspace(y_min, y_max, 100))

    Z = model.predict(np.c_[xx.ravel(), yy.ravel()])
    Z = np.argmax(Z, axis=1)  # Convert one-hot back to class labels
    Z = Z.reshape(xx.shape)

    plt.figure(figsize=(8, 6))
    plt.contourf(xx, yy, Z, alpha=0.3, cmap="coolwarm")
    plt.scatter(X[:, 0], X[:, 1], c=y, cmap="coolwarm", edgecolors="k",
s=100)
    plt.xlabel("Feature 1")
    plt.ylabel("Feature 2")
    plt.title("VQC Decision Boundary")
    plt.grid(alpha=0.5)
    plt.show()
#--------------------------------------------------------------------


def plot_confusion_matrix(vqc, X_test, y_test):
    """ Function to plot confusion matrix """
    y_pred = vqc.predict(X_test)
    y_pred_labels = np.argmax(y_pred, axis=1)  # Convert one-hot back to
labels
    y_test_labels = np.argmax(y_test, axis=1)  # Convert one-hot back to
labels

    fig, ax = plt.subplots(figsize=(6,6))
    ConfusionMatrixDisplay.from_predictions(y_test_labels,
y_pred_labels, cmap="Blues", ax=ax)
    plt.title("Confusion Matrix - VQC")
    plt.grid(False)
    plt.show()
#--------------------------------------------------------------------
```

```python
# Generate synthetic dataset
X, y = make_classification(n_samples=100, n_features=2, n_classes=2,
n_informative=2,
                n_redundant=0, n_clusters_per_class=1, random_state=42)
X_train, X_test, y_train, y_test = train_test_split(X, y,
test_size=0.2, random_state=42)


#-------------------------------------------------------------------
# Convert labels to one-hot encoding
encoder   = OneHotEncoder(sparse=False)
y_train_oh = encoder.fit_transform(y_train.reshape(-1, 1))
y_test_oh = encoder.transform(y_test.reshape(-1, 1))


#-------------------------------------------------------------------
# Define quantum feature map and ansatz
feature_map = ZZFeatureMap(feature_dimension=2, reps=2)
ansatz    = RealAmplitudes(num_qubits=2, reps=1)

# Define quantum kernel for visualization
quantum_kernel = QuantumKernel(feature_map=feature_map,
quantum_instance=Aer.get_backend("aer_simulator"))


#-------------------------------------------------------------------
# Create and train VQC
vqc = VQC(
  feature_map=feature_map,
  ansatz=ansatz,
  optimizer=COBYLA(),
  quantum_instance=Aer.get_backend("aer_simulator"),
)

vqc.fit(X_train, y_train_oh)

# Evaluate VQC
accuracy = vqc.score(X_test, y_test_oh)
print("VQC Accuracy:", accuracy)


#-------------------------------------------------------------------
# Execute visualization functions
plot_decision_boundary(vqc, X, y)
plot_confusion_matrix(vqc, X_test, y_test_oh)
```

8.3.1 *Quantum Feature Mapping*

Quantum feature mapping encodes classical data into high-dimensional quantum states, enabling efficient exploration of complex relationships.

1. **Mathematical Representation**

    Classical data point x is mapped to a quantum state $|\psi(x)\rangle$ using a feature map $\phi(x)$:

$$|\psi(x)\rangle = U_\phi(x)\,|0\rangle$$

    where $U_\phi(x)$ is a unitary operation encoding the data into the quantum state.

2. **Kernel Trick in Quantum Feature Mapping**

    The quantum kernel computes inner products in the quantum feature space, facilitating nonlinear classification:

$$K(x,x') = |\langle\psi(x)\,|\,\psi(x')\rangle|^2$$

Python Code Example: Quantum Feature Map
This example highlights how quantum feature maps can represent data in high-dimensional spaces for machine-learning tasks.

```python
#-----------------------------------------------------------------
# Quantum Feature Map with Visualization
# Chapter 8 in the QUANTUM COMPUTING AND QUANTUM MACHINE LEARNING BOOK
#-----------------------------------------------------------------
# Version 1.0
# Qiskit changes frequently.
# We recommend using the latest version from the book code repository at:
# https://aqtinitiative.org/quantum-computing-for-engineers

# (c) 2025 Jesse Van Griensven, Roydon Fraser, and Jose Rosas
# License: MIT - Citation of this work required
#-----------------------------------------------------------------
import numpy as np
import matplotlib.pyplot as plt

from qiskit import Aer, QuantumCircuit
from qiskit.circuit.library import ZZFeatureMap
from qiskit_machine_learning.kernels import QuantumKernel
```

```python
from qiskit.visualization import circuit_drawer
#-------------------------------------------------------------------

# Visualizing Quantum Circuit
def plot_quantum_circuit(qc):
    """ Function to plot the quantum circuit """
    print("\nQuantum Feature Map Circuit:")
    print(qc.decompose())  # Print circuit decomposition
    circuit_drawer(qc, output='mpl', style={'backgroundcolor':
'white'})
#-------------------------------------------------------------------

# Define a quantum feature map
feature_map = ZZFeatureMap(feature_dimension=2, reps=2)

# Create a quantum kernel
quantum_kernel = QuantumKernel(feature_map=feature_map,
quantum_instance=Aer.get_backend('aer_simulator'))

# Execute visualization
plot_quantum_circuit(feature_map)
```

8.3.2 Applications of QML

QML finds applications in a variety of fields, as shown in Table 8.1.

8.4 Benefits of Quantum Computing for AI and Machine Learning

Quantum computing offers transformative potential for artificial intelligence (AI) and machine learning (ML) by addressing the fundamental limitations of classical computational resources. Quantum properties like superposition, entanglement, and interference provide new paradigms for solving complex problems,

Table 8.1 Summary of QML applications

QML application	Description
Finance	Portfolio optimization using quantum-enhanced algorithms
Healthcare	Accelerating drug discovery with QPCA for molecular feature extraction
Engineering	Optimizing resource allocation and predictive maintenance with VQC

enabling faster algorithms, efficient data handling, and improved optimization techniques.

Quantum computing offers numerous benefits for AI and ML by providing computational speedup, efficient data handling, enhanced learning models, and improved optimization techniques. Algorithms like Grover's search, quantum gradient descent, and QAOA exemplify how quantum systems address classical computational bottlenecks, enabling breakthroughs in machine learning applications. These advancements pave the way for quantum-enabled AI systems with transformative potential across industries.

8.4.1 Computational Speedup

One of the most significant benefits of quantum computing is the ability to solve problems exponentially faster than classical methods in specific scenarios. For example, quantum search algorithms, such as Grover's algorithm, achieve quadratic speedup in unstructured search problems.

Comparison of Time Complexities

For a search problem with dataset size N:

- **Classical Algorithm**: $T_{\text{classical}} = O(N)$, requiring linear time to find the target.
- **Quantum Algorithm**: $T_{\text{quantum}} = O(\sqrt{N})$, reducing the search time quadratically.

Mathematical Formulation of Grover's Algorithm

The state evolution in Grover's algorithm amplifies the amplitude of the desired state. After $\sqrt{N}$ iterations, the probability of measuring the target state approaches 1:

$$|\psi\rangle = \frac{1}{\sqrt{N}} \sum_{i=0}^{N-1} |i\rangle \rightarrow |x_s\rangle$$

where $|x_s\rangle$ is the marked state.

Python Code Example: Grover's Algorithm for Speedup

This example demonstrates a simple implementation of Grover's algorithm, showcasing the computational speedup in quantum search tasks.

```
#-------------------------------------------------------------------
# Grover's Algorithm for Speedup
# Chapter 8 in the QUANTUM COMPUTING AND QUANTUM MACHINE LEARNING BOOK
#-------------------------------------------------------------------
# Version 1.0
# Qiskit changes frequently.
```

```python
# We recommend using the latest version from the book code repository at:
# https://aqtinitiative.org/quantum-computing-for-engineers

# (c) 2025 Jesse Van Griensven, Roydon Fraser, and Jose Rosas
# License: MIT - Citation of this work required
#-----------------------------------------------------------------
import numpy as np
import matplotlib.pyplot as plt

from qiskit import QuantumCircuit, Aer, execute
from qiskit.visualization import plot_histogram, circuit_drawer
from qiskit.circuit.library import GroverOperator
#-----------------------------------------------------------------

def plot_quantum_circuit(qc):
    """ Visualizing Quantum Circuit """
    print("\nGrover's Algorithm Quantum Circuit:")
    print(qc.decompose())  # Print decomposed circuit
    circuit_drawer(qc, output='mpl', style={'backgroundcolor':
'white'})
    return
#-----------------------------------------------------------------

def plot_measurement_results(counts):
    """ Visualization of Measurement Outcomes """
    plt.figure(figsize=(8, 5))
    plot_histogram(counts)
    plt.title("Measurement Results (Grover's Algorithm)")
    plt.show()
    return
#-----------------------------------------------------------------

def create_oracle():
    """ Define a Custom Oracle (Marking |101⟩ as the solution) """
    oracle = QuantumCircuit(3)
    oracle.x(0)  # Apply X-gate to qubit 0
    oracle.h(2)  # Apply Hadamard to qubit 2
    oracle.ccx(0, 1, 2) # Controlled-Controlled-NOT (Toffoli Gate)
    oracle.h(2)  # Hadamard back
    oracle.x(0)  # Undo X-gate
    return oracle
#-----------------------------------------------------------------

# Create a 3-qubit Grover circuit
qc = QuantumCircuit(3)
```

```python
# Apply Hadamard gates to create a superposition
qc.h([0, 1, 2])

# Apply Grover operator (oracle + diffusion)
oracle   = create_oracle()
grover_op = GroverOperator(oracle=oracle)
qc.append(grover_op, [0, 1, 2])

# Measure the qubits
qc.measure_all()

# Simulate the circuit
simulator = Aer.get_backend('aer_simulator')
result   = execute(qc, simulator).result()
counts   = result.get_counts()

# Execute visualizations
plot_quantum_circuit(qc)
plot_measurement_results(counts)
```

8.4.2 Handling Large-Scale Data

Quantum computers can encode massive datasets into quantum states, leveraging the exponentially growing Hilbert space of 2^n dimensions for n qubits. This capability enables efficient representation and manipulation of large datasets.

Quantum State Encoding
Classical data can be embedded into quantum states using amplitude encoding:

$$|\psi\rangle = \sum_{i=0}^{N-1} c_i |i\rangle$$

where c_i are the data coefficients, and $N = 2^n$ represents the total state space.

Advantages
- Efficient Storage: Quantum states compactly encode high-dimensional data.
- Parallel Processing: Quantum operations act on all components of the state simultaneously.

Python Code Example: Quantum Data Encoding
This code example encodes classical data into a quantum state for efficient storage and manipulation.

```python
#-------------------------------------------------------------------
# Quantum Data Encoding with Visualization
# Chapter 8 in the QUANTUM COMPUTING AND QUANTUM MACHINE LEARNING BOOK
#-------------------------------------------------------------------
# Version 1.0
# Qiskit changes frequently.
# We recommend using the latest version from the book code repository at:
# https://aqtinitiative.org/quantum-computing-for-engineers

# (c) 2025 Jesse Van Griensven, Roydon Fraser, and Jose Rosas
# License: MIT - Citation of this work required
#-------------------------------------------------------------------
import warnings
warnings.filterwarnings('ignore')

import numpy as np
import matplotlib.pyplot as plt

from qiskit import QuantumCircuit, Aer, execute
from qiskit.visualization import import circuit_drawer, plot_histogram

#-------------------------------------------------------------------
# Visualization Functions
# Quantum Circuit Diagram:   Shows how data is encoded into quantum
states.
# Statevector Representation: Probability amplitudes of the quantum
states
# Measurement Histogram:    Outcome distribution after measurement
#-------------------------------------------------------------------

# Function to visualize quantum circuit
def plot_quantum_circuit(qc):
    print("\nQuantum Data Encoding Circuit:")
    print(qc)
    circuit_drawer(qc, output='mpl', style={'backgroundcolor':
'white'})

# Function to simulate statevector
def simulate_statevector(qc):
    backend = Aer.get_backend('statevector_simulator')
    job = execute(qc, backend)
    result = job.result()
    statevector = result.get_statevector()
```

```python
  plt.figure(figsize=(8, 5))
  plt.bar(range(len(statevector)), np.abs(statevector),
color='blue', alpha=0.7)
  plt.xlabel("Quantum State")
  plt.ylabel("Probability Amplitude")
  plt.title("Statevector Representation of Encoded Data")
  plt.grid(alpha=0.5)
  plt.show()

# Function to measure qubits and show probability distribution
def measure_and_plot(qc):
  qc_measure = qc.copy()
  qc_measure.measure_all()

  backend = Aer.get_backend('aer_simulator')
  job = execute(qc_measure, backend, shots=1024)
  result = job.result()
  counts = result.get_counts()

  plt.figure(figsize=(8, 5))
  plot_histogram(counts)
  plt.title("Measurement Results of Encoded Data")
  plt.show()
#------------------------------------------------------------------

# Normalize data vector
data = np.array([1, 2, 3, 4])
norm = np.linalg.norm(data)
normalized_data = data / norm

# Create a quantum circuit for data encoding
qc = QuantumCircuit(2)  # 2 qubits for 4 data points
qc.initialize(normalized_data, [0, 1])

# Execute Visualizations
plot_quantum_circuit(qc)
simulate_statevector(qc)
measure_and_plot(qc)
```

8.4.3   Enhanced Learning Models

Quantum parallelism allows simultaneous evaluation of multiple hypotheses in machine learning models, significantly reducing training times. This is particularly beneficial for models with large parameter spaces or complex loss landscapes. Quantum speedup in learning has two main components:

- Quantum Gradient Descent (QGD): Quantum circuits can compute gradients faster using parameter-shift rules:

$$\frac{\partial L}{\partial \theta} = \frac{L(\theta + \pi/2) - L(\theta - \pi/2)}{2}$$

2. Simultaneous Evaluation: Quantum states encode multiple parameter values, enabling simultaneous evaluation of loss functions.

Python Code Example: Variational Circuit for ML
This code example shows a parameterized quantum circuit for ML tasks, demonstrating enhanced parallelism in learning.

```python
#------------------------------------------------------------------
# Quantum Variational Circuit for ML with Visualization
# Chapter 8 in the QUANTUM COMPUTING AND QUANTUM MACHINE LEARNING BOOK
#------------------------------------------------------------------
# Version 1.0
# Qiskit changes frequently.
# We recommend using the latest version from the book code repository at:
# https://aqtinitiative.org/quantum-computing-for-engineers

# (c) 2025 Jesse Van Griensven, Roydon Fraser, and Jose Rosas
# License: MIT - Citation of this work required
#------------------------------------------------------------------
import warnings
warnings.filterwarnings('ignore')

import numpy as np
import matplotlib.pyplot as plt

from qiskit import QuantumCircuit, Aer, execute
from qiskit.visualization import circuit_drawer, plot_histogram

#------------------------------------------------------------------
# Visualization Functions
# Quantum Circuit Diagram:   Shows how data is encoded into quantum
states.
# Statevector Representation: Probability amplitudes of the quantum
states
```

```python
# Measurement Histogram:    Outcome distribution after measurement
#------------------------------------------------------------------

def plot_quantum_circuit(qc):
    """ Function to plot quantum circuit """
    print("\nVariational Quantum Circuit:")
    print(qc)
    circuit_drawer(qc, output='mpl', style={'backgroundcolor':
'white'})


#------------------------------------------------------------------
def simulate_statevector(qc):
    """ Function to visualize the statevector """
    backend = Aer.get_backend('statevector_simulator')
    job = execute(qc, backend)
    result = job.result()
    statevector = result.get_statevector()

    plt.figure(figsize=(8, 5))
    plt.bar(range(len(statevector)), np.abs(statevector),
color='blue', alpha=0.7)
    plt.xlabel("Quantum State")
    plt.ylabel("Probability Amplitude")
    plt.title("Statevector Representation of Variational Circuit")
    plt.grid(alpha=0.5)
    plt.show()


#------------------------------------------------------------------
def measure_and_plot(qc):
    """ Function to measure qubits and show probability distribution """
    qc_measure = qc.copy()
    qc_measure.measure_all()

    backend = Aer.get_backend('aer_simulator')
    job = execute(qc_measure, backend, shots=1024)
    result = job.result()
    counts = result.get_counts()

    plt.figure(figsize=(8, 5))
    plot_histogram(counts)
    plt.title("Measurement Results of Variational Circuit")
    plt.show()


#------------------------------------------------------------------
# Create a variational quantum circuit
```

```python
#----------------------------------------------------------------
def variational_circuit(params):
  qc = QuantumCircuit(2)
  qc.ry(params[0], 0)
  qc.ry(params[1], 1)
  qc.cx(0, 1)
  return qc

# Example parameters
params = [np.pi / 4., np.pi / 3.]
circuit = variational_circuit(params)


#----------------------------------------------------------------
# Execute Visualizations
#----------------------------------------------------------------
plot_quantum_circuit(circuit)
simulate_statevector(circuit)
measure_and_plot(circuit)
```

8.5 Applications in Optimization

Quantum optimization algorithms, such as the quantum approximate optimization
algorithm (QAOA), improve the efficiency of training neural networks and
hyperparameter tuning.

8.5.1 *Quantum Approximate Optimization Algorithm (QAOA)*

QAOA solves combinatorial optimization problems by encoding the cost function
H_C into a Hamiltonian:

$$|\psi\rangle = e^{-i\beta H_M} e^{-i\gamma H_C} |\psi_0\rangle$$

where H_M is the mixing Hamiltonian, β and γ are variational parameters, and $|\psi_0\rangle$ is
the initial state.

Python Code Example: QAOA for Optimization
This example uses QAOA to solve a binary optimization problem, showcasing its
application in tuning machine learning models.

```python
#-----------------------------------------------------------------
# QAOA for Optimization with Visualization
# Chapter 8 in the QUANTUM COMPUTING AND QUANTUM MACHINE LEARNING BOOK
#-----------------------------------------------------------------
# Version 1.0
# Qiskit changes frequently.
# We recommend using the latest version from the book code repository at:
# https://aqtinitiative.org/quantum-computing-for-engineers

# (c) 2025 Jesse Van Griensven, Roydon Fraser, and Jose Rosas
# License: MIT - Citation of this work required
#-----------------------------------------------------------------

import numpy as np
import matplotlib.pyplot as plt

from qiskit import Aer
from qiskit.algorithms import QAOA
from qiskit_optimization import QuadraticProgram
from qiskit_optimization.algorithms import MinimumEigenOptimizer
from qiskit.utils import algorithm_globals
from qiskit.opflow import I, Z
from qiskit.visualization import circuit_drawer


#-----------------------------------------------------------------
# Visualizing the Cost Function
#-----------------------------------------------------------------
def plot_cost_function():
    x1_vals = [0, 1]
    x2_vals = [0, 1]
    costs = {(x1, x2): -x1 - 2*x2 for x1 in x1_vals for x2 in x2_vals}

    plt.figure(figsize=(6, 5))
    plt.bar(range(len(costs)), costs.values(), tick_label=[f"x1={x1},
x2={x2}" for x1, x2 in costs.keys()], color="royalblue")
    plt.xlabel("Binary Variables (x1, x2)")
    plt.ylabel("Cost Function Value")
    plt.title("Cost Function Representation")
    plt.grid(alpha=0.5)
    plt.show()


#-----------------------------------------------------------------
# Visualizing Quantum Circuit of QAOA
#-----------------------------------------------------------------
def plot_qaoa_circuit(qaoa, operator):
```

```python
    """Plot QAOA circuit with required parameters."""
    circuit = qaoa.construct_circuit(param_dict, operator)[0]  #
Extract the first circuit
    print("\nQAOA Quantum Circuit:")
    circuit_drawer(circuit, output='mpl', style={'backgroundcolor':
'white'})


#-----------------------------------------------------------------
# Visualizing Optimization Results
#-----------------------------------------------------------------
def plot_solution(result):
    x_vals = [int(k) for k in result.variables_dict.values()]
    labels = [f"x{i+1}" for i in range(len(x_vals))]

    plt.figure(figsize=(6, 5))
    plt.bar(labels, x_vals, color="green", alpha=0.7)
    plt.xlabel("Variables")
    plt.ylabel("Binary Assignment")
    plt.title("QAOA Optimization Solution")
    plt.ylim(0, 1.2)
    plt.grid(alpha=0.5)
    plt.show()


#-----------------------------------------------------------------
# Define an optimization problem (Binary Quadratic Model)
#-----------------------------------------------------------------
problem = QuadraticProgram()
problem.binary_var('x1')
problem.binary_var('x2')
problem.minimize(linear={'x1': -1, 'x2': -2})  # Minimize -x1 - 2*x2


#-----------------------------------------------------------------
# Define QAOA parameters
#-----------------------------------------------------------------
algorithm_globals.random_seed = 42  # Ensure reproducibility
quantum_instance = Aer.get_backend('aer_simulator')

qaoa = QAOA(optimizer=None, reps=2,
quantum_instance=quantum_instance)
optimizer = MinimumEigenOptimizer(qaoa)

# Convert Quadratic Problem to Qiskit's Operator
operator, offset = problem.tc_ising()
```

```
# Generate a sample set of QAOA parameters
initial_params = np.random.rand(2 * qaoa.ansatz.reps)  # Get reps from
ansatz
param_dict = {p: v for p, v in zip(qaoa.ansatz.parameters,
initial_params)}  # Convert to dictionary

# Solve the problem
result = optimizer.solve(problem)
print("Solution:", result)

#--------------------------------------------------------------------
# Execute Visualizations
#--------------------------------------------------------------------
plot_cost_function()
plot_qaoa_circuit(qaoa, operator)  # Pass required operator
plot_solution(result)
```

8.6 Quantum-Enhanced Data Processing

Quantum computing significantly improves data processing by leveraging unique quantum principles, such as superposition, entanglement, and interference. Techniques like quantum data encoding, quantum Fourier transforms (QFT), and quantum kernel methods enable efficient handling and analysis of large datasets, offering transformative advantages in machine learning and optimization.

Quantum-enhanced data processing revolutionizes traditional approaches to data handling and analysis by leveraging encoding, QFT, and kernel methods. These techniques address computational bottlenecks in classical systems, enabling efficient representation, analysis, and processing of large-scale datasets. With applications spanning signal processing, optimization, and machine learning, quantum data processing is a cornerstone of quantum computing's impact on data-driven fields.

8.6.1 Quantum Data Encoding

Quantum data encoding is a foundational step in quantum data processing, where classical data is embedded into quantum states for further manipulation.

8.6.2   *Mathematical Representation*

Classical data points x_i are mapped to a quantum state $|\psi\rangle$ in a high-dimensional Hilbert space:

$$|\psi\rangle = \sum_{i=1}^{N} c_i \, |x_i\rangle$$

where c_i are coefficients derived from the data values, and $N = 2^n$ for n-qubits. This mapping enables efficient representation and parallel processing of data.

8.6.3   *Methods of Encoding*

The following are descriptions of methods of quantum encoding.

1. **Amplitude Encoding**

   Data values are normalized and directly encoded into the amplitude of the quantum state:

$$|\psi\rangle = \frac{1}{\sqrt{\sum_i x_i^2}} \sum_i x_i \, |i\rangle$$

2. **Basis Encoding**

   Binary representations of data indices are used to encode data into quantum states:

$$x_1 = |00\rangle, \quad x_2 = |01\rangle, \quad x_3 = |10\rangle, \quad x_4 = |11\rangle$$

Python Code Example: Amplitude Encoding
This code demonstrates the "amplitude encoding" of a classical data vector into a 2-qubit quantum state.

```
#------------------------------------------------------------------
# Quantum Amplitude Encoding with Visualization
# Chapter 8 in the QUANTUM COMPUTING AND QUANTUM MACHINE LEARNING BOOK
#------------------------------------------------------------------
# Version 1.0
# Qiskit changes frequently.
# We recommend using the latest version from the book code repository at:
# https://aqtinitiative.org/quantum-computing-for-engineers
```

```python
# (c) 2025 Jesse Van Griensven, Roydon Fraser, and Jose Rosas
# License: MIT - Citation of this work required
#-----------------------------------------------------------------
import warnings
warnings.filterwarnings('ignore')

import numpy as np
import matplotlib.pyplot as plt
from qiskit import QuantumCircuit, Aer, execute
from qiskit.visualization import circuit_drawer, plot_histogram


#-----------------------------------------------------------------
# Visualization Functions
#-----------------------------------------------------------------

def plot_quantum_circuit(qc):
    """ Function to visualize quantum circuit """
    print("\nQuantum Amplitude Encoding Circuit:")
    print(qc)
    circuit_drawer(qc, output='mpl', style={'backgroundcolor':
'white'})


#-----------------------------------------------------------------
def simulate_statevector(qc):
    """ Function to simulate statevector """
    backend = Aer.get_backend('statevector_simulator')
    job = execute(qc, backend)
    result = job.result()
    statevector = result.get_statevector()

    plt.figure(figsize=(8, 5))
    plt.bar(range(len(statevector)), np.abs(statevector),
color='blue', alpha=0.7)
    plt.xlabel("Quantum State")
    plt.ylabel("Probability Amplitude")
    plt.title("Statevector Representation of Amplitude Encoding")
    plt.grid(alpha=0.5)
    plt.show()


#-----------------------------------------------------------------
def measure_and_plot(qc):
    """ Function to measure qubits and show probability distribution """
    qc_measure = qc.copy()
    qc_measure.measure_all()
```

```python
    backend = Aer.get_backend('aer_simulator')
    job = execute(qc_measure, backend, shots=1024)
    result = job.result()
    counts = result.get_counts()

    plt.figure(figsize=(8, 5))
    plot_histogram(counts)
    plt.title("Measurement Results of Amplitude Encoding")
    plt.show()

#----------------------------------------------------------------------
# Normalize data vector
data = np.array([1, 2, 3, 4])
norm = np.linalg.norm(data)
normalized_data = data / norm

#----------------------------------------------------------------------
# Create a quantum circuit for encoding
qc = QuantumCircuit(2)  # 2 qubits for 4 data points
qc.initialize(normalized_data, [0, 1])

#----------------------------------------------------------------------
# Execute Visualizations
#----------------------------------------------------------------------
plot_quantum_circuit(qc)
simulate_statevector(qc)
measure_and_plot(qc)
```

8.7   Quantum Fourier Transform

The quantum Fourier transform (QFT) is the quantum counterpart of the classical fast Fourier transform (FFT), a widely used algorithm in classical computing for analyzing periodic patterns in data. Unlike FFT, which operates on classical signals, QFT acts on quantum states, leveraging the principles of superposition and entanglement to perform efficient transformations.

The quantum Fourier transform (QFT) is a fundamental tool in quantum computing, enabling efficient frequency and periodicity analysis through quantum parallelism. Its time complexity of $O(n^2)$ makes it exponentially faster than the classical FFT for large input sizes.

By leveraging Hadamard gates, controlled phase gates, and qubit swaps, the QFT circuit efficiently transforms quantum states into the Fourier basis. Applications such as Shor's algorithm, quantum phase estimation, and signal processing highlight the transformative potential of QFT in quantum algorithms.

8.7.1   Definition of QFT

The quantum Fourier transform acts on the computational basis states $|j\rangle$ and transforms them into a superposition of states weighted by complex exponential coefficients.

Mathematically, the QFT for a quantum state $|j\rangle$, where $j \in \{0, 1, ..., N-1\}$ and $N = 2^n$ for n-qubits, is defined as:

$$|j\rangle \rightarrow \frac{1}{\sqrt{N}} \sum_{k=0}^{N-1} e^{2\pi ijk/N} |k\rangle$$

Here

- $N = 2^n$ is the total number of quantum states for n-qubits.
- j and k are integers representing computational basis states
- $e^{2\pi ijk/N}$ are the complex exponential weights.

For an arbitrary quantum state $|\psi\rangle$, which is a superposition of basis states:

$$|\psi\rangle = \sum_{j=0}^{N-1} x_j |j\rangle$$

the QFT transforms it as follows:

$$\text{QFT}(|\psi\rangle) = \sum_{k=0}^{N-1} y_k |k\rangle, \quad \text{where} \quad y_k = \frac{1}{\sqrt{N}} \sum_{j=0}^{N-1} x_j e^{2\pi ijk/N}.$$

The QFT coefficients y_k are obtained via an inner product of the input state amplitudes x_j with the Fourier basis.

8.7.2   Quantum Circuit Implementation of QFT

The QFT can be implemented on a quantum circuit using Hadamard gates, controlled phase gates, and swap gates. This is summarized below (Table 8.2).

The QFT circuit for n-qubits has a depth proportional to $O(n^2)$, which is exponentially faster than the classical FFT for large N.

Python Code Example: Step-by-Step QFT Circuit for 3-Qubits
The QFT for 3-qubits can be implemented as follows:

- Apply a Hadamard gate to the first qubit.
- Apply controlled phase rotations (CR_k) to the remaining qubits.

Table 8.2 Gates employed in the quantum Fourier transform algorithm

Gate type in QFT	Action
Hadamard gates (H)	Used to create superposition states
Controlled phase gates	Apply phase shifts proportional to the distance between qubits
Swap gates	Reverse the order of qubits to match the classical Fourier output ordering

- Repeat steps 1 and 2 for the remaining qubits.
- Swap the qubits to reverse their order.

Here is the Qiskit implementation for a 3-qubit QFT:

```python
#-------------------------------------------------------------------
# QFT Circuit for 3-Qubits with Visualization
# Chapter 8 in the QUANTUM COMPUTING AND QUANTUM MACHINE LEARNING BOOK
#-------------------------------------------------------------------
# Version 1.0
# Qiskit changes frequently.
# We recommend using the latest version from the book code repository at:
# https://aqtinitiative.org/quantum-computing-for-engineers

# (c) 2025 Jesse Van Griensven, Roydon Fraser, and Jose Rosas
# License:  MIT - Citation of this work required
#-------------------------------------------------------------------
import warnings
warnings.filterwarnings('ignore')

import numpy as np
import matplotlib.pyplot as plt

from qiskit import QuantumCircuit, Aer, transpile, execute
from qiskit.visualization import circuit_drawer, plot_histogram


#-------------------------------------------------------------------
# Function to implement QFT
#-------------------------------------------------------------------
def qft(circuit, n):
    """Quantum Fourier Transform implementation"""
    for i in range(n):
        circuit.h(i)  # Apply Hadamard gate
        for j in range(i + 1, n):
            angle = 2. * np.pi / 2**(j - i + 1)
            circuit.cp(angle, j, i)  # Apply controlled phase rotation
```

```python
    for i in range(n // 2):
      circuit.swap(i, n - i - 1)  # Reverse qubit order


#----------------------------------------------------------------
# Visualization Functions
#----------------------------------------------------------------


# Function to plot quantum circuit
def plot_quantum_circuit(qc):
  print("\nQFT Quantum Circuit:")
  circuit_drawer(qc, output='mpl', style={'backgroundcolor':
'white'})


# Function to simulate and show statevector
def simulate_statevector(qc):
  backend = Aer.get_backend('statevector_simulator')
  job = execute(qc, backend)
  result = job.result()
  statevector = result.get_statevector()

  plt.figure(figsize=(8, 5))
  plt.bar(range(len(statevector)), np.abs(statevector),
color='blue', alpha=0.7)
  plt.xlabel("Quantum State")
  plt.ylabel("Probability Amplitude")
  plt.title("Statevector Representation of QFT")
  plt.grid(alpha=0.2)
  plt.show()


# Function to measure qubits and show probability distribution
def measure_and_plot(qc):
  qc_measure = qc.copy()
  qc_measure.measure_all()

  backend = Aer.get_backend('aer_simulator')
  job   = execute(qc_measure, backend, shots=1024)
  result  = job.result()
  counts  = result.get_counts()

  plt.figure(figsize=(8, 5))
  plot_histogram(counts)
  plt.title("Measurement Results of QFT Circuit")
  plt.show()
```

```
#---------------------------------------------------------------------
# Create 3-qubit QFT circuit
#---------------------------------------------------------------------
n = 3
qc = QuantumCircuit(n)
qft(qc, n)


#---------------------------------------------------------------------
# Execute Visualizations
#---------------------------------------------------------------------
plot_quantum_circuit(qc)
simulate_statevector(qc)
measure_and_plot(qc)
```

8.7.3   Comparison with FFT

Table 8.3 compares classical and quantum Fourier transforms.

The QFT achieves its efficiency by operating on quantum parallelism, where all computational basis states are processed simultaneously. In contrast, the FFT operates sequentially on classical inputs.

8.7.4   Example: QFT with Input State

To demonstrate the QFT in practice, consider the input state $|\psi\rangle = |000\rangle$, which represents the integer $j = 0$. Applying the QFT on this state results in:

$$\text{QFT}(|000\rangle) = \frac{1}{\sqrt{8}} \sum_{k=0}^{7} |k\rangle$$

The QFT transforms the $|000\rangle$ state into an equal superposition of all basis states.

Table 8.3 Classical (FFT) versus quantum Fourier transforms (QFT)

Property	QFT	FFT
Time complexity	$O(n^2)$	$O(n \log n)$
Input data	Quantum states	Classical signals
Efficiency	Exponentially faster for n-qubits due to superposition	Efficient for classical systems
Parallelism	Leveraged via quantum entanglement	Classical computation only

Python Code Example: QFT

```python
#-----------------------------------------------------------------
# QFT Circuit with Visualization
# Chapter 8 in the QUANTUM COMPUTING AND QUANTUM MACHINE LEARNING BOOK
#-----------------------------------------------------------------
# Version 1.0
# Qiskit changes frequently.
# We recommend using the latest version from the book code repository at:
# https://aqtinitiative.org/quantum-computing-for-engineers

# (c) 2025 Jesse Van Griensven, Roydon Fraser, and Jose Rosas
# License: MIT - Citation of this work required
#-----------------------------------------------------------------
import warnings
warnings.filterwarnings('ignore')

import numpy as np
import matplotlib.pyplot as plt
from qiskit import QuantumCircuit, Aer, transpile, execute
from qiskit.visualization import plot_histogram, circuit_drawer
#-----------------------------------------------------------------

def sprint(Matrix, decimals=4):
  """ Prints a Matrix with real and imaginary parts rounded to 'decimals'
"""
  import sympy as sp
  SMatrix = sp.Matrix(Matrix)  # Convert to Sympy Matrix if it's not
already

  def round_complex(x):
    """Round real and imaginary parts of x to the given number of
decimals."""
    c = complex(x)  # handle any real or complex Sympy expression
    r = round(c.real, decimals)
    i = round(c.imag, decimals)
    # If imaginary part is negligible, treat as purely real
    if abs(i) < 10**(-decimals): return sp.Float(r)
    else: return sp.Float(r) + sp.Float(i)*sp.I

  # Display the rounded Sympy Matrix
  display(SMatrix.applyfunc(round_complex))
  return
```

```python
#-----------------------------------------------------------------------
# Visualization Functions
#-----------------------------------------------------------------------

def plot_quantum_circuit(qc):
    """ Function to plot quantum circuit """
    print("\nQFT Quantum Circuit:")
    display(circuit_drawer(qc, output='mpl', style=
{'backgroundcolor': 'white'}))
    return
#-----------------------------------------------------------------------

def simulate_statevector(qc):
    """ Function to simulate statevector """
    backend    = Aer.get_backend('statevector_simulator')
    job        = execute(transpile(qc, backend), backend)
    result     = job.result()
    statevector = result.get_statevector()

    # Display statevector
    print("\nStatevector after QFT:")
    #print(statevector)
    sprint(statevector)

    # Plot probability amplitudes
    plt.figure(figsize=(8, 5))
    plt.bar(range(len(statevector)), np.abs(statevector),
color='blue', alpha=0.7)
    plt.xlabel("Quantum State")
    plt.ylabel("Probability Amplitude")
    plt.title("Statevector Representation of QFT")
    plt.grid(alpha=0.5)
    plt.show()
#-----------------------------------------------------------------------

# Function to measure qubits and show probability distribution
def measure_and_plot(qc):
    qc_measure = qc.copy()
    qc_measure.measure_all()

    backend = Aer.get_backend('aer_simulator')
    job     = execute(transpile(qc_measure, backend), backend,
shots=1024)
    result = job.result()
    counts = result.get_counts()
```

```python
plt.figure(figsize=(8, 5))
plot_histogram(counts)
plt.title("Measurement Results of QFT Circuit")
plt.show()


#------------------------------------------------------------------
# Function to implement QFT
#------------------------------------------------------------------
def qft(circuit, n):
    """Quantum Fourier Transform implementation"""
    for i in range(n):
        circuit.h(i)  # Apply Hadamard gate
        for j in range(i + 1, n):
            angle = 2. * np.pi / 2**(j - i + 1)
            circuit.cp(angle, j, i)  # Apply controlled phase rotation
    for i in range(n // 2):
        circuit.swap(i, n - i - 1)  # Reverse qubit order


#------------------------------------------------------------------
# Create 3-qubit QFT circuit
#------------------------------------------------------------------
n = 3
qc = QuantumCircuit(n)
qft(qc, n)


#------------------------------------------------------------------
# Execute Visualizations
#------------------------------------------------------------------
plot_quantum_circuit(qc)
simulate_statevector(qc)
measure_and_plot(qc)
```

8.7.5 Applications of QFT

The following are examples of QFT applications:

- **Frequency Analysis in Signal Processing**
  QFT can be used to analyze the frequency components of quantum states, similar to how FFT processes classical signals.
- **Periodicity Detection in Shor's Algorithm**
  Shor's algorithm uses the QFT to identify the periodicity of modular exponentiation, which is critical for integer factorization.
- **Quantum Phase Estimation**

QFT is a central component in quantum phase estimation algorithms, where it extracts phase information from eigenstates of unitary operators.

- **Quantum Simulations**

  In quantum chemistry and material science, QFT is applied to simulate periodic systems and perform Fourier transformations on wave functions.

8.7.6 Inverse Quantum Fourier Transform (IQFT)

The inverse QFT is the reverse operation of the QFT, transforming Fourier basis states back to the computational basis. It is defined as:

$$|k\rangle \rightarrow \frac{1}{\sqrt{N}} \sum_{j=0}^{N-1} e^{-2\pi ijk/N} |j\rangle$$

The IQFT circuit mirrors the QFT circuit with conjugated phase rotations and reverse qubit ordering.

8.8 Applications of Quantum-Enhanced Data Processing

There are literally thousands of applications where quantum can enhance data processing. The following is a short list to serve as examples.

- **Signal Processing**

  QFT accelerates frequency analysis for tasks, such as speech recognition and image processing.
- **Optimization**

  Quantum data encoding simplifies the representation of optimization problems for algorithms like QAOA.
- **Machine Learning**

  Quantum kernels and feature maps improve classification accuracy and enable the analysis of complex datasets.

8.8.1 Examples of Quantum AI Applications

Quantum AI applications span diverse fields, leveraging quantum capabilities to accelerate image processing, enhance NLP tasks, optimize financial models, and create advanced generative models. By addressing classical computational bottlenecks, these applications demonstrate the transformative potential of quantum AI in solving complex real-world problems.

Quantum artificial intelligence (AI) leverages the unique properties of quantum computing to enhance AI applications across various domains. These advancements address computational bottlenecks, enabling novel methods for data processing, optimization, and generative modeling. This section provides an in-depth exploration of quantum AI applications in image processing, natural language processing (NLP), financial modeling, and generative models.

8.9 Quantum Image Processing

Quantum image processing (QIP) is an emerging field that utilizes quantum computing principles to encode, represent, and analyze images. QIP offers the potential to dramatically accelerate image processing tasks such as edge detection, image compression, noise filtering, and object recognition by leveraging the advantages of quantum superposition, entanglement, and quantum parallelism.

Unlike classical image processing methods that operate on pixel-by-pixel computations, QIP works on the entire image encoded as a quantum state. This enables faster processing, particularly for large images, since quantum systems can perform computations on multiple states simultaneously. By utilizing amplitude encoding and quantum parallelism, QIP accelerates tasks, such as edge detection, image compression, and noise filtering.

Quantum edge detection algorithms efficiently compute gradients using amplitude differences, while quantum PCA enables storage-efficient image compression. Although challenges remain, QIP has significant potential to revolutionize classical image processing, particularly in fields requiring high computational efficiency and accuracy.

8.9.1 *Quantum Image Representation*

To process an image on a quantum computer, it must first be represented as a quantum state. Various methods have been proposed for quantum image representation, with amplitude encoding being one of the most widely used techniques due to its efficiency.

Quantum image representation depends on amplitude encoding, where pixel intensity values are encoded into the amplitudes of a quantum state. For an image with N pixels, the quantum state representation is:

$$|\psi\rangle = \sum_{i=1}^{N} c_i \,|i\rangle, \quad \text{where} \quad \sum_{i=1}^{N} |c_i|^2 = 1$$

where c_i is the normalized pixel intensity, and $|i\rangle$ are the computational basis states representing pixel positions.

Normalization is essential since quantum states must satisfy the condition that the sum of squared amplitudes equals 1.

For a grayscale image with pixel values $p_i \in [0, 255]$, normalization is done as follows:

$$c_i = \frac{p_i}{\sqrt{\sum_{j=1}^{N} p_j^2}}.$$

Python Code Example: Encoding a Simple Image

Let us encode a 2×2 grayscale image into a quantum state using amplitude encoding. Assume the pixel values are:

$$P = \begin{bmatrix} 100 & 150 \\ 200 & 50 \end{bmatrix}$$

First, we normalize the pixel values:

$$c_0 = \frac{100}{\sqrt{100^2 + 150^2 + 200^2 + 50^2}}, \quad c_1 = \frac{150}{\sqrt{\cdots}}, \quad \text{etc.}$$

The following is an implementation of encoding a simple image:

```
#-----------------------------------------------------------------
# Quantum Encoding a Simple Image with Visualization
# Chapter 8 in the QUANTUM COMPUTING AND QUANTUM MACHINE LEARNING BOOK
#-----------------------------------------------------------------
# Version 1.0
# Qiskit changes frequently.
# We recommend using the latest version from the book code repository at:
# https://aqtinitiative.org/quantum-computing-for-engineers

# (c) 2025 Jesse Van Griensven, Roydon Fraser, and Jose Rosas
# License: MIT - Citation of this work required
#-----------------------------------------------------------------
import warnings
warnings.filterwarnings('ignore')
```

```python
import numpy as np
import matplotlib.pyplot as plt
from qiskit import QuantumCircuit, Aer, transpile, execute
from qiskit.visualization import circuit_drawer, plot_histogram
#-------------------------------------------------------------------

def sprint(Matrix, decimals=4):
  """ Prints a Matrix with real and imaginary parts rounded to 'decimals'
"""
    import sympy as sp
    SMatrix = sp.Matrix(Matrix)  # Convert to Sympy Matrix if it's not
already

    def round_complex(x):
      """Round real and imaginary parts of x to the given number of
decimals."""
        c = complex(x)  # handle any real or complex Sympy expression
        r = round(c.real, decimals)
        i = round(c.imag, decimals)
        # If imaginary part is negligible, treat as purely real
        if abs(i) < 10**(-decimals): return sp.Float(r)
        else: return sp.Float(r) + sp.Float(i)*sp.I

    # Display the rounded Sympy Matrix
    display(SMatrix.applyfunc(round_complex))
    return

#----------------------------------------------------------------------
# Visualization Functions
#----------------------------------------------------------------------

def plot_quantum_circuit(qc):
  """ Function to plot the quantum circuit """
  print("\nQuantum Image Encoding Circuit:")
  display(circuit_drawer(qc, output='mpl', style=
{'backgroundcolor': 'white'}))

#
def simulate_statevector(qc):
  """ Function to visualize the statevector """
  backend = Aer.get_backend('statevector_simulator')
  job = execute(transpile(qc, backend), backend)
  result = job.result()
  statevector = result.get_statevector()
```

```python
  # Display statevector
  print("\nStatevector after Quantum Encoding:")
  sprint(statevector)

  # Plot probability amplitudes
  plt.figure(figsize=(8, 5))
  plt.bar(range(len(statevector)), np.abs(statevector),
color='blue', alpha=0.7)
  plt.xlabel("Quantum State")
  plt.ylabel("Probability Amplitude")
  plt.title("Statevector Representation of Image Encoding")
  plt.grid(alpha=0.5)
  plt.show()

# Function to simulate measurement outcomes
def measure_and_plot(qc):
  qc_measure = qc.copy()
  qc_measure.measure_all()

  backend = Aer.get_backend('aer_simulator')
  job = execute(transpile(qc_measure, backend), backend, shots=1024)
  result = job.result()
  counts = result.get_counts()

  plt.figure(figsize=(8, 5))
  plot_histogram(counts)
  plt.title("Measurement Results of Image Encoding")
  plt.show()

#-----------------------------------------------------------------------
# Pixel values representing grayscale intensity
#-----------------------------------------------------------------------
pixels = [100, 150, 200, 50]

# Normalize pixel values for amplitude encoding
norm = np.sqrt(sum(p**2 for p in pixels))
amplitudes = [p / norm for p in pixels]

#-----------------------------------------------------------------------
# Quantum circuit for amplitude encoding
#-----------------------------------------------------------------------
qc = QuantumCircuit(2)  # 2 qubits for 4 pixel values
qc.initialize(amplitudes, [0, 1])  # Encode into 2-qubit state
#-----------------------------------------------------------------------
# Execute Visualizations
```

```
#-------------------------------------------------------------------
plot_quantum_circuit(qc)
simulate_statevector(qc)
measure_and_plot(qc)
```

8.9.2 Quantum Edge Detection

Edge detection is a fundamental task in image processing, used to identify boundaries within an image. Classical edge detection methods, such as Sobel and Canny filters, compute gradients of pixel intensity changes. Quantum algorithms achieve edge detection more efficiently by encoding images into quantum states and leveraging amplitude-based computations.

In quantum edge detection, the gradients of pixel intensities are computed using the amplitude differences between adjacent quantum states. Quantum parallelism allows these computations to be performed simultaneously across all pixels.

For a given quantum state $|\psi\rangle$, representing an image, the gradient at position i can be approximated as:

$$G_i = |c_i - c_{i+1}|$$

where c_i and c_{i+1} are amplitudes corresponding to adjacent pixels.

Python Code Example: Quantum Edge Detection
This is a simplified quantum edge detection algorithm using Qiskit. We compare pixel intensities to detect significant changes (edges).

```
#-------------------------------------------------------------------
# Quantum Edge Detection Circuit with Visualization
# Chapter 8 in the QUANTUM COMPUTING AND QUANTUM MACHINE LEARNING BOOK
#-------------------------------------------------------------------
# Version 3.0
# Qiskit changes frequently.
# We recommend using the latest version from the book code repository at:
# https://aqtinitiative.org/quantum-computing-for-engineers

# (c) 2025 Jesse Van Griensven, Roydon Fraser, and Jose Rosas
# License: MIT - Citation of this work required
#-------------------------------------------------------------------

# Import necessary libraries
import numpy as np
import matplotlib.pyplot as plt
from qiskit import QuantumCircuit, Aer, transpile, execute
```

```python
from qiskit.providers.aer import AerSimulator
from qiskit.visualization import circuit_drawer, plot_histogram


#------------------------------------------------------------------
# Step 1: Creates a Quantum Circuit for Edge Detection
# Step 2: Uses Quantum Phase Estimation (QPE) to enhance edges
# Step 3: Applies Quantum Fourier Transform (QFT) to analyze pixel
frequency
#------------------------------------------------------------------
# Define Quantum Edge Detection Circuit
#------------------------------------------------------------------
def create_edge_detection_circuit():
  qc = QuantumCircuit(3, 3)  # 3 qubits, 3 classical bits (for
measurement)

  # Initialize |+> states to create superposition
  qc.h(0)
  qc.h(1)

  # Apply Quantum Phase Estimation (simulating edge detection)
  qc.cp(np.pi / 2, 0, 1)  # Controlled Phase Shift
  qc.cp(np.pi / 4, 1, 2)  # Smaller Phase Shift for enhancement

  # Apply Quantum Fourier Transform (QFT) to analyze frequency
components
  qc.h(2)
  qc.cx(1, 2)
  qc.h(1)

  # Measure the qubits
  qc.measure([0, 1, 2], [0, 1, 2])

  return qc

# Create the quantum edge detection circuit
qc = create_edge_detection_circuit()


#------------------------------------------------------------------
# Visualization Functions
#------------------------------------------------------------------

# Function to plot the quantum circuit
def plot_quantum_circuit(qc):
  print("\nQuantum Edge Detection Circuit:")
```

```
  display(circuit_drawer(qc, output='mpl', style=
{'backgroundcolor': 'white'}))

# Function to simulate measurement outcomes
def measure_and_plot(qc):
  simulator    = AerSimulator()
  transpiled_qc = transpile(qc, simulator)
  job = simulator.run(transpiled_qc, shots=1024)
  result = job.result()
  counts = result.get_counts()

  plt.figure(figsize=(8, 5))
  plot_histogram(counts)
  plt.title("Measurement Results of Quantum Edge Detection")
  plt.show()


#------------------------------------------------------------------
# Execute Visualizations
#------------------------------------------------------------------
plot_quantum_circuit(qc)
measure_and_plot(qc)
```

The above-coded quantum circuit applies Hadamard gates to generate superpositions, which help compute the differences between pixel values. The resulting quantum state can then be analyzed to identify edges.

8.9.3 *Quantum Image Compression*

Quantum image compression exploits quantum superposition and entanglement to reduce the storage requirements of images. Images are encoded into quantum states, and redundant information is minimized through techniques, such as quantum principal component analysis (QPCA).

In quantum PCA, the image is represented as a density matrix ρ, and the principal components are extracted by performing eigenvalue decomposition. The quantum circuit computes the eigenvalues and eigenvectors using the quantum phase estimation (QPE) algorithm. The eigenvectors corresponding to the largest eigenvalues are retained to reconstruct the compressed image. Mathematically, the image compression is formulated as:

$$\rho \approx \sum_{i=1}^{k} \lambda_i \, | v_i \rangle \langle v_i |$$

where λ_i are the largest eigenvalues, $| v_i \rangle$ are the eigenvectors, and $k \ll N$.

8.9.4 Advantages of Quantum Image Processing

The following is a description of the many advantages derived from QIP.

- **Parallelism**
  Quantum algorithms can process all pixel values simultaneously, significantly speeding up computations for large images.
- **Storage Efficiency**
  Quantum systems encode images into quantum states, reducing the memory requirements for high-resolution images.
- **Edge Detection**
  Quantum algorithms can efficiently compute gradients and detect edges using amplitude-based encoding.
- **Compression**
  Quantum PCA allows for efficient image compression while retaining important visual information.
- **Noise Reduction**
  Quantum algorithms can filter noise in images by exploiting quantum properties, such as entanglement and phase estimation.

8.9.5 Challenges in Quantum Image Processing

The following is a list of limitations from the current stages of quantum computers.

- **Noise in Quantum Hardware**
  Current quantum computers are susceptible to errors, which can affect the accuracy of image-processing tasks.
- **Encoding Overhead**
  Encoding classical images into quantum states requires normalization and state preparation, which can be resource-intensive.
- **Limited Qubits**
  Representing large images on NISQ (noisy intermediate-scale quantum) devices is challenging due to limited qubit availability.
- **Measurement Overhead**
  Extracting useful information from quantum states requires repeated measurements, which can increase computational costs.

8.9.6 *Applications of Quantum Image Processing*

The following list contains examples of QIP applications.

- **Medical Imaging**
  Quantum algorithms can accelerate edge detection and noise reduction in medical images, such as MRI scans.
- **Satellite Imaging**
  QIP can process high-resolution satellite images efficiently for edge detection and object recognition.
- **Image Compression**
  Quantum compression techniques reduce the storage and transmission costs for large image datasets.
- **Object Recognition**
  Quantum algorithms can improve object recognition tasks by processing images in high-dimensional quantum spaces.
- **Computer Vision**
  Quantum edge detection and filtering algorithms can enhance image preprocessing for computer vision systems.

Python Code Example: Quantum Edge Detection
This example demonstrates how quantum circuits encode and process image data for edge detection tasks.

```
#----------------------------------------------------------------------
# Quantum Edge Detection
# Chapter 8 in the QUANTUM COMPUTING AND QUANTUM MACHINE LEARNING BOOK
#----------------------------------------------------------------------
# Version 1.0
# Qiskit changes frequently.
# We recommend using the latest version from the book code repository at:
# https://aqtinitiative.org/quantum-computing-for-engineers


# (c) 2025 Jesse Van Griensven, Roydon Fraser, and Jose Rosas
# License: MIT - Citation of this work required
#----------------------------------------------------------------------


from qiskit import QuantumCircuit, Aer, execute
import numpy as np


# Define pixel intensity data for a simple image
image_data = np.array([1, 0, 2, 3])
norm = np.linalg.norm(image_data)
normalized_data = image_data / norm
```

```
# Create a quantum circuit for encoding image data
qc = QuantumCircuit(2)  # 2 qubits for 4 pixels
qc.initialize(normalized_data, [0, 1])

# Apply Hadamard gates for superposition (edge detection preprocessing)
qc.h([0, 1])

# Measure the qubits
qc.measure_all()

# Simulate the circuit
simulator = Aer.get_backend('aer_simulator')
result = execute(qc, simulator).result()
print(result.get_counts())
```

8.9.7   *Natural Language Processing (NLP)*

Quantum systems enhance NLP by leveraging entanglement and superposition to process linguistic structures efficiently. Applications include sentiment analysis, language translation, and question-answering systems.

1. **Quantum Transformers**

   Quantum transformers generalize classical transformer models by leveraging quantum circuits to capture long-range dependencies:

$$|\psi_{\text{output}}\rangle = U_{\text{attention}} |\psi_{\text{input}}\rangle$$

   where $U_{\text{attention}}$ is a unitary operator implementing the attention mechanism.

2. **Advantages in NLP**

   (a) Parallel Processing: Simultaneous evaluation of multiple word embeddings.
   (b) Efficient Representations: Encoding complex linguistic relationships in high-dimensional quantum states.

Python Code Example: Quantum NLP
The following example encodes words and captures their dependencies using quantum gates for NLP tasks.

```
#---------------------------------------------------------------------
# Quantum NLP with Visualization
# Chapter 8 in the QUANTUM COMPUTING AND QUANTUM MACHINE LEARNING BOOK
```

```python
#------------------------------------------------------------------
# Version 1.0
# Qiskit changes frequently.
# We recommend using the latest version from the book code repository at:
# https://aqtinitiative.org/quantum-computing-for-engineers

# (c) 2025 Jesse Van Griensven, Roydon Fraser, and Jose Rosas
# License: MIT - Citation of this work required
#------------------------------------------------------------------
import warnings
warnings.filterwarnings('ignore')

import numpy as np
import matplotlib.pyplot as plt

from qiskit import QuantumCircuit, Aer, transpile, execute
from qiskit.visualization import circuit_drawer, plot_histogram
from qiskit.quantum_info import Statevector
#------------------------------------------------------------------

def sprint(Matrix, decimals=4):
    """ Prints a Matrix with real and imaginary parts rounded to 'decimals'
"""
    import sympy as sp
    SMatrix = sp.Matrix(Matrix)  # Convert to Sympy Matrix if it's not
already

    def round_complex(x):
        """Round real and imaginary parts of x to the given number of
decimals."""
        c = complex(x)  # handle any real or complex Sympy expression
        r = round(c.real, decimals)
        i = round(c.imag, decimals)
        # If imaginary part is negligible, treat as purely real
        if abs(i) < 10**(-decimals): return sp.Float(r)
        else: return sp.Float(r) + sp.Float(i)*sp.I

    # Display the rounded Sympy Matrix
    display(SMatrix.applyfunc(round_complex))
    return


#------------------------------------------------------------------
# Visualization Functions
#------------------------------------------------------------------
```

```python
# Function to plot the quantum circuit
def plot_quantum_circuit(qc):
  print("\nQuantum NLP Encoding Circuit:")
  display( circuit_drawer(qc, output='mpl', style=
{'backgroundcolor': 'white'}) )


# Function to visualize the statevector
def simulate_statevector(qc):
  simulator = Aer.get_backend("statevector_simulator")
  transpiled_qc = transpile(qc, simulator)
  job = simulator.run(transpiled_qc)
  result = job.result()
  statevector = result.get_statevector()

  # Display statevector
  print("\nQuantum Statevector Representation of NLP Encoding:")
  sprint(statevector)

  # Plot probability amplitudes
  plt.figure(figsize=(8, 5))
  plt.bar(range(len(statevector)), np.abs(statevector),
color='blue', alpha=0.7)
  plt.xlabel("Quantum State")
  plt.ylabel("Probability Amplitude")
  plt.title("Statevector Representation of Quantum NLP")
  plt.grid(alpha=0.5)
  plt.show()

# Function to measure qubits and show probability distribution
def measure_and_plot(qc):
  qc_measure = qc.copy()
  qc_measure.measure_all()

  simulator = Aer.get_backend("aer_simulator")
  transpiled_qc = transpile(qc_measure, simulator)
  job = simulator.run(transpiled_qc, shots=1024)
  result = job.result()
  counts = result.get_counts()

  plt.figure(figsize=(8, 5))
  plot_histogram(counts)
  plt.title("Measurement Results of Quantum NLP Encoding")
  plt.show()
```

```
#---------------------------------------------------------------
# Define a quantum circuit for NLP encoding
#---------------------------------------------------------------
qc = QuantumCircuit(3)

# Encode three words into quantum states using parameterized rotations
qc.ry(0.5, 0)
qc.ry(1.0, 1)
qc.ry(1.5, 2)

# Add entanglement to represent contextual dependencies between words
qc.cx(0, 1)
qc.cx(1, 2)


#---------------------------------------------------------------
# Execute Visualizations
#---------------------------------------------------------------
plot_quantum_circuit(qc)
simulate_statevector(qc)
measure_and_plot(qc)
```

8.10 Financial Modeling

Quantum AI provides advanced tools for financial modeling, with applications in portfolio optimization, risk analysis, and market prediction. Financial systems inherently involve high-dimensional optimization problems, where classical approaches often become computationally inefficient as the number of variables increases.

Quantum algorithms exploit quantum parallelism, entanglement, and superposition to process large datasets and solve complex optimization problems efficiently. Specifically, quantum algorithms such as the quantum approximate optimization algorithm (QAOA) and quantum Monte Carlo methods have significant applications in financial modeling. Integrating quantum methods with classical finance models can revolutionize financial decision-making by enhancing speed, accuracy, and scalability.

8.10.1 Quantum Portfolio Optimization

Portfolio optimization is a fundamental problem in finance that involves selecting the optimal allocation of assets to maximize expected returns while minimizing risk. The challenge can be formulated as a quadratic optimization problem, where the covariance matrix of asset returns plays a critical role. Quantum algorithms provide a powerful approach to solving portfolio optimization problems by encoding the cost function into a Hamiltonian and leveraging quantum variational methods.

The classical portfolio optimization problem, based on Markowitz Modern Portfolio Theory, can be expressed as:

$$\text{Minimize} \quad C = \mathbf{x}^T Q \mathbf{x} - \mu^T \mathbf{x}$$

subject to:

$$\sum_{i=1}^{n} x_i = 1 \quad \text{and} \quad x_i \geq 0$$

where

- $\mathbf{x} = [x_1, x_2, \ldots, x_n]^T$ is the vector of asset weights.
- Q is the covariance matrix of asset returns.
- μ is the vector of expected returns,
- C represents the cost function (a balance of risk and return).

To solve the portfolio optimization problem on a quantum computer, we map the cost function C into a Hamiltonian H_C, where quantum states encode the asset allocations.

The cost Hamiltonian for portfolio optimization is defined as:

$$H_C = \sum_{i,j} Q_{ij} x_i x_j$$

where

- Q_{ij} represents the covariance between asset i and asset j, and $x_i \in \{0, 1\}$ are binary variables indicating the inclusion or exclusion of assets.
- The quantum system minimizes H_C to determine the optimal asset weights that balance risk and return.

8.10.2 QAOA Application for Portfolio Optimization

The quantum approximate optimization algorithm (QAOA) is a hybrid quantum-classical algorithm that iteratively minimizes the cost Hamiltonian. The steps of QAOA for portfolio optimization are as follows:

- **Initialization**
  Encode the initial state $|\psi_0\rangle$ as a superposition of all possible asset allocations.
- **Quantum Evolution**
  Apply alternating layers of cost Hamiltonian H_C and mixing Hamiltonian H_M:

$$|\psi(\beta,\gamma)\rangle = e^{-i\beta H_M} e^{-i\gamma H_C} |\psi_0\rangle$$

- **Measurement**

  Measure the final quantum state to obtain the expectation value of the cost Hamiltonian $\langle H_C \rangle$.
- **Classical Optimization**

  Update the variational parameters β and γ using a classical optimizer to minimize $\langle H_C \rangle$.
- **Iteration**

  Repeat the process until convergence is reached at the optimal solution.

Python Code Example: Quantum Portfolio Optimization

Below is an implementation of QAOA for portfolio optimization using Qiskit. This example considers a simple 3-asset portfolio with a predefined covariance matrix.

```python
#-------------------------------------------------------------------
# Quantum NLP with Visualization
# Quantum Portfolio Optimization with Visualization
# Chapter 8 in the QUANTUM COMPUTING AND QUANTUM MACHINE LEARNING BOOK
#-------------------------------------------------------------------
# Version 1.0
# Qiskit changes frequently.
# We recommend using the latest version from the book code repository at:
# https://aqtinitiative.org/quantum-computing-for-engineers


# (c) 2025 Jesse Van Griensven, Roydon Fraser, and Jose Rosas
# License: MIT - Citation of this work required
#-------------------------------------------------------------------
import numpy as np
import matplotlib.pyplot as plt

from qiskit import Aer, QuantumCircuit, transpile, assemble, execute
from qiskit.circuit import Parameter
from qiskit.opflow import I, Z, X, PauliSumOp, StateFn, CircuitSampler
from qiskit.algorithms.optimizers import COBYLA
from qiskit.utils import QuantumInstance
#-------------------------------------------------------------------


def get_cost_hamiltonian(Q):
    """
    Given a covariance matrix Q, build the cost Hamiltonian
    as a PauliSumOp for QAOA.
    """
    num_qubits = len(Q)
    terms = []
```

```python
    for i in range(num_qubits):
        for j in range(num_qubits):
            if Q[i, j] == 0:
                continue

            label = ["I"] * num_qubits
            label[i] = "Z"
            if j != i:
                label[j] = "Z"

            pauli_string = "".join(label)   # e.g. 'ZIZ' for 3 qubits
            coefficient = Q[i, j]

            # (pauli_string, coefficient)
            terms.append((pauli_string, coefficient))

    return PauliSumOp.from_list(terms)


# -----------------------------------------------------------------------
# Create QAOA ansatz with visualization
def qaoa_ansatz(beta, gamma, num_qubits):
    qc = QuantumCircuit(num_qubits)

    # Initialize in superposition
    for i in range(num_qubits):
        qc.h(i)

    # Apply phase-separation unitary
    for i in range(num_qubits):
        qc.rz(2 * gamma * Q[i, i], i)
        for j in range(i + 1, num_qubits):
            qc.cx(i, j)
            qc.rz(2 * gamma * Q[i, j], j)
            qc.cx(i, j)

    # Apply mixing Hamiltonian
    for i in range(num_qubits):
        qc.rx(2 * beta, i)

    return qc


# -----------------------------------------------------------------------
# Define cost function
def cost_function(params):
```

```python
  beta, gamma = params
  qc = qaoa_ansatz(beta, gamma, 3)

  # Uncomment to visualize the QAOA circuit at each iteration
  # print("\nQAOA Circuit:")
  # display(qc.draw(output='mpl'))

  # Calculate expectation value
  measurable_expression = StateFn(H, is_measurement=True) @ StateFn
(qc)
  sampler = CircuitSampler(QuantumInstance(Aer.get_backend
('statevector_simulator')))
  return np.real(sampler.convert(measurable_expression).eval())
#-------------------------------------------------------------------

def track_cost(params):
  val = cost_function(params)
  cost_values.append(val)
  params_list.append(params)
  return val
#-------------------------------------------------------------------

# Define the covariance matrix (Q)
Q = np.array([
  [0.1, 0.02, 0.04],
  [0.02, 0.15, 0.01],
  [0.04, 0.01, 0.12]
])

# Build the Hamiltonian
H = get_cost_hamiltonian(Q)

# Set up optimizer
initial_params = [0.1, 0.1]
optimizer = COBYLA()

# Track cost for visualization
cost_values = []
params_list = []

# Optimization
result = optimizer.minimize(fun=track_cost, x0=initial_params)

# Plot convergence
plt.figure(figsize=(8, 5))
```

```python
plt.plot(range(len(cost_values)), cost_values, marker='o')
plt.title("QAOA Cost Function Convergence")
plt.xlabel("Iteration")
plt.ylabel("Cost")
plt.grid(True)
plt.show()

# Print final result
print(f"Optimal Parameters: {result.x}")

# Run optimized circuit on qasm simulator for measurement distribution
simulator  = Aer.get_backend('qasm_simulator')
qc_optimized = qaoa_ansatz(result.x[0], result.x[1], 3)
qc_optimized.measure_all()

job  = execute(qc_optimized, backend=simulator, shots=1024)
counts = job.result().get_counts()

# Plot measurement distribution
plt.figure(figsize=(8, 5))
plt.bar(counts.keys(), counts.values(), color='navy')
plt.xlabel("Bitstring outcome")
plt.ylabel("Counts")
plt.title("QAOA Output Distribution")
plt.xticks(rotation=45)
plt.show()
```

8.10.3 Risk Analysis Using Quantum Simulations

Risk analysis involves evaluating the uncertainties in asset returns and quantifying their impact on portfolio performance. Quantum Monte Carlo methods can efficiently simulate random variables and probability distributions, providing a faster alternative to classical Monte Carlo simulations.

For a portfolio with n assets, the quantum system can encode the probability distributions of asset returns and compute risk metrics, such as Value at Risk (VaR) and Conditional Value at Risk (CVaR).

8.10.4 Applications in Market Prediction

Quantum machine learning algorithms can be applied to predict stock prices, identify trading patterns, and forecast market trends. For instance, quantum neural networks and quantum-enhanced time-series analysis provide tools for analyzing financial data with improved accuracy.

- **Quantum Neural Networks (QNN)**
  QNNs can be trained on historical market data to predict future trends and optimize trading strategies.
- **Quantum Time-Series Analysis**
  Quantum algorithms can analyze temporal dependencies in financial data, enabling precise predictions of stock price movements.
- **Anomaly Detection**
  Quantum kernels and quantum support vector machines (SVMs) can detect anomalies and outliers in market data, improving fraud detection and error identification.

8.10.5 Advantages of Quantum Financial Modeling

Quantum computing has opened remarkable opportunities in financial modeling, addressing some of the persistent challenges classical computational approaches face. By harnessing the unique properties of quantum systems, such as superposition and quantum parallelism, quantum algorithms significantly improve speed, accuracy, and scalability. These advantages are particularly beneficial for complex financial operations, including large-scale optimization tasks, risk assessment, and portfolio management. As the financial industry increasingly demands faster and more precise modeling techniques to handle vast datasets and intricate scenarios, quantum computing emerges as a transformative tool capable of redefining financial decision-making processes. The list below further expands on these advantages:

- **Speed**
  Quantum algorithms solve large-scale optimization problems faster than classical algorithms, particularly as the number of assets grows.
- **Accuracy**
  Quantum methods provide higher risk analysis and portfolio optimization precision by leveraging quantum parallelism.
- **Scalability**
  Quantum systems efficiently handle high-dimensional financial datasets, enabling optimization for large portfolios.

8.10.6 Challenges in Quantum Financial Modeling

While quantum computing promises revolutionary advancements for financial modeling, several practical challenges must be addressed to realize its full potential. Quantum financial modeling today primarily relies on noisy intermediate-scale quantum (NISQ) hardware, characterized by limitations in qubit counts and high error rates, restricting the complexity and size of solvable problems. Additionally, encoding classical financial data into quantum states remains resource-intensive and computationally demanding, posing significant implementation hurdles. Moreover, integrating quantum and classical systems in hybrid approaches introduces complexities that can lead to performance overhead and operational inefficiencies. These challenges include:

- **NISQ Hardware**
  Current quantum computers are limited by noise and qubit count, impacting their ability to solve large-scale problems.
- **Data Encoding**
  Encoding financial data into quantum states remains a resource-intensive process.
- **Classical-Quantum Integration**
  Hybrid quantum-classical approaches require seamless integration, which can introduce overhead in practical implementations.

8.11   Quantum Generative Models

Quantum generative adversarial networks (QGANs) generate high-quality data samples, advancing fields like drug discovery, materials science, and simulations (Table 8.4).

8.11.1   QGAN Architecture

QGANs consist of two components:

- A quantum generator (G) that produces synthetic data samples.
- A quantum discriminator (D) that distinguishes between real and synthetic samples.

The objective function is:

Table 8.4 Quantum generative adversarial network applications

QGAN application	Description
Drug discovery	Generate molecular structures for targeted drug design
Materials science	Simulate new materials with desired properties
Data augmentation	Generate synthetic datasets for training machine learning models

$$\min_{G} \max_{D} \mathbb{E}_{x \sim p_{\text{data}}}[\log D(x)] + \mathbb{E}_{z \sim p_z}[\log(1 - D(G(z)))]$$

where p_{data} is the data distribution, and p_z is the latent space distribution.

Python Code Example: QGAN

This example code showcases a QGAN implementation for generating synthetic data samples.

```python
#------------------------------------------------------------------
# QGAN
# Chapter 8 in the QUANTUM COMPUTING AND QUANTUM MACHINE LEARNING BOOK
#------------------------------------------------------------------
# Version 1.0
# Qiskit changes frequently.
# We recommend using the latest version from the book code repository at:
# https://aqtinitiative.org/quantum-computing-for-engineers

# (c) 2025 Jesse Van Griensven, Roydon Fraser, and Jose Rosas
# License: MIT - Citation of this work required
#------------------------------------------------------------------
import numpy as np
import matplotlib.pyplot as plt

from qiskit import Aer, QuantumCircuit, transpile, assemble, execute
from qiskit.circuit import Parameter
from qiskit.opflow import I, Z, X, PauliSumOp, StateFn, CircuitSampler
from qiskit.algorithms.optimizers import COBYLA
from qiskit.utils import QuantumInstance
#------------------------------------------------------------------
import numpy as np
import matplotlib.pyplot as plt

from qiskit_machine_learning.algorithms import QGAN
from qiskit_machine_learning.datasets import gaussian
from qiskit import Aer
from qiskit.utils import QuantumInstance
#------------------------------------------------------------------

# 1) Generate training data (2D)
training_size = 100
test_size = 0
n_features = 2
```

```python
training_data, training_labels, _, _ = gaussian(training_size,
test_size, n_features)
data = training_data


# 2) Compute data bounds
min_x, max_x = np.min(data[:, 0]), np.max(data[:, 0])
min_y, max_y = np.min(data[:, 1]), np.max(data[:, 1])
bounds = np.array([[min_x, max_x], [min_y, max_y]])


# 3) Initialize QGAN with positional arguments
qgan = QGAN(
    data,   # your real training data (shape (100, 2))
    bounds, # shape (2,2)
    [2, 2], # num_qubits: list of 2 ints for 2 features
    10,    # batch_size
    50     # epochs (or num_epochs)
)


# 4) Create a QuantumInstance and Train QGAN
quantum_instance = QuantumInstance(
    backend=Aer.get_backend('qasm_simulator'),
    shots=1024
)


generated_samples, gen_params = qgan.run
(quantum_instance=quantum_instance)
print("Trained QGAN Discriminator and Generator.")


# 5) Plot training losses (if available)
loss_d = qgan._ret.get('loss_d', [])
loss_g = qgan._ret.get('loss_g', [])
epochs_range = range(len(loss_d))


plt.figure(figsize=(6, 4))
plt.plot(epochs_range, loss_d, 'r-o', label='Discriminator Loss')
plt.plot(epochs_range, loss_g, 'b-o', label='Generator Loss')
plt.title("QGAN Training Losses")
plt.xlabel("Epoch")
plt.ylabel("Loss")
plt.grid(True)
plt.legend()
plt.show()


# 6) Compare Real vs. Generated Data
plt.figure(figsize=(6, 4))
```

```
plt.title("Real vs. Generated Data (After Training)")
plt.scatter(data[:, 0], data[:, 1], alpha=0.7, label="Real data")
plt.scatter(generated_samples[:, 0], generated_samples[:, 1],
alpha=0.7, label="Generated data")
plt.legend()
plt.grid(True)
plt.show()
```

8.12   Challenges in Quantum AI

Quantum AI holds immense potential to revolutionize computational capabilities, but it faces significant challenges that hinder its widespread adoption. These challenges are rooted in the limitations of current quantum hardware, the nascent state of quantum machine learning (QML) algorithms, and the complexities of data encoding. Overcoming these obstacles is essential for realizing quantum AI's full potential in practical applications.

8.12.1 Hardware Limitations

In the current stage, quantum computers are quite limited. This section highlights the two main limitations of NISQ quantum computers.

1. **Limited Qubit Count and Connectivity**
   Quantum computers currently have a limited number of qubits, which restricts the scale of problems they can address. Additionally, the physical layout and connectivity of qubits in a quantum processor often impose constraints on gate operations, leading to increased circuit depth and error rates.
2. **High Error Rates and Decoherence**
   Quantum systems are prone to errors due to decoherence, noise, and imperfections in gate implementations. The fidelity of quantum gates, represented by the probability of error-free operation, often falls short for large-scale computations.
3. **Error Correction Challenges**
   Quantum error correction (QEC) schemes, such as the surface code, require a significant overhead in physical qubits to encode a single logical qubit. The relationship between logical and physical qubits is given by:

$$n_{\text{physical}} = k \cdot (1 + r)$$

where k is the number of logical qubits, and r is the redundancy ratio for error correction.

Python Code Example: Simulating Hardware Noise
The following code demonstrates how noise affects quantum computations. It applies a depolarizing noise model to simulate errors during quantum operations.

```python
#------------------------------------------------------------------
# Simulating Hardware Noise
# Chapter 8 in the QUANTUM COMPUTING AND QUANTUM MACHINE LEARNING BOOK
#------------------------------------------------------------------
# Version 1.0
# Qiskit changes frequently.
# We recommend using the latest version from the book code repository at:
# https://aqtinitiative.org/quantum-computing-for-engineers

# (c) 2025 Jesse Van Griensven, Roydon Fraser, and Jose Rosas
# License: MIT - Citation of this work required
#------------------------------------------------------------------

from qiskit import QuantumCircuit, Aer, execute
from qiskit.providers.aer.noise import NoiseModel,
depolarizing_error

# Define a quantum circuit
qc = QuantumCircuit(2)
qc.h(0)  # Apply a Hadamard gate
qc.cx(0, 1)  # Entangle the qubits
qc.measure_all()

# Simulate the circuit with noise
noise_model = NoiseModel()
noise_model.add_all_qubit_quantum_error(depolarizing_error(0.01,
1), ['h'])
noise_model.add_all_qubit_quantum_error(depolarizing_error(0.02,
2), ['cx'])

simulator = Aer.get_backend('aer_simulator')
result = execute(qc, simulator, noise_model=noise_model).result()
print(result.get_counts())
```

8.12.2   Algorithmic Maturity

Most quantum algorithms are fairly new. For example, new quantum algorithms for factoring large numbers are faster and require less qubits than the historical Shor's.

1. **Early Development Stage**

   Many QML algorithms are still in their infancy and require extensive testing to demonstrate advantages over classical counterparts. Algorithms such as the quantum support vector machine (QSVM) and variational quantum classifier (VQC) show promise but often lack robustness when scaled.

2. **Scalability Issues**

   The performance of QML algorithms often deteriorates with increasing problem size due to circuit depth, noise, and decoherence. Ensuring scalability is a critical research area.

3. **Lack of Standardization**

   There is no universal framework for benchmarking quantum algorithms. Metrics such as accuracy, runtime, and resource efficiency need standardization to compare quantum and classical methods fairly.

Python Code Example: Variational Circuit Optimization

This code demonstrates the iterative nature of variational circuits, highlighting the need for robust optimization methods.

```
#-----------------------------------------------------------------
# Variational Circuit Optimization
# Chapter 8 in the QUANTUM COMPUTING AND QUANTUM MACHINE LEARNING BOOK
#-----------------------------------------------------------------
# Version 1.0
# Qiskit changes frequently.
# We recommend using the latest version from the book code repository at:
# https://aqtinitiative.org/quantum-computing-for-engineers

# (c) 2025 Jesse Van Griensven, Roydon Fraser, and Jose Rosas
# License: MIT - Citation of this work required
#-----------------------------------------------------------------

from qiskit import QuantumCircuit, Aer
import numpy as np

# Define a parameterized quantum circuit
def variational_circuit(params):
    qc = QuantumCircuit(2)
    qc.rx(params[0], 0)
    qc.ry(params[1], 1)
    qc.cx(0, 1)
    return qc
```

```python
# Simulate a variational circuit with random parameters
params = np.random.rand(2)
qc = variational_circuit(params)
qc.measure_all()

simulator = Aer.get_backend('aer_simulator')
result = execute(qc, simulator).result()
print(result.get_counts())
```

8.12.3   Data Encoding

In quantum computing, data encoding is required to transform the classical data into quantum states.

1. **Computational Complexity of Encoding**

   Encoding classical data into quantum formats is a non-trivial task. For amplitude encoding, normalizing data and preparing quantum states require $O(2^n)$ resources for n-qubits:

$$|\psi\rangle = \frac{1}{\sqrt{\sum_i |x_i|^2}} \sum_i x_i |i\rangle$$

2. **Loss of Information**

   Data truncation during encoding can lead to the loss of significant information, particularly for high-dimensional datasets. Ensuring that data encoding preserves the fidelity of the original dataset is an ongoing challenge.

3. **Hardware Constraints**

   Preparing quantum states often requires sophisticated hardware and precise control over operations, which increases resource requirements and execution time.

Python Code Example: Data Encoding for QML

This code example demonstrates the computational overhead involved in preparing quantum states for machine learning tasks.

```python
#----------------------------------------------------------------------
# Data Encoding for QML
# Chapter 8 in the QUANTUM COMPUTING AND QUANTUM MACHINE LEARNING BOOK
#----------------------------------------------------------------------
# Version 1.0
# Qiskit changes frequently.
```

```python
# We recommend using the latest version from the book code repository at:
# https://aqtinitiative.org/quantum-computing-for-engineers

# (c) 2025 Jesse Van Griensven, Roydon Fraser, and Jose Rosas
# License: MIT - Citation of this work required
#----------------------------------------------------------------------

from qiskit import QuantumCircuit
import numpy as np

# Normalize classical data
data = np.array([1, 2, 3, 4])
norm = np.linalg.norm(data)
normalized_data = data / norm
# Encode data into a quantum circuit
qc = QuantumCircuit(2)
qc.initialize(normalized_data, [0, 1])

print("Quantum Circuit for Data Encoding:")
print(qc)
```

8.12.4   Overcoming Quantum AI Challenges

Addressing the existing challenges in quantum AI requires coordinated progress across multiple fronts, from hardware advancements to algorithmic innovations. By systematically targeting key areas, such as hardware capabilities, algorithmic efficiency, data encoding methods, and benchmarking practices, the field can move closer to practical, real-world implementations. Current research and development efforts are focused on four primary strategies:

1. **Improved Hardware Design**

   Advances in qubit fidelity, connectivity, and error correction will mitigate hardware limitations. Efforts to develop fault-tolerant quantum computers are crucial for scaling quantum AI.

2. **Algorithmic Innovations**

   Research into more robust and scalable QML algorithms will enhance their applicability. Hybrid quantum-classical methods provide a practical pathway by combining the strengths of both paradigms.

3. **Efficient Data Encoding**

   Novel encoding techniques, such as tensor-network-based encodings, aim to reduce the computational complexity of mapping classical data into quantum states.

4. **Standardized Benchmarks**

Establishing standardized metrics for evaluating quantum algorithms will enable fair comparisons with classical approaches and foster the adoption of best practices.

8.12.5 Final Notes on Quantum AI

Quantum AI faces substantial challenges in hardware capabilities, algorithm development, and data encoding. Addressing these limitations requires interdisciplinary efforts in quantum engineering, algorithm research, and computational methods. Progress in these areas will unlock the potential of quantum AI, enabling breakthroughs in computation, optimization, and machine learning.

Exercise Questions: Introduction to Quantum AI
- **Foundations of Quantum Machine Learning**

    (a) Define quantum machine learning (QML) and explain how it combines quantum computing with classical machine learning techniques.
    (b) Discuss how quantum properties like superposition and entanglement can improve machine learning tasks, such as classification and optimization.

- **Quantum-Enhanced Algorithms**

    (a) Write the mathematical expression for a quantum kernel in a quantum support vector machine (QSVM) and explain its role in nonlinear classification.
    (b) Describe the main steps in quantum principal component analysis (QPCA) for dimensionality reduction and explain its advantage over classical PCA.

- **Hybrid Quantum-Classical Approaches**

    (a) Explain how variational quantum classifiers (VQCs) combine quantum circuits with classical optimization.
    (b) Using the loss function $L(\theta) = \sum_{i=1}^{n} (y_i - \mathsf{Pred}_\theta(x_i))^2$, explain how classical gradient-based methods optimize quantum circuit parameters.

- **Quantum Feature Mapping**

    (a) Describe the process of encoding classical data into quantum states using quantum feature maps.
    (b) Explain how the kernel trick in quantum feature mapping facilitates nonlinear classification tasks.

- **Benefits of Quantum AI**

  (a) Compare the time complexities of classical and quantum algorithms for unstructured search problems, highlighting the advantages of Grover's algorithm.

  (b) Discuss how quantum parallelism reduces the training time for machine learning models with large parameter spaces.

- **Applications of QML**

  (a) Provide an example of how QML is used in healthcare for drug discovery.

  (b) Explain how quantum approximate optimization algorithm (QAOA) can optimize resource allocation in engineering.

- **Quantum Data Encoding**

  (a) Write the mathematical expression for amplitude encoding and explain how it represents classical data in a quantum state.

  (b) Describe the trade-offs between amplitude encoding and basis encoding for large datasets.

- **Quantum Fourier Transform (QFT)**

  (a) Write the mathematical formula for the QFT and explain its role in periodicity detection.

  (b) Compare the efficiency of QFT with the classical fast Fourier transform (FFT) in terms of time complexity.

- **Quantum Generative Models**

  (a) Define a quantum generative adversarial network (QGAN) and explain its components: the quantum generator and discriminator.

  (b) Discuss one application of QGANs in materials science or drug discovery.

- **Quantum Image Processing**

  (a) Explain how quantum edge detection algorithms leverage amplitude encoding for analyzing image gradients.

  (b) Describe the process of encoding pixel intensity values into a quantum state for image processing tasks.

- **Challenges in Quantum AI**

  (a) Identify two major hardware limitations in quantum AI and propose potential solutions to address them.

  (b) Explain the difficulties associated with data encoding for quantum machine learning and discuss novel encoding techniques to mitigate these challenges.

- **Future Directions in Quantum AI**

  (a) Discuss how hybrid quantum-classical methods are bridging the gap between current quantum hardware limitations and practical AI applications.
  (b) Propose a framework for standardizing benchmarks in quantum AI to enable fair comparisons with classical methods.

Additional Bibliography

1. S. Aaronson, *Quantum Computing Since Democritus* (Cambridge University Press, 2013)
2. F. Arute et al., Quantum supremacy using a programmable superconducting processor. Nature **574**, 505–510 (2019)
3. A. Aspuru-Guzik, A. Aspuru-Guzik, *Quantum Chemistry and Machine Learning* (Wiley, 2021)
4. P. Benioff, *The Computer as a Physical System* (Springer, 2016)
5. J. Biamonte, P. Wittek, *Quantum Machine Learning: An Introduction* (Cambridge University Press, 2017)
6. J. Biamonte et al., Quantum machine learning. Nature **549**(7671), 195–202 (2017)
7. R. Cleve, *Quantum Algorithms Revisited* (Elsevier, 2001)
8. D. Deutsch, *The Fabric of Reality: The Science of Parallel Universes—and Its Implications* (Penguin Books, 1997)
9. E. Farhi, J. Goldstone, *A Quantum Approximate Optimization Algorithm* (Springer, 2014)
10. E. Farhi et al., A quantum approximate optimization algorithm. *arXiv preprint arXiv:1411.4028* (2014).
11. L.K. Grover, *Quantum Mechanics Helps in Searching for a Needle in a Haystack* (MIT Press, 1996)
12. L. K. Grover, A Fast Quantum Mechanical Algorithm for Database Search. *Proceedings of the 28th Annual ACM Symposium on Theory of Computing (STOC)* (1996), pp. 212–219.
13. V. Havlíček et al., Supervised learning with quantum-enhanced feature spaces. Nature **567**(7747), 209–212 (2019)
14. P. Kaye, R. Laflamme, M. Mosca, *An Introduction to Quantum Computing* (Oxford University Press, 2007)
15. S. Lloyd, *Programming the Universe: A Quantum Computer Scientist Takes on the Cosmos* (Vintage, 2006)
16. S. Lloyd et al., Quantum algorithms for machine learning. Science **273**(5278), 1073–1078 (2013)
17. S. Majidy, C. Wilson, R. Laflamme, *Building Quantum Computers – A Practical Introduction* (Cambridge Press, 2025)
18. A. Montanaro, *Quantum Algorithms: An Overview* (Springer, 2021)
19. M. A. Nielsen, I. L. Chuang. *Quantum Computation and Quantum Information*. 10th Anniversary ed., Cambridge University Press (2010)
20. J. Preskill, *Quantum Computing in the NISQ Era and Beyond* (Cambridge University Press, 2020)
21. C. Rigetti, *Practical Quantum Computing for Developers* (Addison-Wesley Professional, 2020)
22. M. Schuld, F. Petruccione, *Supervised Learning with Quantum Computers* (Springer, 2018)
23. M. Schuld et al., Circuit-centric quantum classifiers. Phys. Rev. A **101**(3), 032308 (2020)
24. P.W. Shor, *Algorithms for Quantum Computation: Discrete Logarithms and Factoring* (IEEE, 1994)
25. P.W. Shor, Polynomial-time algorithms for prime factorization and discrete logarithms on a quantum computer. SIAM J. Comput. **26**(5), 1484–1509 (1997)

26. A. Vaswani et al., Attention is all you need. Adv. Neural Inf. Proces. Syst. **30**, 5998–6008 (2017)
27. W.K. Wootters, W.H. Zurek, *Quantum Theory and Measurement* (Princeton University Press, 1983)
28. W.H. Zurek, *Decoherence and the Transition from Quantum to Classical* (Oxford University Press, 2003)
29. W.H. Zurek, Decoherence, einselection, and the quantum origins of the classical. *Rev. Mod. Phys.* **75**(3), 715–775 (2003)

Chapter 9
Quantum Neural Networks (QNNs)

This chapter highlights how quantum computing can redefine neural networks by offering new capabilities and addressing existing challenges. Engineers can leverage these innovations for advanced applications, such as optimization, generative modeling, and quantum-specific problem-solving. Future research and technological advancements will continue to shape the role of QNNs in engineering and beyond.

QNNs aim to leverage quantum phenomena such as superposition, entanglement, and quantum parallelism to process high-dimensional data and accelerate machine learning tasks.

9.1 The Motivation for Quantum Neural Networks

This section discusses the motivation behind QNNs, their theoretical evolution, and the fundamental concepts underpinning their architecture, such as quantum perceptrons and quantum neurons.

Despite their success, classical neural networks (NNs) face significant challenges as the complexity of tasks and datasets increases. These challenges are summarized in Table 9.1.

9.2 Benefits of Quantum Computing on Neural Nets

Quantum computing provides a natural framework to enhance neural networks by leveraging uniquely quantum properties, such as superposition, entanglement, and unitary operations. Rather than directly comparing classical and quantum neural networks, Fig. 9.1 illustrates various capabilities, limitations, and hybrid integration points that define quantum neural network (QNN) architectures. The left-hand categories highlight essential features of QNNs, while the right-hand columns

J. Van Griensven Thé et al., *Quantum Computing and Quantum Machine Learning for Engineers and Developers*, https://doi.org/10.1007/978-3-031-98245-3_9

Table 9.1 Limitations of classical neural networks

Factor	Description
Computational complexity	Training deep neural networks requires significant computational power, particularly when dealing with large datasets or complex architectures. The time complexity of training scales poorly for massive models
Energy consumption	Classical hardware consumes enormous energy during training, limiting the sustainability and scalability of deep learning systems
Optimization challenges	Many machine learning problems require solving non-convex optimization problems with exponentially large solution spaces, which are computationally intractable for classical computers
High-dimensional data representation	Representing and processing high-dimensional data efficiently is challenging in classical systems, leading to data redundancy and resource bottlenecks

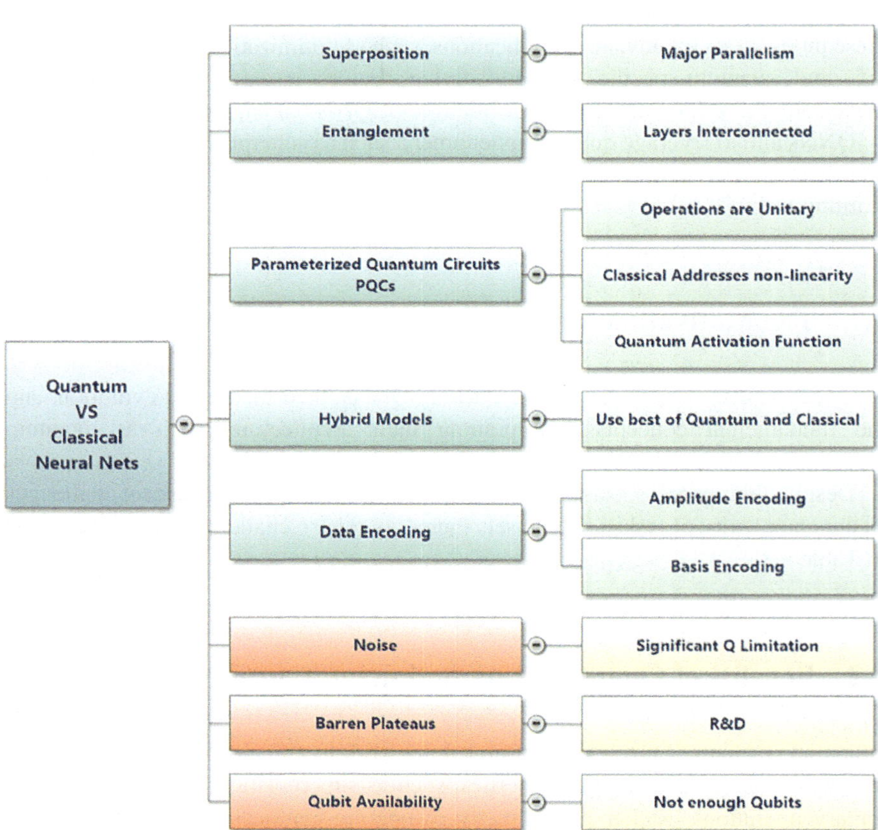

Fig. 9.1 Main aspects between classical and quantum NNs. (By the authors)

provide their associated implications and connections to classical neural network properties or challenges.

Quantum neural networks (QNNs) offer several advantages that can significantly enhance machine learning performance:

1. Exponential speedups for specific computational tasks compared to classical methods. Problems demanding extensive classical resources may be solved more efficiently on quantum hardware.
2. Representing and processing high-dimensional state spaces. A quantum state of n qubits can represent 2^n dimensions simultaneously, enabling efficient processing of complex data.
3. Exploitation of quantum properties such as superposition and entanglement, which allow QNNs to:

   (a) Encode and process multiple states in parallel (quantum parallelism).
   (b) Build correlations across quantum neurons efficiently.

4. **Parameterized quantum circuits** form the hidden layers of QNNs, typically using quantum rotation gates (e.g., R_x, R_y, and R_z) and entanglement to create trainable quantum models.
5. Optimization of these circuits occurs via updates to rotation gate parameters and can be executed using either classical or quantum algorithms.

QNNs embody a hybrid quantum-classical paradigm, leveraging quantum mechanics to accelerate learning, increase scalability, and reduce energy consumption in solving complex machine-learning tasks.

9.3 Evolution of Quantum Neural Networks

The concept of quantum neural networks has evolved over time, with contributions from both theoretical and experimental research.

1. **Early Theoretical Models**

   (a) Initial models attempted to mimic classical perceptrons using quantum mechanical systems.
   (b) Early explorations included quantum-inspired algorithms and theoretical frameworks for quantum learning.

2. **Modern Developments**

   (a) The introduction of variational quantum circuits (VQCs) has enabled the design of quantum models that can learn parameters through optimization processes.
   (b) Algorithms like the quantum approximate optimization algorithm (QAOA) provide quantum-enhanced optimization for combinatorial problems.

(c) Hybrid quantum-classical neural networks integrate quantum layers into classical deep learning architectures, achieving quantum advantages where feasible.

3. **Key Contributors and Breakthroughs**

Significant contributions to QNNs have come from both academia and industry, with support from platforms like:

(a) IBM Quantum
(b) Google Quantum AI
(c) Rigetti Computing
(d) D-Wave Systems

These platforms provide cloud-accessible quantum processors for implementing and testing QNNs in practical scenarios.

9.4 Quantum Perceptron and Quantum Neurons

The building blocks of quantum neural networks are quantum perceptrons and quantum neurons, which are analogous to classical perceptrons and neurons but operate using quantum principles.

9.4.1 Quantum Perceptron

The classical perceptron works as follows:

1. It takes inputs x_i, multiplies them with corresponding weights w_i, and computes a weighted sum:

$$z = \sum_i w_i x_i$$

2. An activation function $f(z)$ is applied to introduce nonlinearity and produce the output.

Quantum perceptrons are designed to achieve a similar function but in the quantum framework.

1. **Input Representation**

Classical inputs x_i are encoded into quantum states $|\psi\rangle$ using techniques like amplitude encoding or phase encoding.

2. **Weight Application**

Quantum gates, such as rotation gates R_x, R_y, or R_z, are used to apply weights to the input qubits. For example:

$$R_y(\theta) = \begin{bmatrix} \cos(\theta/2) & -\sin(\theta/2) \\ \sin(\theta/2) & \cos(\theta/2) \end{bmatrix}$$

where θ represents the weight applied to the qubit.

3. **Activation Function**

Implementing nonlinear activation functions in quantum systems is challenging because quantum operations must be unitary (reversible). Therefore:

(a) Activations can be approximated using parameterized rotations, employing rotation gates (e.g., R_x, R_y, and R_z).

(b) Measurement followed by classical post-processing introduces nonlinearity by mapping quantum states to classical outputs, enabling expressive decision boundaries.

9.5 Quantum-Friendly Activation Functions

To accommodate the unitary nature of quantum operations, activation functions in QNNs are often defined as quantum gates. For example:

1. **Rotation Activation**

Quantum rotation gates like $R_y(\theta)$ serve as activations.

2. **Measurement-Based Nonlinearity**

After applying quantum operations, the measurement outcome can be mapped to a classical activation function.

For instance, if the measurement produces probabilities p_0 and p_1 for states $|0\rangle$ and $|1\rangle$, a classical activation function f can be applied as:

$$f(\theta) = p_0 - p_1$$

9.6 Quantum Neurons

Quantum neurons extend the concept of quantum perceptrons by leveraging quantum entanglement and state evolution.

1. **Weight Representation**

Weights are encoded using quantum gates applied to input qubits. For example, the combination of R_y gates and entangling Controlled-NOT (CNOT) gates introduces correlations between qubits.

2. **Entanglement**

Entanglement allows quantum neurons to correlate data efficiently. For a two-qubit system, entanglement can create a Bell state:

$$| \Phi^+ \rangle = \frac{1}{\sqrt{2}} (| 00 \rangle + | 11 \rangle)$$

3. **Quantum Threshold Gates**

   Quantum threshold models mimic classical decision-making processes. For instance, the measurement outcomes of qubits can act as thresholds to determine the output class.

9.7   Impact of Superposition and Entanglement in Neurons

The parameterized quantum circuits representing each neural net layer require the following constituents:

1. **Superposition**

   Quantum neurons can encode multiple inputs simultaneously due to superposition. This enables QNNs to process vast amounts of data in parallel.
2. **Entanglement**

   Entanglement introduces strong correlations between quantum neurons, allowing them to model intricate patterns in data more effectively than classical neurons.

9.8   Mapping Neural Networks to Quantum Circuits

Quantum neural networks (QNNs) bridge the gap between classical deep learning and quantum computing, introducing new possibilities for efficient computation and enhanced learning models. This section provides a detailed exploration of how traditional neural network architectures are mapped to quantum circuits.

Mapping neural networks to quantum circuits involves encoding classical data into quantum states, designing parameterized quantum layers, and training these layers using hybrid optimization techniques. By leveraging quantum analogs of classical components, QNNs introduce powerful capabilities for data representation and learning, paving the way for advancements in AI and quantum computing.

Quantum neural networks employ primarily two types of entanglement topologies. They differ in how qubits are interconnected (entangled), impacting computational efficiency and expressive capability. These two topologies are:

1. **Linear entanglement** is generally recommended for simpler tasks or initial exploration, especially when quantum hardware resources or coherence time are limited.
2. **Circular entanglement** is preferable when the computational task benefits from higher correlation complexity and expressivity, such as modeling complex interactions or cyclic dependencies.

Table 9.2 Comparison between linear and circular entanglement topologies

Aspect	Linear entanglement	Circular entanglement
Connectivity	Sequential, neighbors only	Sequential plus loop link
Circuit complexity	Lower	Slightly higher
Quantum correlation	Moderate, localized	Enhanced, global
Quantum hardware compatibility	Better for NISQ (near-term) hardware	Slightly more demanding
Task suitability	Problems with linear or sequential structure	Problems requiring global or cyclic correlations

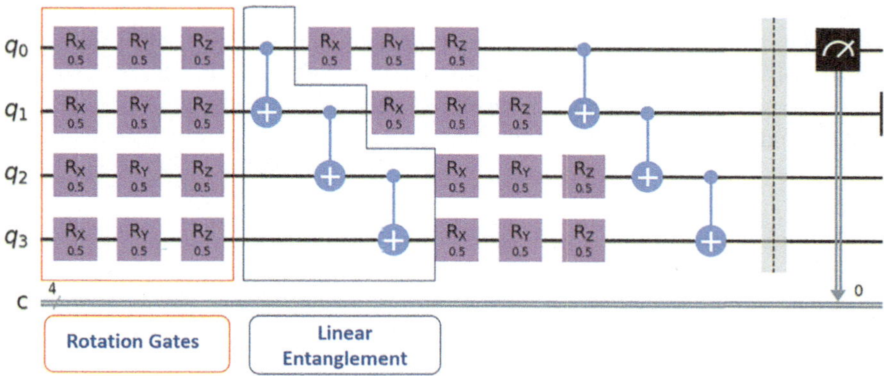

Fig. 9.2 Quantum neural network with linear entanglement. (By the authors)

The differences between these two topologies are highlighted in Table 9.2.
Figure 9.2 presents a quantum neural network circuit with linear entanglement.

9.9 Quantum Encoding of Neural Networks

Quantum neural networks utilize quantum principles for data representation and
computation.

1. **Encoding Classical Data**

   (a) **Amplitude Encoding:** Encodes classical data into quantum state amplitudes:

   $$|\psi\rangle = \sum_i \frac{x_i}{\sqrt{\sum_j x_j^2}} |i\rangle$$

   (b) **Basis Encoding:** Maps binary representations of data to computational basis
       states.

2. **Mapping Weights and Biases**

Weights and biases are translated into quantum gates. Parameterized quantum circuits (PQCs) serve as trainable quantum layers.

3. **Variational Ansatz for Quantum Circuits**

QNNs rely on a "variational form," also known as an ansatz, where unitary operations $U(\theta)$ represent trainable parameters:

$$|\psi(\theta)\rangle = U(\theta_1)U(\theta_2)...U(\theta_n)|0\rangle$$

Python Code Example: Parameterized Quantum Circuit

This code defines a simple parameterized quantum circuit (PQC), often used as a variational ansatz to generate quantum states that depend on tunable parameters. These circuits are fundamental to variational quantum algorithms, such as QNNs, QAOA, and VQE.

```python
#------------------------------------------------------------------
# Parameterized Quantum Circuit
# Chapter 9 in the QUANTUM COMPUTING AND QUANTUM MACHINE LEARNING BOOK
#------------------------------------------------------------------
# Version 1.0
# Qiskit changes frequently.
# We recommend using the latest version from the book code repository at:
# https://aqtinitiative.org/quantum-computing-for-engineers

# (c) 2025 Jesse Van Griensven, Roydon Fraser, and Jose Rosas
# License: MIT - Citation of this work required
#------------------------------------------------------------------
import numpy as np
import matplotlib.pyplot as plt

from qiskit import QuantumCircuit
from qiskit.visualization import plot_circuit_layout, circuit_drawer
#------------------------------------------------------------------

# Create a parameterized quantum circuit
def variational_ansatz(params):
  qc = QuantumCircuit(2)
  qc.ry(params[0], 0)
  qc.ry(params[1], 1)
  qc.cx(0, 1)
  return qc
#------------------------------------------------------------------
```

```python
# Example parameters
params = [np.pi / 4., np.pi / 3.]
qc = variational_ansatz(params)

# Display the quantum circuit
print("Quantum Circuit:")
print(qc)
qc.draw('mpl')  # Draws the circuit as a matplotlib plot

# Plot the parameter values
plt.figure(figsize=(6, 4))
plt.bar(['Theta_0', 'Theta_1'], params, color=['blue', 'orange'])
plt.title("Parameterized Rotation Angles")
plt.ylabel("Angle (radians)")
plt.grid(axis='y', linestyle='--', alpha=0.6)
plt.show()
```

9.10   Quantum Analogs of Neural Layers

QNNs replace classical layers with quantum layers, introducing new mechanisms for computation.

9.10.1   Quantum Perceptrons

Quantum perceptrons replace activation functions with quantum operations, such as controlled rotations or measurements.

9.10.2   Unitary Transformations

Unitary gates simulate both linear and nonlinear transformations, enabling the quantum analog of matrix multiplication and nonlinear activation.

9.10.3   Circuit Depth and Expressivity

The depth of a quantum circuit determines its capacity to approximate complex functions. A balance must be struck between expressivity and trainability to avoid issues like overfitting or vanishing gradients.

Python Code Example: Quantum Perceptron

```python
#----------------------------------------------------------------------
# Quantum Perceptron
# Chapter 9 in the QUANTUM COMPUTING AND QUANTUM MACHINE LEARNING BOOK
#----------------------------------------------------------------------
# Version 1.0
# Qiskit changes frequently.
# We recommend using the latest version from the book code repository at:
# https://aqtinitiative.org/quantum-computing-for-engineers

# (c) 2025 Jesse Van Griensven, Roydon Fraser, and Jose Rosas
# License: MIT - Citation of this work required
#----------------------------------------------------------------------
import numpy as np
import matplotlib.pyplot as plt

from qiskit import QuantumCircuit, Aer, execute
from qiskit.visualization import import plot_histogram
#----------------------------------------------------------------------

# Quantum perceptron with controlled rotation
qc = QuantumCircuit(2)
qc.h(0)  # Create superposition
qc.crx(np.pi / 4., 0, 1)  # Controlled rotation
qc.measure_all()

print("Quantum Perceptron Circuit:")
print(qc)

# Visualize the quantum circuit
qc.draw('mpl')  # High-quality circuit diagram

# Simulate the circuit
simulator = Aer.get_backend('qasm_simulator')
result   = execute(qc, simulator, shots=1000).result()

# Extract and visualize the measurement results
counts = result.get_counts(qc)
print("Measurement Results:", counts)

# Plot the results as a histogram
plot_histogram(counts)
plt.title("Quantum Perceptron Measurement Results")
plt.show()
```

9.11 Training Quantum Neural Networks

Training QNNs involves optimizing parameters in quantum circuits using hybrid quantum-classical approaches. QNN's workflow is presented in Fig. 9.3.

9.11.1 Quantum Gradient Descent (QGD)

QGD uses parameter-shift rules to compute approximate gradients:

$$\frac{\partial L}{\partial \theta} = \frac{L(\theta + \pi/2) - L(\theta - \pi/2)}{2}$$

9.11.2 Cost Function Evaluation

The cost function is evaluated using the expectation value of quantum observables:

$$C(\theta) = \langle \psi(\theta) \,|\, H \,|\, \psi(\theta) \rangle$$

where H is a Hamiltonian representing the problem.

9.11.3 Mitigating Vanishing Gradients

Vanishing gradients occur in deep circuits due to reduced variance in parameter shifts. Techniques like layer-wise learning and adaptive optimizers mitigate this issue.

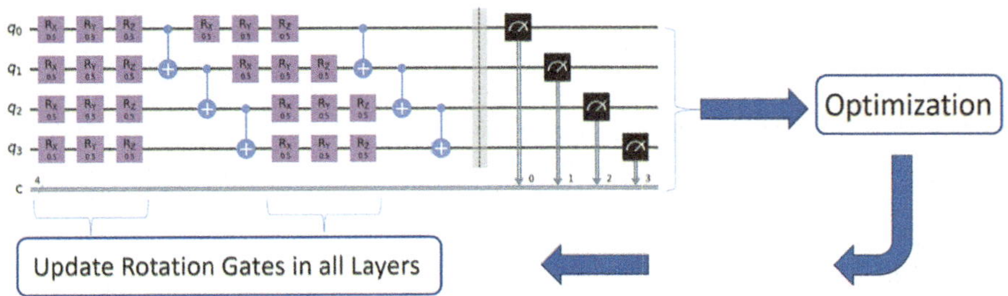

Fig. 9.3 QNN workflow where optimization can be classical or quantum-based. (By the authors)

Python Code Example: Training a QNN

This code demonstrates the parameter-shift rule for computing gradients in QNN training.

```python
#------------------------------------------------------------------
# Training a QNN
# Chapter 9 in the QUANTUM COMPUTING AND QUANTUM MACHINE LEARNING BOOK
#------------------------------------------------------------------
# Version 1.0
# Qiskit changes frequently.
# We recommend using the latest version from the book code repository at:
# https://aqtinitiative.org/quantum-computing-for-engineers

# (c) 2025 Jesse Van Griensven, Roydon Fraser, and Jose Rosas
# License: MIT - Citation of this work required
#------------------------------------------------------------------

from qiskit import Aer, execute, QuantumCircuit
from qiskit.circuit import Parameter
import numpy as np
import matplotlib.pyplot as plt
#------------------------------------------------------------------

# Define cost function
def cost_function(param_value):
  circuit = qc.bind_parameters({theta: param_value})
  result = execute(circuit, backend, shots=1000).result()
  counts = result.get_counts()
  expectation = counts.get('0', 0) / sum(counts.values())
  return expectation
#------------------------------------------------------------------

# Define a parameterized circuit
qc   = QuantumCircuit(1)
theta = Parameter('θ')
qc.rx(theta, 0)
qc.measure_all()

# Simulate circuit for parameter shift
backend = Aer.get_backend('aer_simulator')

# Parameter shift for gradient computation
param = np.pi / 4.
grad = (cost_function(param + np.pi / 2.) - cost_function(param - np.pi
```

```
/ 2.)) / 2
print("Gradient:", grad)

# Visualize the cost function
param_values = np.linspace(0, 2 * np.pi, 100)
cost_values = [cost_function(p) for p in param_values]

plt.figure(figsize=(10, 7))
plt.plot(param_values, cost_values, label="Cost Function",
color="blue")
plt.axvline(param, color="red", linestyle="--", label=f"Gradient at
θ={param:.2f}")
plt.title("Cost Function for Quantum Neural Network")
plt.xlabel("Parameter θ (radians)")
plt.ylabel("Expectation Value")
plt.legend()
plt.grid(True)
plt.show()
```

9.12   Quantum Cost Functions

In quantum neural networks (QNNs), cost functions play a central role in training, analogous to loss functions in classical neural networks. They measure the performance of the network by quantifying the difference between predicted outcomes and target values. Unlike classical cost functions that directly evaluate numerical outputs, quantum cost functions are computed based on quantum measurements and quantum state properties.

By combining quantum cost functions, hybrid architectures, and probabilistic outputs, quantum neural networks provide a robust framework for tackling machine learning tasks using quantum computing. These components allow QNNs to exploit quantum advantages while mitigating current hardware limitations, paving the way for future advancements in quantum-enhanced learning systems.

9.12.1   Defining Cost Functions in QNNs

A quantum cost function evaluates the performance of a quantum neural network by using information extracted from the quantum states through measurements. These cost functions share conceptual similarities with classical loss functions, such as mean squared error (MSE), cross-entropy, or Hinge loss in classical machine learning.

In QNNs:

(a) Input data is encoded into quantum states $|\psi\rangle$ using quantum encoding schemes.
(b) Quantum circuits, parameterized by learnable variables θ, process the input state to produce an output state $|\psi_{\text{out}}(\theta)\rangle$.
(c) Measurements on the output state produce probabilities or other relevant values to compute the cost function.

General Form of a Quantum Cost Function

The cost function $C(\theta)$ can be defined as:

$$C(\theta) = \left\langle \psi_{\text{out}}(\theta) \,|\, \widehat{O} \,|\, \psi_{\text{out}}(\theta) \right\rangle$$

where

(a) $|\psi_{\text{out}}(\theta)\rangle$ is the output quantum state dependent on the circuit parameters θ.
(b) $\widehat{O}$ is a Hermitian operator (observable) acting on the quantum state space, used to compute measurable quantities. The "hat" notation denotes that this is a quantum operator.

The choice of observable depends on the problem:

(a) For regression problems, it might represent the expectation value of an observable.
(b) For classification problems, it might relate to the probabilities of measured states.

9.12.2   Examples of Quantum Cost Functions

Similar to classical neural nets, QNNs require a cost function to be minimized, two of which are described below:

1. **Fidelity-Based Cost Functions**

   Fidelity measures the similarity between two quantum states. If $|\psi_{\text{target}}\rangle$ is the target quantum state and $|\psi_{\text{out}}(\theta)\rangle$ is the output state, the fidelity-based cost function is:

$$C_{\text{fidelity}}(\theta) = 1 - |\langle \psi_{\text{target}} \,|\, \psi_{\text{out}}(\theta)\rangle|^2$$

   The goal is to minimize this cost, ensuring the output state matches the target state as closely as possible.

2. **Overlap-Based Cost Functions**
   Overlap cost functions are defined based on the inner product between quantum states. Given a set of basis states $|z\rangle$, the cost can be computed as:

$$C_{\text{overlap}}(\theta) = \sum_z p(z) \cdot C(z)$$

   where $p(z)$ is the probability of measuring $|z\rangle$ from $|\psi_{\text{out}}(\theta)\rangle$, and $C(z)$ is a classical cost associated with the basis state $|z\rangle$.

   This formulation is particularly useful for tasks like optimization and sampling.

9.12.3 Challenges with Quantum Cost Functions

QNNs have similar problems to deep neural nets (DNNs) during minimization:

1. Barren Plateaus
   Barren plateaus are regions in the parameter space where the gradient of the cost function becomes exponentially small, leading to vanishing gradients. This phenomenon arises in large quantum circuits and prevents effective training.

   (a) Mathematically, the gradient $\nabla_\theta C(\theta)$ vanishes as the circuit depth increases:

$$\mathbb{E}[|\nabla_\theta C(\theta)|] \sim \frac{1}{2^n}$$

   where n is the number of qubits.
   (b) Mitigation Strategies:

   - Shallow circuits with fewer entangling operations.
   - Layer-wise training methods.
   - Designing cost functions with smaller parameter spaces.

2. Quantum Noise
   In real quantum hardware, noise arising from decoherence and gate imperfections can impact the evaluation of cost functions. Quantum measurements may produce unreliable results, leading to noisy gradients. Some solutions to this problem are:

   (a) Error mitigation techniques, such as measurement averaging or zero-noise extrapolation.
   (b) Hybrid quantum-classical training, where classical optimizers compensate for noisy quantum evaluations.

9.13 Hybrid Quantum-Classical Neural Networks

Quantum computers currently operate in the noisy intermediate-scale quantum (NISQ) era, where

1. Quantum devices have limited qubits.
2. Quantum gates are prone to noise and decoherence.
3. Fully quantum neural networks remain challenging to implement.

To overcome these limitations, hybrid quantum-classical architectures integrate quantum computing into classical machine learning workflows. These hybrid models leverage the strengths of both paradigms:

1. Classical systems handle preprocessing, optimization, and data post-processing.
2. Quantum circuits process specific tasks, such as feature learning or optimization.

9.13.1 Architecture of Hybrid QNNs

A typical hybrid QNN architecture comprises the following components:

1. **Classical Input Layer**
   Input data is pre-processed and encoded into quantum states using suitable encoding schemes (e.g., angle or amplitude encoding).
2. **Quantum Circuit (Quantum Layer)**
   The encoded quantum state is processed by parameterized quantum circuits (e.g., variational quantum circuits), which apply quantum gates to transform the input state into an output state based on trainable parameters.
3. **Classical Output Layer**
   The quantum state is measured, and the resulting classical information is passed to a classical neural network layer or an optimization algorithm for post-processing or decision-making.
4. **Architecture Flow**
   Classical Input $\rightarrow$ Quantum Encoding $\rightarrow$ Quantum Circuit $\rightarrow$ Quantum Measurement $\rightarrow$ Classical Output

9.13.2 Advantages of Hybrid Models

Due to limitations in qubit availability and coherence times, adopting a hybrid quantum neural network (QNN) architecture is currently advantageous. Here are the main advantages:

1. **Computational Efficiency**

   Classical systems handle computationally intensive tasks, while quantum layers tackle specific problems where quantum speedup is possible.

2. **Optimization**

   Hybrid models use classical optimizers (e.g., Adam and gradient descent) to train quantum circuits.

3. **Flexibility**

   Hybrid QNNs can be implemented on existing noisy intermediate-scale quantum (NISQ) hardware, making them practical for near-term quantum devices.

9.13.3   Use Cases of Hybrid QNNs

In the NISQ era, most QNN use cases benefit from a hybrid architecture. Here are some use cases that help substantiate this point:

1. **Quantum-Enhanced Classifiers**

   Quantum layers extract high-dimensional features, improving classification accuracy for complex datasets.

2. **Quantum Variational Algorithms**

   Hybrid architectures use algorithms like the variational quantum eigensolver (VQE) or quantum approximate optimization algorithm (QAOA) for machine learning tasks, such as clustering and regression.

9.13.4   Quantum States as Neural Network Outputs

In quantum neural networks, the outputs of quantum circuits are quantum states. To extract meaningful classical results, the quantum states are measured, and the measurement outcomes are interpreted for downstream tasks.

9.13.4.1   Quantum State Measurement

Quantum states are probabilistic in nature, meaning measurements yield outcomes based on probabilities defined by the quantum state amplitudes. For a quantum state $|\psi\rangle$, the probability of measuring a basis state $|z\rangle$ is:

$$p(z) = |\langle z | \psi \rangle|^2$$

Measurements collapse the quantum state to a specific outcome, and repeated measurements provide statistical estimates of these probabilities.

9.13.4.2   Representing Outputs as Probability Distributions

QNN outputs can be represented as probability distributions over measurement outcomes. For example:

(a) In classification tasks, class probabilities are encoded in quantum amplitudes.
(b) Measured outcomes p_0, p_1, ..., p_n are interpreted as the likelihood of each class.

9.13.4.3   Examples of Quantum Outputs

In quantum machine learning, interpreting the outputs of quantum models requires translating quantum measurement outcomes into meaningful predictions. Quantum systems encode information in probability amplitudes, and their measurements yield inherently probabilistic outcomes. These outcomes can be directly utilized for standard machine learning tasks, such as binary and multi-class classification. The essential quantum principle of state collapse upon measurement naturally aligns with classical decision-making processes, allowing quantum predictions to be intuitively mapped to conventional classification outputs. Key examples of quantum outputs include:

1. **Binary Classification**

   For a two-class problem, the output probabilities p_0 and p_1 (corresponding to measurement outcomes $|0\rangle$ and $|1\rangle$) determine the class prediction:

$$Prediction = \arg\max_i p_i$$

2. **Multi-Class Classification**

   For multi-class problems, amplitudes are mapped to class probabilities, and the most probable outcome is selected.

3. **Quantum State Collapse**

   The collapse of the quantum state after measurement represents the model's prediction, analogous to an argmax operation in classical networks.

9.14   Quantum Boltzmann Machines

Quantum Boltzmann machines (QBMs) extend classical Boltzmann machines by incorporating quantum mechanics, enabling exploration of a broader solution space through quantum superposition and entanglement. This approach holds promise for solving complex optimization and generative modeling problems, particularly in fields like machine learning and quantum chemistry.

Quantum Boltzmann machines extend classical Boltzmann machines into the quantum domain, leveraging superposition and entanglement for enhanced generative modeling and optimization. Applications span diverse fields, from materials science to financial modeling, but challenges such as hardware limitations and decoherence necessitate innovative solutions, including hybrid architectures. QBMs hold the potential to unlock new possibilities in data-driven quantum applications.

Boltzmann machines are energy-based models that utilize statistical mechanics principles to learn patterns in data.

Energy-Based Learning

In classical Boltzmann machines, the energy of a system determines the probability of a specific state s:

$$P(s) = \frac{e^{-E(s)}}{Z}$$

where $E(s)$ is the energy of state s, and Z is the partition function:

$$Z = \sum_s e^{-E(s)}$$

The energy function for a state s is typically defined as:

$$E(s) = -\sum_{i,j} w_{ij} s_i s_j - \sum_i b_i s_i$$

where w_{ij} are the weights and b_i are the biases.

Restricted Boltzmann Machines (RBMs)

RBMs simplify Boltzmann machines by restricting connections to only those between visible and hidden layers, making inference and learning computationally tractable. RBMs are widely used for feature learning and dimensionality reduction.

Python Code Example: RBM Implementation

```
#-------------------------------------------------------------
# Restricted Boltzmann Machines
# Chapter 9 in the QUANTUM COMPUTING AND QUANTUM MACHINE
LEARNING BOOK
#-------------------------------------------------------------
# Version 1.0
# Qiskit changes frequently.
# We recommend using the latest version from the book code
repository at:
#      https://aqtinitiative.org/quantum-computing-for-
engineers
# (c) 2025 Jesse Van Griensven, Roydon Fraser, and Jose
Rosas
# License: MIT - Citation of this work required
```

```python
#-----------------------------------------------------------
import numpy as np
import matplotlib.pyplot as plt
#-----------------------------------------------------------
# Sigmoid activation function
def sigmoid(x):
return 1 / (1 + np.exp(-x))
# Forward pass: Compute hidden activations
def sample_hidden(visible, weights, hidden_bias):
activation = np.dot(visible, weights) + hidden_bias
probabilities = sigmoid(activation)
return probabilities
#-----------------------------------------------------------
# Define RBM parameters
num_visible = 6
num_hidden = 3
weights = np.random.randn(num_visible, num_hidden)  #
Initialize weights
visible_bias = np.zeros(num_visible)
hidden_bias = np.zeros(num_hidden)
# Example visible layer input
visible_input = np.array([1, 0, 1, 0, 1, 0])
hidden_probs = sample_hidden(visible_input, weights,
hidden_bias)
print("Hidden Layer Probabilities:", hidden_probs)
# Visualization: Weight matrix as a heatmap
plt.figure(figsize=(8, 5))
plt.imshow(weights, cmap='coolwarm', aspect='auto')
plt.colorbar(label="Weight Value")
plt.title("RBM Weight Matrix Heatmap")
plt.xlabel("Hidden Units")
plt.ylabel("Visible Units")
plt.xticks(range(num_hidden), [f"H{i+1}" for i in range
(num_hidden)])
plt.yticks(range(num_visible), [f"V{i+1}" for i in range
(num_visible)])
plt.grid(False)
plt.show()
# Visualization: Hidden layer activation probabilities
plt.figure(figsize=(8, 5))
plt.bar(range(1, num_hidden + 1), hidden_probs,
color='blue', alpha=0.7)
plt.title("Hidden Layer Activation Probabilities")
plt.xlabel("Hidden Units")
plt.ylabel("Probability")
```

```
    plt.xticks(range(1, num_hidden + 1), [f"H{i}" for i in
range(1, num_hidden + 1)])
    plt.ylim(0, 1)
    plt.grid(axis='y', linestyle='--', alpha=0.6)
    plt.show()
```

9.14.1 QBM Quantum Extension

QBMs extend RBMs into the quantum domain by using quantum states to represent probability distributions.

Quantum States as Probability Distributions

The quantum analog of a probability distribution is the density matrix ρ, defined using the quantum Hamiltonian H:

$$\rho = \frac{e^{-\beta H}}{\mathrm{Tr}(e^{-\beta H})}$$

where

$\beta = 1/k_B T$ is the inverse temperature
k_B is Boltzmann's constant
T is the temperature
Tr is the trace operator

Quantum Gibbs Sampling

QBMs utilize quantum Gibbs sampling to generate samples from the quantum distribution encoded by ρ. This process involves evolving a quantum system to thermal equilibrium under the influence of Hamiltonian H.

Quantum Hamiltonian in QBMs

The Hamiltonian H typically includes both classical and quantum components:

$$H = H_{\text{classical}} + H_{\text{quantum}}$$

where H_{quantum} introduces off-diagonal terms to enable superposition and tunneling.

9.14.2 Applications of QBMs

The following examples describe use cases for QBMs:

1. **Learning Complex Data Distributions**

   QBMs can capture intricate patterns in data distributions, making them suitable for applications in materials science, quantum chemistry, and financial modeling.

2. **Quantum-Enhanced Generative Modeling**

   QBMs serve as powerful generative models capable of producing high-quality samples for data augmentation and synthetic dataset generation.

Python Code Example: Quantum Gibbs Sampling

This example approximates quantum Gibbs sampling using entanglement and controlled rotations.

```python
#----------------------------------------------------------------------
# Quantum Gibbs Sampling
# Chapter 9 in the QUANTUM COMPUTING AND QUANTUM MACHINE LEARNING BOOK
#----------------------------------------------------------------------
# Version 1.0
# Qiskit changes frequently.
# We recommend using the latest version from the book code repository at:
# https://aqtinitiative.org/quantum-computing-for-engineers

# (c) 2025 Jesse Van Griensven, Roydon Fraser, and Jose Rosas
# License: MIT - Citation of this work required
#----------------------------------------------------------------------
from numpy import pi
import matplotlib.pyplot as plt

from qiskit import QuantumCircuit, Aer, execute
from qiskit.visualization import plot_histogram
#----------------------------------------------------------------------

# Define a simple Hamiltonian H = Z1 + Z2 + X1X2
qc = QuantumCircuit(2)

# Apply thermal relaxation (approximation)
qc.h(0)        # Create superposition
qc.cx(0, 1)    # Entangle qubits
qc.rz(pi / 4., 0) # Apply rotation to simulate inverse temperature
effect
qc.measure_all()

# Print the quantum circuit
print("Quantum Gibbs Sampling Circuit:")
```

```
print(qc)
qc.draw('mpl')  # Graphically display the circuit

# Simulate the circuit
backend = Aer.get_backend('aer_simulator')
result = execute(qc, backend, shots=1000).result()
counts = result.get_counts()

# Print results
print("Quantum Gibbs Sampling Result:", counts)

# Visualize the measurement results
plot_histogram(counts)
plt.title("Quantum Gibbs Sampling Probability Distribution")
plt.show()
```

9.14.3 QBM Implementation Challenges

Similar to QNNs, quantum Boltzmann machines, in the NISQ era, suffer from the
following challenges:

1. **Hardware Limitations**
   Current quantum computers are limited by qubit count, coherence times, and
   gate fidelities, restricting the scale of QBMs.
2. **Quantum Decoherence**
   Noise and decoherence impact the accuracy of quantum Gibbs sampling,
   leading to errors in state preparation.
3. **Hybrid Architectures**
   To mitigate hardware limitations, hybrid approaches combine classical sampling with quantum optimization. For example, the classical components handle
   large-scale sampling, while quantum systems optimize the energy landscape.

Python Code Example: Hybrid QBM
This hybrid QBM approach uses classical sampling to generate visible states and
quantum circuits to optimize the energy function.

```
#-------------------------------------------------------------------
# Hybrid QBM
# Chapter 9 in the QUANTUM COMPUTING AND QUANTUM MACHINE LEARNING BOOK
#-------------------------------------------------------------------
# Version 1.0
# Qiskit changes frequently.
```

```python
# We recommend using the latest version from the book code repository at:
# https://aqtinitiative.org/quantum-computing-for-engineers

# (c) 2025 Jesse Van Griensven, Roydon Fraser, and Jose Rosas
# License: MIT - Citation of this work required
#------------------------------------------------------------------
import numpy as np
import matplotlib.pyplot as plt

from qiskit.algorithms.optimizers import COBYLA
from qiskit.circuit import Parameter, QuantumCircuit
from qiskit import Aer, execute
from qiskit.visualization import plot_histogram
#------------------------------------------------------------------

# Classical sampling function
def classical_sampling(num_samples, weights, biases):
    samples = []
    for _ in range(num_samples):
        visible = np.random.choice([0, 1], size=len(biases))
        samples.append(visible)
    return np.array(samples)
#------------------------------------------------------------------

# Objective function
def cost_function(param):
    simulator = Aer.get_backend('aer_simulator')
    circuit = qc.bind_parameters({theta: float(param[0])})  # FIXED:
Ensure param is a scalar
    result = execute(circuit, simulator, shots=1000).result()
    counts = result.get_counts()
    expectation = counts.get('0', 0) / sum(counts.values())
    return -expectation  # Minimize energy
#------------------------------------------------------------------

def cost_tracking(param):
    value = cost_function(param)
    optimization_steps.append((float(param[0]), value))  # FIXED:
Extract scalar value
    return value
#------------------------------------------------------------------

# Parameters for classical sampling
num_samples = 1000
num_visible = 5
```

```python
weights = np.random.randn(num_visible)  # Randomized weights
biases  = np.random.randn(num_visible)  # Randomized biases

# Generate classical samples
classical_samples = classical_sampling(num_samples, weights, biases)

# Plot classical sampling results
plt.figure(figsize=(8, 5))
plt.hist(classical_samples.sum(axis=1), bins=np.arange(-0.5,
num_visible + 1.5, 1), color='blue', alpha=0.7)
plt.xlabel("Sum of Visible Units")
plt.ylabel("Frequency")
plt.title("Classical Sampling Distribution")
plt.grid(axis='y', linestyle='--', alpha=0.6)
plt.show()

# Quantum optimization using Qiskit
theta = Parameter('θ')
qc = QuantumCircuit(1)
qc.rx(theta, 0)
qc.measure_all()

# Optimize using COBYLA
optimizer = COBYLA(maxiter=100)
optimization_steps = []

result = optimizer.optimize(num_vars=1,
objective_function=cost_tracking, initial_point=[0.5])
optimal_param = result[0]
print("Optimal Parameter:", optimal_param)

# Plot optimization process
params, values = zip(*optimization_steps)
plt.figure(figsize=(8, 5))
plt.plot(params, values, marker='o', linestyle='-', color='red')
plt.xlabel("Parameter θ")
plt.ylabel("Objective Function Value")
plt.title("Quantum Optimization Progress")
plt.grid(True)
plt.show()
```

9.15 Quantum Generative Adversarial Networks

Quantum generative adversarial networks (QGANs) extend the concept of classical GANs into the quantum domain. By leveraging quantum computing's superposition, entanglement, and exponential state space, QGANs offer powerful capabilities for generative modeling across diverse applications, including synthetic data generation, quantum system simulations, and advanced AI training.

QGANs combine the strengths of classical GANs and quantum computing to achieve significant advancements in generative modeling. They are particularly impactful in applications requiring high-quality synthetic data, enhanced simulation capabilities, and accurate quantum system modeling. Addressing challenges such as convergence issues and input preparation will pave the way for the widespread adoption of QGANs across scientific and engineering domains.

9.15.1 Overview of Classical GANs

Classical generative adversarial networks (GANs) consist of two neural networks: the generator and the discriminator. These networks compete in a zero-sum game, where the generator aims to produce data resembling the real dataset, and the discriminator evaluates whether the input is real or fake.

1. **Architecture**

    (a) Generator G: Maps random noise $z \sim p_z$ to a data space. Its objective is to create data indistinguishable from real samples.
    (b) Discriminator D: Distinguishes between real data $x \sim p_{\mathsf{data}}$ and fake data $G(z)$.

2. **Training Objective**
    The GAN optimization problem is formulated as follows:

$$\min_{G} \max_{D} \mathbb{E}_{x \sim p_{\mathsf{data}}}[\log D(x)] + \mathbb{E}_{z \sim p_z}[\log(1 - D(G(z)))]$$

where G and D are trained iteratively to achieve a Nash equilibrium.

Python Code Example: Classical GAN Architecture
This example constructs the generator and discriminator models for a classical GAN.

```
#-------------------------------------------------------------------
# Classical GAN Architecture
# Chapter 9 in the QUANTUM COMPUTING AND QUANTUM MACHINE LEARNING BOOK
#-------------------------------------------------------------------
# Version 1.0
```

```python
# Qiskit changes frequently.
# We recommend using the latest version from the book code repository at:
# https://aqtinitiative.org/quantum-computing-for-engineers

# (c) 2025 Jesse Van Griensven, Roydon Fraser, and Jose Rosas
# License: MIT - Citation of this work required
#-------------------------------------------------------------------
import numpy as np
import matplotlib.pyplot as plt

import tensorflow as tf
from tensorflow.keras import layers
#-------------------------------------------------------------------

def create_generator():
  """ Generator model """
  model = tf.keras.Sequential([
    layers.Dense(128, activation='relu', input_dim=100),
    layers.Dense(256, activation='relu'),
    layers.Dense(784, activation='sigmoid'),
    layers.Reshape((28, 28))  # Reshape to 28x28 images
  ])
  return model
#-------------------------------------------------------------------

def create_discriminator():
  """ Discriminator model """
  model = tf.keras.Sequential([
   layers.Flatten(input_shape=(28, 28)), # Ensure correct input shape
    layers.Dense(256, activation='relu'),
    layers.Dense(128, activation='relu'),
    layers.Dense(1, activation='sigmoid')
  ])
  return model
#-------------------------------------------------------------------

def generate_noise(batch_size=5, latent_dim=100):
  """ Generate random noise """
  return np.random.normal(0, 1, (batch_size, latent_dim))
#-------------------------------------------------------------------

def generate_and_plot_images(generator, num_images=5):
  """ Generate sample images from random noise """
  noise = generate_noise(num_images)
  generated_images = generator.predict(noise)
```

```python
plt.figure(figsize=(10, 2))
for i in range(num_images):
  plt.subplot(1, num_images, i + 1)
  plt.imshow(generated_images[i], cmap='gray')
  plt.axis('off')
plt.suptitle("Generated Images from GAN Generator", fontsize=12)
plt.show()
#-------------------------------------------------------------------

# Instantiate models
generator     = create_generator()
discriminator = create_discriminator()
print("Generator and Discriminator Created.")

# Generate and plot images from the generator
generate_and_plot_images(generator)
```

9.15.2 GAN Quantum Extension

QGANs adapt the GAN architecture for quantum systems, replacing classical
components with quantum circuits.

1. **Quantum Generator**

    The quantum generator is a parameterized quantum circuit (PQC) that gener-
    ates quantum states $|\psi_G(\theta)\rangle$, where θ are the trainable parameters:

$$|\psi_G(\theta)\rangle = U(\theta)\,|0\rangle$$

    with $U(\theta)$ representing a sequence of quantum gates.

2. **Quantum Discriminator**

    The quantum discriminator evaluates the output of the generator using quan-
    tum measurements. Its objective is to maximize the difference in responses to real
    and generated quantum states.

3. **Quantum Discriminator Loss Function**

    The discriminator's loss function evaluates the quality of real and fake
    samples:

$$L_D = \mathbb{E}_{\rho_{\text{real}}}[\log\langle D\,|\,\rho_{\text{real}}\,|\,D\rangle] + \mathbb{E}_{\rho_{\text{fake}}}[\log(1 - \langle D\,|\,\rho_{\text{fake}}\,|\,D\rangle)]$$

    where

ρ_{real}: Density matrix representing a real quantum state drawn from training data.

ρ_{fake} : Density matrix representing a generated (fake) quantum state from the
    quantum generator.

$|D\rangle$: Measurement basis state associated with the discriminator's readout.

$\langle D \mid \rho \mid D \rangle$: Probability of measuring the state ρ and obtaining outcome $|D\rangle$, i.e., the Born rule.

$\mathbb{E}_{\rho_{real}}[\cdot], \mathbb{E}_{\rho_{fake}}[\cdot]$: Expectation over the real and fake quantum data distributions, respectively.

$\log(\cdot)$: Natural logarithm, applied to measurement outcome probabilities.

Python Code Example: Quantum Generator
This code defines a simple parameterized quantum generator circuit.

```python
#-------------------------------------------------------------------
# Quantum Generator
# Chapter 9 in the QUANTUM COMPUTING AND QUANTUM MACHINE LEARNING BOOK
#-------------------------------------------------------------------
# Version 1.0
# Qiskit changes frequently.
# We recommend using the latest version from the book code repository at:
# https://aqtinitiative.org/quantum-computing-for-engineers

# (c) 2025 Jesse Van Griensven, Roydon Fraser, and Jose Rosas
# License: MIT - Citation of this work required
#-------------------------------------------------------------------
import numpy as np
import matplotlib.pyplot as plt

from qiskit import QuantumCircuit, Aer, execute
from qiskit.circuit import Parameter
from qiskit.visualization import import plot_histogram
#-------------------------------------------------------------------

def quantum_generator(params):
    """ Define a quantum generator circuit """
    qc = QuantumCircuit(2)
    qc.ry(params[0], 0)
    qc.ry(params[1], 1)
    qc.cx(0, 1)
    qc.measure_all()
    return qc
#-------------------------------------------------------------------

# Example parameters
params = [0.5, 1.0]
qc = quantum_generator(params)
```

```python
# Display the quantum circuit
print("Quantum Generator Circuit:")
print(qc)
qc.draw('mpl')  # Graphically display the circuit

# Simulate the circuit
backend = Aer.get_backend('aer_simulator')
result  = execute(qc, backend, shots=1000).result()
counts  = result.get_counts()

# Print measurement results
print("Quantum Generator Measurement Results:", counts)

# Visualize the measurement results
plot_histogram(counts)
plt.title("Quantum Generator Probability Distribution")
plt.show()
```

9.15.3   QGAN Applications

The following use cases exemplify applications of QGANs:

1. **Synthetic Data Generation for AI Models**
   QGANs generate synthetic datasets for training AI models, particularly when
   real-world data is scarce or sensitive. Quantum-enhanced generation ensures
   higher diversity and realism.
2. **Enhancing Resolution in Engineering Simulations**
   QGANs improve the resolution of engineering simulations by generating
   high-fidelity data points. Applications include computational fluid dynamics
   (CFD) and materials design.
3. **Modeling Quantum Systems and Materials**
   QGANs simulate quantum systems and generate potential energy surfaces for
   complex molecular systems, aiding quantum chemistry and material science
   research.

Python Code Example: Training QGAN
This code trains a QGAN on a synthetic dataset using Qiskit's QGAN
implementation.

```python
# -----------------------------------------------------------------------
# Training QGAN
# Chapter 9 in the QUANTUM COMPUTING AND QUANTUM MACHINE LEARNING BOOK
```

```python
#----------------------------------------------------------------------
# Version 1.0
# Qiskit changes frequently.
# We recommend using the latest version from the book code repository at:
# https://aqtinitiative.org/quantum-computing-for-engineers

# (c) 2025 Jesse Van Griensven, Roydon Fraser, and Jose Rosas
# License: MIT - Citation of this work required
#----------------------------------------------------------------------
import numpy as np
import matplotlib.pyplot as plt

from qiskit import Aer
from qiskit.utils import QuantumInstance
from qiskit_machine_learning.algorithms import QGAN
from qiskit_machine_learning.datasets import ad_hoc_data
#----------------------------------------------------------------------

# Generate training data using ad_hoc_data (compatible with Qiskit ML
0.3.1)
training_features, training_labels, _, _ = ad_hoc_data(
   training_size=100, test_size=0, n=2, gap=0.3, plot_data=False,
one_hot=False
)

# Convert dataset into required format
data = np.array(training_features)

# Define bounds for QGAN
bounds = np.array([
   [np.min(data[:, 0]), np.max(data[:, 0])],
   [np.min(data[:, 1]), np.max(data[:, 1])]
])

# Initialize QGAN (without 'epochs' argument)
qgan = QGAN(data, bounds=bounds, num_qubits=[2, 2], batch_size=10)

# Set epochs separately (fixing the previous error)
qgan.epochs = 50

# Define a Quantum Instance
qi = QuantumInstance(Aer.get_backend('qasm_simulator'), shots=1024)
```

```python
# Train QGAN using the Quantum Instance (fixing the backend error)
generated_samples, _ = qgan.run(quantum_instance=qi)
print("Trained QGAN Discriminator and Generator.")

# Extract loss values for visualization
loss_d = qgan._ret.get('loss_d', [])
loss_g = qgan._ret.get('loss_g', [])
epochs_range = range(len(loss_d))

# Plot training losses
plt.figure(figsize=(8, 5))
plt.plot(epochs_range, loss_d, 'r-o', label='Discriminator Loss')
plt.plot(epochs_range, loss_g, 'b-o', label='Generator Loss')
plt.title("QGAN Training Losses")
plt.xlabel("Epoch")
plt.ylabel("Loss")
plt.grid(True)
plt.legend()
plt.show()

# Compare Real vs. Generated Data
plt.figure(figsize=(8, 5))
plt.title("Real vs. Generated Data (After Training)")

# Real data scatter plot
plt.scatter(data[:, 0], data[:, 1], alpha=0.7, label="Real Data")

# Generated data scatter plot
plt.scatter(generated_samples[:, 0], generated_samples[:, 1],
alpha=0.7, label="Generated Data")

plt.legend()
plt.grid(True)
plt.show()
```

9.15.4   QGAN Challenges and Solutions

Quantum generative adversarial networks (QGANs) combine the generative power
of classical GANs with quantum computing, offering potential improvements in
efficiency and capability. However, similar to their classical counterparts, QGANs
face several practical challenges, particularly concerning training stability and efficient data handling. Common difficulties include convergence issues, such as oscillations and mode collapse, as well as the computational complexity associated with

preparing quantum states from classical data. Addressing these challenges is essential for realizing the full potential of QGANs, and various solutions have been proposed:

1. **Convergence Issues**

   QGANs, like classical GANs, suffer from convergence problems. Oscillations and mode collapse are common issues where the generator fails to learn a diverse data distribution. Some solutions for these issues include:

   (a) Gradient Clipping: Limits the magnitude of gradients during training to stabilize updates.
   (b) Alternate Training Strategies: Train the discriminator and generator alternately to ensure balanced progress.

2. **Efficient Preparation of Input States for the Generator**

   Preparing input states for the quantum generator is computationally expensive, especially for high-dimensional data. Some solutions for these issues include:

   (a) Hybrid Approaches: Combine classical preprocessing with quantum generation.
   (b) Feature Reduction: Use techniques like PCA to reduce data dimensions before quantum encoding.

Python Code Example: Hybrid QGAN Workflow

This hybrid approach preprocesses data classically before quantum generation.

```
#-----------------------------------------------------------------------
# Hybrid QGAN Workflow
# Chapter 9 in the QUANTUM COMPUTING AND QUANTUM MACHINE LEARNING BOOK
#-----------------------------------------------------------------------
# Version 1.0
# Qiskit changes frequently.
# We recommend using the latest version from the book code repository at:
# https://aqtinitiative.org/quantum-computing-for-engineers

# (c) 2025 Jesse Van Griensven, Roydon Fraser, and Jose Rosas
# License: MIT - Citation of this work required
#-----------------------------------------------------------------------
import numpy as np
import matplotlib.pyplot as plt
from sklearn.decomposition import PCA

from qiskit import Aer, execute, QuantumCircuit
from qiskit.visualization import import plot_histogram
#-----------------------------------------------------------------------
```

```python
# Classical preprocessing: Perform PCA on input data
data = np.array([[1.2, 0.7], [0.9, 1.5], [1.8, 1.0]])
pca = PCA(n_components=1)
reduced_data = pca.fit_transform(data)

# Quantum generator for reduced data
qc = QuantumCircuit(1)
qc.ry(reduced_data[0][0], 0)
qc.measure_all()

# Display the quantum circuit
print("Quantum Generator Circuit:")
print(qc)
qc.draw('mpl')

# Simulate the quantum generator
backend = Aer.get_backend('aer_simulator')
result = execute(qc, backend, shots=1000).result()
counts = result.get_counts()

# Plot PCA results
plt.figure(figsize=(6, 4))
plt.scatter(data[:, 0], data[:, 1], label="Original Data",
color='blue')
plt.scatter(reduced_data, np.zeros_like(reduced_data),
label="Reduced Data (PCA)", color='red')
plt.axhline(0, color='black', linestyle="--", alpha=0.6)
plt.xlabel("Feature 1")
plt.ylabel("Feature 2")
plt.title("PCA Dimensionality Reduction")
plt.legend()
plt.grid(True)
plt.show()

# Print and visualize measurement results
print("Hybrid QGAN Simulation Result:", counts)
plot_histogram(counts)
plt.title("Quantum Generator Measurement Results")
plt.show()
```

9.16 Advantages and Limitations of Quantum Neural Networks

Quantum neural networks (QNNs) leverage the principles of quantum computing to extend the capabilities of traditional neural networks. While they hold immense promise for addressing complex computational challenges, they are not without significant limitations. This section explores the advantages, limitations, and future directions for QNN development.

Quantum neural networks offer exponential speedups, higher capacity for modeling complex systems, and potential energy efficiency gains. However, they are constrained by hardware limitations, training challenges, and scalability issues. Hybrid architectures, quantum-specific learning paradigms, and seamless integration into existing workflows represent promising directions for advancing QNN applications. By addressing these limitations, QNNs have the potential to transform fields ranging from AI to quantum chemistry and beyond.

9.16.1 Exponential Speedup

QNNs have the potential to solve problems intractable for classical neural networks by exploiting quantum parallelism and entanglement. For certain problems, such as unstructured search and quantum chemistry simulations, QNNs offer an exponential speedup over classical approaches. The quantum advantage is evident in tasks with complexity reductions:

$$T_{\text{quantum}} = O\left(\sqrt{N}\right), \quad T_{\text{classical}} = O(N)$$

Python Code Example: Grover's Speedup Simulation
For a dataset with N elements, a classical algorithm may require $O(N)$ computations, whereas a quantum algorithm such as Grover's search reduces this to $O(\sqrt{N})$.

This code demonstrates the basic principle behind Grover's search, highlighting the efficiency of quantum parallelism.

```
#------------------------------------------------------------
# Grover's Speedup Simulation
# Chapter 9 in the QUANTUM COMPUTING AND QUANTUM MACHINE LEARNING BOOK
#------------------------------------------------------------
# Version 1.0
# Qiskit changes frequently.
# We recommend using the latest version from the book code repository at:
# https://aqtinitiative.org/quantum-computing-for-engineers
```

```python
# (c) 2025 Jesse Van Griensven, Roydon Fraser, and Jose Rosas
# License: MIT - Citation of this work required
#-------------------------------------------------------------------
import numpy as np
import matplotlib.pyplot as plt

from qiskit import QuantumCircuit, Aer, execute
from qiskit.visualization import plot_histogram
#-------------------------------------------------------------------


# Quantum search with Grover's algorithm
qc = QuantumCircuit(2)

# Apply Hadamard to create superposition
qc.h([0, 1])

# Phase inversion (Oracle step)
qc.cz(0, 1)

# Diffusion operator (Amplification step)
qc.h([0, 1])
qc.measure_all()

# Display the quantum circuit
print("Grover's Quantum Circuit:")
print(qc)
qc.draw('mpl')

# Simulate the circuit
simulator = Aer.get_backend('aer_simulator')
result = execute(qc, simulator, shots=1000).result()
counts = result.get_counts()

# Print and visualize measurement results
print("Quantum Speedup Simulation Result:", counts)
plot_histogram(counts)
plt.title("Grover's Algorithm Measurement Results")
plt.show()
```

9.16.2   Higher Capacity

QNNs can model complex quantum systems, directly leveraging quantum states, to capture the intricate behaviors of quantum systems. This is particularly useful in fields like quantum chemistry, where classical simulations of molecular systems become computationally expensive.

Example

QNNs efficiently solve Schrödinger's steady-state equation for large molecular systems:

$$H \,|\, \psi \rangle = E \,|\, \psi \rangle$$

where H is the Hamiltonian, $|\psi\rangle$ is the quantum state, and E is the energy eigenvalue.

9.16.3 Energy Efficiency

Quantum computations, by their nature, can be more energy-efficient than classical computations for specific tasks. Unlike classical GPUs that require substantial energy for matrix multiplications, quantum gates perform operations with minimal energy, E, dissipation in cryogenic environments.

For a quantum operation:

$$E_{\text{quantum}} < E_{\text{classical}} \quad \text{(for certain tasks)}$$

9.17 QNN Limitations

Currently, quantum neural nets have many limitations caused by the NISQ era of quantum computers. The most important ones are listed below.

1. **Hardware Constraints**

   The limited number of qubits and the presence of noise in quantum gates pose significant challenges for implementing large-scale QNNs. Quantum error correction, though promising, requires additional qubits and increases system complexity.

2. **Error Rates**

   The fidelity, F, of quantum gates affects the accuracy of QNNs:

$$F = |\, \langle \psi_{\text{ideal}} \,|\, \psi_{\text{real}} \rangle \,|^2$$

3. **Mitigation Strategies**

   (a) Layer-wise training of QNNs.

   (b) Adaptive optimizers that adjust learning rates dynamically.

4. **QNN Scalability**

   Scaling QNN architectures to solve large, practical problems remains a challenge due to qubit constraints and decoherence. The number of qubits required for realistic applications often exceeds the capacity of current quantum hardware.

5. **Resource Requirements**

$$n_{\text{qubits}} = \log_2(N)$$

   where N is the dataset size.

6. **QNN Training Complexity**

   QNNs often face barren plateaus, where the gradient of the cost function diminishes exponentially with the number of parameters, hindering optimization. For a parameterized quantum circuit:

$$\frac{\partial L}{\partial \theta} \to 0 \quad \text{(as the number of parameters increases)}.$$

Python Code Example: Noise Simulation

The following simulation highlights the impact of noise on quantum gate operations.

```python
#-------------------------------------------------------------------
# Quantum Noise Simulation
# Chapter 9 in the QUANTUM COMPUTING AND QUANTUM MACHINE LEARNING BOOK
#-------------------------------------------------------------------
# Version 1.0
# Qiskit changes frequently.
# We recommend using the latest version from the book code repository at:
# https://aqtinitiative.org/quantum-computing-for-engineers

# (c) 2025 Jesse Van Griensven, Roydon Fraser, and Jose Rosas
# License: MIT - Citation of this work required
#-------------------------------------------------------------------
import numpy as np
import matplotlib.pyplot as plt

from qiskit import QuantumCircuit, Aer, execute
from qiskit.providers.aer.noise import NoiseModel,
depolarizing_error
from qiskit.visualization import plot_histogram
#-------------------------------------------------------------------

# Define a quantum circuit with noise
qc = QuantumCircuit(1)
```

```python
qc.h(0)  # Apply Hadamard gate
qc.measure_all()

# Add noise model
noise_model = NoiseModel()
noise_model.add_all_qubit_quantum_error(depolarizing_error(0.1,
1), ['h'])

# Simulate with noise
simulator = Aer.get_backend('aer_simulator')
result   = execute(qc, simulator, noise_model=noise_model,
shots=1000).result()
counts   = result.get_counts()

# Display the quantum circuit
print("Quantum Circuit with Noise:")
print(qc)
qc.draw('mpl')

# Print and visualize measurement results
print("Noisy Circuit Result:", counts)
plot_histogram(counts)
plt.title("Quantum Noise Simulation Results")
plt.show()
```

9.18   Hybrid Classical-Quantum QNN Architectures

Hybrid architectures leverage the strengths of classical and quantum systems. Classical components handle data preprocessing and large-scale optimization, while quantum circuits address high-dimensional feature extraction and optimization subtasks.

Python Code Example: A Simple Hybrid QNN
This example code combines classical preprocessing with quantum encoding for a hybrid workflow.

```python
#--------------------------------------------------------------------
# A Simple Hybrid QNN
# Chapter 9 in the QUANTUM COMPUTING AND QUANTUM MACHINE LEARNING BOOK
#--------------------------------------------------------------------
# Version 1.0
# Qiskit changes frequently.
```

```python
# We recommend using the latest version from the book code repository at:
# https://aqtinitiative.org/quantum-computing-for-engineers

# (c) 2025 Jesse Van Griensven, Roydon Fraser, and Jose Rosas
# License: MIT - Citation of this work required
#-------------------------------------------------------------------
import numpy as np
import matplotlib.pyplot as plt

from sklearn.preprocessing import StandardScaler
from qiskit import QuantumCircuit, Aer, execute
from qiskit.visualization import plot_histogram
#-------------------------------------------------------------------

# Classical preprocessing
data  = np.array([[2.1, 3.4], [1.5, 2.2], [0.8, 1.7]])
scaler = StandardScaler()
normalized_data = scaler.fit_transform(data)

# Quantum encoding
qc = QuantumCircuit(2)
qc.ry(normalized_data[0][0], 0)
qc.ry(normalized_data[0][1], 1)
qc.measure_all()

# Simulate the quantum circuit
simulator = Aer.get_backend('aer_simulator')
result  = execute(qc, simulator, shots=1000).result()
counts  = result.get_counts()

# Display the quantum circuit
print("Hybrid Quantum Circuit:")
print(qc)
qc.draw('mpl')

# Plot normalized data
plt.figure(figsize=(6, 4))
plt.scatter(data[:, 0], data[:, 1], label="Original Data",
color='blue')
plt.scatter(normalized_data[:, 0], normalized_data[:, 1],
label="Normalized Data", color='red')
plt.axhline(0, color='black', linestyle="--", alpha=0.6)
plt.axvline(0, color='black', linestyle="--", alpha=0.6)
plt.xlabel("Feature 1")
plt.ylabel("Feature 2")
```

```
plt.title("Data Normalization for Hybrid QNN")
plt.legend()
plt.grid(True)
plt.show()


# Print and visualize measurement results
print("Quantum Circuit Measurement Results:", counts)
plot_histogram(counts)
plt.title("Hybrid QNN Quantum Encoding Results")
plt.show()
```

Python Code Example: A More Complete Hybrid QNN Implementation

This code example is a hybrid quantum-classical architecture, where a quantum circuit acts as a parameterized layer, and classical optimization techniques are used to train the QNN. This example uses the variational quantum classifier (VQC) concept, where the quantum circuit processes input data, and the output is optimized to solve a simple binary classification problem.

This example code is complex and requires additional explanation as follows:

1. **Quantum Circuit**

    (a) A 2-qubit parameterized quantum circuit is created with trainable parameters (θ).
    (b) The circuit applies initial Hadamard gates to create superposition, followed by parameterized rotations and a CNOT gate to introduce entanglement.

2. **EstimatorQNN**

    Qiskit's EstimatorQNN is used to define a quantum neural network. It maps input quantum circuits to expected values of observables.

3. **Dataset**

    A simple dataset with two features and binary labels (0 or 1) is created for demonstration purposes.

4. **Neural Network Classifier**

    (a) This classifier integrates the QNN with a classical optimizer (COBYLA) to minimize the loss function.
    (b) Cross-entropy is used as the loss function to train the QNN.

5. **Training and Prediction**

    (a) The QNN is trained on the provided dataset using the classical optimizer.
    (b) Predictions are made on test data.

6. **Visualization**

    The training and test points are plotted to visualize the QNN's classification performance.

```python
#------------------------------------------------------------------
# Complete Hybrid QNN
# Chapter 9 in the QUANTUM COMPUTING AND QUANTUM MACHINE LEARNING BOOK
#------------------------------------------------------------------
# Version 1.0
# (c) 2025 Jesse Van Griensven, Roydon Fraser, and Jose Rosas
# License: MIT - Citation of this work required
#------------------------------------------------------------------

import numpy as np
import matplotlib.pyplot as plt
from qiskit import Aer, QuantumCircuit
from qiskit.circuit import Parameter
from qiskit.utils import QuantumInstance, algorithm_globals
from qiskit.algorithms.optimizers import COBYLA
from qiskit.opflow import Z, StateFn, CircuitStateFn
from qiskit_machine_learning.neural_networks import CircuitQNN
from qiskit_machine_learning.algorithms.classifiers import
NeuralNetworkClassifier

# Set random seed for reproducibility
algorithm_globals.random_seed = 42

# Step 1: Define the Quantum Neural Network (QNN)
def create_quantum_circuit():
    """Create a parameterized quantum circuit for the QNN."""
    circuit = QuantumCircuit(2)  # 2-qubit quantum circuit

    # Define trainable parameters (weights)
    theta_0 = Parameter('θ0')
    theta_1 = Parameter('θ1')
    theta_2 = Parameter('θ2')

    # Initial Hadamard gates to create superposition
    circuit.h(0)
    circuit.h(1)

    # Parameterized rotations (trainable layers)
    circuit.ry(theta_0, 0)
    circuit.ry(theta_1, 1)
    circuit.cx(0, 1)  # Entanglement using a CNOT gate
    circuit.ry(theta_2, 1)

    return circuit
```

```python
quantum_circuit = create_quantum_circuit()
print("Quantum Circuit:")
print(quantum_circuit.draw())

# Step 2: Define the QNN with the CircuitQNN Class
quantum_instance = QuantumInstance(Aer.get_backend
('aer_simulator'), shots=1024)

qnn = CircuitQNN(
    circuit=quantum_circuit,
    input_params=[], # No classical input encoding for now
    weight_params=quantum_circuit.parameters,
    quantum_instance=quantum_instance, # Required in your Qiskit
version
)

# Step 3: Prepare a Simple Dataset for Binary Classification
X_train = np.array([[0, 1], [1, 0], [1, 1], [0, 0]]) # Input features
y_train = np.array([0, 1, 1, 0]) # Binary labels: 0 or 1

# Step 4: Use NeuralNetworkClassifier to Train the QNN
classifier = NeuralNetworkClassifier(
    neural_network=qnn,
    optimizer=COBYLA(maxiter=50),
    loss='cross_entropy',
    one_hot=False
)

# Train the QNN on the dataset
classifier.fit(X_train, y_train)

# Step 5: Test the QNN
X_test = np.array([[0, 1], [1, 0]])
predictions = classifier.predict(X_test)

print("\nTest Predictions:")
for i, prediction in enumerate(predictions):
    print(f"Input: {X_test[i]} → Predicted Label: {prediction}")

# Step 6: Visualization of Results
plt.figure(figsize=(6, 5))
plt.title("QNN Predictions")
plt.scatter(X_train[:, 0], X_train[:, 1], c=y_train, marker='o',
label='Train Data', edgecolors='black')
```

```
plt.scatter(X_test[:, 0], X_test[:, 1], c=predictions, marker='x',
label='Test Data', edgecolors='black')

plt.xlabel('Feature 1')
plt.ylabel('Feature 2')
plt.legend()
plt.grid()
plt.show()
```

This code output is complex and requires additional explanations as follows:

1. **Quantum Circuit**
   The circuit is displayed, showing trainable rotation gates and entanglement.
2. **Predictions**
   The QNN predicts binary labels (0 or 1) for test input data.
3. **Visualization**
   A scatter plot visualizes the training and test data, showing the QNN's decision boundaries.

9.19   Quanvolutional Neural Networks

Quanvolutional neural networks (QuanvNNs) represent a novel quantum-enhanced adaptation of classical convolutional neural networks (CNNs), leveraging the unique computational strengths of quantum mechanics to improve feature extraction and representation. These hybrid quantum-classical models aim to address challenges arising in classical deep learning systems, particularly for computationally intensive tasks like image classification, object detection, and pattern recognition.

Classical CNNs have revolutionized the field of computer vision by effectively capturing hierarchical spatial features from images. Through convolutional layers, they progressively extract features, such as edges, textures, and objects, enabling robust performance across diverse tasks. However, as datasets grow larger and the complexity of visual recognition tasks increases, classical architectures encounter significant limitations:

1. Scalability Issues: Large models require enormous computational power and memory. Training on high-dimensional datasets becomes resource-intensive and time-consuming.
2. Feature Limitations: Classical models may struggle to discover complex or global relationships within data, particularly in noisy or ambiguous scenarios.
3. Data Bottlenecks: While classical models excel at low-dimensional feature extraction, capturing intricate features in high-dimensional data remains challenging.

Quanvolutional neural networks address these bottlenecks by integrating quantum computing principles into the CNN architecture. Quantum systems offer properties like superposition, entanglement, and quantum parallelism, enabling the efficient representation and processing of information in high-dimensional spaces. These unique characteristics make quantum-enhanced neural networks particularly well-suited to tasks requiring large-scale computation and the identification of complex patterns.

9.19.1   The Role of Quantum Computing in Convolution

At the heart of a QuanvNN lies the ability to leverage quantum computations in tandem with classical models. Convolutional operations, which are the foundation of CNNs, involve applying filters (small matrices) to sections of input data (e.g., image patches). In classical CNNs, these filters learn meaningful patterns through iterative optimization. However, this process becomes computationally expensive when dealing with higher-dimensional data.

Quantum computing enhances CNN by encoding information into quantum states and performing operations within quantum circuits. Specifically:

1. **High-Dimensional Representations**

   Quantum systems naturally represent data in Hilbert spaces, which grow exponentially with the number of qubits. For instance, an n-qubit quantum system can encode 2^n dimensions of information, allowing quantum models to efficiently process high-dimensional features.

2. **Quantum Feature Maps**

   Quantum feature maps enable the transformation of classical data into quantum states that capture richer, more complex features. These quantum-encoded features can represent global relationships that are otherwise difficult to capture using classical filters.

3. **Quantum Parallelism**

   Quantum gates operate on superpositions of states, allowing simultaneous evaluation of multiple inputs. This property accelerates computations and enhances feature extraction efficiency.

In QuanvNNs, quantum operations replace or complement classical convolutional filters. Small regions of input data, such as patches from an image, are encoded into quantum states, processed through quantum circuits, and decoded back into classical outputs. The resulting quantum-enhanced features are then passed to subsequent layers in the neural network.

9.19.2   The Architecture of Quanvolutional Neural Networks

A QuanvNN architecture integrates quantum operations into the classical CNN pipeline. The core components of the pipeline include:

1. Data Encoding

   The input image is divided into smaller patches, typically corresponding to the receptive fields of classical convolutional layers. These patches are encoded into quantum states using techniques such as:

   (a) Angle Encoding: Information is encoded into the rotation angles of qubits.
   (b) Amplitude Encoding: The pixel intensities are mapped to the amplitudes of a quantum state.
   (c) Mathematically: Given a classical vector $x \in \mathbb{R}^n$, the quantum state encoding can be represented as:

$$|x\rangle = \frac{1}{\|x\|} \sum_{i=1}^{n} x_i \, |i\rangle$$

   where x_i are the pixel values.

2. Quantum Circuit Processing

   Once encoded, the quantum states are processed using a small quantum circuit. The quantum operations typically consist of the following:

   (a) Parameterized Quantum Gates: Quantum gates, such as rotation gates R_x, R_y, and R_z, are applied to manipulate the qubits.
   (b) Entanglement Operations: Multi-qubit gates, like the Controlled-NOT (CNOT) gate, introduce entanglement, enabling the circuit to explore correlations between pixels.

   The quantum circuit acts as a learnable filter, and its parameters are optimized during training. The output is a transformed quantum state, which encodes a processed feature map.

3. Measurement and Classical Output

   The processed quantum state is measured, and the outcomes are converted back into classical data. For example, the measurement result could be the expectation value of an observable O:

$$\langle O \rangle = \langle \psi | O | \psi \rangle$$

   where $|\psi\rangle$ is the output quantum state. These classical outputs serve as feature maps for downstream layers in the neural network.

4. Classical Post-Processing

   After quantum processing, the resulting feature maps are passed through subsequent classical convolutional layers, pooling layers, and fully connected layers, as in a standard CNN. The quantum-enhanced features provide a richer representation of the input data.

9.19.3   Benefits of Quanvolutional Neural Networks

QuanvNNs combine the strengths of quantum computing and classical deep learning, offering several advantages:

1. **Efficient High-Dimensional Computation**
   Quantum circuits operate in exponentially large Hilbert spaces, enabling efficient representation and processing of high-dimensional data. This allows QuanvNNs to capture complex relationships within data with fewer resources.
2. **Enhanced Feature Extraction**
   Quantum operations can produce nonlinear transformations that classical filters may struggle to replicate. Entanglement introduces correlations between features, enhancing the network's ability to recognize intricate patterns.
3. **Reduction in Computational Overhead**
   By leveraging quantum parallelism, QuanvNNs can process multiple feature maps simultaneously, reducing the overall computational cost for feature extraction.
4. **Hybrid Quantum-Classical Approach**
   QuanvNNs integrate quantum circuits as modular components within classical architectures. This hybrid approach allows them to be implemented on near-term quantum hardware while leveraging the robustness of classical deep learning frameworks.

9.19.4   Challenges and Future Directions

While quanvolutional neural networks hold significant promise, there are several challenges to address:

1. **Noisy Quantum Hardware**
   Current quantum computers are prone to noise and errors, which can impact the performance of quantum circuits. Error mitigation techniques are critical for reliable execution.
2. **Limited Qubit Count**
   Near-term quantum hardware has a limited number of qubits, restricting the size of quantum circuits that can be implemented.
3. **Data Encoding Overhead**
   Mapping classical data into quantum states can be computationally expensive, particularly for high-resolution images.
4. **Scalability**
   Scaling QuanvNNs to larger datasets and deeper architectures remains an open challenge due to hardware constraints.

9.20   The Architecture of Quanvolutional Neural Networks

The architecture of quanvolutional neural networks (QuanvNNs) combines quantum and classical computational layers to form a hybrid system that capitalizes on the strengths of both paradigms. By embedding quantum operations into the classical convolutional neural network (CNN) pipeline, QuanvNNs can extract richer, high-dimensional features while maintaining the versatility of classical deep learning architectures.

The QuanvNN architecture consists of four primary components which are systematically integrated to ensure seamless information flow and quantum-classical collaboration:

1. The input layer
2. The quantum convolutional layer
3. The classical processing layer
4. The output layer

9.20.1   Input Layer

The input layer serves as the interface between classical data and the quantum components of the network. The key steps at this stage include:

1. **Classical Input Data**

   The input data, such as a two-dimensional image, is divided into smaller overlapping or non-overlapping patches. Each patch represents a localized region of the input that will be processed independently. For a $N \times N$ image and patch size $k \times k$, the total number of patches is determined by:

$$\text{Number of patches} = \left(\frac{N - k}{s} + 1 \right)^2$$

   where s is the stride length determining how far the window moves.

   For example, a 28×28 image with 3×3 patches and stride $s = 1$ results in $26 \times 26 = 676$ patches.

2. **Quantum State Encoding**

   Each selected patch $P(i,j)$ is transformed into a quantum state using one of the quantum encoding schemes:

   (a) Amplitude Encoding: Maps pixel values into the amplitudes of a quantum state.

   (b) Basis Encoding: Represents the data as binary quantum states.

   (c) Phase Encoding: Converts pixel intensities into phase angles of qubits.

   (d) Mathematically: For a patch $x \in \mathbb{R}^k$, the quantum encoding can be written as:

$$|\psi_{i,j}\rangle = \mathcal{E}(P(i,j)) \quad \text{where} \quad \mathcal{E}: P \rightarrow |\psi\rangle$$

The resulting quantum state $|\psi_{i,\,j}\rangle$ represents the encoded patch data, ready for processing by the quantum convolutional layer.

9.20.2   *Quantum Convolutional Layer*

The quantum convolutional layer serves as the quantum equivalent of classical convolutional layers. It applies quantum circuits to the encoded quantum states (patches), acting as quantum filters that extract expressive features. The key steps in the quantum convolutional layer are listed below:

1. **Quantum Circuit as a Filter**

    Each patch $|\psi_{i,\,j}\rangle$ is processed by a parameterized quantum circuit (PQC), which applies a sequence of quantum gates. The PQC is defined as:

$$U(\theta) = U_L(\theta_L)...U_2(\theta_2)U_1(\theta_1)$$

    where θ represents learnable parameters optimized during training.

2. **Measurement and Quantum Feature Extraction**

    After the quantum circuit processes the state, measurements are performed to extract classical outputs. For an observable O, the quantum feature value, f_{quantum}, is obtained by computing the expectation value of the observable with respect to the output state:

$$f_{\text{quantum}}(P(i,j)) = \langle \psi_{\text{out}} | O | \psi_{\text{out}} \rangle$$

    where $|\psi_{\text{out}}\rangle = U(\theta)|\psi_{i,j}\rangle$, and $U(\theta)$ is the parameterized quantum circuit.

    These expectation values are computed for each patch and then aggregated to form a quantum feature map, which serves as the output of the quantum convolutional layer.

Python Code Example: Quantum Circuit as a Filter

An example of a quantum circuit acting on a 2-qubit system. The circuit combines single-qubit gates (e.g., R_y, R_z) and entangling gates (e.g., CNOT) to manipulate the quantum state and capture both local and global correlations within the patch.

```
#------------------------------------------------------------------
# Quantum Circuit as a Filter
# Chapter 8 in the QUANTUM COMPUTING AND QUANTUM MACHINE LEARNING BOOK
#------------------------------------------------------------------
```

```
# Version 1.0
# Qiskit changes frequently.
# We recommend using the latest version from the book code repository at:
# https://aqtinitiative.org/quantum-computing-for-engineers

# (c) 2025 Jesse Van Griensven, Roydon Fraser, and Jose Rosas
# License: MIT - Citation of this work required
#-------------------------------------------------------------------

# Example of a quantum circuit acting on a 2-qubit system

from qiskit import QuantumCircuit
from qiskit.circuit import Parameter
theta = Parameter('θ')
qc = QuantumCircuit(2)

# Apply rotations and entanglement
qc.ry(theta, 0)
qc.ry(theta, 1)
qc.cx(0, 1)
qc.draw()
```

9.20.3   Classical Processing Layer

The quantum feature maps generated in the previous layer are passed into the classical processing layer for further refinement and analysis. This stage leverages traditional CNN components, ensuring compatibility with existing deep learning frameworks. The key components of the classical processing layer are:

1. **Pooling Layers**

   Quantum feature maps can be down sampled using pooling operations (e.g., max pooling and average pooling) to reduce the spatial dimensions and computational complexity.

2. **Dense Layers (Fully Connected Layers)**

   The down-sampled feature maps are flattened and passed through fully connected layers, where weights are optimized to learn complex mappings from quantum features to output predictions.

   Mathematically, a fully connected layer applies:

$$\text{Output} = \sigma\left(W \cdot f_{\text{quantum}} + b\right)$$

   where W and b are learnable parameters.

   σ is a classical activation function (e.g., ReLU and softmax).

f_{quantum} is the quantum feature vector, where each component corresponds to an expectation value of a quantum observable measured on the output quantum state after quantum convolution (as defined in Sect. 9.20.2).

9.20.4   Output Layer

The final layer generates predictions based on the refined quantum-classical feature maps. The output can vary depending on the task:

1. **Classification**
   Assigns class probabilities using a SoftMax activation function.
2. **Regression**
   Predicts continuous values using a linear activation.
3. **Object Detection**
   Outputs bounding boxes and class labels for objects in an image.
   For a classification task with C classes, the output layer computes:

$$\text{Prediction}(x) = \text{softmax}(W_{\text{out}} \cdot f_{\text{final}} + b_{\text{out}})$$

where W_{out} and b_{out} are trainable parameters, and f_{final} is the output from the dense layers.

9.20.5   Workflow of a Quanvolutional Neural Network

The end-to-end workflow of a QuanvNN can be summarized as follows:

1. **Input Image**: The input image is provided in classical form.
2. **Patch Selection**: The image is divided into small patches.
3. **Quantum Encoding**: Each patch is encoded into a quantum state using amplitude, phase, or basis encoding.
4. **Quantum Circuit Processing**: Parameterized quantum circuits process the encoded states, generating quantum features.
5. **Quantum Measurements**: Quantum states are measured to extract classical feature maps.
6. **Quantum Feature Map**: The measured outputs form the quantum feature map.
7. **Classical Layers**: The quantum feature maps are passed through classical pooling, dense, and activation layers.
8. **Output Prediction**: The final output is generated for tasks like classification or regression.

9.20.6 Advantages of the QuanvNN Architecture

The main advantages of the QuanvNN Architecture are presented below:

1. **Expressive Feature Extraction**

   Quantum circuits can capture complex patterns and correlations that classical filters might miss.
2. **Hybrid Quantum-Classical Design**

   Integrating quantum operations into classical pipelines allows QuanvNNs to leverage the best of both worlds.
3. **Parallelism and Efficiency**

   Quantum operations can process high-dimensional data efficiently through superposition and entanglement.

9.21 Applications of Quanvolutional Neural Networks

Quanvolutional neural networks (QuanvNNs) represent an exciting advancement in deep learning by integrating quantum computing principles into classical convolutional neural networks. The hybrid quantum-classical architecture enables efficient processing of high-dimensional data, improved feature extraction, and the modeling of intricate correlations that classical systems may struggle to capture. These capabilities position QuanvNNs for transformative applications in fields like image classification, pattern recognition, time-series analysis, medical diagnostics, remote sensing, cybersecurity, and robotics.

9.21.1 QuanvNNs Enhanced Image Classification

QuanvNNs are particularly effective for image classification tasks due to their quantum convolutional layers, which act as advanced feature extractors. Instead of relying solely on classical filters, quantum circuits process localized image patches, capturing fine-grained details, correlations, and hidden relationships that classical filters may overlook.

Mathematically, for an image patch x, the quantum feature extraction involves:

$$| \psi_{\text{out}}(x) \rangle = U(\theta) | \psi_{\text{encoded}}(x) \rangle$$

where $U(\theta)$ is a parameterized quantum circuit applied to the encoded state $| \psi_{\text{encoded}} \rangle$.

1. **Advantages Over Classical CNNs**

   (a) Efficient High-Dimensional Feature Extraction: Quantum circuits can efficiently represent complex relationships in data due to superposition and entanglement.
   (b) Faster Convergence: In certain scenarios, QuanvNNs achieve faster convergence due to their enhanced representational power.

2. **Applications**

   (a) Medical Imaging: Quantum feature maps can help detect tumors in MRI and CT scans with higher accuracy.
   (b) Facial Recognition Systems: Improved robustness in identifying faces under varying conditions.
   (c) Satellite Image Classification: Faster and more precise classification of geographical regions for environmental monitoring.

Python Code Example: A QuanvNN Implementation

```python
#----------------------------------------------------------------------
# Quanvolution NN
# Chapter 9 in the QUANTUM COMPUTING AND QUANTUM MACHINE LEARNING BOOK
#----------------------------------------------------------------------
# Version 1.0
# Qiskit changes frequently.
# We recommend using the latest version from the book code repository at:
# https://aqtinitiative.org/quantum-computing-for-engineers


# (c) 2025 Jesse Van Griensven, Roydon Fraser, and Jose Rosas
# License: MIT - Citation of this work required
#----------------------------------------------------------------------


from qiskit import QuantumCircuit, Aer, transpile
import numpy as np


# Example: Quantum filter for image patches
def quantum_filter(theta):
    qc = QuantumCircuit(2)
    qc.h(0)  # Apply Hadamard gate for superposition
    qc.ry(theta[0], 0)  # Rotation gate with learnable parameter
    qc.ry(theta[1], 1)
    qc.cx(0, 1)  # Introduce entanglement
    qc.measure_all()
    return qc
```

```
theta = [np.pi / 4, np.pi / 3]
qc = quantum_filter(theta)
print(qc.draw())
```

9.21.2 QuanvNNs Pattern Recognition

Pattern recognition involves identifying textures, edges, and geometric shapes within data. QuanvNNs exploit quantum entanglement to model intricate correlations, enabling improved detection of structured and unstructured patterns.

1. **Real-World Use Cases**

   (a) Handwriting Recognition: Quantum feature maps improve accuracy in recognizing handwritten digits or characters, such as those in the MNIST dataset.
   (b) Traffic Sign Detection: Real-time pattern recognition enhances autonomous vehicle vision systems.

2. **Example**
   For handwriting recognition, patches from handwritten digits are processed through quantum circuits, and features extracted via quantum measurements are fed into classical dense layers for classification.

9.21.3 QuanvNNs Advantage in Time-Series Data

While QuanvNNs are primarily designed for spatial data, they can be adapted for time-series analysis by treating time windows as patches. Quantum layers excel at modeling temporal dependencies and correlations within these patches. Some time-series applications Use cases are:

1. **Financial Forecasting**
   QuanvNNs analyze historical stock market data to predict trends using quantum-enhanced time-window processing.
2. **Sensor Data Analysis**
   Predictive maintenance in IoT systems based on time-series sensor inputs.
3. **Energy Load Forecasting**
   Modeling temporal patterns in energy consumption for smart grid systems.

QuanvNNs Time-Series Example Workflow:

1. Divide time-series data into overlapping time windows.
2. Encode each window into quantum states.
3. Process quantum states through parameterized quantum circuits to extract quantum features.
4. Use classical layers to predict trends.

9.21.4 QuanvNNs Enhanced Medical Diagnostics

QuanvNNs are particularly promising for analyzing large-scale medical images, where detecting subtle anomalies is critical. Quantum circuits improve the feature extraction process, enabling the identification of minute differences that might be overlooked by classical filters. Some use case examples are presented below, in the list and in Fig. 9.4:

1. **Cancer Detection**: Early detection of tumors in MRI and CT scans through quantum-enhanced analysis.
2. **Organ Segmentation**: Improved segmentation accuracy for tissues or organs using quantum-enhanced convolution.
3. **QuanvNN Example Pipeline**: Input MRI scan → Patch selection → Quantum encoding → Quantum circuit processing → Quantum feature extraction → Classical classification for tumor detection.

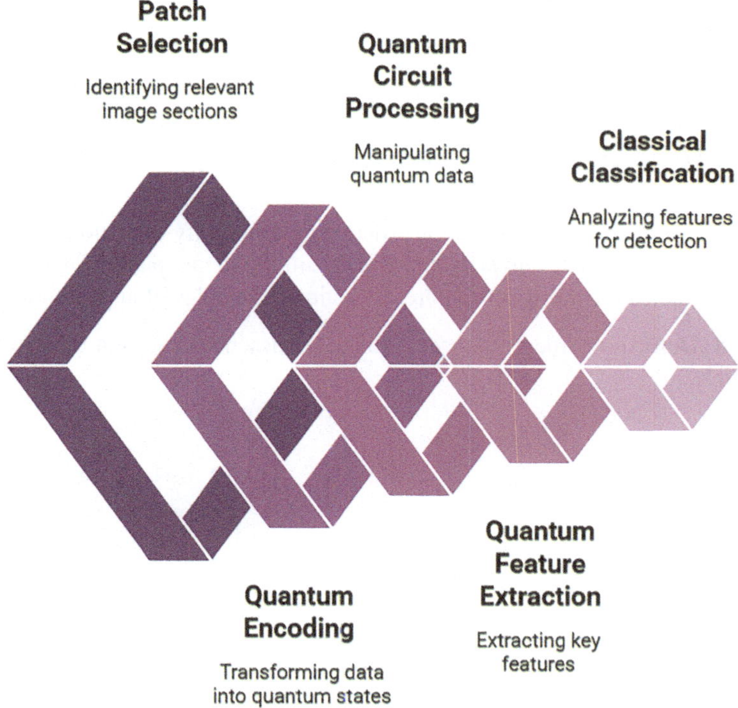

Fig. 9.4 Quanvolutional neural net workflow. (By the authors)

9.21.5   QuanvNNs Remote Sensing

QuanvNNs enhance the processing of high-resolution satellite imagery by applying quantum convolutional operations to identify patterns and changes in geographical features. Some of the applications for remote sensing are:

1. **Climate Change Modeling**: Quantum-processed Earth observation data aids in detecting deforestation patterns, glacier melting, and rising sea levels.
2. **Disaster Response**: Faster identification of wildfires, floods, and other natural disasters through quantum-enhanced feature detection.
3. **Resource Management**: Monitoring water bodies, forests, and land use with improved accuracy.

The QuanvNN Remote Sensing Workflow is basically the steps shown in the text below and visually in Fig. 9.5.

1. High-resolution satellite image
2. Divide into regions
3. Quantum feature extraction
4. Classical interpretation and decision-making.

9.21.6   QuanvNNs Cybersecurity and Anomaly Detection

Cybersecurity systems rely on anomaly detection to identify deviations in network traffic or financial data. QuanvNNs improve sensitivity to rare patterns by leveraging quantum-enhanced learning for feature extraction. Examples of applications are:

1. **Fraud Detection:** Identifying fraudulent transactions in real-time financial systems.

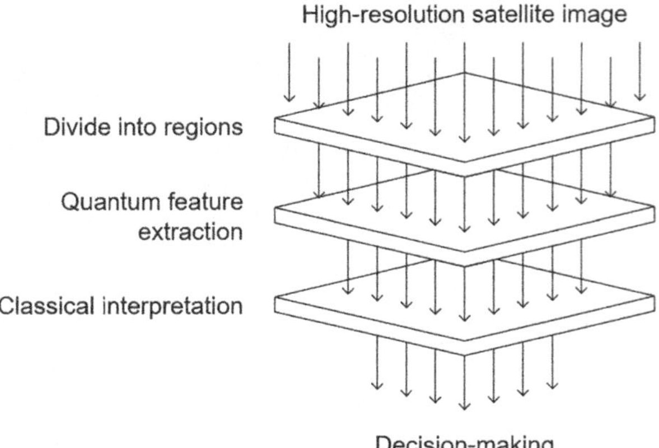

Fig. 9.5 QuanvNN remote sensing workflow. (By the authors)

2. **Intrusion Detection**: Detecting anomalies in network traffic to flag potential cyberattacks.
3. **Data Breach Prevention**: Monitoring large datasets to spot unauthorized data access or modifications.

Python Code Example: Anomaly Detection

```python
#----------------------------------------------------------------
# Quantum Anomaly Detection
# Chapter 9 in the QUANTUM COMPUTING AND QUANTUM MACHINE LEARNING BOOK
#----------------------------------------------------------------
# Version 1.0
# Qiskit changes frequently.
# We recommend using the latest version from the book code repository at:
# https://aqtinitiative.org/quantum-computing-for-engineers

# (c) 2025 Jesse Van Griensven, Roydon Fraser, and Jose Rosas
# License: MIT - Citation of this work required
#----------------------------------------------------------------
import numpy as np
import matplotlib.pyplot as plt

from qiskit import QuantumCircuit, Aer, execute
from qiskit.visualization import import plot_histogram
#----------------------------------------------------------------

def quantum_anomaly_filter(theta):
    """ the Quantum Anomaly Detection Circuit """
    qc = QuantumCircuit(3)

    # Apply Hadamard gates to create superposition
    qc.h(range(3))

    # Apply parameterized rotations
    for i in range(3):
        qc.ry(theta[i], i)

    # Introduce entanglement using CNOT gates
    qc.cx(0, 1)
    qc.cx(1, 2)

    # Measure all qubits
    qc.measure_all()
```

```python
  return qc
#-----------------------------------------------------------------

# Define sample anomaly detection parameters - Example angles
theta_values = [np.pi / 4., np.pi / 3., np.pi / 6.]

# Create and display the quantum circuit
qc = quantum_anomaly_filter(theta_values)
print("Quantum Anomaly Detection Circuit:")
print(qc)
qc.draw('mpl')

# Simulate the circuit
simulator = Aer.get_backend('aer_simulator')
result   = execute(qc, simulator, shots=1000).result()
counts   = result.get_counts()

# Convert measurement results into a probability distribution
bitstrings   = list(counts.keys())
probabilities = np.array(list(counts.values())) / 1000.0

# Sort results for visualization
sorted_indices     = np.argsort(probabilities)[::-1]
sorted_bitstrings  = np.array(bitstrings)[sorted_indices]
sorted_probabilities = probabilities[sorted_indices]

# Print and visualize measurement results
print("Quantum Anomaly Detection Results:", counts)
plot_histogram(counts)
plt.title("Quantum Anomaly Detection Output Distribution")
plt.show()

# Plot the anomaly detection probability distribution
plt.figure(figsize=(8, 5))
plt.bar(sorted_bitstrings, sorted_probabilities, color='purple',
alpha=0.7)
plt.xlabel("Measured Bitstring")
plt.ylabel("Probability")
plt.title("Quantum Anomaly Detection - Probability Distribution")
plt.grid(axis='y', linestyle="--", alpha=0.6)
plt.show()
```

9.21.7 QuanvNNs Enhanced Robotics and Edge Computing

QuanvNNs can enhance object detection and environmental understanding for robotics and drones. Quantum convolutional layers enable real-time processing of visual data, making them suitable for edge computing applications where low latency is essential. Here are some applications:

1. **Autonomous Navigation**: Improved vision for robots and drones operating in dynamic environments.
2. **Industrial Automation**: Real-time object tracking for warehouse management, logistics, and smart factories.
3. **Edge Computing**: Deploying QuanvNNs on edge devices with limited computational resources.

Exercise Questions: Quantum Neural Networks (QNNs)
1. **Mapping Neural Networks to Quantum Circuits**

    (a) Explain the process of mapping a classical neural network architecture to a quantum circuit.
    (b) Using the equation $z = \sum w_i x_i + b$, describe how weights and biases are translated into parameterized quantum gates in QNNs.

2. **Quantum Encoding of Neural Networks**

    (a) Compare amplitude encoding and basis encoding for representing classical data in quantum states.
    (b) Derive the normalized quantum state for a dataset $x = [2, 1, 3]$ using amplitude encoding.

3. **Quantum Analogs of Neural Layers**

    (a) What are quantum perceptrons, and how do they replace classical activation functions?
    (b) Explain how unitary transformations in quantum circuits enable both linear and nonlinear operations.

4. **Training Quantum Neural Networks**

    (a) Describe the parameter-shift rule for computing gradients in a QNN. Provide its mathematical formulation.
    (b) Write Python code to compute the gradient of a simple parameterized quantum circuit using the parameter-shift rule.

5. **Quantum Gradient Descent**

    (a) Define the cost function $C(\theta) = \langle \psi(\theta) | H | \psi(\theta) \rangle$ and explain its role in QNN optimization.

   (b) Discuss how vanishing gradients, or barren plateaus, affect the training of
       QNNs and propose strategies to mitigate this issue.

6. **Quantum Boltzmann Machines (QBMs)**

   (a) Explain the concept of energy-based learning in classical and quantum
       Boltzmann machines. Provide the mathematical expression for the energy
       function.
   (b) Discuss the role of quantum Gibbs sampling in QBMs and its advantages
       over classical sampling techniques.

7. **Quantum Generative Adversarial Networks (QGANs)**

   (a) Compare the architecture of classical GANs with QGANs, highlighting the
       role of quantum circuits in the generator and discriminator.
   (b) Describe an application of QGANs in synthetic data generation for AI model
       training.

8. **Applications of QNNs**

   (a) How do QNNs address complex problems in quantum chemistry, such as
       solving the Schrödinger equation?
   (b) Discuss how QNNs improve generative modeling in computational fluid
       dynamics (CFD) or material design.

9. **Hybrid Quantum-Classical Architectures**

   (a) Explain how hybrid architectures combine classical preprocessing with
       quantum circuits for QNN training.
   (b) Write Python code to demonstrate a hybrid workflow where classical data is
       preprocessed and encoded into a quantum circuit.

10. **Advantages and Limitations of QNNs**

    (a) Discuss the potential exponential speedup offered by QNNs for tasks like
        unstructured search and molecular simulations.
    (b) Identify two significant hardware constraints in implementing QNNs and
        propose solutions to overcome them.

11. **Future Directions for QNNs**

    (a) What are quantum-native learning paradigms, and how do they differ from
        adapting classical methods for quantum systems?
    (b) Propose how QNNs can be integrated into real-time optimization workflows
        for engineering applications.

12. **Python Programming for QNNs**

    (a) Write Python code to create a simple quantum generator circuit using Qiskit.
    (b) Modify the circuit to include noise simulation and analyze its impact on the
        generator's output.

Additional Bibliography

1. D. Abbott et al., Dreams versus reality: plenary debate session on quantum computing. Quantum Inf. Process **9**(4), 401–428 (2010)
2. S. Aaronson, *Quantum Computing Since Democritus* (Cambridge University Press, 2013)
3. C.H. Bennett, G. Brassard, *Quantum Cryptography: Public Key Distribution and Coin Tossing* (*Proceedings of IEEE International Conference on Computers, Systems and Signal Processing*, 1984), pp. 175–179
4. G. Benenti et al., *Principles of Quantum Computation and Information: A Comprehensive Textbook* (World Scientific Publishing, 2004)
5. C. Bernhardt, *Quantum Computing for Everyone* (MIT Press, 2019)
6. D. Bouwmeester, A.K. Ekert, A. Zeilinger (eds.), *The Physics of Quantum Information: Quantum Cryptography* (Quantum Teleportation, Quantum Computation. Springer, 2000)
7. R. Cleve et al., Quantum algorithms revisited. Math. Program. **97**(1), 17–25 (2003) Springer
8. D. Deutsch, Quantum theory, the Church-Turing Principle, and the universal quantum computer. Proc. R. Soc. Lond. A. Math. Phys. Sci. **400**(1818), 97–117 (1985)
9. D.P. DiVincenzo, The physical implementation of quantum computation. Fortschritte der Physik **48**(9–11), 771–783 (2000)
10. S. S. Gill et al. Quantum computing: vision and challenges. *arXiv preprint arXiv:2403.02240* (2024)
11. J. Gruska, *Quantum Computing* (McGraw-Hill, 1999)
12. J.D. Hidary, *Quantum Computing: An Applied Approach* (Springer, 2019)
13. C. Hughes et al., *Quantum Computing for the Quantum Curious* (Springer, 2021)
14. G. Jaeger, *Quantum Information: An Overview* (Springer, 2007)
15. E.R. Johnston et al., *Programming Quantum Computers: Essential Algorithms and Code Samples* (O'Reilly Media, 2019)
16. P. Kaye et al., *An Introduction to Quantum Computing* (OUP Oxford, 2007)
17. A.Y. Kitaev et al., *Classical and Quantum Computation* (American Mathematical Society, 2002)
18. P. Krantz et al., A quantum engineer's guide to superconducting qubits. Appl. Phys. Rev. **6**(2), 021318 (2019)
19. S. Majidy, C. Wilson, R. Laflamme, *Building Quantum Computers – A Practical Introduction* (Cambridge Press, 2025)
20. N.D. Mermin, *Quantum Computer Science: An Introduction* (Cambridge University Press, 2007)
21. A. Montanaro, Quantum algorithms: an overview. npj Quan. Inform. **2**, 15023 (2016)
22. M.A. Nielsen, I.L. Chuang, *Quantum Computation and Quantum Information: 10th Anniversary Edition* (Cambridge University Press, 2010)
23. J. Preskill, Quantum computing and the entanglement frontier. Proc. Natl. Acad. Sci. **111**(15), 5759–5763 (2014)
24. P.W. Shor, Polynomial-time algorithms for prime factorization and discrete logarithms on a quantum computer. SIAM J. Comput. **26**(5), 1484–1509 (1997)
25. D. R. Simon. On the power of quantum computation. *Proceedings of the 35th Annual Symposium on Foundations of Computer Science*, pp. 116–123 (1994)
26. A.M. Steane, Error correcting codes in quantum theory. Phys. Rev. Lett. **77**(5), 793–797 (1996)
27. L. Susskind, A. Friedman, *Quantum Mechanics: The Theoretical Minimum* (Basic Books, 2014)
28. V. Vedral, *Decoding Reality: The Universe as Quantum Information* (Oxford University Press, 2010)
29. T. Wong, *Introduction to Classical and Quantum Computing* (Rooted Grove, 2017)
30. B. Zeng et al., *Quantum Information Meets Quantum Matter* (Springer, 2019)
31. W.H. Zurek, Decoherence, einselection, and the quantum origins of the classical. Rev. Mod. Phys. **75**(3), 715–775 (2003)

Index

The manufacturer's authorised representative in the EU is Springer
Nature Customer Service Centre GmbH, Europaplatz 3, 69115 Heidelberg,
Germany. If you have any concerns regarding our products, please
contact ProductSafety@springernature.com

Printed and bound by CPI Group (UK) Ltd, Croydon, CR0 4YY
19/05/2026
02113616-0016